# High Energy Physics with Polarized Beams and Polarized Targets
(Argonne, 1978)

**AIP Conference Proceedings**
Series Editor: Hugh C. Wolfe
**Number 51**
Particles and Fields Subseries, No. 17

# High Energy Physics with Polarized Beams and Polarized Targets
(Argonne, 1978)

Editor
**G.H. Thomas**
Argonne National Laboratory

**American Institute of Physics**
New York                1979

Copying fees: The code at the bottom of the first page of each article in this volume gives the fee for each copy of the article made beyond the free copying permitted under the 1978 US Copyright Law. (See also the statement following "Copyright" below). This fee can be paid to the American Institute of Physics through the Copyright Clearance Center, Inc., Box 765, Schenectady, N.Y. 12301.

Copyright © 1979 American Institute of Physics

Individual readers of this volume and non-profit libraries, acting for them, are permitted to make fair use of the material in it, such as copying an article for use in teaching or research. Permission is granted to quote from this volume in scientific work with the customary acknowledgment of the source. To reprint a figure, table or other excerpt requires the consent of one of the original authors and notification to AIP. Republication or systematic or multiple reproduction of any material in this volume is permitted only under license from AIP. Address inquiries to Series Editor, AIP Conference Proceedings, AIP.

L.C. Catalog Card No. 79-64565
ISBN 0-88318-150-9
DOE CONF- 781095

## FOREWORD

This year is in some sense a milestone in polarization physics. The polarized beam facility at Argonne will be closing down, along with the ZGS, and one has the chance to evaluate the output from that program. Moreover, polarized beams and targets have been used in lepton interactions and these results indicate the great potential of studying the spin dependence of fundamental interactions. Finally, it is now apparent that the techniques used at the ZGS for making a polarized beam can be extended to strong focusing machines. Therefore, the general physics community has an interest in determining whether such technology can be used to answer questions of fundamental interest.

These considerations played a role in the organization of the polarization symposium held in October 1978 at Argonne. We hope these proceedings of that symposium carry some of the flavor of the numerous contributed papers, invited talks and general summary talks which were given. An attempt was made to make the conference truly international, and this can be seen reflected in the number of foreign speakers and participants. It is the hope of the organizing committee that the tradition of this symposium will continue; the plans are to have the next one in Lausanne, Switzerland in 1980.

I would like to take this opportunity to express my gratitude to the many people who have contributed to the organization of the symposium. First to the organizing committee of O. Chamberlain, E. D. Courant, G. Fidecaro, J. D. Jackson, L. Michel, L. Soloviev, S. Suwa, G. A. Voss and their energetic chairman A. D. Krisch for providing general guidance and help in establishing the overall balance of the symposium. In working out the detailed program, I was fortunate to have the suggestions and help of a hard-working Program Committee consisting of R. N. Cahn, D. G. Crabb, H. E. Miettinen, and C. K. Sinclair. The conference secretary K. Novak performed an outstanding job in taking care of correspondence and the several mailings which were made prior to the symposium. I would also like to especially thank her, as well as M. Ambats, B. Angelos and J. Day for their help as conference secretariat. Without their vigilance, the symposium could not have gone as smoothly. Also, we are indebted to the professional services of the Conference Planning and Management Staff, M. Holden and D. Burdzinski, for their logistical help at all stages of the symposium.

To the speakers I am grateful, for they are responsible for the scientific excellence of the program. They are to be especially thanked for producing a written record of their talks for these proceedings. The session chairman should also be commended for accomplishing the difficult task of keeping physicists from talking too long.

My final thanks go to Argonne, AUA, The University of Chicago, The University of Michigan, and The ZGS Users group who provided special funding for the symposium. On behalf of all the participants we are especially appreciative to the AUA for the lovely meal in the Sullivan Room of the Chicago Art Institute, and the special tour of the Pompeii exhibit which followed.

G. Thomas
Argonne National Laboratory
February, 1979

TABLE OF CONTENTS

1. POLARIZED TARGETS

   A. Abragam: Polarized Targets in High Energy and Elsewhere............................................. 1

   R. C. Fernow: Report of the Workshop on Polarized Target Materials...................................... 15

   J. Button-Shafer: The University of Massachusetts Spin Refrigerator and Strange Particle Physics with the Brookhaven Multiparticle Spectrometer........... 41

   D. H. Saxon: An Axially Polarized Proton Target for the Measurement of A and R Parameters in the Reaction $\pi^- p \to K^\circ \Lambda^\circ$ in the Momentum Range 1.34-2.24 GeV/c................................................. 48

   T. O. Niinikoski: Recent Developments in Polarized Targets at CERN........................................ 62

2. POLARIZED LEPTON BEAMS

   V. N. Baier: Polarized Electron Beams................... 70

   R. F. Schwitters: Experimental Review of Beam Polarization in High Energy $e^+e^-$ Storage Rings............ 91

   J. R. Johnson: Measurements of SPEAR Beam Polarization Using a Back-Scattered Lasar Technique............... 110

   R. Rossmanith: Measurement of Beam-Polarization in the Storage Ring "PETRA"........................... 115

   W. von Drachenfels: Acceleration of Polarized Electrons in the 2.5 GeV Synchrotron at Bonn.................. 120

   B. W. Montague: Polarized $e^\pm$ Beams in LEP?............... 129

3. LEPTON-HADRON AND WEAK INTERACTIONS

   J. J. Sakurai: Weak Electromagnetic Interference......... 138

   V. W. Hughes: Lepton Hadron Scattering................... 171

   C. Y. Prescott: Parity Violations in Inelastic Scattering of Polarized Electrons........................ 202

   K. P. Schüler: Polarized Electron Polarized Proton Scattering Experiments at SLAC...................... 217

   D. E. Nagle: Parity Violations in the Scattering of 15 MeV Protons by Hydrogen......................... 224

   D. E. Nagle: Search for Parity Violation in Polarized Proton Scattering at 6 GeV/c....................... 231

   F. E. Paige: Estimates of W Production with Polarized Protons............................................ 235

   P. Langacker: Final State Polarization in Neutrino Induced Reactions................................... 241

   L. Michel: Comment on Contribution of P. Langacker....... 247

4. POLARIZED HADRON BEAMS

L. C. Teng: Acceleration of Polarized Ions in Synchrotrons.................................... 248

W. Haeberli: Polarized Ion Sources................... 269

E. F. Parker: Polarized Ion Source Development at Argonne National Laboratory..................... 290

Ya. S. Derbenev: On the Possibilities of Obtaining High Energy Polarized Particles in Accelerators and Storage Rings........................... 292

E. D. Courant: Possibility of Polarized Beams at the AGS.................................................. 307

D. G. Underwood: Hyperon Beams as a Source of Polarized Protons...................................... 318

S. Suwa: Studies on Possibility of Accelerating Polarized Protons at KEK............................. 325

R. M. Beurtey: Polarized Particles at SATURNE.......... 330

P. W. Schmor: The 200 nA Variable Energy Polarized Proton Beam at TRIUMF.............................. 341

A. Turrin: Depolarization During Acceleration.......... 347

A. A. Belushkina: The Cryogenic Source "CRYPOL 2" of Polarized Hydrogen and Deuterium Atoms.......... 351

5. MEDIUM ENERGY HADRONIC INTERACTIONS

D. V. Bugg: NN Scattering Below 1 GeV................. 362

H. M. Spinka: Pure Spin Total Cross Sections.......... 382

N. Hoshizaki: pp Phase Shifts and Dibaryon Resonances... 399

H. B. Willard: Proton-Proton Elastic Scattering Spin Correlative Parameter $A_{nn}(\theta)$ at 643 and 796 MeV..... 420

W. Leo: Measurements of the Polarization A and the Spin-Correlation $A_{nn}$ in Elastic p-p Scattering Between 400 and 600 MeV........................... 424

T. A. Mulera: Measurement of the Energy Dependence of $\Delta\sigma_T$ and $A_{nn}$ for pp Scattering in the 1-3 GeV/c Region............................................ 428

6. HADRONIC INTERACTIONS

M. Fidecaro: Polarized Target Experiments at CERN....... 434

A. M. Jonckheere: Polarization Measurements in $\pi^{\pm}p$ and pp Elastic Scattering at 100 and 300 GeV........ 439

A. C. Irving: Implications of Nucleon-Nucleon Spin-Polarization Measurements........................ 445

P. Kroll: Interpretation of the Spin Structure in Nucleon-Nucleon Scattering........................ 451

P. Kyberd: Polarization in P-N Elastic Scattering at 24 GeV/c.................................................. 455

S. W. Gray: Inclusive $\pi^-$ Asymmetries Near $x = 0$........... 457

K. Rybicki: A Model Independent Partial Wave Analysis of the $\pi^+\pi^-$ System Produced at Low Four-Momentum Transfer in the Reaction $\pi^-p_\uparrow \to \pi^+\pi^-n$ at 17.2 GeV/c... 461

J. T. Donohue: Polarization Measurements and a Narrow $\varepsilon(750)$.......................................... 476

L. Van Rossum: $K^+N$ Charge Exchange at 6 and 12 GeV/c..... 478

M. Svec: Tensor Exchange Amplitudes in $K^+N$ Charge Exchange Reactions......................................... 491

N. Van Hieu: Asymptotic Theorems and Polarization Phenomena................................................. 496

R. L. Kelly: Comment on $Z_1^*(1800)$ and Spin-Rotation Parameter Measurements................................... 501

7. HIGH $P_T$ PHENOMENA

D. Sivers: High $P_T$ Theory in a Spin...................... 505

K. M. Terwilliger: Spin Measurements in Hadronic High Momentum Transfer Scattering............................. 521

K. Heller: Polarization in Inclusive Lambda Production... 542

R. O. Polvado: Proton Polarization in Inclusive Processes at 100, 200, 300 and 400 GeV....................... 549

J. Pumplin: Transverse Quark Polarization in Large $P_T$ Reactions and $e^+e^-$ Jets.................................. 554

D. E. Soper: Production of Dimuons From High Energy Polarized Proton-Proton Collisions........................ 559

C. K. Chen: Spin Correlation Asymmetries of Large Angle Elastic NN Scattering and the Nature of Partons................................................... 563

8. SYMPOSIUM SUMMARY

M. G. Borghini............................................ 567

PROGRAM....................................................... 573

LIST OF PARTICIPANTS.......................................... 577

## POLARIZED TARGETS IN HIGH ENERGY AND ELSEWHERE

A. Abragam

CEA-Saclay, B.P. n° 2 - 91 190 Gif-sur-Yvette, France

The subject of polarized targets had been covered extensively in two international conferences that took place at Saclay in 1965 and Berkeley in 1971. Proceedings have been published for both.

One of my problems as an invited speaker at the present Conference is to avoid repeating what I have said at the previous two. This goal is best achieved by speaking of things that happened long before the first, or after the second.

With respect to the latter period there is a second problem : namely that I have no first-hand knowledge of what happened, in the field of polarized targets for High Energy Physics since 1971. This is why the second part of this talk will be dealing with "elsewhere".

### The Overhauser Effect

To come back to the first part we go back to 1953 and to Overhauser's proposal to polarize nuclei in metals by saturating the ESR resonance of conduction electrons.

This proposal presented at a meeting of the APS did not meet, to say the least, with unanimous approval. For a suitable description of the reactions of the audience I would like to borrow that given to me a long time ago by the late Van de Graaf à propos of the PhD thesis examination of Louis de Broglie, that he had attended in Paris in 1923, an historical event.

"Never has so much gone over the heads of so many".

In spite of the general incredulity two brave men, T.R.Carver and C.P.Slichter decided to do just what Dr Overhauser had ordered, saturate the ESR of conduction electrons of sodium metal and see what happens. What happened is shown in Fig.1, an NMR signal of $^7$Li, quite invisible in the absence of electron saturation. Overhauser had also predicted that the ESR line should be displaced by the field produced by the polarized nuclei and "seen" by the conduction electrons. This is shown in Fig.2 (Saclay 1960), where the ESR line so displaced, floats back (Overhauser had said snaps back, an overstatement), with a time constant equal to the nuclear relaxation time.

To predict his results Overhauser had made three unnecessarily restrictive assumptions

    a) Fermi statistics for the conduction electrons
    b) Equal number of electrons and nuclei
    c) A scalar coupling of the form $a\underset{\sim}{I}.\underset{\sim}{s}$ between electronic and nuclear spins.

ISSN: 0094-234X/79/510001-14$1.50 Copyright 1979 American Institute of Physics

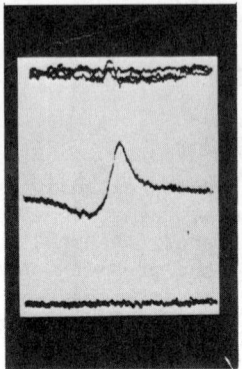 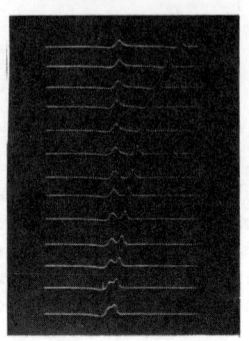

Fig.1 - Enhanced signal of $^7$Li in lithium metal.

Fig.2 - Displaced ESR line in lithium metal.

It was rapidly realized that neither of these three assumptions was necessary. Fig.3 shows a large Overhauser enhancement of the NMR signal of $^{29}$Si (Saclay 1958), observed in semi-conducting silicon where conduction electrons are far less numerous than the nuclei of $^{29}$Si and are known to obey Boltzmann statistics, in contradiction with assumptions a) and b).

Fig.4 (Saclay 1958) shows Overhauser enhancement of nuclear signals of protons in water where dissolved free radicals with unpaired electronic spins play the part of conduction electrons. The electron-nuclear coupling is known to be dipolar rather than scalar. In a paper published in 1955 under the title "The Overhauser effect in non-metals" I had shown that dipolar electron-nucleus coupling led to an Overhauser nuclear polarization enhancement of a sign opposite to that predicted for a scalar coupling, in agreement with Fig.4.

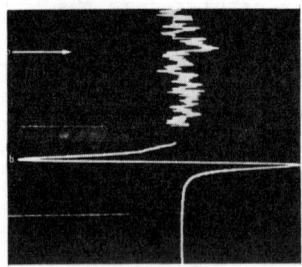 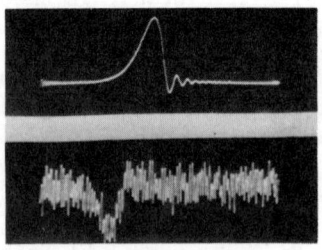

Fig.3 - Enhanced signal of $^{29}$Si in semi-conducting silicon.

Fig.4 - Enhanced signal of protons in water containing free radicals.

In lectures, although never in print, and to call attention to the negative sign of the enhancement I had referred to it as the Underhauser effect. This had the unfortunate consequence of misleading a very respectable colleague who later wrote a large textbook on magnetic resonance where Underhauser appears in the Name Index.

After disbelieving Overhauser's predictions until Carver and Slichter proved dramatically their validity, there was a swing in the opposite direction, a widespread belief that the saturation of an ESR resonance in any type of bulk matter should lead to an enhanced nuclear polarization.

## The Solid Effect

In my paper of 1955 I had analyzed the situation in solid insulators and shown that saturation of an ESR resonance due to fixed paramagnetic impurities with dipolar coupling to surrounding nuclei should *not* lead to an Overhauser enhancement. It was three years later in 1958 that it occurred to me (and quite independently to Carson Jeffries at Berkeley) that, in that case, driving the forbidden transitions, sum or difference of electronic and nuclear frequencies, could lead to a large nuclear polarization parallel or antiparallel to that of the electrons.

Fig.5 (Saclay 1958) shows how this phenomenon occurs in LiF where the rare isotope $^6$Li with a small magnetic moment plays the part of the nucleus and the large moment of $^{19}$F that of the electron. Driving the frequency $\Omega = \omega(^{19}F) \pm \omega(^6Li)$ yields for $^6$Li a polarization enhanced by $\mp \gamma(^{19}F)/\gamma(^6Li) \simeq \mp 6$.

I cannot understand why it took me, (or anybody else who took the trouble to read the paper), three long years to discover something which was really implicit in my analysis of 1955 but there it is.

In magnetic resonance there is a fundamental difference with far-reaching consequences between solids where the relative positions of the spins are fixed and liquids where they move rapidly with respect to each other. To distinguish the Overhauser effect where, be it in metals, semiconductors, or liquids, electronic and nuclear spins move rapidly with respect to each other, from the new effect where the relative positions of electronic and nuclear spins are fixed I had called the new effect the "solid effect".

## Polarized Targets

The discovery of the "solid effect" started research on "dynamic nuclear polarization" and then on polarized targets in various laboratories with Saclay and Berkeley being by far the most active in the field.

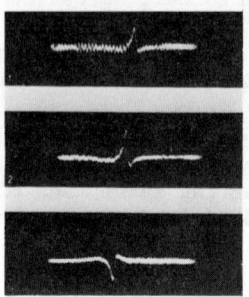 

Fig.5 - Enhanced signal of $^6$Li obtained by "solid effect" in LiF.

Fig.6 - Thin target of LMN for low energy protons.

The first experiment using a polarized proton target and a low energy polarized proton beam was performed at Saclay in 1962. The first high energy experiment with a polarized target and a high energy pion beam, or at least what at the time was considered as a high energy beam, was performed at Berkeley in 1963. Fig. 6,7,8, of historical interest now, show some details of the first Saclay experiment, the target, a thin crystal of LMN (short for Lanthanum Magnesium Nitrate) doped with cerium, the lay-out of the experiment and a photograph of the target assembly.

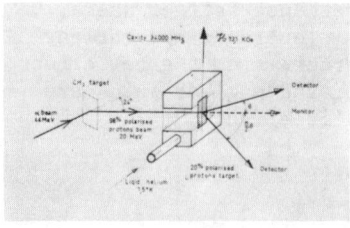 

Fig.7 - Scattering of low energy protons by a polarized LMN target experimental lay-out. a)

Fig.8 - Scattering of low energy protons by a polarized LMN target experimental lay-out. b)

In my talk at the Saclay conference of 1965 I had given a simple-minded description of the well resolved solid effect as resulting from forced flips-flops or flip-flips of an electron-nucleus pair followed by a relaxation back flip of an electron-spin. This model describes well the situation in a substance such as LMN in high field, doped with neodymium, where the electronic line width is much smaller than the nuclear Larmor frequency. This is illustrated in Fig.9 which shows that enhancements of opposite sign of the nuclear polarization do indeed occur for microwave frequencies $\Omega = \omega_e \pm \omega_n$. For several years, say from 1962 to 1966, LMN was the main proton-target material.

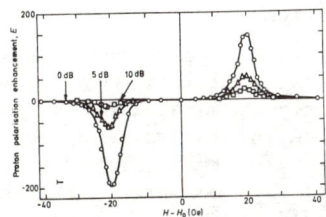

Fig.9 - Enhancement of proton NMR signal in LMN by solid effect.

Its main attractions were the ease with which single crystals could be grown and above all the high proton polarizations, 70 % or more, that it yielded.

Its main defect was the small ratio of free polarized protons to all the protons of the target, of the order of $f \approx 0.06$ and there was a growing demand from high energy physicists, for target materials with higher free protons content.

The most obvious choice solid orthohydrogen $H_2$, did not work because of the strong leakage relaxation of the protons due to their intramolecular coupling. Such a coupling is not effective in its ground state $J = 0$ in ortho hydrogen free solid HD, which for that reason looked more promising, and was studied at Saclay for over a year to no avail. The snag was the very long electronic relaxation time of the electronic centres, namely hydrogen (and deuterium) atoms produced by irradiation of the sample with $\gamma$ rays.

More modestly, the solution chosen in the late sixties was organic substances, alcohols (butanol) doped with free-radicals (porphyrexides), and diols (propanediol) doped with $Cr^V$ complexes, which provided ratios f of the order of 0.24 for butanol and 0.19 for propanediol.

On the experimental side a great step forward was the move to lower temperatures with the use of $^3$He cryostats and later on of dilution refrigerators.

At present proton-polarizations nearing 100 % are obtained for instance in propane-diol doped with $Cr^V$.

## Spin Temperature Theory of DNP

The process of dynamic polarization in these substances is somewhat different from the simple scheme of the well-resolved solid effect and cannot be understood without introducing the concept of spin-temperature. I am going to give of that process a description somewhat different from that given usually, because I shall make use of it in the second part of my talk. It was well accepted some time ago by low temperature physicists ; whether it agrees with particle physicists remains to be seen. The description is in three steps :

a) Adiabatic demagnetization

It has been known for many years that a system of interacting magnetic moments in thermal equilibrium in a high magnetic field $H_o$ could be cooled efficiently by adiabatic demagnetization. Reducing the applied field from the initial value $H_o$ to zero, results in a temperature reduction $T_i/T_f$ of the order of $H_o/H_L$ where the constant $H_L$, much smaller than $H_o$, is a measure of the local field, "seen" by a given moment and produced by its neighbours.

b) Adiabatic demagnetization in high field

Given a system of interacting spins in high field it is also possible to achieve their adiabatic demagnetization while remaining all the time in a high magnetic field. There is no contradiction either in terms since the word demagnetization means reduction of the magnetization, not of the field, or in fact, for such a demagnetization can be easily achieved in practice, using an rf field. The principle of this operation called ADRF (adiabatic demagnetization in the rotating frame) is best understood using the familiar concept of the rotating frame.

The behaviour of a system of spins in a large dc field $H_o$ and a small nearly resonant rf field of amplitude $H_1$, rotating at a frequency $\omega$ in the neighbourhood of the Larmor frequency $\gamma H_o$ of the spins, in a plane perpendicular to $H_o$, is described most simply in a frame rotating around $H_o$ with the angular velocity $\omega$. In that frame the nuclear spins behave as if they were "seeing" static fields only: a dc field parallel to $H_o$ but of amplitude $\Delta H = H_o - \omega/\gamma$ and a dc field of amplitude $H_1$ at right angle to it.

The dipolar Hamiltonian of the spins which can be written

$$H_D = \gamma^2 \hbar^2 \sum_{i<j} \{\underline{I}_i \cdot \underline{I}_j - 3(\underline{\hat{r}}_{ij} \cdot \underline{I}_i)(\underline{\hat{r}}_{ij} \cdot \underline{I}_j)\} r_{ij}^{-3} \qquad (1)$$

is replaced in the rotating frame by the so called truncated Hamiltonian where all elements of (1) which do not commute with the Zeeman Hamiltonian :

$-\gamma\hbar \underset{\sim}{H}_o \cdot \underset{\sim}{I} = -\gamma\hbar H_o I_z$ are thrown out :

$$H'_D = \frac{1}{2}\gamma^2\hbar^2 \sum_{i<j}(1-3\cos^2\theta_{ij})\{3I_{iz}I_{jz}-\underset{\sim}{I}_i\cdot\underset{\sim}{I}_j\} r_{ij}^{-3} \qquad (2)$$

where $\theta_{ij}$ is the angle of the unit vector $\hat{r}_{ij}$ with the dc field $H_o$. In a crystal the truncated Hamiltonian $H'_D$ depends on the orientation of the dc field $H_o$ with respect to the crystal axes. The Hamiltonian of the system can then be written :

$$H = -\gamma\hbar\Delta H I_z - \gamma\hbar H_1 I_x + H'_D \quad . \qquad (3)$$

This is the Hamiltonian of a system of spins in a static field $\Delta H\hat{z} + H_1\hat{x}$, interacting through the truncated Hamiltonian $H'_D$ of eq. (2). It is plausible and has been verified experimentally that applying the rf field far off resonance, then sweeping the frequency (or the dc field) so as to reduce $\Delta H$ to zero, and finally suppressing the rf field itself, leads to a cooling of the dipolar energy $H'_D$ in a ratio $H_o/H'_L$, comparable to, but somewhat different from, $H_o/H_L$, (the truncated Hamiltonian $H'_D$ yields a somewhat different value $H'_L$ for the local field). Besides leaving the dipolar interaction cold in high field, and allowing a wealth of Hamiltonians $H'_D$ by changing the orientation of the field $H_o$ with respect to the crystal axes, the ADRF procedure has yet a third advantage : by choosing at will the initial sign of $\Delta H = (H_o-\omega/\gamma)$, the effective field before the ADRF, negative as well positive temperatures can be given to the truncated dipolar energy $H'_D$. Once the ADRF is over, the rf field $H_1$ can be suppressed.

c) DNP

Paramagnetic electronic impurities introduced in the sample, because of their short spin-lattice relaxation time, have no difficulty in reaching the temperature of the lattice (and of the refrigerator) which is of the order of a kelvin or so. The electronic spins interact with each other through a truncated electronic dipolar interaction $H'^e_D$ which corresponds to an internal electronic magnetic field $H^e_L$ of the order of say twenty gauss. Actually this concept of local electronic field for a randomly dilute system of spins should be used with caution. For most pairs of electronic spins it will be much less and for a few it would be a good deal more. The main point is that because of the large electronic gyromagnetic factor $\gamma_e$, the spread of the energy spectrum of $H'^e_D$, of the order of $\gamma_e H^e_L/2\pi \approx 50$MHz will have wings reaching into the nuclear Larmor frequency of the order of a hundred MHz. The nuclear Zeeman energy $Z_n$ and the electronic truncated dipolar $H'^e_D$ are thus "on speaking terms" and at the same spin temperature.

If now we perform an ADRF on the *electronic* spins with a microwave generator near the *electronic* Larmor frequency which is in the range of a hundred GHz, we should cool the electronic dipolar energy by a factor of the order of $H_o/H^e_L$ that is three orders of magnitude. Since $H'^e_D$ and $Z_n$ are on "speaking terms" this results in a cooling of the nuclear Zeeman energy $Z_n$ but by a far smaller amount

because of the heat capacity of $Z_n$, much higher than that of $H'^e_D$.
However once the ADRF is over, because of the short electronic relaxation time, one can repeat it again and again until $Z_n$ has reached a temperature smaller than that of the lattice by a factor of the order of $H_o/H^e_L$ that is reaching into the millikelvin range. In practice, instead of repeating the electronic ADRF many times one obtains a comparable result by applying the microwave field continuously at a distance from the electronic resonance of the order of $H^e_L$. Depending on the side of the resonance at which this microwave is applied, temperatures of either sign are obtained for $Z_n$.

Thus it is literally a cooling of nuclear spins that occurs in the DNP process. This is examplified by the fact that if there is more than one nuclear species in the sample with different magnetic moments and or spins, they reach the same nuclear spin temperature $T_s$ rather than the same polarization. The different nuclear polarizations resulting from this common spin temperature are easily computed by the classical formulae $P = B_I(\gamma_I \hbar H_o/k_B T_s)$ where $B_I$ is the well-known Brillouin function for spin I. Fig.10 shows the inverse temperatures reached by proton and deuteron spins in a substance containing both, plotted against time (CERN 1974). The fact that proton and deuteron points fall on a single smooth curve is convincing evidence of the existence of a single nuclear spin temperature.

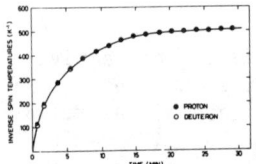
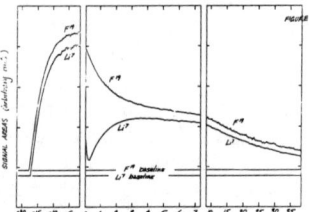

Fig.10 - Growth of inverse spin temperature for protons and deuterons during dynamic polarization.

Fig.11 - Thermal coupling of $^7$Li and $^{19}$F in dynamic polarization.

In Fig.11 (Saclay 1973) the following experiment was performed in dynamically polarized LiF. The polarization of $^7$Li is suddenly reduced by a strong rf field. Its polarization then grows at the expense of the $^{19}$F polarization at a rate which expresses the coupling between the two nuclear species. Neither one is directly on "speaking terms" with the other because of their widely different Larmor frequencies but they communicate indirectly through the electronic spin-spin energy. Having thus reached the same spin temperature, both species relax together at a much slower rate towards the lattice temperature.

The existence of a common spin temperature explains why in a sample where the proton polarization is within experimental error 100 %, the deuteron polarization does not exceed 45 %. This follows directly from the fact that $\gamma(D) \ll \gamma(^1H)$.

We pass onto the second part of this talk, "polarized targets elsewhere".

### Nuclear Antiferromagnetism and Neutron Diffraction

It is well known that interacting magnetic moments can order themselves below a certain critical temperature $T_c$ to form a new phase. The long range arrangement of moments in this ordered phase can be ferromagnetic, antiferromagnetic or even more complicated.

For the last 25 years the study of electronic antiferromagnetism has been dominated by neutron diffraction. An antiferromagnetic structure contains successive lattice planes with opposite magnetizations. Thanks to its own magnetic moment a neutron can "tell" a magnetic moment up from a moment down. It will "see" two successive lattice planes in an antiferromagnetic structure as different and the magnetic period of the lattice as twice that of the crystal.

Neutron diffraction can then give rise to extra Bragg reflections at an angle different from that for the normal lattice.

These extra Bragg lines, called superstructure lines or antiferromagnetic lines appear below the critical temperature.

We had set as our goals at Saclay a) the production of antiferromagnetic structures of nuclear spins originating in their mutual dipolar interactions b) the observation of antiferromagnetic Bragg peaks.

Both goals appear at first sight as rather formidable.

a) Production of nuclear antiferromagnetism

The critical temperature $T_c$ for a nuclear antiferromagnetic transition is given in order of magnitude by $k_B T_c \sim \mu H_L$ where $H_L$ is the nuclear local field and $\mu H_L$ is the magnetic energy of a nuclear magnetic moment $\mu$ in that field. By taking for $H_L$ a few gauss, which is the right order of magnitude for a system of nuclear spins and for $\mu$ the magnetic moment of, say, a proton a temperature $T_c$ of a microkelvin is obtained. It may be amusing for high energy physicists to remark that if one takes as a central energy 10 eV, average ionization energy of an atom, $\mu H_L \simeq 10^{-10}$ eV, is as far below it, as one TeV, energy planned in the Fermilab energy doubler, is above.

Such temperatures can in fact be obtained in a straightforward manner using DNP. As explained earlier, the dynamic polarization reduces the temperature of the nuclear Zeeman energy by a factor of the order of $H_o/H_L^e$, that is, something like three orders of magnitude. A second reduction of similar magnitude is then obtained by the demagnetization of the nuclear spins themselves. If the lattice temperature is a fraction of a Kelvin, this yields a final dipolar spin temperature in the desired microkelvin range.

The first nuclear antiferromagnetic structure was produced and observed (although not by neutron diffraction for reasons to appear shortly) 9 years ago at Saclay in a crystal of $CaF_2$ where the nuclei of $^{19}F$ form a simple cubic lattice.

b) Observation by neutron diffraction

One might fear that because the nuclear magnetic moments are three to four orders of magnitude smaller than electronic moments, neutron diffraction by an antiferromagnetic nuclear structure would be a very weak phenomenon. This is not so.

This is due to the fact that besides magnetic interactions there are sizeable *nuclear* interactions between the spin of the nucleus and that of the neutron which enable the neutron to "tell" a nuclear spin up from a nuclear spin down. The order of magnitude of these interactions is described conveniently by assigning to each nuclear species a hypothetical pseudo-magnetic moment $\mu^*$ which would provide a magnetic scattering amplitude equal to the actual nuclear spin-dependent scattering amplitude, of that nucleus. The pseudomagnetic moments of most nuclei were very poorly known and original methods, for measuring these moments have been developed at Saclay. The general name of nuclear pseudomagnetism has been proposed for these studies. It was a great disappointment that $\mu^*$ ($^{19}F$) has turned out to be very small (0.017 $\mu_B$). Otherwise neutron diffraction studies of nuclear antiferromagnetism would have been reported at least five years earlier.

Neutron Diffraction Study of Antiferromagnetism in Lithium Hydride

The possibility of performing a neutron diffraction study on a nuclear antiferromagnet is limited to substances whose nuclei have a sufficiently large pseudomagnetic moment $\mu^*$. Furthermore, for a first study it is advisable to choose a system as simple as possible where the prediction of antiferromagnetism has a good chance to be correct. These considerations have led to the choice of lithium hydride : the pseudomagnetic moment of the proton is the largest of all nuclei, $\mu^*(^1H) = 5.4~\mu_B$, and that of lithium 7 is $\mu^*(^7Li) = -0.62~\mu_B$. As for the crystalline structure of LiH, it is of the NaCl type, i.e. it consists of two intercalated f.c.c. lattices of $^1H$ and $^7Li$.

Fig.12 shows the antiferromagnetic structure expected at negative spin temperature when the external field is parallel to a four-fold crystalline axis. This is the field orientation to which the neutron diffraction study has been limited so far.

The whole experimental procedure : dynamic polarization and ADRF has to be performed in a high homogeneous field and at low temperature, that is in a superconducting magnet and with a dilution refrigerator. At the same time, free access must be provided to the incident and diffracted neutron beams : an unacceptable absorption would result from their crossing the coïl or an excessive thickness of dilute $^3He$.

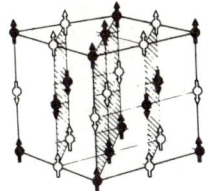

Fig.12 - Antiferromagnetic nuclear structure in lithium hydride LiH.

The superconducting magnet is of the split-coil type, which allows neutron-diffraction experiments in a plane perpendicular to the (vertical) magnetic field. The sample of LiH, in the form of a platelet of typical dimensions 5 × 4 × 0.5 mm is placed vertically in the cryostat. The thickness of the 6 % - $^3$He solution is limited to 0.5 mm on each crystal face, as a compromise between the necessities of cooling the crystal and reducing neutron absorption. The whole cryostat can be rotated around a vertical axis, so as to bring the crystal to the Bragg-reflexion orientation.

The lithium hydride sample contains of the order of $2.10^{19}$ paramagnetic F-centres per/cc. They are created by irradiation at liquid nitrogen temperature with a beam of 3-MeV electrons. The magnetic field is 6.5 Teslas, and the frequency of the microwave irradiation used for DNP is 185 GHz. During the polarization period, the temperature of the bath is around 200 mK. The maximum polarization, after 2 to 3 days of irradiation, corresponds to :

$p_H \simeq 95$ % ; $p_{Li} \simeq 80$ %.

The ADRF is performed on both $^1$H and $^7$Li spins, by using two rf fields of carefully adjusted frequencies, (of the order of 280 and 110 MHz, respectively).

Fig.13 shows the antiferromagnetic diffraction line 110 observed immediately after an ADRF with initial polarizations $p_H$ = 0.95 and $p_{Li}$ = 0.80. On the same Figure is shown for comparison the crystalline diffraction line 220 observed with the unpolarized sample. The reality of a long-range nuclear antiferromagnetic order whose structure is precisely that expected from theory, is thus directly established. The diffraction of neutrons by this nuclear antiferromagnet is a large effect, as anticipated from the large value of the proton pseudo-magnetic moment. The width at half-intensity of the antiferromagnetic line, equal to 0.65°, is much larger than that of the crystalline line 220 (∼ 0.2°). It is experimentally the same whatever the initial polarizations and the antiferromagnetic line intensity. This broadening is attributed to the existence of antiferromagnetic domains.

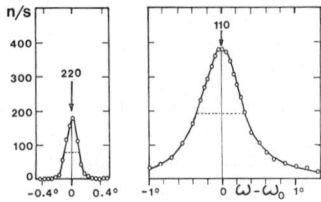

Fig.13 - Antiferromagnetic Bragg peak for the 110 reflection and normal Bragg peak for the normal 220 reflection.

This experiment, performed at Saclay last summer is the first direct proof of the long range order which establishes itself among nuclear spins solely because of their tiny dipolar interactions. The fact that the temperature obtained for nuclear spins is of the order of a microkelvin is of no importance per se : the significant fact is that at a temperature as low as this, something still happens, namely a thermodynamic phase transition.

Return to High Energy

A few remarks can be made with respect to the relevance of this to a discussion on polarized targets.

Strictly speaking the target of antiferromagnetic LiH is not polarized since the polarizations of $^7$Li and $^1$H are both zero and if its scattering of slow neutrons is so dramatically different from that of a spin-disordered target, it is because of the collective character of the Bragg scattering of each neutron by all the nuclei of the target.

Still, one may point out that even for fast particles with a wavelength much smaller than the lattice spacing, antiferromagnetic LiH is not equivalent to a spin-disordered LiH. This is because nuclei of $^7$Li (spin I = 3/2), belonging to two antiferromagnetic sublattices, although of opposite polarizations, have the same, non-vanishing alignment. Suppose for argument's sake that the polarization of each sublattice is unity in absolute value. The alignment

$3I_z^2 - I(I+1)$ will be 3 for all the $^7$Li nuclei of the target and will be perceived by a fast incoming particle.

The second remark concerns the value of LiH as a polarized proton target before the nuclear demagnetization. The high polarization $p_H = 0.95$ and the high free proton content $f = 0.25$ are attractive. Among its weaknesses one should emphasize the length of the polarization times, 2 to 3 days in our experiment.

A potentially interesting target material is $^6$Li D. At first sight it might appear that the polarization expected for D should be very much smaller than the 95 % observed for $^1$H in $^7$Li, 40 % at most, because of the spin-temperature law in DNP. This is not necessarily so. In contrast with the targets currently used in high energy physics the linewidth in lithium hydride is determined by the hyperfine couplings of the paramagnetic impurities (F centers) with neighbouring nuclei (mainly lithium). The magnetic moment of $^7$Li is roughly 4 times that of $^6$Li and one may expect in $^6$Li D an ESR line 4 times narrower than in $^7$Li H and possibly a nuclear spin temperature much lower also. This could compensate to some extent for the smaller $\gamma$ of D and $^6$Li. D and $^6$Li have the same spin, and the same magnetic moment within 4 %. They would thus carry nearly equal polarizations p which hopefully could exceed the 45 % observed until now. It is rather remarkable that $\mu(^6\text{Li}) \cong \mu(D)$ and also

$$\mu(^6\text{Li}) \cong \frac{\mu(H_e^3) + \mu(^3H)}{2} \quad \text{within 5 \%.}$$

If we accept for the wave function of $^6$Li the picture : $^6$Li $\sim$ D + $^4$He (or $^6$Li $\sim$ $^3$He + $^3$H) we are led to the conclusion that in a sample of $^6$Li D where both nuclei have polarization p the average neutron (or proton) spin polarization including bound nucleons is $p/2$, that is a factor $f = 0.5$ for both neutrons and protons.

All this is highly speculative but amusing. Among other problems, one would have to lay hands on $^6$Li D which, if one is not in the thermonuclear business, may prove difficult.

## References

Considering the introductory and at least in part historical nature of this talk it has appeared that references to books or review articles would be more suitable than references to the original papers (with a single exception for the most recent and yet unpublished work)

A. Abragam, 1961 "The Principles of Nuclear Magnetism" (Oxford : Clarendon Press)
M. Goldman, 1970 "Spin Temperature and Nuclear Magnetic Resonance in Solids", (Oxford : Clarendon Press)
C.D. Jeffries, 1963 "Dynamic Nuclear Orientation" (New-York Interscience)
A. Abragam and M. Borghini, 1964, Prog.Low.Temp.Phys. $\underline{4}$ chap.8
C.D. Jeffries, 1972 chap.3 "in Electron Paramagnetic Resonance Geschwind S (ed) (New-York Plenum)

M.Goldman, 1977 "Nuclear Magnetic Ordering" (Phys.Rep. 32 C 1)
A.Abragam and M.Goldman, 1978 (Reports on Progress in Physics 41,3) "Principles of Dynamic Nuclear Polarization"
Y.Roinel, V.Bouffard, G.L.Bacchella, M.Pinot, P.Mériel, P.Roubeau, O.Avenel, M.Goldman and A.Abragam, "First Study of Nuclear Antiferromagnetism by Neutron Diffraction", (Phys. Rev. Lett. to be published).

REPORT OF THE WORKSHOP ON POLARIZED TARGET MATERIALS*

Argonne National Laboratory
Argonne, Illinois 60439

October 23-24, 1978
and
October 30-31, 1978

| | |
|---|---|
| G. R. Court | Liverpool University |
| D. G. Crabb | University of Michigan |
| R. C. Fernow | Brookhaven National Laboratory |
| D. H. Fitzgerald | University of California at Los Angeles |
| S. W. Gray | Indiana University |
| D. A. Hill | Argonne National Laboratory |
| J. J. Jarmer | Los Alamos Scientific Laboratory |
| A. D. Krisch | University of Michigan |
| M. Krumpolc | University of Illinois, Circle Campus |
| T. O. Niinikoski | CERN |

---

*Work supported by the U.S. Department of Energy.

ISSN: 0094-243X/79/510015-26$1.50 Copyright 1979 American Institute of Physics

## OUTLINE

1. Introduction

2. Survey of "Clean" Target Materials

   2.1 $H_2$     2.6 $B_2H_6$, $LiH_4B$

   2.2 HD     2.7 $CH_4$

   2.3 $D_2$     2.8 $C_2H_6$

   2.4 $He^3$     2.9 $NH_3$, $NH_2CH_3$

   2.5 $Li^6D$

3. Radiation Damage in Polarized Target Materials

   3.1 Current Problems

   3.2 Damage Mechanisms

   3.3 DNP With Radiation Induced Free Radicals

   3.4 Optical Bleaching

   3.5 Effects on DNP

4. Dynamic Polarization with Stable $Cr^V$ Complexes

5. Miscellaneous Topics

   5.1 Figure of Merit

   5.2 Theoretical Remarks on DNP

   5.3 Cooling Power and Heat Transfer

   5.4 Choice of Magnetic Field for DNP

   5.5 EPR Measurements in a Polarized Target

6. Summary

   Acknowledgements

   References

1. INTRODUCTION

During the last two decades there has been an increasing interest in spin effects in high energy physics. The polarized proton targets which were developed some 20 years ago allowed the first precise experiments on spin effects. However, the field is now coming to the point where present polarized target materials are no longer totally adequate.

Recently large spin effects in high energy strong interactions have been observed in:
1. Large $P_\perp^2$ processes [1]
2. Inclusive processes [2]

It seems clear that better polarizable materials might significantly improve the data in these processes for the following reasons:
1. Large $P_\perp^2$ requires high beam intensity which reduces target polarization because of radiation damage
2. Inclusive data is seriously contaminated by events from the heavy C and O atoms in PPT materials

It thus seemed appropriate to reexamine the polarized target materials currently in use. An earlier examination of many of these materials was made by Borghini.[3] The workshop concentrated on an examination of
- radiation damage in polarized target materials
- a survey of "clean" target materials
- dynamic polarization results with the new stable Cr(V) complexes

In addition to the normal polarized target experts with backgrounds in high energy physics, low temperature physics and solid state physics, we included scientists with strong backgrounds in various areas of chemistry and radiation damage physics, as these areas were quite crucial to our goals. However, it is clear that much closer collaboration with experts in these areas will be necessary to find polarized target materials that allow more precise experiments on high $P_\perp^2$ processes and inclusive processes.

2. SURVEY OF "CLEAN" TARGET MATERIALS

Since the Berkeley Conference[4] of 1971 there have not been many advances in the polarizable nucleon content of materials used for polarized targets. In fact, the experimental situation with respect to some potential target materials has not changed greatly since Borghini's review of 1966.[3] In part this appears to reflect the inherent difficulty of working with highly hydrogen-rich substances. Also, most of the active groups have been kept busy doing the many scattering experiments which could exploit the hydrocarbon targets that came into wide use around 1970.

There are a number of light molecular species which have a polarizable nucleon content appreciably greater than that of currently used target materials. In what follows we review previously obtained results for each species and discuss the prospects for improvement, with emphasis on those species that either appear more tractable or are inherently more attractive from the point of view of polarizable nucleon content.

In general materials in which neutrons and protons can be independently polarized are the most useful. However, there are scattering experiments such as $\pi^\circ$ inclusive and jet production, in addition to the obvious ones in which recoils are identified, for which targets containing only polarizable pn pairs are not necessarily as a disadvantage. Thus, species such as $D_2$ and $Li^6$ are well worth considering.

## 2.1 $H_2$

"Brute force" polarization of the order of 10% may well be possible in $H_2$ with current magnet and refrigeration technology, although this has not been demonstrated. The attainment of larger polarization is made improbable by fundamental limitations on the heat transfer at the solid-liquid boundary of the ortho-para heat of converstion. In any case, such a target would have severe limitations in the areas of magnet aperture, polarization reversal, and very high field at the target. Dynamic polarization in $H_2$ is hindered because paramagnetic impurities catalyze the conversion of nearby molecules to the non-magnetic para form.

## 2.2 HD

"Brute force" polarization in HD is eminently feasible. Bozler and Graf[5] have obtained 40% proton polarization in a field of 10T with a dilution refrigerator at 23 mK, in a sample doped with a small amount of ortho-$H_2$ to relax the nuclear spins. However, a target of this sort would again suffer from restricted aperture, difficulty of polarization reversal, and high ambient field. Also, at these temperatures beam heating effects would already be appreciable at a charged particle flux of $10^6/cm^2sec$.

Honig[6] has proposed a method for surmounting most of the above problems by exploiting the ortho-para conversion of the $H_2$ dopant. Hereafter we refer to this method as the "modified brute force" (MBF) method. New problems arise in this method, to be discussed later.

Dynamic nuclear polarization (DNP) in HD has been a tantalizing possibility for better than a decade, but to date results have been sparse and not highly encouraging. Nevertheless, there are a few hopeful hints, and the stakes are high enough to warrant a vigorous pursuit of this possibility in the near future.

DNP in HD would presumably be based on a "solid effect" acting on a dopant of paramagnetic H-atoms. Other types of center can certainly be induced in HD, but H-atoms are the most straightforward. We consider four different methods for producing H-atoms dispersed in HD.

A.  IRRADIATION

Rebka and Solem[7] have obtained a proton polarization of 3.7% and a deuteron polarization of $\sim$0.4% in HD at 1.2T and 1.2K, by the "solid effect" on one of the two EPR transitions of H-atom impurities. The H-atoms were induced by irradiation of solid HD in a bremsstrahlung beam of 60 MeV maximum energy. Total dose was $\sim 10^7$R which resulted in $\sim 10^{-4}$ H-atoms per HD molecule. By raising the field to 2.5T, lowering the temperature to 0.4K, and saturating both "forbidden" EPR transitions, one (naively) might expect to attain $P_H \sim 2 \times 3 \times 2 \times 3.7 = 44\%$.

Experiments of this sort are difficult and tedious. They require access to an electron accelerator and work in a high-radiation area. Worse, irradiation, probably inevitably, produces "foreign" paramagnetic species and $H_2$, $D_2$ recombination products, which are depolarizing and complicate the physics. One can wait for the $H_2$ to convert to para-$H_2$, but this requires a long wait after irradiation, $\sim$10 days for Rebka and Solem.

B.  TRITIUM SPECIES DOPING

Here one mixes HT or $T_2$ with HD, freezes the mixture and waits for decay and ionization products to build up. One has good control over initial concentrations, and requirements for special apparatus are minimal. However, the physics is again rather complicated. This technique has been used in $D_2$, with limited success.[8]

C.  ATOMIC BEAM DOPING

Rebka[9] has proposed that a relatively clean and controllable method of doping HD would be to mix an RF discharge-produced atomic-H beam with a molecular beam or gas stream of HD, and then freeze the mixture. This method has the advantage that $H_2$ and unwanted paramagnetic species could be minimized.

D.  SPARK DISCHARGE DOPING

Here HD gas passes through a spark discharge, or, $H_2$ gas passes through the discharge and is added to a stream of HD gas, before freezing. This is similar to method C, perhaps somewhat simpler but less controllable.

Any target utilizing HD, polarized by whatever means, will have certain advantages and disadvantages relative to current targets, inherent in the material.

Advantages

1. The NMR linewidths are fairly small:

$$\Delta f_H \sim 19 \text{ kHz}, \Delta f_D \sim 1.3 \text{ kHz}.$$

This makes thermal equilibrium (TE) calibrations easier and more reliable.

2. Target insertion and removal can in principle be quite rapid, at cryostat temperatures which never have to exceed 30K.

Disadvantages

1. Spatial density of H is only $\sim 1/2$ that of current targets.
2. The target is destroyed whenever the cryostat temperature rises much above 5K. Since currently visualized HD targets would have lengthy "maturation" periods, this is a serious reliability problem. However, this period could be usefully used for unpolarized "dummy" running.
3. The density of the target may be difficult to determine and reproduce accurately.
4. In order to take full advantage of the low background of HD, cryostats with less material intercepting the beam must be used.
5. So far, high purity (>> 99%) HD is not readily available, from commercial sources.

The following are lists of advantages and disadvantages peculiar to MBF or DNP, listed roughly in order of decreasing importance.

HD-MBF Advantages

1. Simpler physics than HD-DNP.
2. Target preparation method is already fairly clear.
3. No homogeneous field is required, except for dynamically polarized D.
4. No microwaves are required. But strong RF is needed for polarized D.

HD-MBF Disadvantages and Problems

1. Magnet failure destroys target.
2. Have to hold stable low temperature for $\gtrsim$ 2 weeks.
3. Target not available during this period.
4. Probably much more radiation-sensitive than HD-DNP. The maximum tolerable flux of charged particles is $\sim 10^7/cm^2$-sec.
5. Target or polarizing magnet <u>must</u> be movable.
6. Homogeneous field required for D-DNP.
7. Polarization reversal: adiabatic passage or reversal of the holding field are the only possibilities.
8. Need high-field magnet, $\sim$ 10T.
9. TE calibration unavailable for long periods.

HD-DNP Advantages

1. Some DNP attainable immediately following target preparation. In Solem's experiments, pol. $\sim 1/3$ of maximum, immediately after irradiation, rising to maximum with $\sim$ 5 day time constant.

2. Magnet failure does not destroy target.
3. Probably much less radiation-sensitive than HD-MBF.
4. Continuous very low temperatures are not required.
5. Pol. reversal dynamically.
6. Does not require high-field magnet.
7. No need to move either target or magnet.

HD-DNP Disadvantages and Problems

1. How to prepare target without use of bremsstrahlung radiation.
2. Physics more complicated than HD-MBF.
3. The EPR linewidth went from $\sim$ 10G to $\sim$ 20G in going from 4.2 to 1.2K, in Solem's case.
4. D and H polarizations occur at different microwave frequencies. (This could also be an advantage if H and D polarization is wanted separately.)
5. The apparent necessity of $O_2$-impurity means another parameter to control, and complicates the radiation chemistry.
6. Requires two sources or a modulator for two microwave transitions? Separation is $\sim$ 1.4 GHz.

It is clear that the problems with the MBF method are relatively greater and that a DNP approach would be far more flexible. However, the MBF method has one big advantage in that the essential features have all been proven.

## 2.3 $D_2$

Rebka and Waine[10] achieved $\sim$ 1% polarization at 0.8T and 1.2K in solid $D_2$ containing D-atoms. Procedures for doping similar to those for HD may be possible.

## 2.4 $He^3$

$He^3$ would make an attractive neutron target. Optical pumping has yielded polarizations in excess of 50% in low pressure $He^3$ gas.[11] Although it is possible to preserve polarization while compressing the gas into the NTP range,[12] attempts to liquify the polarized gas meet with severe difficulties.[13] Attempts to polarize solid $He^3$ by the use of Pomeranchuk cooling have met with more success, yielding polarization of at least 47% at 5.4T at temperatures of less than 5mK.[14] In addition to a few other problems with this method, beam heating would probably cause the overall figure of merit of such a target to be less than that of conventional deuterated targets.[15] The beam flux would be limited to $\sim 10^5/cm^2$-sec.

## 2.5 $Li^6D$

Abragam and co-workers[16] have demonstrated polarizations of $P_H \sim$ 95%, $P_{Li7} \sim$ 80% in $Li^7H$. The sample was a crystal 5 x 5 x .5mm which was irradiated by $\sim 10^{18}/cm^2$ of $\sim$ 3MeV electrons. Dynamic cooling on the resulting centers was carried out in a dilution refrigerator at $\sim$ .5K in a 5T field. Equal spin temperatures of H and $Li^7$ were attained, with a polarizing time constant of $\sim$ 2 days.

While the figure of merit of a $Li^7H$ target would not be better than existing targets, the above result, along with other considerations, suggests that $Li^6D$ might be highly polarized. The nuclear ground-state of $Li^6$ is predominantly $^3S_1$ and the magnetic moment is nearly equal to that of the deuteron, which implies that $Li^6D$ may be thought of as a deuteron plus a "bound deuteron", both polarizable. In other words, four nucleons out of a total of eight may be polarized.

2.6  $B_2H_6$, $LiH_4B$

These materials, while interesting from the point of view of hydrogen content, are considered to be too unstable and reactive to be serious candidates.

2.7  $CH_4$

Chemical doping of methane might be achieved by mixing it with free radical vapor at room temperature, and then freezing the mixture (see $C_2H_6$). A more likely possibility would be to create $CH_3$-radicals or H atoms in the gas phase before freezing, or to irradiate solid methane. Methane is sufficiently hydrogen-rich as to make these possibilities well worth exploring. A previous doubt about the equilibrium spin-state of solid methane now appears to be clarified.[17-19]

2.8  $C_2H_6$

Methods of doping already discussed might be tried on ethane. Also, Hwang[20] has mixed DTBN vapor with ethane and observed the presence of the $N^{14}$ hyperfine structure (hfs) in the EPR of the solid. This at least indicates that a good dispersion of the radical in the solid can probably be achieved by such a method.

2.9  $NH_3$, $NH_2CH_3$

Scheffler[21] obtained sizeable polarization in solid ammonia doped with glycerol: $Cr^V$, but the time required for polarization reversal was many hours. Scheffler attributed the long time constant to "clustering" of the glycerol: $Cr^V$ in the ammonia matrix. When pure $Cr^V$ species are used with no third additive such as glycerol, the results are more extreme: very small dynamic polarization and very long nuclear time constant (see Sec. 4). These results seem best explained by the ease with which ammonia tends to crystallize and exclude foreign materials from the crystals. If this is true, there is probably a similar problem with most hydrogen-rich substances, which tend to have small, easily ordered molecules.

It should be noted that crystalline materials may have another difficulty in addition to this reluctance to accept impurities into the lattice. That is, dopants which do manage to become incorporated may themselves show a preferred orientation correlated with the orientation of the crystallite. Dopants such as $Cr^V$ which depend on g-anisotropy broadening for much of their polarizing ability will then not be effective.

It might be possible to find some additive which would enable ammonia to form a glassy solid. Or, one may consider the asymmetric molecule of methylamine ($CH_3NH_2$) to be a more likely candidate for glass formation.

## 3  RADIATION DAMAGE IN POLARIZED TARGET MATERIALS

### 3.1  Current Problems

It is well known that radiation damage steadily reduces the polarization in all the currently used polarized target materials.[22-25] The proton polarization decays roughly exponentially with accumulated radiation flux $\Phi$

$$P = P_o e^{-\Phi/\Phi_A} \qquad (1)$$

where $\Phi_A$ is the characteristic flux for the polarization to decay by 1/e. We use the subscript A to indicate that much of this damage can be repaired by annealing. In general there is a different characteristic flux for positive and negative enhancement signals. The standard butanol-$H_2O$/porphyrexide (PX) and diol/Cr(V) targets have $\Phi_A$ values of $1 - 4 \cdot 10^{14}$ minimum ionizing particles/$cm^2$. In addition the nuclear spin lattice relaxation time $T_{1N}$ is observed to decrease roughly as one over the dose. This loss of polarization entails additional running time to measure some asymmetry with a fixed error. Still more time is lost since the material must be frequently annealed and changed.

It is also found[22,25] that the polarization after an anneal slowly declines from the starting polarization. This indicates that there is probably more than one type of damage mechanism occuring in the material. The polarization following annealing appears to follow the empirical relation

$$P_o(ann) = P_{asym} + (P_o - P_{asym}) e^{-\Phi/\Phi_{NA}} \qquad (2)$$

where $P_{asym}$ and $\Phi_{NA}$ are material and enhancement dependent parameters.

In the diol-Cr(V) system the plus enhancement state appears to be more resistant to the annealable type of damage but less resistant to the non-annealable type (see Table 3.1). The damaged sample has a higher effective g value with the greatest shift occuring in the negative enhancement peak.

Table 3.1 Properties of Current Polarized Target Materials (.5K, 2.5T)[25,26]

| Material | $P_o^+$ | $P_o^-$ | $T_{ann}[K]$ | $\phi_A^+$ | $\phi_A^-$ | $\phi_{NA}^+$ | $\phi_{NA}^-$ | $\frac{P_a^+}{P_o^+}$ | $\frac{P_a^-}{P_o^-}$ | (77K) $G(e_{trap})$ | $G(Rad)$ |
|---|---|---|---|---|---|---|---|---|---|---|---|
| ethanediol-Cr(V) | 82 | -82 | 160 | 2.3±.3 | 1.8±.2 | .8±.2 | .8±.2 | .69 | .89 | 1.5 | ~4 |
| 1,2 propanediol-Cr(V) | 82 | -82 | 180 | 2.0±.3 | 1.6±.2 | .8±.3 | 1.0±.3 | .73 | .90 | | |
| n-butanol/$H_2O$-PX | 68 | -62 | 130 | 2.2±.5 | 4.5±1.0 | | | | | 0.7* | |

*In pure butanol.

In the butanol-$H_2O$-PX system the negative enhancement state is more resistant to the annealable type of damage. The optimum microwave frequency for positive polarization decreases somewhat, while that for negative remains essentially constant.[27] When the last point is looked at in more detail by doing a complete plot of polarization against microwave frequency on a heavily damaged sample, it is seen that the line shape is asymmetrically broadened on the lower frequency side compared with the same plot for undamaged material which is essentially symmetric. Thus, both materials suffer an optimum frequency shift in the enhancement closest to g = 2. These same enhancements have the most annealable-type radiation damage.

Butanol can be annealed to extend its working life by heating it to approximately 120K. After each anneal the proton relaxation time is typically 5% longer than its previous value and the polarization is slightly smaller. This overall increase in relaxation time, and hence polarization reversal time, and decrease in polarization mean that the practical total life of a target involves typically not more than ten anneals.

3.2 Damage Mechanisms

The mechanisms by which the beam damages the target material are complicated and material dependent. Polarized target materials consist of paramagnetic centers such as prophyrexide or Cr(V) embedded in an irregular lattice of alcohol molecules. Irradiation of the material with high energy beams results in ionization energy being deposited in the material. This energy results in partial disassociation of the solid matrix and creation of new species of free radicals, ion fragments, and molecules.

Radical formation in aliphatic alcohols RHOH is described in Ref. 26. The major paramagnetic products are the free radicals ROH, RH, H, and trapped electrons, ion fragments such as $RHOH_2^+$ and $RHO^-$, and molecular $H_2$ and $H_2O$. Other species are present at intermediate steps in the reaction. However, EPR spectra show that only a few major radicals are produced in a given material.

One species of radicals ROH is formed by the elimination of an α hydrogen atom. Thus the hydroxyalkyl radical (·) $CHOHCH_2OH$ is formed in ethanediol and $C_3H_7C$(·)HOH in n-butanol. The hydrogen atoms in the -OH group do not appear to be significant in the EPR hyperfine structure. The EPR spectra for these radicals at 77K and 600 MHz are centered near g = 2 with a width of 100 gauss for ethanediol and 140 gauss for n-butanol.[28] The second major species of free radical is hydrogen atoms. Their EPR spectra show a doublet centered around g = 2 and with a separation of 508 gauss. The remaining major radical is trapped electrons. These cause a singlet line near g = 2 in the EPR spectrum with widths 8-20 gauss. They have an optical absorption peak at visible wavelengths and cause an intense coloration in the damaged material.

In addition to radical formation the beam may also cause damage to the desired paramagnetic species. For example, paramagnetic Cr(V) may be converted to Cr(IV), Cr(VI), or paramagnetic Cr(III) by reacting with radiation products. There is also evidence that porphyrexide is depleted during the course of irradiation.

Thermal annealing successfully depletes the trapped electron concentration by liberating the electrons and allowing them to recombine with position ions. In addition radical concentrations may be greatly reduced by heating to 0.6 - 0.7 of the melting temperature. Trapped electrons are freed below this temperature.

## 3.3 DNP With Radiation Induced Free Radicals

In order to better understand the effects of the produced free radicals tests have been made with a sample of undoped butanol by irradiating it in a 4 GeV bremsstrahlung beam.[27] The measurements were made at a field of 2.53T and a temperature of 1K. During the irradiation, attempts were made to polarize the sample. It was found that the polarization increased linearly with dose, reaching values of +7 and -6% after a dose equivalent to eight times the dose which would reduce the polarization to 0.75 of its initial value in a normally doped material. The total absorbed dose was $\sim \Phi_A$. Over this irradiation period the proton relaxation time fell from an initial value of 1500 secs. to 550 secs. The material was then annealed in the normal way. The proton relaxation time was found to have returned approximately to its initial value while the polarization had not appreciably changed.

A further small irradiation was then made before the experiment had to be terminated because no further beam time was available. The maximum polarization reached was +8% and it still appeared to be rising linearly with dose.

A polarization against microwave frequency plot carried out immediately after the anneal showed a shape and width very similar to that for an undamaged phorprexide sample but shifted $\sim$ 0.2% lower in frequency. This shows that the non-annealable centers have a g value approximately 2.001. When the microwave plot from the damaged undoped sample was added to the microwave plot for the undamaged porphyrexide sample the result was in excellent agreement with that obtained for the damaged normally doped sample both with respect to shape and absolute size (see Fig. 3.1).

When the undoped sample was finally removed from the cryostat it was observed carefully while it was allowed to warm up. The beads, which when initially loaded had been completely colorless were seen to have a strong grey-purple color which disappeared quite suddenly just before they melted.

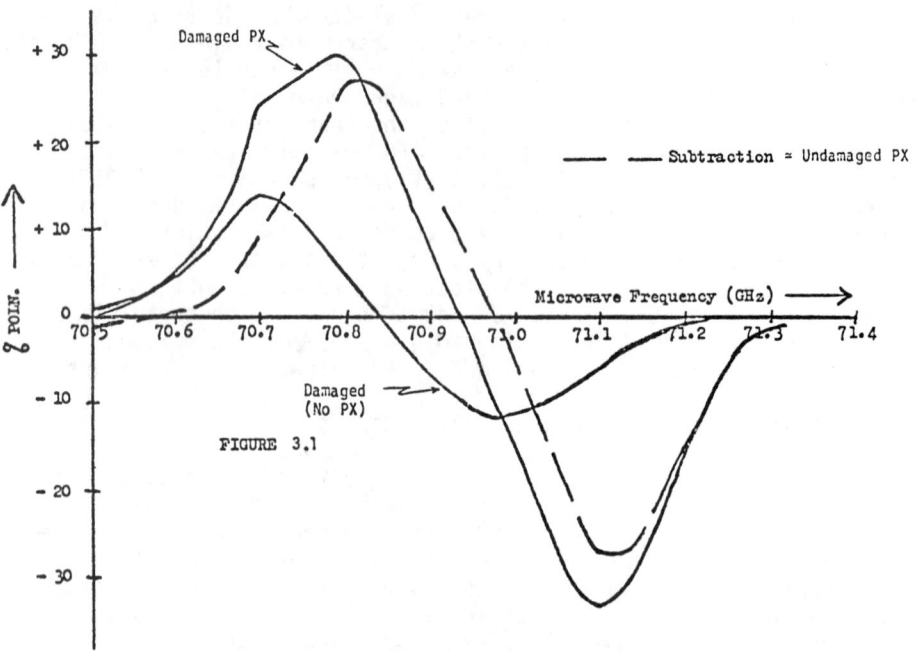

FIGURE 3.1

## 3.4 Optical Bleaching

We have seen that at least some of the annealing centers are associated with color centers. It should be noted that the formation of red purple color centers in butanol and other alcohols by radiation damage at 77K has been reported elsewhere.[28] In these measurements it was generally observed that irradiated material showed a broad absorption spectrum in the range 200 to 500 nm peaked at 500 nm with a second peak in the ultra violet. The visible region absorption could be removed either by annealing the material at temperatures $\approx$ 110K or by exposing the material to white or monochromatic light of wavelength 540 nm (optical bleaching). The ultra violet absorption line was unaffected by annealing but could be partially bleached by exposure to radiation of 250 nm wavelength. The EPR singlet line was effectively removed. The presence of other paramagnetic centres associated with the ultra violet absorption was demonstrated by EPR.

This immediately suggests two further developments. Firstly it may be possible to optically bleach a target material. This could have very important operational advantages as the time now employed for heating the material to 120K and cooling it back to operating temperature is a significant part of the running time. It is also quite conceivable that, because the energy input required to bleach the centers is unlikely to be very much higher and could be much less than that required to produce them, a process of continuous optical bleaching would be possible. The technological problems involved do not look too serious as targets have already been used with light guides for scintillators operating close to target temperatures.

Secondly if radiation damage centers are actually used to produce the polarization then once the optimum concentration has been obtained, continuous optical bleaching could be used to maintain this optimum, again possibly giving a very long life time. Alternatively if more than one type of center is produced then the unwanted ones could perhaps be removed by selective bleaching. Again the technology would not appear to present serious problems as high power tunable dye lasers are currently available.

One possible drawback is that in optical bleaching, unlike thermal annealing in damaged alcohols, the freed trapped electron interacts to form a new radical of the ROH type. Thus the total number of induced paramagnetic centers is not reduced. The experiment would be worthwhile however because it isn't clear at this point which radicals are most deleterious to DNP.

## 3.5 Effects on DNP

In the preceding sections we have described the loss of dynamic polarization with radiation damage and various products formed in the radiolysis of the target material. The results cited there rest on at least some experimental evidence. In this section we become more speculative and consider how the radiation damage may cause the loss of polarization.

The results of section 3.2 show that there are at least 3 species of radiation induced radicals present besides the original paramagnetic dopant. These are the trapped electron, $e_{trap}$, hydrogen atom $\dot{H}$, and hydroxyalkyl radical $\dot{R}OH$. The data in section 3.3 show that at least one of the radicals produced in undoped butanol can participate in DNP. When the sample was annealed the polarization remained and $T_{1n}$ increased back to its starting value. Since the $e_{trap}$ are annealed (or bleached) easily and there is no evidence for $\dot{H}$ at 77K we can probably assume that these are the species that anneal and are responsible for the shortened relaxation time. Since the polarization didn't change after the anneal we assume that the $\dot{R}OH$ radical is responsible for the DNP, does not anneal at 120K, and that it does not appreciably shorten $T_{1n}$. Presumably then long-term non-annealable damage results from the build up of the $\dot{R}OH$ or other radicals and destruction of the original paramagnetic dopant.

It is not clear at this point precisely how the presence of these new paramagnetic centers affects DNP. The most likely possibilities are (1) increasing the leakage factor by shortening $T_{1n}$, (2) modification of the EPR line shape, (3) destruction of the original paramagnetic entity and (4) short-circuiting the original spin temperature reservoirs to the lattice. To resolve these questions it would be extremely helpful to have an EPR spectrometer connected to a working (irradiated) polarized target. Then one could observe the growth of new paramagnetic centers with radiation damage. One could also perform careful annealing studies by warming up until specific radicals anneal out and then recooling to see how that radical affected DNP. Unfortunately problems with attenuation in the long waveguides and the lack of a tuned cavity make this direct approach very difficult. However, the same EPR information can be obtained from the NMR signal using the NEDOR[16] technique (see Sec. 5.5). These experiments could be very significant in helping to elucidate the effects of radiation damage on DNP. Coarse information on radical production can also be obtained using a bolometer in the target cavity and sweeping the magnetic field.

Further experiments with radiation induced radicals in <u>undoped</u> materials are clearly desirable. Targets doped in this way would eliminate the Cr(V)/porphyrexide destruction problem and at least part of the EPR interference problem. The big questions are how large a $P_n$ can be achieved this way and how much more radiation resistant the target would be. This experiment could be coupled to a test of optical bleaching. In this case trapped electrons would provide even more of the ROH radicals which are presumably responsible for the polarization.

Other useful experiments could include a study of the effects of the addition of electron scavengers to the target material to change the annealing properties and an investigation of how $\Phi_A$ depends on T and H. One should investigate the addition of naphthalene and related compounds to butanol to improve its radiation resistance, and DNP in isopropyl-benzene which should freeze to a suitable glass.

Finally we mention the interesting possibility that HD may have superior radiation resistance properties. It has low density, the D$^{\bullet}$ radical decays away at low temperature[7], and it probably has a low Kapitza resistance and high thermal conductivity. It would be quite remarkable if the same substance simultaneously solved the problems of high hydrogen content and radiation resistance.

4.  DYNAMIC POLARIZATION WITH STABLE Cr(V) COMPLEXES

Recently prepared chromium (V) complexes[29,30] of tertiary α-hydroxy acids possess physical and chemical properties which are of considerable importance for polarized target experiments. They are remarkably stable and soluble in solvents with high hydrogen content (water, alcohols, diols, ammonia), and are easily accessible in pure form.

The chromium (V) complexes of the general structure

$$Na^+ \left[ \begin{array}{c} R_2\backslash \phantom{C}R_1 \\ \phantom{R}C - O \\ | \\ C - O \\ O \phantom{=} \end{array} \begin{array}{c} O \\ \| \\ \phantom{/}Cr \\ / \phantom{Cr}\backslash \end{array} \begin{array}{c} O - C{=}O \\ | \\ O - C\phantom{/}R_1 \\ \phantom{O - C}\backslash R_2 \end{array} \right]^-$$

(1) $R_1 = CH_3$, $R_2 = C_2H_5$
(2) $R_1 = R_2 = CH_3$
(3) $R_1 = R_2 = C_2H_5$
(4) $R_1 = R_2 = C_4H_9$
(5) $R_1, R_2 = (CH_2)_4$
(6) $R_1, R_2 = (CH_2)_5$
(7) $R_1 = CH_3$, $R_2 = C_6H_5$

were prepared in 60-90% yields by the reaction of anhydrous sodium dichromate in acetone with the following hydroxy acids: 2-hydroxy-2-methylbutyric (HMBA), 2-hydroxy-2-methylpropionic (HMPA), 2-ethyl-2-hydroxybutyric (EHBA), 2-butyl-2-hydroxyhexanoic (BHHA), 1-hydroxycyclopentanecarboxylic (HCpCA), 1-hydroxylcyclohexane carboxylic (HCCA), and 2-hydroxy-2-phenylpropionic (HPPA) acids. The synthesis is carried out at room temperature and can be accomplished in two to three days. The stoichiometry of the reaction is $Na_2Cr_2O_7 + 5\ R_1R_2C(OH)COOH = 2\ Na[OCr(O_2COCR_1R_2)_2] + R_1R_2CO + CO_2 + 5\ H_2O$.

Some physical and chemical properties are summarized in Table 4.1, while a summary of dynamic polarization results achieved to date is given in Table 4.2.

Table 4.1 Solubility and Stability of Chromium(V) Complexes Of Tertiary α-Hydroxy Acids $R_1R_2C(OH)COOH$ [29,30]

| Complex | Solubility[a] Water | Acetone | Degree of Decomposition[b](%) In $H_2O$ | Thermal Decomposition (°C) |
|---|---|---|---|---|
| Cr(V)-HMBA (1) | 290 | 103 | 58 | 180 |
| Cr(V)-HMPA (2) | 270 | 4.1 | 81 | 180 |
| Cr(V)-EHBA (3) | 190 | 203 | 28 | 180 |
| Cr(V)-BHHA (4) | 16 | 87 | 27 | 180 |
| Cr(V)-HCpCA(5) | 204 | 730 | 100 | 170 |
| Cr(V)-HCCA (6) | 340 | 710 | 100 | 170 |
| Cr(V)-HPPA (7) | 170 | 810 | 100[d] | 140 |

a - g/L, 25°C
b - 0.01 M solutions of chromium(V) complexes followed over a period of 24 hr at 25°C.

Table 4.2
Summary of Stable Cr(V) Results

| Material[1] | Complex | [$10^{19}$ spins/ml] | $P_H$ | $P_D$ | $\nu_N$ [MHz] | f± [GHz] | Tin at T [M] | T [K] | τ(70%) at Q at T [M] | [mW/g] | [K] | Refrig. | Degas[6] | Ref. |
|---|---|---|---|---|---|---|---|---|---|---|---|---|---|---|
| EG | HMBA | 10 | 76 ± 2, -71 ± 2 | | 108 | Std Cr(V) | 9 | .89 | 8 | 5 | ~.7 | He[3] | Y | ANL[31] |
| EG | HMBA | 7 | 77 ± 2, -78 ± 2 | | 108 | Std Cr(V) | 12 | .89 | 9 | 5 | ~.7 | He[3] | Y | ANL[31] |
| $NH_3$ | HMBA | 7 | ~ ± 1.5 | | 108 | Std Cr(V) | ~160[7] | .89 | 60 ± 30 | 5 | ~.7 | He[3] | | ANL |
| PeOH | HMBA | 2.6 | 51 ± 3, -59 ± 3 | | 106.8 | 69.31, 69.74 | 160 ± 30 | .48 | | | | He[3] | | MICH[32] |
| PeOH | HMBA | (sat.) | 50 ± 3, -58 ± 3 | | 106.8 | | 103 ± 28 | .48 | | | | He[3] | | MICH[32] |
| PeOH | EHBA | 2.7 | 61 ± 3, -63 ± 3 | | 106.8 | 69.26, 69.68 | 244 ± 60 | .48 | | | | He[3] | | MICH[32] |
| PD | EHBA | 10 | ±82 ± 4 | | 106 | | 20 | .58 | | | | He[3] | Y | KEK[33] |
| BuOH | EHBA | 3 | 75, -80 | | 106.45 | | 21 | .78 | 15 | 3 | .48 | He[3] | Y | CERN[34] |
| PD-$d_6$[2] | EHBA | 10 | (±97)[3] | ±40 ±2 | 106.50 | 69.23, 69.71 | 3.5 | .98 | 10 | 5 | .4 | DR | Y | CERN[34] |
| EG-$d_6$[4] | EHBA | 10 | | 24 ±3 | 106 | | | | | | | He[3] | | KEK[33] |
| BuOD-$d_{10}$[5] | BHHA | 10 | (±94)[3] | ±30 ±3 | 106.50 | 69.13, 69.69 | 1 | .98 | 5(D) | 5 | .4 | DR | Y | CERN[34] |

(1) Materials
  EG: ethylene glycol
  PD: propanediol
  BuOH: 1-butanol
  PeOH: 1-pentanol

(2) $CD_2OHCDOH^CD_3$ Rapid spin-temperature equilibrium (STE) seen between species.
(3) These large $P_H$ cannot be obtained in undeuterated samples.
(4) $(CD_2OD)_2$
(5) Sample contained 2% H STE observed.
(6) Dissolved oxygen removal.
(7) Non-exponential.

# 5. MISCELLANEOUS TOPICS

## 5.1 Figure of Merit

There are many factors which may enter into the figure of merit of a polarized target. The following list includes factors that are frequently of interest. The relative importance of each factor varies with the type of scattering experiment. Reports concerned with potential target materials should ideally contain enough information to determine these factors.

1. target polarization(s)
2. $\Sigma p_\uparrow/\Sigma N$, $\Sigma n_\uparrow/\Sigma N$, $\Sigma N_\uparrow/\Sigma N$
3. radiation resistance, $\Phi_A$ and $\Phi_{NA}$
4. density
5. polarization reversing time
6. independent polarizability of $n_\uparrow, p_\uparrow$
7. atomic numbers and weights present
8. duty cycle, "dead" times
9. magnet field and homogeneity requirements
10. existence of a suitable dummy
11. susceptibility to beam heating

## 5.2 Theoretical Remarks on DNP

Many publications cover this subject.[16,35,36] We only wish to point out here the differences between DNP in LMN (and many other crystalline anisotropic solids) in which the "solid effect" is the dominant mechanism, and isotropic, non-crystalline organic solids doped with paramagnetic inpurities, in which the major source of EPR broadening is due to the g-tensor anisotropy (dynamic cooling of nuclear spins).

The main difference is that the "solid effect" can be understood in terms of the response of isolated electron-nuclear spin pairs without any consideration of their interactions among other spins of their species, whereas the response of g-tensor broadened EPR to microwaves can only be handled when these interactions are taken into account. Another important difference is that in the solid effect the transitions induced by microwaves are not "allowed", but in dynamic cooling they are. Two consequent differences follow from these considerations: (1) power absorption in the "solid effect" is very weak but in "dynamic cooling" it is strong (2) response of LMN-type material is "transition probabilistic" whereas in organic targets it is entirely collective and can only be handled by quantum statistics (i.e., thermodynamics). This is why we speak of spin temperature and dynamic cooling in the latter case.

## 5.3 Comparison of Cooling by Evaporation and Dilution of $^3$He

Specifying a target's performance by coolant (helium) temperature is all right if a comparison is not being made between different cooling systems. As is well known, what matters for the end result in DNP is the material temperature, not the coolant temperature. We wish to point out here how evaporation and dilution cooling systems differ from each other and why the <u>heat transfer</u> is more important than the coolant temperature in most practical cases.

The cooling powers and heat transfer characteristics[37] are given below for the two systems.

|  | COOLING POWER | HEAT TRANSFER |
|---|---|---|
| Dilution | $\dot{Q}_m \simeq 12.5 \, \sigma S T_m^4$ | $\dot{Q}_m \equiv \sigma_t S_t (T_L^4 - T_m^4)$ <br> (Due to Kapitza Resistance) |
| He$^3$ Evaporation | $\dot{Q}_m = \dfrac{\dot{V}_{pump}}{RT_{pump}} \, pL_3^o \sim \exp^{-L_3^o/kT}$ | $\Delta T \simeq c \cdot \dfrac{\dot{Q}_m}{\sigma_t}$ <br> (Due to Nucleate Boiling Resistance) |

$\sigma, \sigma_t$ = heat exchanger and target surface area.

$S, S_t$ = surface boundary conductance in the heat exchanger and target.

$\dot{Q}_m$ = load into the mixing chamber, or $^3$He evaporator applied to the target by microwaves.

$L_3^o$ = $^3$He latent heat of evaporation.

Using the above we can draw the target temperature $T_L$ as a function of applied load $\dot{Q}_m$:

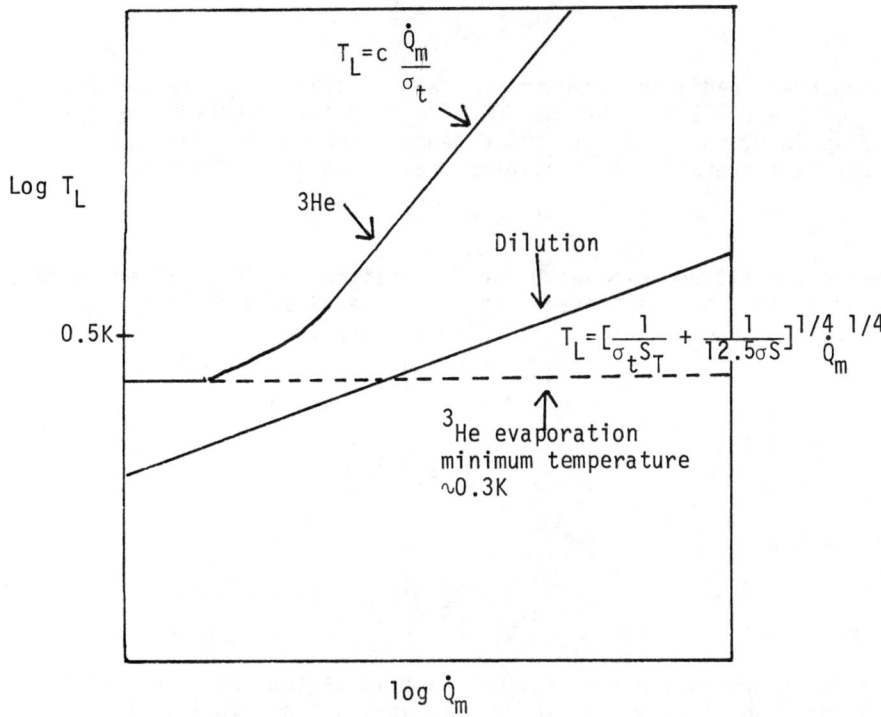

We conclude that it may well be that the target in a dilution refrigerator is colder at any heat load. The polarization results support this picture.

## 5.4 Choice of Magnetic Field for DNP

Currently almost every laboratory uses a 2.5T magnetic field for DNP in polarized targets. This is mainly due to the availability of microwave sources from the LMN-era. However, it is by no means clear that 2.5T would give optimum results for the maximum polarization or reversal speed. In fact, it would seem rather natural that there would be an optimum field for DNP in each substance.

To illustrate this, let us consider the spin-temperature model, which gives

$$\max \{P\} = \tanh \left[\max \{\frac{\overline{\mu} \cdot \overline{B}}{kT_S}\}\right]$$

for spin $\frac{1}{2}$ (or a corresponding Brillouin function for spin I). The spin temperature can be related to the lattice temperature $T_L$ by

$$T_S = \cdot T_L \frac{\Delta_e}{f_e},$$

when the irradiation frequency is at a distance $\Delta_e$ from the center $f_e$ of the EPR line. If the EPR line is mainly broadened by g-tensor anisotropy, then $\Delta_e/fe$ is not dependent on magnetic field. The lattice temperature in a dilution refrigerator is given by

$$T_L \sim \dot{Q}_{sat}^{1/4},$$

where the microwave power $\dot{Q}$ for full saturation is supposed to be proportional to the electron spin lattice relaxation rate $T_{1e}^{-1}$. This, on the other hand, has a field dependence

$$T_{1e}^{-1} \sim H^{+5},$$

leading to

$$T_L \sim (H^{+5})^{1/4} = H^{5/4}.$$

Therefore

$$\frac{\mu B}{kT_S} \sim \frac{H}{H^{5/4}} = H^{-1/4}$$

and so P probably has a maximum which is determined by the region where one or more of the above relations are not valid. It is suggested that this maximum could possibly be located for some of the well-polarizing materials in a dilution refrigerator.

Advantages of lower field

1. Magnet is less expensive and simpler to construct.
2. Microwave sources more powerful.
3. Less cooling required per gram of material.
4. A lower magnetic field permits detection of lower momentum charged particles. Some experiments may not be possible or at best extremely difficult because of the severe bending in the 2.5T field.
5. A lower field on the target may allow spectrometer magnets to be placed closer resulting in an increase in solid angle acceptance and access to the target.
6. Polarization reversal times might be made shorter. The technique of rapidly reversing the target polarization by rapid reversal of the target field would be simpler for reduced fields.

Some experience with low field targets has been gained with the CERN "frozen spin" targets.[38] The experimental method involves cooling the target sample by means of a dilution refrigerator. The target is dynamically polarized at about 0.5K and 2.5T. Then the microwave power is turned off and the temperature reduced. At the lower temperature the nuclear spin lattice relaxation time is longer and the polarization is frozen in. At this point the magnetic field can be reduced to 1T or less. In this configuration the target can be operated for several hours with only a relatively small loss in polarization.

A recent development[39] along these lines is the proposed Saclay frozen spin polarized target being designed for nucleon nucleon scattering experiments at SATURNE II. The polarizing solenoid and correction coils can achieve a relative uniformity of $\pm 5 \cdot 10^{-5}$ over the volume of the target at 2.5T. Experiments are performed in much smaller holding fields of 0.5 or 1T provided by freely orientable superconducting solenoids.

## 5.5 EPR Measurements in a Polarized Target

Large polarized targets are rather unsuitable for conventional EPR measurements because of the multimode cavities used and because of the very strong absorption of the good target materials. However, useful and relevant information can be obtained of the electron spin system by simply placing a resistance thermometer in contact with the microwave field of the cavity and deducing the microwave field strength from the resistance as a function of magnetic field at constant frequency. The bolometer shows clearly the transition of the target material from the non-absorbing state to complete absorption. This transition in a uniform field is rather sharp and has been used[40] successfully for determining the best polarizing frequencies in propanediol-Cr(V) and in butanol-PX at CERN. The same information could also be obtained by measuring the reflected power from the cavity, but one is then bound to use rather high microwave power, in contrast to the bolometric method whose sensitivity increases at low power (in a dilution refrigerator).

The Nucleon Electron Double Resonance (NEDOR) method[16] operates on the principle that the NMR line changes its shape and position, when the EPR is saturated. In particular, when the EPR is fully saturated (whole line), then the shift of the NMR line is given by $\Delta H = H_S$, where

$$H_S = \frac{4\pi}{3} M_S \xi ;$$

$H_S$ is the internal field caused by the electron magnetization $M_S$ and $\xi$ is a factor of the order of unity depending on lattice symmetry. The magnetization $M_S$ can be related to the number $N_S$ of electron spins in the system

$$M_S = N_S \gamma_S \hbar P_S .$$

If the microwaves are rapidly switched off, then the time constant of the recovery of the NMR frequency can be related to the various electronic relaxation times, mainly with $T_{1s}$ when it is slow. Complicated cases may be obtained when $T_{1s}$ is rather fast.

We might speculate that NEDOR could be used, in addition to measurements of spin concentration and shape of EPR, for distinguishing between different electron spin species, such as free radicals produced by radiation damage. This is clearly straightforward in the case of resolved EPR lines, but might also be applied to overlapping EPR lines whose spin lattice relaxation times differ significantly and which do not couple strongly with each other. For example, the H atom doublet line probably can be resolved, but trapped electrons may overlap with some broad radical lines; however, trapped electrons relax slowly.

We propose to study both of these in-situ methods in order to see their potential in trying to understand radiation damage in polarized targets.

Abragam and his coworkers have already used NEDOR in irradiated LiH.[16] At CERN the changes in proton NMR lineshape have been observed but frequency shifts have not yet been measured.

## 6. SUMMARY

It was apparent at this workshop that polarized target physics is still an active and fertile field. Stable Cr(V) complexes have been developed. In particular, n-butanol-EHBA-Cr(V) targets with their fast polarization time, high polarization in $He^3$ cryostats, higher hydrogen content than the diols, and increased stability compared to PX should become the standard target material once people gain experience in preparing the compound. Further experiments with the "dream" materials HD and $CH_4$ are being contemplated with some optimism and $Li^6D$ appears to have some promise. Further understanding of the radiation damage problem could come from incorporating NEDOR techniques into working polarized targets. Radiation induced DNP in undoped material offers a promising approach to a radiation resistant target especially when used in conjunction with optical bleaching.

In order to have a continuing program for investigating new polarized target materials and doping agents and their performance when exposed to ionizing radiation, it is necessary to have a dedicated polarized target facility near a high intensity source of radiation. Such a facility should also have a NEDOR or ESR capability to look for radicals which are produced during irradiation to try to better understand the processes which lead to depolarization.

Two institutions which currently have facilities for the study of DNP in the laboratory are Saclay and CERN though, at present, neither facility is located near an intense radiation source. It might be possible to install these systems in a beam at SATURNE or the CERN SC. Another possibility would be to obtain a small betatron or linac to use as a dedicated source of particles.

Such a facility would make possible a much more systematic investigation of target material than hitherto. Several groups with polarized target systems at various accelerators have made studies in a piecemeal way, but generally such measurements have been in accord with the physics program being conducted and existing accelerator schedules. This situation is not likely to change significantly and only specific measurements which can be fit into a particular schedule are likely to be made in the near future. Possible measurements are discussed below and could be made with PPT V at Argonne which routinely operates in a beam of greater than $10^{10}$ protons per pulse.

6.1 Irradiate Undoped Samples

G. Court et al.[27] have shown that it is possible to obtain a build-up of DNP in undoped butanol irradiated by photons. This work should be continued with butanol and other materials to measure what levels of polarization are possible with this method. The effect of annealing at various temperatures should also be studied. Such measurements could be done in a period when the physics experiment is being tuned up and when there is no need for the target to be polarized. After the cryostat measurements the sample could be removed for conventional ESR analysis. Doped samples could also be tested for comparison.

6.2 Study of New Complexes

Recent measurements of DNP at CERN[34] using butanol doped with EHBA show that polarizations of 75-80% are obtainable. This is to be compared with $\sim$ 65% normally obtained with butanol doped with porphyrexide. On the other hand the butanol-porphyrexide mixture is more resistant (by a factor of $\sim$ 3) to radiation damage when compared with ethanediol or propanediol doped with the more conventional $Cr^V$ complexes. It is important to establish whether the butanol-EHBA mixture retains this resistance to radiation damage and what the annealing properties are. These measurements can easily be done during a normal physics run using the material as a target and polarizing it.

6.3 Tests of New Complexes

DNP tests of materials such as butanol and pentanol doped with new complexes can be carried out as part of the normal physics program. So long as reasonable polarizations are obtained, samples can be used as targets during the course of a physics experiment and their properties examined.

## ACKNOWLEDGEMENTS

The PPT Workshop Committee would like to express our gratitude to the following individuals for valuable suggestions and discussion.

| | |
|---|---|
| I. P. Auer | Argonne National Laboratory |
| W. R. Ditzler | Argonne National Laboratory |
| M. J. Eder | |
| P. H. Hansen | University of Michigan |
| P. Kyberd | Oxford University |
| J. D. Lesikar | Rice University |
| H. E. Miettinen | Rice University |
| J. R. Miller | Argonne National Laboratory |
| A. Penzo | INFN - Trieste |
| A. Perlmutter | University of Miami |
| L. Ratner | Argonne National Laboratory |
| J. Roček | University of Illinois |
| G. L. Salmon | Oxford University |
| S. Suwa | KEK |
| G. Theodosiou | Argonne National Laboratory |
| A. D. Trifunac | Argonne National Laboratory |
| D. G. Underwood | Argonne National Laboratory |
| L. VanRossum | CEN - Saclay |

We would like to give special thanks to A. Abragam of the College de France for enlightening and entertaining discussions on many of the topics discussed in this report.

# REFERENCES

1. D. G. Crabb et al., Phys. Rev. Lett. 41, 1257 (1978).
2. K. Heller et al., Phys. Rev. Lett. 41, 607 (1978); K. M. Terwilliger, these Proceedings.
3. M. Borghini, "Choice of Substances for Polarized Proton Targets", CERN 66-3, Jan. 1966.
4. Proc. 2nd Int. Conf. on Polarized Targets (ed. G. Shapiro; LBL 500, UC-34 Physics, Nat. Tech.Inf. Serv., Springfield, Va., 1972).
5. H. M. Bozler et al., Bull. Am. Phys. Soc. 18, 545 (1973).
6. A. Honig, Phys. Rev. Lett. 19, 1009 (1967); H. Mano and A. Honig, Nucl. Instr. and Meth. 124, 1(1975); A. Honig and H. Mano, Phys. Rev. B14, 1858 (1976); A. Honig, Proceedings of Symp. on Experiments Using Enriched Antiproton, Polarized-Proton, and Polarized-Antiproton Beams (ed. A. Yokosawa; ANL-HEP-CP-77-45, Argonne, Ill., 1977)p. 186.
7. J. C. Solem, thesis, Yale, 1968; J. C. Solem, Nucl. Instr. and Meth. 117, 477 (1974); J. C. Solem and G. A. Rebka, Phys. Rev. Lett. 21, 19 (1968).
8. M. Sharnoff and R. V. Pound, Phys. Rev. 132, 1003 (1963).
9. G. A. Rebka, private communication.
10. G. A. Rebka and M. Waine, Bull. Am. Phys. Soc. 7, 538 (1962).
11. R. L. Gamblin and T. R. Carver, Phys. Rev. 138, A946 (1965).
12. R. S. Timsit et al., Bull. Am. Phys. Soc. 15, 761 (1970).
13. H. H. McAdams and G. K. Walters, Phys. Rev. Lett. 18, 436 (1967).
14. R. T. Johnson et al., J. Low Temp. Phys. 10, 35 (1973).
15. E. Byckling et al., "A Polarized $^3$He Target, Proposal for a Feasibility Study", Otaniemi, 1974.
16. Y. Roinel and V. Bouffard, J. Phys. 38, 817 (1977); A. Abragam and M. Goldman, Rep. Prog. Phys. 41, 395 (1978).
17. O. Runolfsson and S. Mango, Phys. Lett. 28A, 254 (1968).
18. P. VanHecke et al., Phys. Lett. 33A, 379 (1970).
19. A. J. Nijman and A. J. Berlinsky, Phys. Rev. Lett. 38, 408 (1977).
20. C. F. Hwang, private communication.
21. K. Scheffler, in ref. 4, p. 271.
22. M. Borghini et al., Nucl. Instr. and Meth. 84, 168 (1970).
23. C. C. Morehouse et al., Phys. Rev. Lett. 25, 835 (1970).
24. H. Petri and G. Abshire, Nucl. Instr. and Meth. 119, 205 (1974).
25. R. C. Fernow, Nucl. Instr. and Meth. 148, 311 (1978), and to be published.
26. S. Pshezetskii et al., EPR of Free Radicals in Radiation Chemistry (J. Wiley, New York, 1974).
27. B. Craven, Thesis, Univ. of Liverpool (1973).
28. R. Alger et al., J. Chem. Phys. 30, 695 (1959).
29. M. Krumpolc, B. G. DeBoer, and J. Roček, J. Am. Chem. Soc. 100, 145 (1978).

30. M. Krumpolc and J. Roček, to be published.
31. D. A. Hill et al., Nucl. Instr. and Meth. 150, 331 (1978).
32. R. C. Fernow, to be published in Nucl. Instr. Meth.
33. A. Masaike et al., KEK preprint, 1978.
34. T. O. Niinikoski, these proceedings.
35. A. Abragam, Proc. 1st Int. Conf. on Polarized Targets, Saclay, 1966.
36. M. Borghini, in ref. 4, p. 1.
37. T. O. Niinikoski, Proc. 6th Cryogenic Eng. Conf., Grenoble.
38. T. O. Niinikoski and F. Udo, Nucl. Instr. and Meth. 134, 219 (1976).
39. S. Bréhin et al., to be published.
40. T. O. Niinikoski, Proc. Conf. on High Energy Phys. with Polarized Beams and Targets (ed. M. Marshak; AIP Conf. no. 35, Argonne, 1976) p. 458.

# THE U. MASS. SPIN REFRIGERATOR
# AND STRANGE-PARTICLE PHYSICS
# WITH THE BROOKHAVEN MULTIPARTICLE SPECTROMETER

J. Button-Shafer
University of Massachusetts, Amherst, MA 01003

## ABSTRACT

Proton polarization up to 80% has been achieved with the University of Massachusetts "spin refrigerator" in a nonuniform 1.07-T field at 1.25 K. The free protons in $(Yb,Y)(C_2H_5SO_4)_3 \cdot 9H_2O$ are polarized through sample rotation at approximately $100$ rps. Results represent the first confirmation of high polarization predicted by a single-Yb-ion interaction model since the original proposals of Jeffries and Abragam. Performance of the target system in a BNL test beam has been excellent. The target, with thickness equivalent to 9 cm $LH_2$, is now scheduled for use in strange-particle experiments at the Brookhaven MPS: the first studies will involve inelastic $K^-p$ reactions with emphasis on the final state $Y^*(1385) + \pi$.

## INTRODUCTION

At the Saclay Conference on "Polarized Targets and Ion Sources" in 1966, C. D. Jeffries stated, "The expense of the microwave system, helium consumption, uniform magnetic fields, and required operator skill and patience lead naturally to the question: isn't there a simpler method? One suggestion was given by myself and by Abragam several years ago: one can polarize nuclei simply by rotating a suitable crystal in a magnetic field at low temperatures."[1,2]

The proposed technique involves cyclically transferring a large paramagnetic-ion polarization to nuclear (proton) spins without the use of microwaves and with no constraint on the uniformity of the magnetic field. It requires a very anisotropic g factor for the

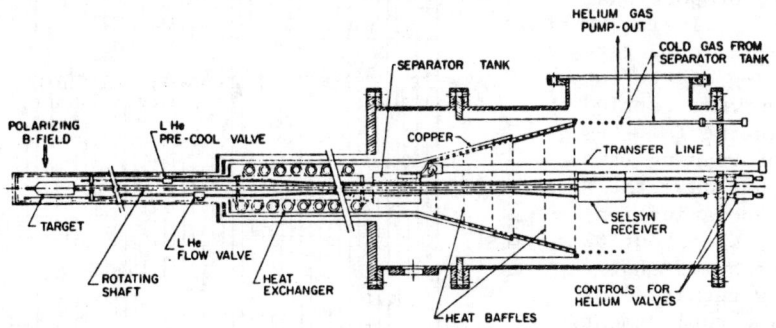

Fig. 1. The University of Massachusetts polarized target. Target length is 3 in.; cryostat length, 60 in.

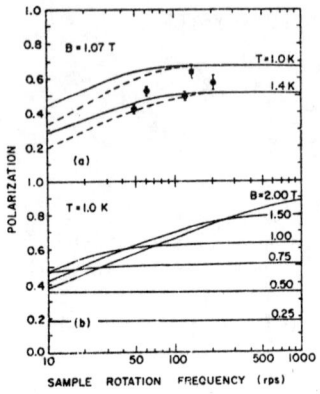

Fig. 2, (a) Early measurements of proton polarization vs. rotation frequency. Calculated curves represent Yb-ion polarization and maximum expected proton polarization (dashed curves).
(b) Calculation of Yb-ion polarization vs. frequency at various magnetic fields.

paramagnetic impurity,

Fig. 1 shows the target system which I have developed with my group at the University of Massachusetts (my major associates in recent years being R. L. Lichti in hardware and S. Dhar in physics planning). The project would likely not have been undertaken without the advice and encouragement of C. D. Jeffries. It represents the melding of solid state technology with that of continuous-flow cryostats. (W. H. Potter of U. C. Davis has been a major consultant on the former, and ANL personnel R. Niemann and D. Hill have given very valuable advice on the latter.)

## U. MASS. RESULTS

The U. Mass. results obtained two years ago (Fig. 2) demonstrated conclusively that the early predictions of high proton polarization are realized with ytterbium concentration of $\leq 0.04$ at.% in $Y(C_2H_5SO_4)_3 \cdot 9H_2O$. The failure of extensive prior efforts to achieve polarizations higher than 35% was caused by poorly understood effects of higher Yb-ion concentration (or the inability to achieve low enough temperatures).

In recent months, polarizations up to 80% have been obtained since establishing the target system in a test beam at BNL. Following construction of a rather elaborate system of counter hodoscopes, spark chamber, proportional chamber, analyzing magnet, and on-line computer system, a π-p scattering experiment was carried out. In the last few weeks of running (two shifts per day) the average polarization achieved was about 62%; only one brief shutdown was required for a bearing replacement. Time consumed in polarizing was about 10%. (Relaxation time was not maximized because of the need to keep the field at

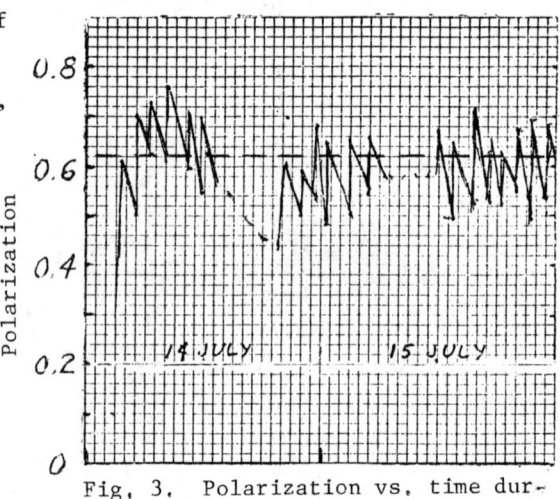

Fig. 3. Polarization vs. time during last two days of test-beam run,

≤ 2 kG.)

The spin refrigerator as constructed at U. Mass. is remarkably simple; only a few physicists and technicians have been involved and the financial investment has been small. (Some time was consumed in attempting a turbine drive system, which was eventually abandoned.) High-energy experiments have been carried on at national laboratories simultaneously with the target development at home.

Two groups from other institutions, one in solid-state physics and the other in nuclear, have consulted with us during the past year concerning the constructing of systems similar to ours.

Relaxation times measured in recent months have been as long as thirty to fifty hours. (Fig. 4.) At low magnetic field, the times are much shorter than theory predicts. It is likely that very small amounts of impurities are preventing the attainment of the long predicted relaxation times (and we found that crystal-growing in teflonware gave improvement).

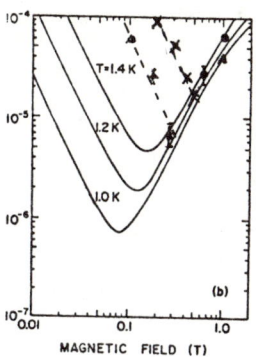

Fig. 4. Relaxation rate vs. B field. Dashed curve on the left indicates best results at low fields (for sample grown in teflon-ware). Solid curves are theoretical. (Rates in sec$^{-1}$)

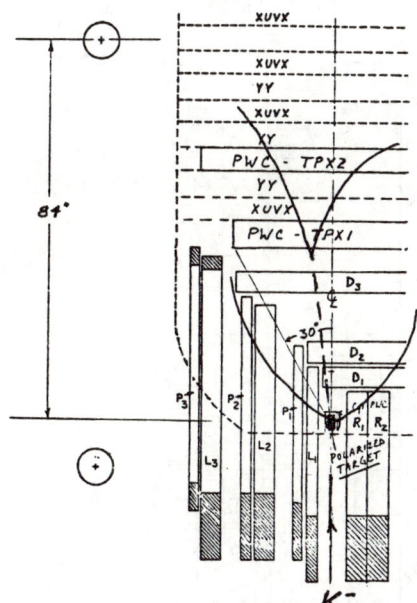

Fig. 5. Plan view of upstream half of the MPS at Brookhaven

## MECHANISM OF POLARIZING

In the presence of the external magnetic field, the Yb ions populate the lowest of four Kramers doublets. This doublet has an anisotropic $g$ factor, $g(\theta) = (g_{\parallel}^2 \cos^2\theta + g_{\perp}^2 \sin^2\theta)^{1/2}$, where $\theta$ is the angle between the applied field and the crystal c axis. With the effective spin taken as 1/2, the Yb $g_{\perp}$ is as small as the proton g value of 0.00302, while $g_{\parallel}$ = 3.33. The Yb-ion relaxation rate varies approximately as $\sin^2\theta \cos^2\theta$. If the crystal is rotated with respect to the magnetic field at a rate greater than the Yb-ion relaxation rate, then the Yb-ion polarization assumes an almost constant value corresponding roughly to $g(45°)$, or tanh ($2.4\mu B/2kT$). As $\theta$ approaches $90°$, the energy-level difference of the doublet greatly diminishes so that it nearly matches the equilibrium energy-splitting of the protons.

At 90° a dipole-dipole interaction may cause the Yb spin orientation to be exchanged with that of a nearby proton. The proton relaxation rate drops rapidly beyond 90°. The polarization process continues with further rotation. The (nonequilibrium) polarization enhancement of the protons is equivalent to reducing their temperature by several orders of magnitude in this "spin refrigerator."

## PLANS FOR USE

The U. Mass. target was planned for use with a facility (bubble chamber, streamer chamber, or spark-chamber complex) which is capable of studying complex, unstable-particle final states. It is ideally suited for use in an inhomogeneous magnetic field, which need not be very high. The target material has a free-to-total proton ratio of more than 10%.

With the Brookhaven MPS, the continuous-flow cryostat will be employed in a near-horizontal orientation, with the MPS main field serving as the polarizing field. Helmholtz coils will be used to enhance the MPS field to about 14 kG for polarizing; they will also serve to reduce the field (for maximum relaxation time) during data-taking, and to reverse the field **for reverse-polarization** running. The hysteresis-type synchronous motor (which has replaced the selsyn receiver of Fig. 1) will be shielded from the MPS field.

The Helmholtz coils' separation is well-matched to the acceptance of the MPS detectors.

Fig. 5 shows a plan view of the polarized target with the MPS apparatus. Besides triggering PWC's, capacitive-readout spark chambers are used in the target region and magnetostrictive spark chambers are presently employed downstream. Only the upstream half of the MPS is shown, the main detecting volume being 4 ft high X 6 ft wide X 15 ft long (with a 10-kG central field).

## PHYSICS

The purpose of the low-momentum $K^-$ experiment which has been scheduled for the polarized target in the MPS is to extend the understanding of formation and final-state $Y^*$ res-

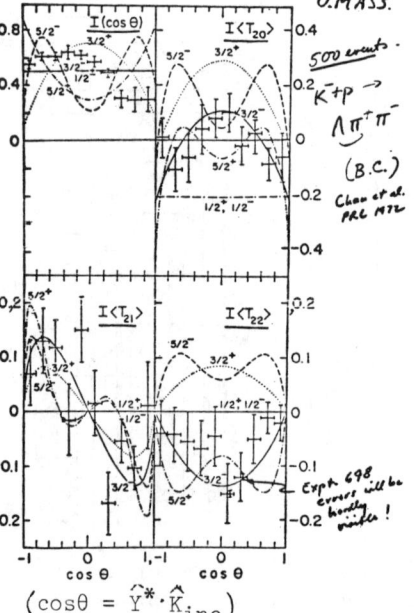

---

Fig. 6. Intensity I and $I\langle T_{LM}\rangle$ for the $Y^*_1(1385)$ from a U. Mass. 500-event study at 1520 MeV (c.m.). Curves are predictions for various $J^P$ intermediate states. The MPS experiment with polarized target will have errors 1/6 as great.

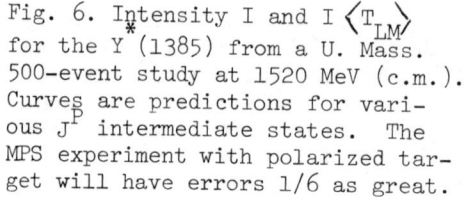

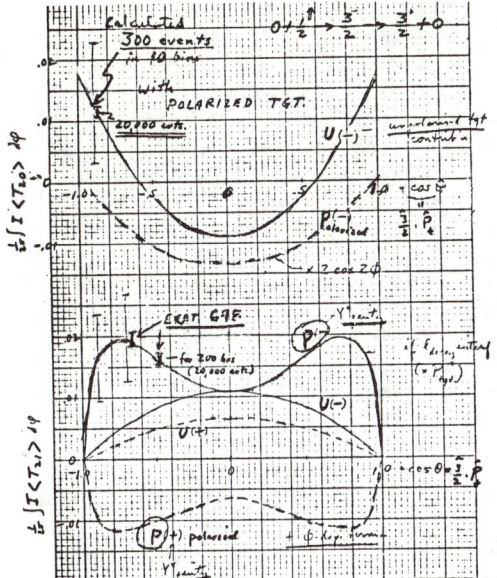

Fig. 7. Predictions of $Y^*$ spin-state parameters vs. $\hat{Y}^* \cdot \hat{P}_{tgt}$. Unpolarized contributions are marked "U," whereas those depending on target polarization are marked "P." Errors expected in the 20,000-event MPS experiment will be nearly invisible.

Rates will be excellent with the MPS. At 2 GeV/c, some 20,000 events per inelastic channel will be produced in the $K^-$ study. Besides the many new parameters afforded by the polarized target, errors will be much less than in the typical bubble-chamber experiment.

A test run done with a target mock-up (in a bubble-chamber) leads us to expect background suppression to be excellent for the anticipated $K^-$ experiment.

Fig. 6 shows measured parameters in a U. Mass. bubble-chamber study of formation resonance decay into the $Y^*(1385)$. Errors for the MPS polarized-target experiment will be much smaller. In addition, the target will provide $\phi$-dependence, more complex

onances, especially through study of inelastic channels. Resonance studies for the sake of quark models can be much enhanced by the use of a polarized target; e.g., the process $K^- p \to \Lambda \pi^+ \pi^-$ will yield valuable physics knowledge (more so than would $\pi p \to \Delta \pi$). (See Petersen and Rosner for $SU(6)_W$ discussion[4] and Button-Shafer for treatment of cascading resonance decays.[5])

A second anticipated phase of polarized-target study is that of high-momentum associated production, $\pi^- p \to K^0 \Lambda$ and $K^{*0} \Lambda$. R and A parameters for the former and many polarization, alignment, and correlation quantities for the latter would be measured.

For any process, the polarized target makes available at least <u>three</u> times the usual number of measurable parameters and hence imposes much stronger constraints on theoretical models.

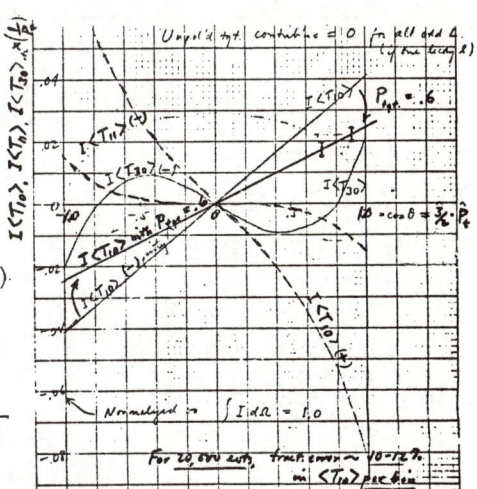

Fig. 8. Predictions for some odd-L $Y^*$ spin parameters vs. $\hat{Y}^* \cdot \hat{P}_{tgt}$. Errors again are very small.

θ-dependence, imaginary parts of $\langle T_{LM} \rangle$, and odd-L $\langle T_{LM} \rangle$. Figs. 7 and 8 show predictions for unpolarized and polarized contributions to some of the even-L and odd-L I$\langle T_{LM} \rangle$ characterizing the final-state $Y^*$, with the assumption of a 3/2 intermediate state. Errors will be <u>very</u> small for 20,000 events. Also, there are some interesting sign flips which depend on the relative parity between intermediate and final-state baryon systems, these being similar to the Bilenky-Bohr sign flip for polarized-target production of spins 0 + 1/2.[6]

If one does not have a pure intermediate state, other types of calculation are possible. One is a partial-wave treatment yielding, for the various I$\langle T_{LM} \rangle$, sums of bilinear combinations of amplitudes.

Another is a general "S-matrix" treatment which is based on the formalism of N. Byers and S. Fenster[7] and on later work of Byers at CERN. Summaries of our calculations follow; they demonstrate that eight measurable parameters become 24 with a polarized target. Also, predictions of errors may be made by using real $Y^*$ data from unpolarized-target experiments. For example, the $b_{00}$ coefficient depends on $\langle T_{10} \rangle$ and $\langle T_{30} \rangle$ ($a_{10}$ and $a_{30}$ below) and might be expected to have a 5% error for 20,000 events; this is to be contrasted with $\geq 30\%$ errors in $\langle T_{LM} \rangle$ measurements in the most sensitive $Y^*$ bubble-chamber

ang. distrib. of decay (strong)

$$I(\theta,\varphi) = \sum_{L,m,M} n'_{L0}\, a_{LM}\, Y^*_{LM}(\theta,\varphi)$$
$$- P_t \sin\varphi \sum_{L,m,M} n_{L0}\, b_{LM}\, Y^*_{LM}(\theta,\varphi)$$
$$+ P_t \cos\varphi \sum_{L,m,M} n_{L0}\, c_{LM}\, Y^*_{LM}(\theta,\varphi)$$

$$I\vec{P}_\Lambda \cdot \hat{\lambda} = \sum_{L,m,M} n\, a_{LM} Y_{LM} - P_t\sin\varphi \sum n\, b\, Y + P_t\cos\varphi \sum n\, c$$

$$I\vec{P}_\Lambda \cdot \hat{z} = \gamma\, Re \sum_{L,m,M} \left(\frac{2J+1}{J(J+1)}\right) n_{L0} \sqrt{\frac{2L+1}{4\pi}}\, \mathcal{D}^{L^*}_{M1}(\varphi,\theta,0)$$
$$\times [a_{LM} - P_t\sin\varphi\, b_{LM} + P_t\cos\varphi\, c_{LM}]$$

(±1 for $J = J - \frac{1}{2}$, $J + \frac{1}{2}$)

$$I\vec{P}_\Lambda \cdot \hat{y} = \gamma\, Im \sum_{L,m,M} (\text{same terms as } I\vec{P}\cdot\hat{z})$$

Final-state
$$Y^* \to \Lambda + \pi.$$

With polarized target,
$$\langle T_{LM} \rangle \to a_{LM},\ b_{LM},\ \text{and}\ c_{LM}$$

Unpolarized contributions:

$$a_{00} = \tfrac{1}{2}\left[S_{3+}^2 + S_{1-}^2 + S_{1+}^2 + S_{3-}^2\right]$$
$$a_{10} = \tfrac{1}{2}\tfrac{1}{\sqrt{15}}\left[3S_{3+}^2 + S_{1-}^2 - S_{1+}^2 - 3S_{3-}^2\right]$$
$$a_{20} = \tfrac{1}{2\sqrt{5}}\left[S_{3+}^2 - S_{1-}^2 - S_{1+}^2 + S_{3-}^2\right]$$
$$\genfrac{}{}{0pt}{}{Re}{Im}\, a_{22} = \genfrac{}{}{0pt}{}{Re}{Im}\,\tfrac{\sqrt{2}}{2\sqrt{5}}\left[S_{3+}^* S_{1+} + S_{1-}^* S_{3-}\right]$$
$$a_{30} = \tfrac{1}{2\sqrt{35}}\left[S_{3+}^2 - 3S_{1-}^2 + 3S_{1+}^2 - S_{3-}^2\right]$$
$$\genfrac{}{}{0pt}{}{Re}{Im}\, a_{32} = \genfrac{}{}{0pt}{}{Re}{Im}\,\tfrac{\sqrt{2}}{2\sqrt{7}}\left[S_{3+}^* S_{1+} - S_{1-}^* S_{3-}\right]$$

Polarized contributions:

$$b_{00} = \tfrac{1}{2}\left[S_{3+}^2 - S_{1-}^2 + S_{1+}^2 - S_{3-}^2\right]$$ — in $I(\theta,\varphi)$, but $\sum$ containing polari- ($\propto$ odd-L $a_{LM}$) info

$$b_{10} = \tfrac{1}{2\sqrt{15}}\left[3S_{3+}^2 - S_{1-}^2 - S_{1+}^2 + 3S_{3-}^2\right]$$ — in $I\vec{P}(\theta,\varphi)$, contain even-L $a_{LM}$ but yield $2J+1$ and parity!

$$b_{20} = \tfrac{1}{2\sqrt{5}}\left[S_{3+}^2 + S_{1-}^2 - S_{1+}^2 - S_{3-}^2\right]$$
$$\genfrac{}{}{0pt}{}{Re}{Im}\, b_{22} = \genfrac{}{}{0pt}{}{Re}{Im}\,\tfrac{\sqrt{2}}{2\sqrt{5}}\left[S_{3+}^* S_{1+} - S_{1-}^* S_{3-}\right]$$
$$b_{30} = \tfrac{1}{2\sqrt{35}}\left[S_{3+}^2 + 3S_{1-}^2 + 3S_{1+}^2 + S_{3-}^2\right]$$
$$\genfrac{}{}{0pt}{}{Re}{Im}\, b_{32} = \genfrac{}{}{0pt}{}{Re}{Im}\,\tfrac{\sqrt{2}}{2\sqrt{7}}\left[S_{3+}^* S_{1+} + S_{1-}^* S_{3-}\right]$$

$$\genfrac{}{}{0pt}{}{Re}{Im}\, c_{11} = \genfrac{}{}{0pt}{}{Re}{Im}\,\tfrac{-\sqrt{2}}{2\sqrt{15}}\left[\sqrt{3}\, S_{3+}^* S_{1-} - 2 S_{1-}^* S_{1+} + \sqrt{3}\, S_{1+}^* S_{3-}\right]$$
$$\genfrac{}{}{0pt}{}{Re}{Im}\, c_{21} = \genfrac{}{}{0pt}{}{Re}{Im}\,\tfrac{\sqrt{2}}{2\sqrt{5}}\left[-S_{3+}^* S_{1-} + S_{1+}^* S_{3-}\right]$$
$$\genfrac{}{}{0pt}{}{Re}{Im}\, c_{31} = \genfrac{}{}{0pt}{}{Re}{Im}\,\tfrac{\sqrt{7}}{2\sqrt{35}}\left[-S_{3+}^* S_{1-} - \sqrt{3}\, S_{1-}^* S_{1+} - S_{1+}^* S_{3-}\right]$$
$$\genfrac{}{}{0pt}{}{Re}{Im}\, c_{33} = \genfrac{}{}{0pt}{}{Re}{Im}\,\tfrac{\sqrt{7}}{2\sqrt{7}}\left[-S_{3+}^* S_{3-}\right]$$

study to establish its spin and parity, with 895 events.[8]

In summary, the spin-refrigerator target makes possible some very pretty physics. We hope it will be useful physics as well.

## REFERENCES

1. C. D. Jeffries, Cryogenics 3, 41 (1963).
2. A. Abragam, Cryogenics 3, 42 (1963).
3. J. Button-Shafer, R. L. Lichti, and W. H. Potter, Phys. Rev. Letters 39, 677 (1977).
4. W. P. Petersen and J. L. Rosner, Phys. Rev. D6, 820 (1972), and private communication from J. L. Rosner at the Aspen Center for Physics.
5. J. Button-Shafer, Phys. Rev. 139, B607 (1965).
6. J. Button-Shafer, Phys. Rev. 150, 1308 (1966).
7. N. Byers and S. Fenster, Phys. Rev. Letters 11, 52 (1963) (and also unpublished appendix).
8. J. B. Shafer and D. O. Huwe, Phys. Rev. 134, B1372 (1964).
9. S-b. Chan et al., Phys. Rev. Letters 28, 256 (1972).

AN AXIALLY POLARISED PROTON TARGET FOR THE MEASUREMENT
OF A AND R PARAMETERS IN THE REACTION $\pi^- p \to K^° \Lambda^°$
IN THE MOMENTUM RANGE 1.34 - 2.24 GeV/C

R. W. Newport, M. Ball, P. H. T. Banks, G. Brewer, P. T. M. Clee,
D. A. Cragg, N. H. Cunliffe, J. Simkin, J. H. Swain, R. Wigley
(To be presented by Dr. D. H. Saxon)
Rutherford Laboratory, Chilton, Didcot, Oxfordshire, OX11 0QX, England

INTRODUCTION

In this paper we describe an axially polarised proton target, designed and built at the Rutherford Laboratory.

The target was required for use with the RL Associated Production Spectrometer to investigate the A and R spin rotation parameters of the interaction

$$\pi^- p \to \Lambda^° K^°$$

in the momentum range $1.34 \to 2.24$ GeV/C.

Apart from the requirement for polarisation in the beam direction a large exit cone (half angle 60°) was requested in order to collect the decay products for the neutral secondary particles.
In addition good access is also required on the inlet side of the target for monitoring the incident particles. These requirements coupled with the normal polarised target demands of a good uniformity 2.5 Tesla field, led to the choice of a superconducting magnet and some novel features in the design of the target cryostat.

The use of a superconducting magnet and a large capacity He-3 liquefier indicated a high consumption of liquid helium -4 so that from the beginning, for economic and operational reasons, the design included the use of a closed circuit helium-4 liquefier.

Despite the complexity and scale of the equipment high reliability was achieved.

SPECIFICATION: After preliminary discussions with the particle physics group design specifications were agreed as follows:

ISSN: 0094-243X/79/510048-14$1.50 Copyright 1979 American Institute of Physi

| | | |
|---|---|---|
| Target: | Volume | 35 cm$^3$ |
| | Material | propanediol with CR V complexes |
| | Polarisation | >75% |
| | Polarisation Measurements | ± 2% (long term) |
| Magnet: | Field | 2.5 Tesla |
| | Homogeneity over 35 cm$^3$ | ±1 parts in 10$^4$ |
| | Exit cone | >60° half angle |
| | Stability | 5 parts in 10$^5$ over 8 hrs. |
| He-3 Cryostat: | Operating Temperature | 0.5K |
| | Capacity | 35 mwatts at 0.5K |
| | | 150 mwatts at 0.55K |

## GENERAL DESIGN FEATURES

Figure 1 shows the main components of the system, diagramatically. Figure 2 shows the system during assembly.

The last downstream element is the superconducting magnet which allows complete access to the target via the exit cone of the magnet, for particle detection. The nose of the target cryostat is inserted into the room temperature bore of the magnet from the upstream side.

The complete He-3 cryostat is in two parts one containing the target and the other containing the He-3 liquefier. This separation of functions allows a common liquefier design to be used for different targets with freedom to choose a target geometry best suited to the particular experiment. The two parts are joined by a short pipe through which He-3 liquid is passed from the vertical liquefier cryostat to the target cryostat. In this system the target cryostat is mainly at 35° with respect to the vertical but has a short horizontal section containing the target. The two sections are mounted together in a frame which has rollers to allow the target to be moved into the bore of the magnet.

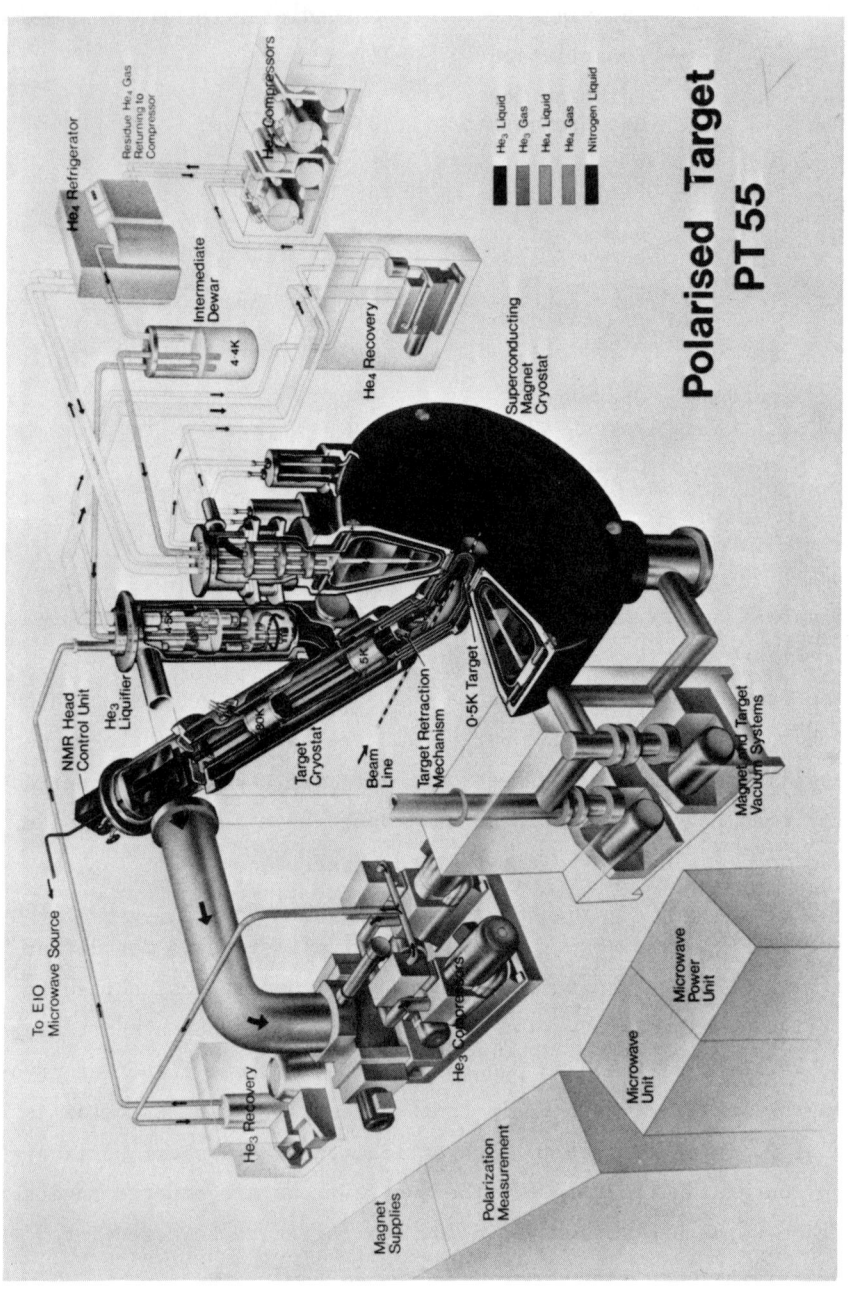

FIGURE 1   SCHEMATIC DIAGRAM OF THE TARGET SYSTEM

FIGURE 2  THE TARGET BEING ASSEMBLED IN THE TEST AREA

Service access to the two sections of the He-3 cryostat and the magnet is mainly from the top so that control panels, the He-4 liquefier and He-3 storage and purification system are mounted close to the cryostats but on a platform above the beam line. This not only makes for ease of operation but leaves space for beam line elements and particle detectors and avoids the worst problems of the large stray field associated with the unshielded superconducting magnet. Only the He-4 compressors, He-3 pumps and insulating vacuum pumps are mounted at floor level.

A separate control room adjacent to the cryogenic controls and on the same platform contains the microwave and nuclear magnetic resonance systems, the magnet power supplies and controls and a general alarm panel. It is from this room that the target is normally operated.

The complete system was built in modules so that installation in the experimental area after testing in a separate assembly area could be quickly and reliably carried out and also to allow re-use of the equipment in other experiments where the layout may be quite different.

To minimise the effect on the homogeneity of the magnetic field, frames and components near to the magnet are constructed of aluminium alloys. Also to avoid damage to the components from magnetic missiles an area around the magnet was enclosed by fences and was not normally accessible.

Because of complexity of the system and the need to minimise the use of operating staff many of the systems are controlled automatically. Not surprisingly this has resulted in very consistent and reliable performance.

## THE SUPERCONDUCTING MAGNET (FIG.3)

The magnet consists of two pairs of symmetrically placed superconducting coils in an aluminium cryostat and vacuum tank to give angular access over a cone of half angle 60° with respect to the centre of the magnet.

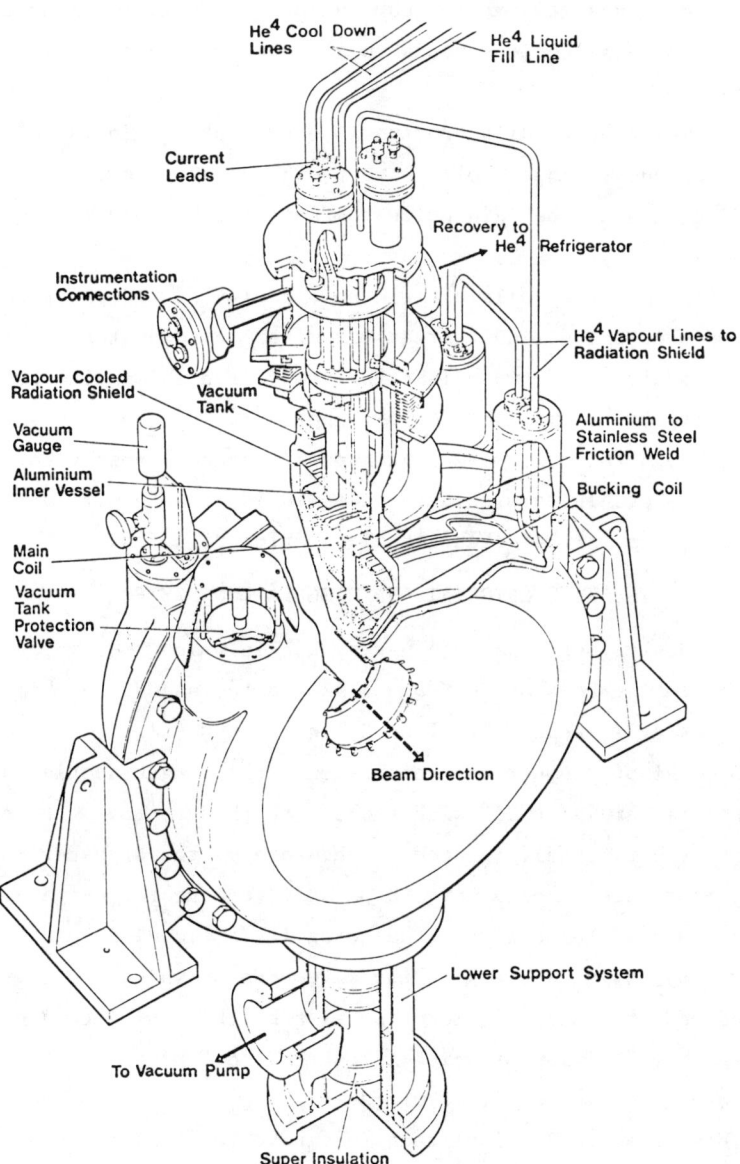

SUPERCONDUCTING MAGNET FOR USE WITH PT55 POLARISED TARGET

FIGURE 3

The four coils are connected in series with the main power supply, but each has its own low current power supply to allow trimming of the individual coil currents to achieve the required field uniformity.

The main outer coils provide in excess of 2.5 Tesla at the centre and the secondary inner coils have their current flowing in the opposite direction to the main coils to flatten the field in the target region.

Using only the coils a field homogeneity of $\sim\pm 2$ parts in $10^4$ was achieved over the nominal target volume of 3cm diameter by 5cm long.

The addition of two cylindrical soft iron shims allowed the homogeneity to be improved to $\sim\pm 0.5$ parts in $10^4$ over the same volume.

Although the coils are continuously powered, tests showed the field to be stable to within $\pm 5$ parts in $10^6$ over a period of 16 hours after an initial stabilisation period of a few hours.

## THE He-3 CRYOSTAT AND CIRCULATION SYSTEM

The He-3 cryostat has two parts, one in which He-3 is cooled and condensed using He-4 at 4.2K and 1.5K and the second in which the target cavity is kept full of He-3 liquid at 0.48K.

Because of geometrical requirements the target section of the cryostat is mainly at 35° with respect to the vertical with the target cavity horizontally placed in the bore of the cryostat. This arrangement made it necessary to be able to move the target cavity within the cryostat before withdrawing it for loading. This is achieved by having a break in the waveguide on which the cavity is mounted and attaching the lower section of this guide to a guided carriage the position of which is controlled from outside the cryostat through stainless steel wires.

This arrangement also leads to a novel design for supplying He-3 to the cavity. A small bore pipe from the Joule-Thompson valve in the main cryostat is fixed into the target cryostat with its open end facing backwards towards the cavity. As the cavity is moved forward into the working position a canopy over a rear compartment of the target receives the tube so that the two-phase mixture of He-3 is caught in this compartment. The liquid level is measured and maintained in this compartment and the temperature is also measured here. He-3 vapour leaves through the opening in the canopy and liquid He-3 flows through a small tube (microwave trap) into the bottom of the front of the target which is the microwave cavity and which contains the beads of propanediol in a perforated FEP container. This container is suspended from the copper lid of the cavity which is perforated with many small holes (microwave traps) to allow helium-3 vapour to escape freely. Two NMr coils are suspended from this lid. The complete cavity is supported from underneath by the waveguide with its longer side vertical.

Although the lid of the cavity is of copper the rest is made of thin stainless steel plated with copper and gold, this material is not only rigid but its low thermal conductivity eases construction and repairs.

The cavity is protected from heat leaks by various shields, cooled by liquid He-4 at 4.2K and liquid nitrogen and is enclosed in a vacuum of $2 \times 10^{-6}$ torr. (Figure 4 shows the target insert, with cavity).

A further feature of the target system is the use of an aluminium alloy nose cone for the He-3 containment at 0.5K. This is attached to the stainless steel tube, which forms most of the cryostat, by an aluminium - stainless steel friction weld.

Returning to the overall system, the helium-3 is circulated by a combination of sealed Roots and rotary pumps with a throughput of 3000 $m^3$/hr. At this temperature the available refrigeration is 50 milliwatts over a static heat inleak of ~80 milliwatts. The refrigeration capacity is shown in Figure 5.

## He-4 SYSTEM

The total consumption of liquid helium at 4.2K arising from the magnet and the He-3 cryostat was estimated to be 15 litres/hour. This led to an early decision to purchase a commercial helium liquefier, namely a CTI-1400 with a capacity of 20 litres/hour.

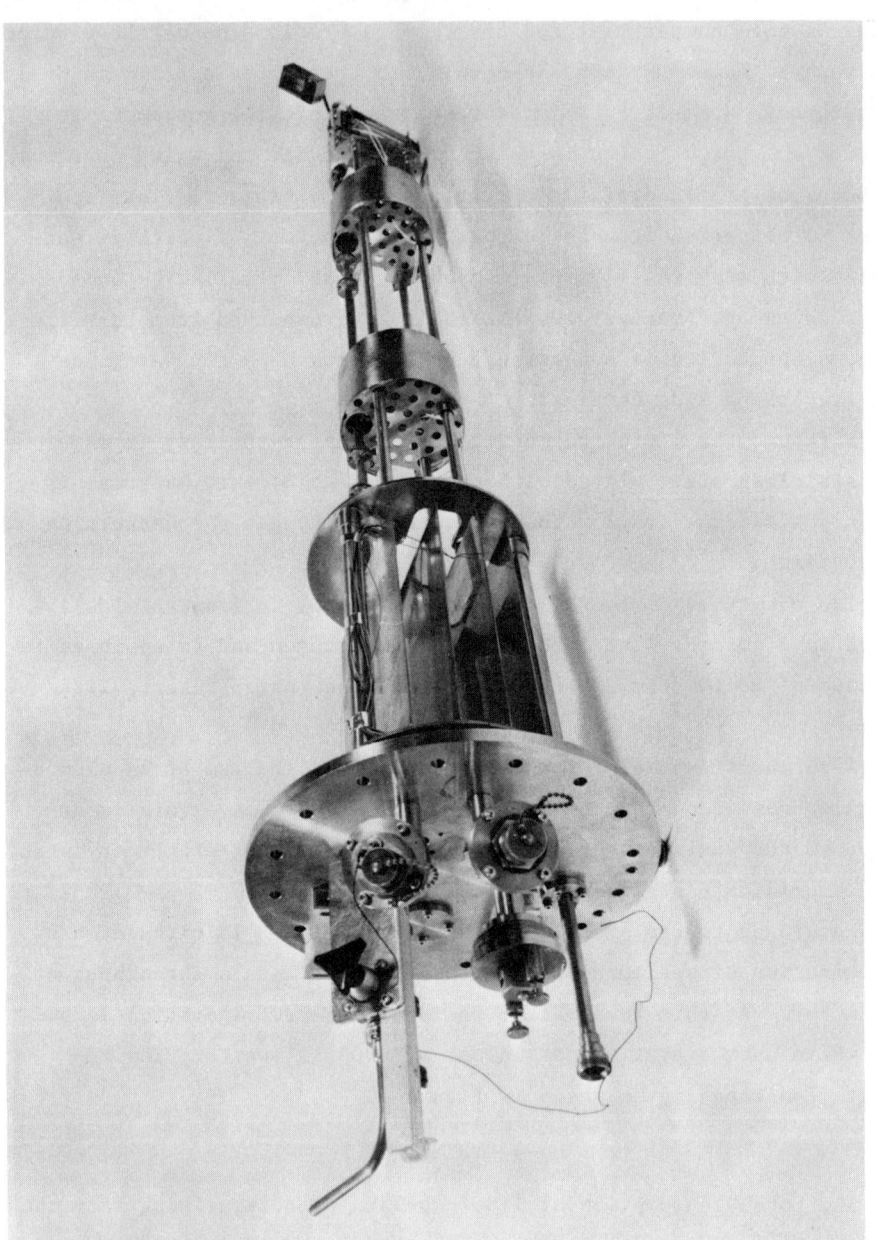

FIGURE 4 TARGET INSERT, WITH CAVITY

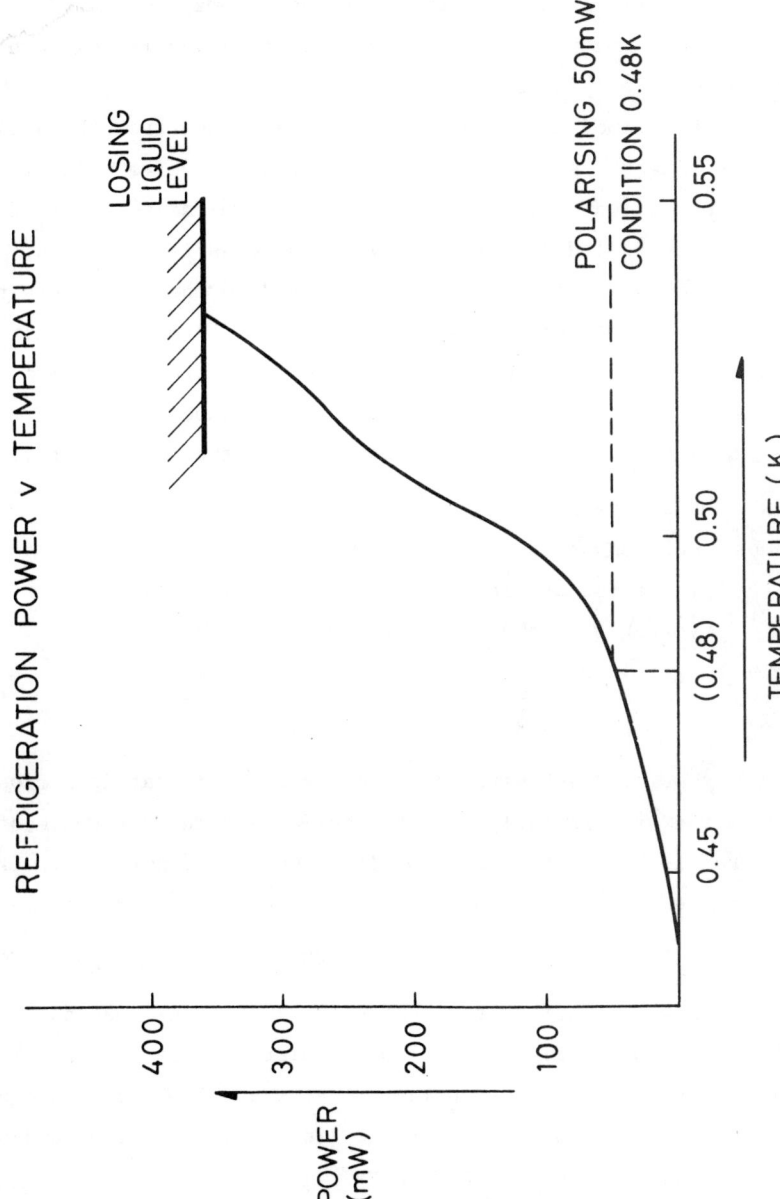

FIGURE 5   AVAILABLE REFRIGERATION POWER

The output of this machine is used to maintain the liquid level in a special dewar with a capacity of 30 litres from which the magnet and the 4.2K baths in the He-3 cryostat are automatically supplied with He-4 liquid via automatically controlled valves in response to superconducting level sensors.

The 1.5K bath of the He-3 cryostat is supplied from the 4.2K bath of the same cryostat via an internal control valve which is automatically controlled from a superconducting level sensor in that bath.

All helium boiled off from the various baths is recovered, compressed to in excess of 400 psig and returned to the system via the internal purifier of the CTI-1400. The boil-off from the magnet is split into two streams to allow vapour cooling of the radiation shields and of the current leads with the neck tube.

The boil-off from the various sources was typically as follows:

| | |
|---|---|
| Intermediate dewar | 1.1 ℓ/hr |
| Magnet neck and current leads | 6.4 ℓ/hr |
| radiation shields | 3.4 ℓ/hr |
| He-3 cryostat 4.2K and 1.5K baths | 3.6 ℓ/hr |
| He-3 target insert | 2.2 ℓ/hr |
| TOTAL | 16.7 ℓ/hr |

After some problems with water in the early operating periods the system ran very reliably over the normal accelerator cycle of three weeks plus one extra week for cooldown and preparation. It also ran for eight weeks when two consecutive cycles were used.

## THE MICROWAVE SYSTEM

The microwave power source chosen for this target was an extended interaction oscillator (E.1.0) manufactured by Varian. The model used was capable of producing 28 watts at 70 GHz so could be situated some 20 metres from the target in the control room and still provide more than adequate r.f. power.

The device which has mechanical tuning over ±2GHz and electronic tuning over ±0.2 GHz was controlled automatically at either of the two frequencies corresponding to maximum positive and negative polar-

isation. The desired frequency is selected at a two position switch
and stabilised with respect to a reference frequency stable to within
±5 ppm. Having optimised the microwave power level to the cavity at
50 milliwatts this level was maintained throughout all operations.

Very reliable and consistent operation was obtained and after
5000 hours operation the E10 shows no signs of deterioration.

## THE POLARISATION MEASURING SYSTEM

Instead of using one of the conventional Q' meter circuits to
measure polarisation, a balanced linear absorption meter circuit was
developed. This system gives good thermal equilibrium signal to
noise ratios. It also eliminates the quadrature reactance and dispersion errors associated with conventional Q' meter systems. Thus
producing a linear base line which is readily removed from NMR signal
automatically.

The area of the NMR signal is then obtained by integration to
obtain a measurement directly proportional to polarisation. After
suitable scaling this measurement is displayed as a digital number
to read percentage polarisation.

Drift errors were reduced to very low levels by incorporating
several automatic compensation systems. Thus the system which is
simple to set up requires no operation during normal running. In
order to measure thermal equilibrium polarisation it is only necessary to switch the gain of one amplifier by a fixed amount.

Daily checks were carried out by means of a built in calibration
unit, which switched standard signals into various parts of the system. This enables us to make accurate assessment of the measurement
systems performance with respect to long term measurement accuracy,
this was found to be 0.2% per month. The performance of the system
was excellent as indicated by Figure 6 which shows a typical thermal
equilibrium signal taken at 0.67K. Over a period of 12 months the
thermal equilibrium calibrations gave a standard deviation of ±0.3%.

Two coils were installed into the cavity, with their axis perpendicular to reduce coupling for the purpose of assessing the uniformity of polarisation throughout the volume of the target. The two

60

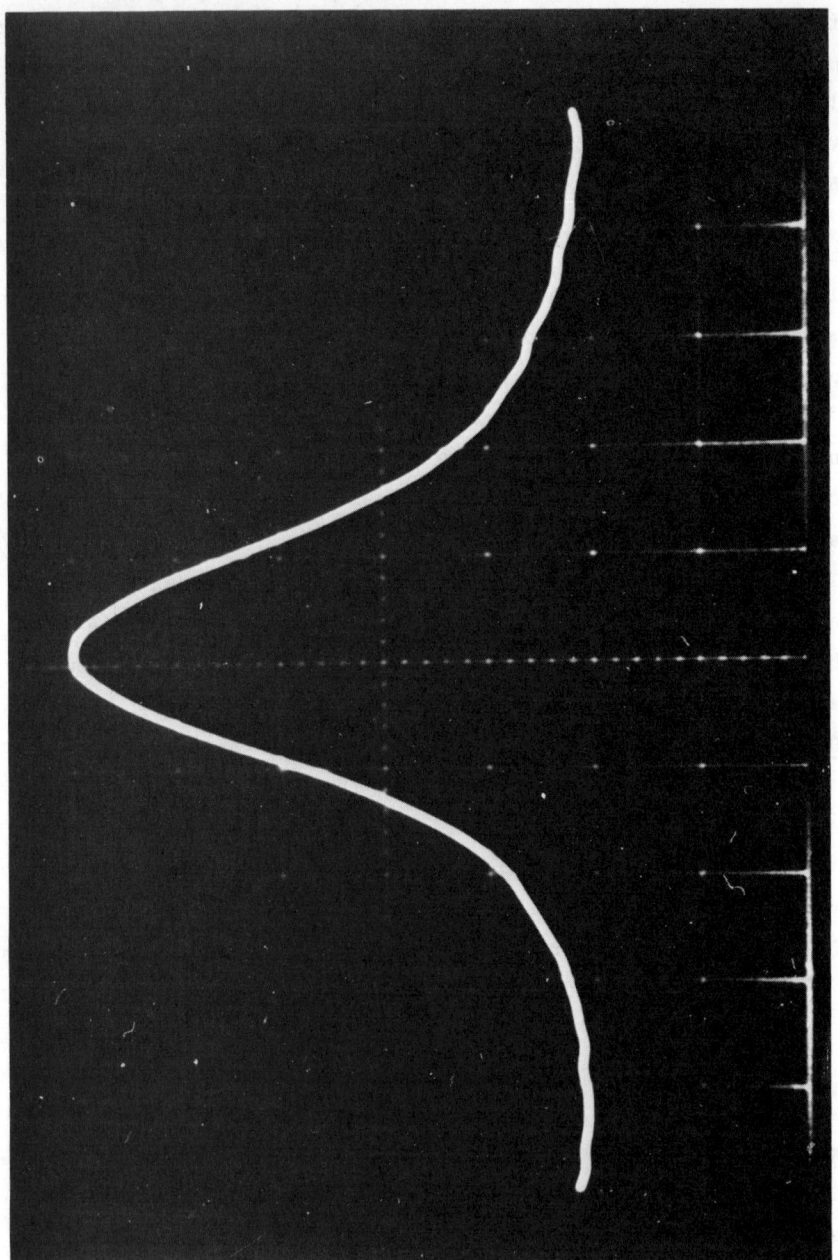

FIGURE 6  THERMAL EQUILIBRIUM SIGNAL AT 0.67K

coils gave measurements of polarisation which agreed with each other to within ±0.2%.

## MATERIAL

The material used in the target was 1,2 propanediol doped with Cr.V complexes to give an electron spin density of approximately $1.5 \times 10^{20}$ per cubic cm beads of 1-2 mm diameter made by an electrostatic bead making device were dry loaded into the cavity at 78K.

## TARGET PERFORMANCE

Polarisations of close to 85% were obtained consistently over a period of one year during which particle physics data was taken. The actual operating time was 21 weeks during which the target was available for over 95% of the time of which data taking was possible.

TABLE 1. Summary of Target Performance

| | |
|---|---|
| Polarisation | ±83% (stability ±0.7% over 24 hours) |
| Polarisation accuracy | 1% |
| Operating temperature | 0.48K |
| Target volume | $35 cm^3$ |
| Field homogeneity at 2.5T | $\pm 5$ in $10^5$ |
| Magnet stability | 4 parts in $10^5$ (over 30 days) |
| Thermal equilibuium S/N ratio | >45db for 10m sec. sweep |
| Polarisation difference between NMR coils | ±0.2% |

## ACKNOWLEDGEMENTS

Thanks are due to the following - D. J. Nicolas & F. M. Russell who participated in the initial discussions of the target and magnet designs, S. F. J. Read and J. L. Thomas who developed and prepared the material, the drawing office and workshops of Technology Division who played their usual roles and A. Jones, J. H. Craig, J. Chauhan, W. Lester, M. Sedwards and B. Thompson who not only operated the system but helped to construct and test it.

# RECENT DEVELOPMENTS IN POLARIZED TARGETS AT CERN

T.O. Niinikoski
CERN, Geneva, Switzerland

## ABSTRACT

This report describes some of the development work done at CERN on polarized targets during past two years. The main emphasis is given to target material research, but some technical developments are also described.

## INTRODUCTION

The major handicaps in today's polarized targets are their low hydrogen content and poor resistance to radiation, and sometimes the infrequent reversal due to slow polarization build-up. Although in the long term dramatic improvements in these parameters can be expected[1], present users of polarized targets might profit from some more classical improvements which could possibly be implemented straight-away. In view of this we have tried to reintroduce the old butanol doped with the new Cr(V) complexes[2], and give some preliminary results here. Some tests have been done with deuterated butanol and propanediol as well.

Instrumental developments may also improve the figure of merit and availability of a given polarized target. In particular, optimization of the microwave parameters and automation of reversal are discussed with a view to shortening the reversal period. Cooling of the target by evaporation or dilution of $^3$He are compared, and a novel "easy loading" dilution refrigerator is briefly described. This refrigerator is now being used in our target material research.

## DEVELOPMENT OF POLARIZABLE MATERIALS

The results of our recent work on 1-butanol alcohol doped with porphyrexide or a new stable Cr(V) complex BHHA-Cr(V) [3] are summarized in Table 1. We have also performed one measurement with deuterated propanediol PD-$D_6$ doped with another new stable complex EHBA-Cr(V) [3]; the results are also given in Table 1. The advantages of the above materials are i) the direct chemical doping by dissolving a stable paramagnetic impurity allows large quantities of uniform material to be prepared, and ii) the effective chemical composition of BuOH is $CH_{1.88}$ in comparison with propanediol which is equivalent to $CH_{1.41}$, i.e. BuOH has 1.33 times more hydrogen per carbon atom than PD. The second advantage might be compensated by the somewhat lower polarization so far obtained in BuOH, but we hope to find better ways of preparing solid solutions of the new complexes. In the following we shall briefly discuss the results of Table 1.

BuOH-PR: The two samples were prepared in the usual way by adding 5% water to BuOH and dissolving the complex in it after having deoxygenated the solution by bubbling dry $N_2$ through the liquid. Porphyrexide decays with a time constant of some tens of minutes in a

Table I

| Material | Dopant and concentration (spin/ml) or (%) a) | $f^+/f^-$ (GHz) | $\frac{\max P_H^+}{\max P_H^-}$ (%) | $\frac{\max P_D^+}{\max P_D^-}$ (%) | $\tau_{0.7}$ (min) b) | $\tau$ | T (K) | Refr. c) |
|---|---|---|---|---|---|---|---|---|
| BuOH +5% $H_2O$ | PR "½" | 70.235 / 70.580 | 67 / -65 | — | 25 | 20 min | 0.91 | DR |
|  |  |  |  |  |  | 700 h | 0.04 |  |
| BuOH +5% $H_2O$ | PR "2" | 70.235 / 70.580 | 80 / -78 | — | 10 | 11 min | 0.98 | DR |
| BuOD($D_{10}$) no water | BHHA-Cr(V) $10^{20}$ | 69.180 / 69.685 | (±94) d) | +30 / -30 | 5 | 59 sec | 0.98 | DR |
| PD($D_6$) no additive | EHBA-Cr(V) $10^{20}$ | 69.230 / 69.708 | (±97) d) | +40 / -40 | 10 | 3.5 min | 0.98 | DR |
| BuOH no water | EHBA-Cr(V) $3 \times 10^{19}$ |  | +75 / -77 |  | 15 | 21 min | 0.78 | $^3$He |

a) Abbreviations: PR = porphyrexide; BHHA-Cr(V) = chromium (V) complex with 2-butyl-2-hydroxy-hexanoix acid (see Refs. 2); EHBA-Cr(V) = chromium (V) complex with 2-ethyl-2-hydroxybutyric acid (see Refs. 2).

b) $\tau_{0.7}$ means the time necessary for reaching $0.7 \times (\max P)$ starting from opposite maximum polarization.

c) DR = dilution refrigerator; $^3$He = helium three evaporation refrigerator.

d) These numbers are in parentheses because they refer to free proton polarization of a deuterated substance, which is normally not used in a proton target.

solution thus prepared, which gives time to make beads from quantities of the order of some tens of millilitres. The first of our samples was taken out of the last 25% of beads prepared from a 100 ml solution; the estimated effective concentration is indicated as ½%, referring to the concentration equivalent to that of a freshly prepared solution. The second sample was prepared rapidly into beads after having dissolved the nominal amount of 2% of porphyrexide. The polarization results show a marked difference between these two samples, both of which were cooled in the mixing chamber of a large dilution refrigerator[3]) at 25 kG field.

The optimum microwave frequencies observed for the two signs of polarization correspond rather well to those observed earlier. Also, the relaxation and polarization times roughly agree with earlier work. Contrary to the case of PD-Cr(V), the lower temperature available in the dilution refrigerator does not seem to greatly improve the polarizations which have already been reached in $^3$He evaporation refrigerators. This is in agreement with the predictions of Borghini[4]). Another point of comparison is that the optimum frequency in BuOH-PR is much flatter than that in PD-Cr(V); in the above samples, little polarization change was observed in the interval of ±10 MHz around the optimum frequencies.

  BuOD($D_{10}$)-BHHA-Cr(V): The butanol sample was 98% deuterated and no water was added to it, because the BHHA-Cr(V) complex was soluble in it at room temperature. Upon solidification, however, the complex may have crystallized to form clusters. This is supported by the line shape of the residual hydrogen nuclei in the sample, which is shown in Fig. 1. The narrow line corresponds to the usual hydrogen NMR frequency in butanol; the broad line is supposed to belong to the hydrogen in the BHHA-Cr(V) molecules, which have formed clusters and therefore see an effective field smaller than the weakly magnetized BuOD matrix. Also, the maximum polarization $P_D$ = ±30% and the large distance (about 500 MHz) between the optimum positive and negative frequencies suggest that the material does not have a uniform distribution of paramagnetic spins. However, the spin temperature equilibrium was attained rapidly under microwave irradiation. The spin-lattice relaxation was unusually rapid in this sample.

  PD($D_6$)-EHBA-Cr(V): The sample was prepared by dissolving the complex directly into the partly deuterated propanediol $CD_2OHCDOHCD_3$. The stability of the complex in the deuterated solution at room temperature and exposed to usual atmosphere was far better than the stability in an undeuterated solution. The somewhat higher polarization obtained in this sample compared with the BuOD($D_{10}$) might be explained by a better distribution of the complex in the material; no second hydrogen line could be distinguished. However, the maximum polarizations of +40% and -40% are worse than those in PD($D_6$)-Cr(V) mixed with PD($D_8$), where +44% and -47% maximum deuteron polarizations have been measured in a large target[5]).

  BuOH-EHBA-Cr(V): This test was done with a large sample (40 cm$^3$) cooled in the copper cavity of a $^3$He evaporation refrigerator. No water was added to the butanol. The polarization results suggest that this material might be a potentially good replacement for the standard PD-Cr(V). The advantages of propanediol -- routinely about 90% polarization and a fast polarization speed (60% in 10 min) --

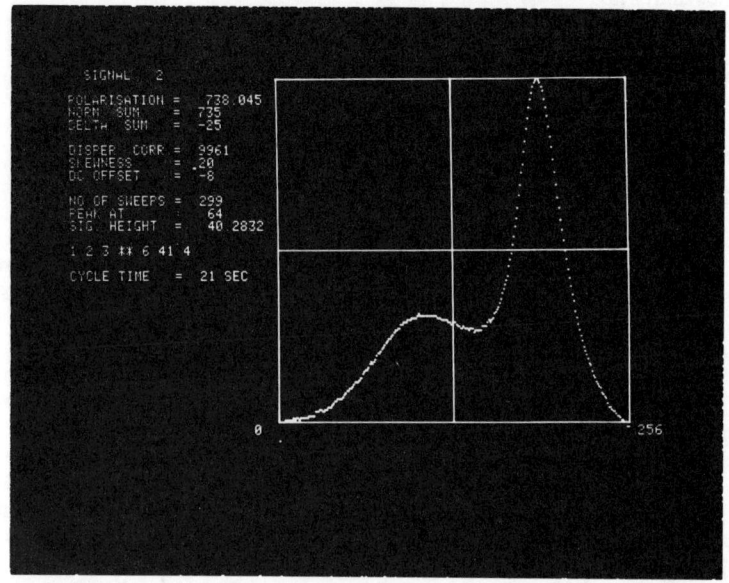

Fig. 1. The proton NMR line shape in 98% deuterated butanol alcohol. The narrow line occurs at the usual frequency in pure butanol and is interpreted as belonging to the 2% unsubstituted protons. The broad line is supposed to be due to clusters of the complex molecules which have a larger magnetization and therefore a shifted resonance frequency.

might well be compensated in some experiments by the higher hydrogen content of butanol.

The work with the new complexes is in progress, and we expect that improvements can be made quite soon in the preparation of these materials.

## INSTRUMENTAL DEVELOPMENTS

<u>Microwave optimization</u>: It has been noted on several occasions that the optimum microwave frequency in PD-Cr(V) changes with the accumulating dose of radiation in the target. If a computer is used for deducing the polarization from a digitized NMR signal, then it is possible to control the microwave frequency in such a way that one always stays at the optimum. This was implemented at CERN by F. Udo; the microwave frequency was "modulated" synchronously with the target polarization measurement cycle, and the time derivative of the polarization was then "demodulated" at the same frequency in the computer. This demodulated signal was fed back to the frequency control of the microwave supply. In this was the system searches out the frequency at which the polarization time derivative is maximum, ending in a steady state with maximum polarization. The microwave power was stabilized at the cavity by feedback from the $^3$He vapour pressure to a motorized attenuator.

The same scheme could be applied to maximizing the polarization growth speed in conditions where the target polarization reversal should take place as rapidly as possible.

<u>Rapid polarization reversal</u>: Several schemes could be considered, of which the adiabatic fast passage and rotating polarization are not discussed here. We focus our attention on two very simple and easily realizable cases: rotating field with frozen spin technique, and combined field zeroing and strong microwave irradiation. The first case is relevant to experiments where one does not have high beam intensity, say below $10^7$ charged particles/(sec cm$^2$); the latter is compatible with high beam intensity.

<u>Rotating field technique</u>: The target polarization always follows the direction of a slowly rotating field, which can be obtained for example with a rotating magnet or with orthogonal fields. The latter case has been implemented at CERN by winding a thin superconducting solenoid axially around the vacuum jacket of the CERN frozen spin target[3]. The additional wall thickness around the target increased by 0.4 mm, including the wire, isolation, and bonding epoxy. This solenoid can carry 20 A current, giving a field of 6.5 kG. The solenoid can be used either for holding the target polarization axially to the beam, or for helping to rotate a transverse field to the opposite direction without passing through zero. In the former case, the main (polarizing) field can be used for rotating the polarization to the opposite axial position without passing through zero.

The asymmetry measured in an experiment where the target polarization is rotated together with field will be at least partly masked by the varying field. This can be taken into account by reversing the target polarization with respect to the magnetic field with regular intervals. Statistically, two experiments instead of one have to be done, and the somewhat increased counting time has to be compared with the suppression of drifts in the apparatus and beam gained by frequent reversals of the polarization. It is clear that the false asymmetry created by the rotating field is relatively smaller in the case of polarization axial to the beam.

Field zeroing technique: Some magnets can be swept rather rapidly to zero field and back up to a stable 25 kG field. This offers a rapid means of zeroing the target polarization, and therefore a significant speeding up of the time necessary for reaching the polarization value, say 60%, at which the data-taking will be started. The gain in the reversal time would be rather insignificant if the polarization build-up would be exponential. This is rarely the case, and therefore the time spent from 0 to 60% (for example) depends strongly on the initial polarization at the moment when one switches on the microwaves at the frequency of the opposite polarization. Some physical insight into the situation can be gained by first noting that some parts of the target polarize slower than others, depending on the effective value of the magnetic field. Another cause of non-uniform polarization growth might be material inhomogeneity, in particular in large targets. Furthermore, microscopic polarization non-uniformities will definitely arise when the spin diffusion is needed for spreading the polarization around paramagnetic centres or clusters, a typical case being ammonia doped with PD-Cr(V) [6].

Combined with microwave optimization, the field zeroing technique brings the reversal time from 25 min to 10 min in a large (40 cm$^3$) propanediol target cooled in a $^3$He refrigerator. With microwave optimization, some further gain in reaching 60% could be expected, and if the experimenter would be happy to start counting at 40% polarization, only a few minutes are spent in the reversal.

Dilution refrigerators: It is a rather common belief that dilution refrigeration does not bring about any significant improvement in the dynamic polarization in comparison with the $^3$He evaporation refrigeration, because at the power levels needed, the helium temperatures are almost equal in both cases. This is true: as an example we might compare our $^3$He refrigerators and dilution refrigerators, which use the same size of pumps and cool 50 mW load at 485 and 305 mK temperatures, correspondingly. The ratio of these two temperatures is 0.63. However, these are cooling fluid temperatures, which are not directly relevant in dynamic polarization. The important parameter is the target material lattice temperature $T_L$.

The essential difference between these two cooling systems lies in the fact that the heat transfer mechanism in them is fundamentally different: in evaporating $^3$He the dominating heat resistance is the so-called nucleate boiling resistance[7], whereas in the dilution refrigerator it is the Kapitza resistance. In the former case the rise of the lattice temperature $T_L$ can be determined from the boiling correlations, which have never been studied systematically for $^3$He and very little for $^4$He at low pressure. Generally, in nucleate boiling the temperature rise is proportional to the heat flux, giving

$$T_L = T_3 + \Delta T = T_3 + \frac{\dot{Q}_m}{\sigma_t \cdot c}, \tag{1}$$

where $T_3$ is the $^3$He temperature ($T_3 = L_3^0/(k \log \dot{Q}_m/A)$), $\dot{Q}_m$ is the microwave power dissipated in the target material, and c is a coefficient which relates the heat flux to the temperature rise $\Delta T$ in nucleate boiling; c may depend weakly on the vapour pressure, which we omit in the following. In a dilution refrigerator the lattice temperature can be expressed by[8]

$$T_L = \left(\frac{1}{\sigma_t S_t} + \frac{1}{12.5\sigma S}\right)^{\frac{1}{4}} \dot{Q}_m^{\frac{1}{4}}, \qquad (2)$$

where $\sigma$ is the effective surface area of the heat exchanger[8] and S the Kapitza surface conductivity constant. Taking the heat input 50 mW of the above example and applying it to a 40 cm$^3$ target made in the form of 1.5 mm beads, we find lattice temperatures of 800 mK in the $^3$He refrigerator and 365 mK in the dilution refrigerator, by taking c = 0.15 mW cm$^{-2}$ K$^{-1}$ and $S_t$ = 2.5 × 10$^{-3}$ W K$^{-4}$ cm$^{-2}$. The ratio of these temperatures is now 0.46, illustrating well the large spin temperature differences that can be found particularly in the dynamic polarization of deuterium nuclear spins.

Easy-loading dilution refrigerator: As can be deduced from Eq. (2), with a limited target surface area $\sigma_t$ it is not necessary that the dilution refrigerator runs at a temperature much below, say, 0.3 K. Such a refrigerator can theoretically run without isolating vacuum separating the mixing chamber and the still, allowing very easy loading of the target material into the mixing chamber. We have constructed a dilution refrigerator insert[9] replacing an old $^3$He refrigerator; in this dilution refrigerator the heat exchanger and mixing chamber are covered with a simple teflon jacket, which just holds the helium mixture inside. The creeping superfluid film on the outside surface of this jacket, together with other residual heat leaks, limits the ultimate low temperature of this refrigerator to about 140 mK, probably preventing its use in a frozen spin target, but allowing, for example, high deuteron polarizations to be reached in the continuous mode. This refrigerator can be loaded in about 30 min, and material tests can be performed in it on a basis of one sample per two days, calibration included. The maximum cooling power is 20 mW, allowing rather sizeable targets to be polarized. For details we refer to the original publication[9].

## ACKNOWLEDGEMENTS

I wish to thank my collaborators J.-M. Rieubland and J.-C. Soulié for their intense efforts in the course of the work described here, which was mainly executed alongside our normal tasks. Thanks are also due to Drs. G. Court and M. Krumpolc for their contributions in the material development projects.

## REFERENCES

1) G.R. Court et al., Report of the workshop on polarized target materials, these proceedings.

2) M. Krumpolc and J. Roček, Synthesis of stable chromium (V) complexes of tertiary hydroxy acids, to be published in J. Amer. Chem. Soc.
M. Krumpolc and J. Roček, J. Amer. Chem. Soc. $\underline{100}$, 145 (1978).
J. Amer. Chem. Soc. $\underline{99}$, 137 (1977).
J. Amer. Chem. Soc. $\underline{98}$, 872 (1976).
3) T.O. Niinikoski and F. Udo, Nucl. Instrum. Methods $\underline{134}$, 219 (1976).
4) M. Borghini, Proc. 2nd Internat. Conf. on Polarized Targets, Berkeley, 1971 (Ed. G. Shapiro) (National Technical Information Service, Springfield, Virginia, 1972), LBL 500, UC 34 Physics, p. 1.
5) T.O. Niinikoski, Proc. AIP Conference on High-Energy Physics with Polarized Beams and Targets, Argonne, 1976 (AIP Conf. Proceedings No. 35), p. 458 (1976).
6) K. Scheffler, Proc. 2nd Internat. Conf. on Polarized Targets, Berkeley, 1971 (Ed. G. Shapiro) (National Technical Information Service, Springfield, Virginia, 1972, LBL 500, UC 34 Physics, p. 271.
7) See, for example, I.R. McDougall, Cryogenics $\underline{11}$, 260 (1971) and references therein.
8) T.O. Niinikoski, Proc. ICEC 6, Grenoble, 1976, p. 102.
9) T.O. Niinikoski and J.-M. Rieubland, Proc. IIR Commission A 1-2, Zurich, 1978 (International Institute of Refrigeration, Paris, 1978), p. 181.

# POLARIZED ELECTRON BEAMS

V.N.Baier *

Institute of Nuclear Physics, Novosibirsk 630090, USSR

## ABSTRACT

Different methods of production of polarized electrons - radiative separation of the beam, radiative polarization and acceleration of polarized particles - are discussed. Recent status of polarization measurement is considered.

## INTRODUCTION

During past years the research with polarized electron beams have been significantly extended. Among of the already obtained results it should be mentioned the observation of asymmetry in electron scattering on protons and deutrons due to interference of electromagnetic and weak interactions at SLAC, measurement of azimuthal asymmetry in the inclusive cross section of hadron production and at jet production in the process $e^+e^- \longrightarrow$ hadrons on SPEAR, metrological measurement on the VEPP-2M storage ring at INP, Novosibirsk.

Polarized electron beams can be produced by the following methods:

a) Polarization by radiative separation of the particles with opposite spin projections in an electron beam at superhigh energies (hundred GeVs) using megagauss magnetic fields [1-3]. This method is suitable for produicng the polarized electrons in the secondary beams of the biggest proton accelerators (FNAL, CERN II).

b) Radiative polarization on electron storage rings, starting with an energy of hundred MeVs and higher. At present this mechanism is widely employed (see the reviews [4-7]).

c) Direct acceleration of polarized electrons with

_____
* Presented by Ya. Derbenev

appropriate sources[8]. This problem is solved on linear accelerators at SLAC. Considerable interest is of the solution of this problem for cyclic accelerators. The polarized particle production is closely associated with the problem of stability of the polarization.

d) In all the cases a problem of measuring the polarization is of importance. The problems listed above are discussed in the present review.

## ELECTRON POLARIZATION BY RADIATIVE SEPARATION

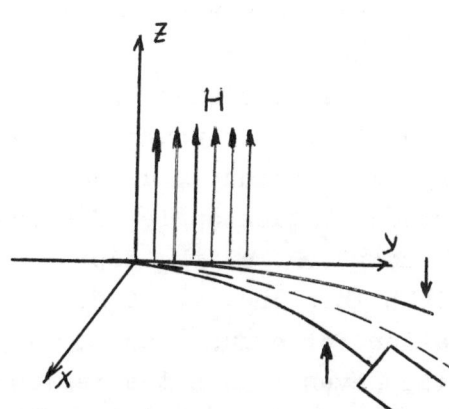

Fig. 1. A picture of electron polarization by radiative separation

When the electron is passing through a magnetic field, the radiation probability depends on a spin projection on the direction of magnetic field $\zeta$. Due to this, the energy losses and, hence, the deviation angle of electron do so, that finally leads to spatial separation of electrons with different projections of $\zeta$ (see Fig.1).

This effect becomes already noticeable at $\chi \sim 0.05$. The parameter

$$\chi = \chi(\varepsilon) = \frac{H}{H_0} \frac{\varepsilon}{m} \tag{1}$$

where H is the magnetic field, $H_0 = \frac{m^2 c^3}{e\hbar} = 4.41 \cdot 10^{13}$ Oe, $\varepsilon$ is the energy of electron, m is its mass, serves for describing the quntum properties of the magnetic bremsstrahlung. Note, that the spin term in the expression for radiation probability contains an additional power of parameter $\chi$. At an electron energy of 250 GeV (secondary beams of the accelerators FNAL, CERN II) the magnetic field H = 4 MG is required to obtain $\chi \sim 0.05$. Some methods are known for producing the megagauss magnetic fields at short times (not shorter than the transit-time of particles) and in small volumes (not exceeding a few cm³), in particular, the explosive method. The devices with megagauss magnetic fields (magnetic converters) are proposed to produce hard photons (at $\chi \lesssim 1$ a maximum of spectral distribution lies at $\omega \sim \varepsilon \chi$). It is clear that the method proposed here essentially broadens the possibilities of magnetic converters since both hard photons and secondary polarized electrons can be used.

To analyse the polarization effect, one should consider the problem of passing the electron through the magnetic field with radiation taken into account, depending on the spin projection $\zeta$, the deviation angle of electron in a field $\varphi$ and at an energy $\varepsilon$ at the depth t (the velocity of light c = 1). To this end, it is necessary to solve the kinetic equation for a function $\varrho_\zeta(\varepsilon,\varphi,t)$ The general form of this equation even within the regions $\chi \lesssim 1$ where the pair production is negligible, is rather complicated and its solution in an analytical form seems to be impossible. However, in the case important for application when

$$\chi_o \equiv \chi(\varepsilon_o) = \frac{H}{H_o}\frac{\varepsilon_o}{m} \ll 1$$

one can expand all the entering quantities in powers of $\chi \leqslant \chi_o$ ( $\varepsilon_o$ is the initial electron energy). In a first approximation, which may be considered as a Fokker-Plank approximation, one keeps only the leading terms of expanxion in powers of $\chi_o$ in the coefficients of equation. Then

$$\left(\frac{\partial}{\partial z} + \frac{\beta}{\tau}\frac{\partial}{\partial \varphi}\right)\varrho_\zeta(\tau,\varphi,z) = \frac{\partial}{\partial \tau}\left[\tau^2(1+a_\zeta\tau)\varrho_\zeta\right]+ \quad (2)$$

$$+ \frac{1}{2}\frac{\partial^2}{\partial \tau^2}(\tau^4 b_2 \varrho_{\zeta})$$

where

$$\tau = \frac{\varepsilon}{\varepsilon_0}, \quad z = \frac{I_c(\varepsilon_0)t}{\varepsilon_0}, \quad I_c(\varepsilon_0) = \frac{2}{3}\alpha m^2 \chi_0^2$$

$$\beta = \frac{eH}{I_c(\varepsilon_0)}, \quad a_2 = -(6c + \frac{3}{2}\zeta)\chi_0, \quad b_2 = 2c\chi_0, \tag{3}$$

$$c = 55/32\sqrt{3} = 0{,}992$$

The set of equations (2) ($\zeta = \pm 1$) will be solved under the initial condition

$$\varrho_{\zeta}(\varepsilon, \varphi, 0) = \delta(\varepsilon - \varepsilon_0)\delta(\varphi - \varphi_0) \tag{4}$$

It is evident that for the initial condition of a general form $\varrho_{\zeta}^{(g)}(\varepsilon, \varphi, 0) = \varrho_0^{(g)}(\varepsilon, \varphi)$ this solution is

$$\varrho_{\zeta}^{(g)}(\varepsilon, \varphi, z) = \int \varrho_{\zeta}(\varepsilon, \varphi, z; \varepsilon_0, \varphi_0) \varrho_0^{(g)}(\varepsilon_0, \varphi_0) d\varepsilon_0 d\varphi_0 \tag{5}$$

Everywhere this is permissible we shall assume that $\varphi_0 = 0$.

From the system (2) follows directly that in this approximation the number of particles with a given $\zeta$ remains the same since this system is diagonal over variable $\zeta = \pm 1$.

If one introduces a new variable x connected to $\tau$ by the following relation

$$x = x(\tau) = \frac{1}{\tau} - 1 + A_2 \ln \tau; \quad \frac{1}{\tau} \simeq 1 + x + A_2 \ln(1+x); \tag{6}$$

$$A_2 = a_2 + b_2;$$

and a new function

$$F(x, \varphi, z) = \varrho(\tau(x), \varphi, z) \left|\frac{d\tau}{dx}\right| \varepsilon_0 \beta \tag{7}$$

then, retaining the leading over $\chi_0$ terms, we have the following equation for $F(x, \varphi, z)$ from (2):

$$\frac{\partial F}{\partial z} + \beta(1+x+A_1 \ln(1+x))\frac{\partial F}{\partial \varphi} = \frac{b_1}{2}\frac{\partial^2 F}{\partial x^2} - \frac{\partial F}{\partial x} \qquad (8)$$

with the initial condition $F(x,\varphi,0) = \delta(x)\delta(\varphi/\beta)$

If one puts $A_1 = b_1 = 0$ in (8), then the equation derived will be satisfied by the following function

$$F = F\left(x-z, \varphi - \beta\left(z + \frac{z^2}{2} + z(x-z)\right)\right)$$

that gives the classical solution for the presented initial condition:

$$F_c(x,\varphi,z) = \delta(x-z)\delta\left(\frac{\varphi}{\beta} - z - \frac{z^2}{2}\right) \qquad (9)$$

Therefore, in the classical limit

$$x = z, \quad \varphi = \varphi_c = \beta\left(z + \frac{z^2}{2}\right) \qquad (10)$$

It is clear that at $\chi_0 \to 0$ a quantity $x - z \to 0$, i.e. when solving the equation (8) it is possible to use the expansion in powers of $u = x - z$. Retaining the leading terms of this expansion we have from (8):

$$\frac{\partial F}{\partial z} + \beta(u+1+z+A_1\ln(1+z))\frac{\partial F}{\partial \varphi} = \frac{b_1}{2}\frac{\partial^2 F}{\partial u^2} \qquad (11)$$

Let us introduce now a new variable in (11):

$$\Psi = \frac{\varphi}{\beta} - \int_0^z [1+t+A_1\ln(1+t)]\,dt \qquad (12)$$

then we obtain a final form of equation

$$\frac{\partial F}{\partial z} + u\frac{\partial F}{\partial \Psi} = \frac{b_1}{2}\frac{\partial^2 F}{\partial u^2}, \quad F(u,\Psi,0) = \delta(u)\delta(\Psi) \qquad (13)$$

The solution of this equation has the form:

$$F(u,\Psi,z) = \frac{1}{\pi\Delta\delta}\exp\left(-\frac{2u^2}{\Delta^2} - \frac{2\Psi^2}{\delta^2} + \frac{2\sqrt{3}}{\Delta\delta}u\Psi\right) \qquad (14)$$

where
$$u = x - z, \quad \psi = \frac{\varphi}{\beta} - z - \frac{z^2}{2} - \mathcal{A}_\zeta \left[ (1+z)\ln(1+z) - z \right],$$
$$\Delta^2 = \ell_z z, \quad \delta^2 = \ell_z \frac{z^3}{3} \tag{15}$$

Integrating the function (14), one can find a distribution over range, or deviation angle:
$$F(u,z) = \int_{-\infty}^{+\infty} F(u,\psi,z) d\psi = \frac{1}{\sqrt{2\pi\Delta^2}} e^{-\frac{u^2}{2\Delta^2}}, \quad F(\psi,z) = \frac{e^{-\frac{\psi^2}{2\delta^2}}}{\sqrt{2\pi\delta^2}} \tag{16}$$

Thus, in the approximation under consideration the angular distribution and energy one are Gaussian. This is just the form in which the angular distribution function has been restored by the known distribution moments in ref.[1]. The mean deviation angle is (see (16)):

$$\langle \varphi \rangle_\zeta = \beta \left[ z \left( 1 + \frac{z}{2} \right) + \mathcal{A}_\zeta g(z) \right], \quad \psi = \frac{\varphi - \langle \varphi \rangle_\zeta}{\beta} \tag{17}$$

where
$$g(z) = (1+z)\ln(1+z) - z \tag{18}$$

so that the particles with different spin projections deviate at a different angle, so

$$\langle \varphi \rangle_{\zeta=-1} - \langle \varphi \rangle_{\zeta=1} = 3\beta X_0 g(z) \tag{19}$$

From above it follows that the particles with different $\zeta$ are spatially separated. If the initial beam has been unpolarized, then after its transit through the magnetic field, it becomes, generally speaking, partially polarized, if one chooses from it the particles within some interval of angles and energies, the polarization degree essentially depends on a choice of these intervals. Selecting the particles in the interval of deviation angles, let us introduce

$$dN_{\vec{\zeta}} = d\varphi \frac{N}{2} \int g_{\vec{\zeta}}(\varepsilon,\varphi,z) d\varepsilon \qquad (20)$$

where N is the total number of particles in the beam. There will be $N_{\vec{\zeta}}(\alpha_1,\alpha_2)$ particles in the angular interval $\alpha_1 \div \alpha_2$:

$$N_{\vec{\zeta}}(\alpha_1,\alpha_2) = \int_{\alpha_1}^{\alpha_2} dN_{\vec{\zeta}} \qquad (21)$$

the polarization degree of these particles being

$$\xi(\alpha_1,\alpha_2) = \frac{N_{\vec{\zeta}=-1}(\alpha_1,\alpha_2) - N_{\vec{\zeta}=1}(\alpha_1,\alpha_2)}{N_{\vec{\zeta}=-1}(\alpha_1,\alpha_2) + N_{\vec{\zeta}=1}(\alpha_1,\alpha_2)} \qquad (22)$$

The polarization, due to radiative separation, is an effect of the first order over $\chi$. The consideration above has been carried out to the same accuracy. The account of the terms of a higher order over $\chi$ is of some interest. This enables one not only to obtain a much more accurate result, but to define an accuracy and limits of this approach. In the second order over $\chi^2$ the beam polarization as a whole arises due to action of the radiative polarization mechanism

$$\xi_R = \frac{3\chi_0 z}{2(1+z)} \qquad (23)$$

This results in decreasing the number of electrons with $\vec{\zeta}$ = 1 and in increasing the number of electrons with $\vec{\zeta}$ = -1. Besides, in the second order over $\chi$ the distribution function differs from Gaussian one and becomes asymmetric, the number of particles with either polarization in the region of large deviation angles increasing, and both the difference of mean deviation angles $\langle\varphi\rangle_{\vec{\zeta}=-1} - \langle\varphi\rangle_{\vec{\zeta}=1}$ and the dispersions of distributions decreasing.

Fig.2 presents the dependence of the polarization degree $\xi(\alpha,\infty), \xi(-\infty,\alpha)$ on the angle $\alpha$ which is measured

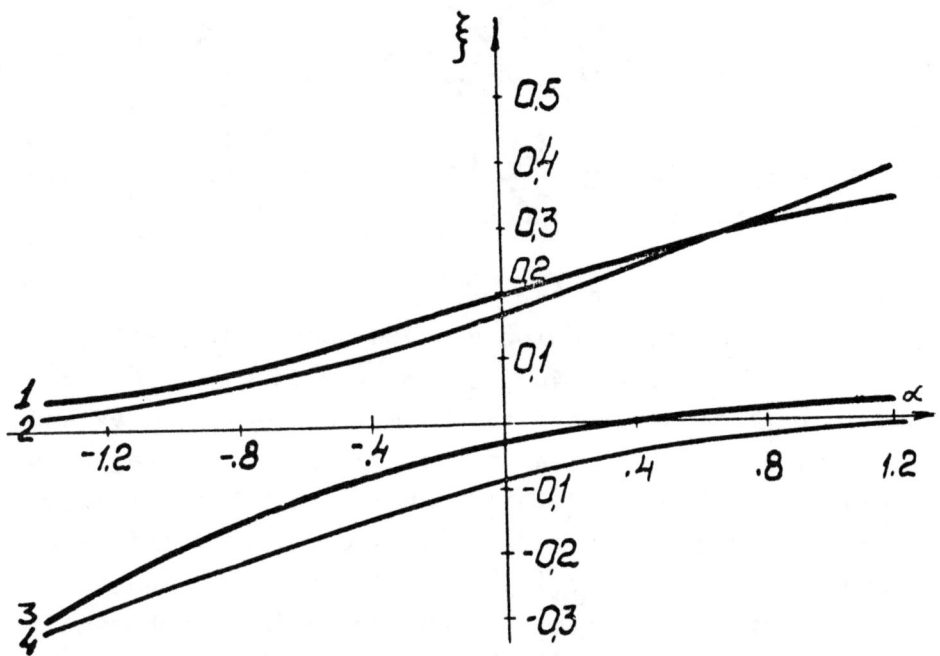

Fig.2 Polarization degree $\xi(\alpha,\infty)$ (curve 1), $\xi(-\infty,\alpha)$ (curve 3) and corresponding quantities $\xi^{(1)}(\alpha,\infty)$ (2) and $\xi^{(1)}(-\infty,\alpha)$ (4), calculated in the first order over $\chi$, all these at $\chi_o = 0.05$ and $z = 1$.

in units (see (10))

$$\frac{1}{\sqrt{2\delta^2}} \frac{(\varphi - \varphi_c)}{\beta} \quad (24)$$

For comparison the quantities $\xi^{(1)}(\alpha,\infty), \xi^{(1)}(-\infty,\alpha)$ are given which have been calculated in the first order over $\chi$. To this end, it is necessary to substitute the distribution $F(\psi,z)$ (16) into formulae (20)-(22). In this case, the polarization is calculated without any difficulties since it is defined by a simple function (22) of Gaussian integrals of the form (see (16), (17)):

$$N_{\zeta}(\alpha_1,\alpha_2) = \frac{N}{2}\int_{\alpha_1}^{\alpha_2}\frac{d\varphi}{\sqrt{2\pi}\beta\delta}\exp\left[-\left(\frac{\varphi-\langle\varphi\rangle_{\zeta}}{\sqrt{2}\beta\delta}\right)^2\right] \quad (25)$$

As a function of range z (note that the electron energy at the range z is $\varepsilon = \frac{\varepsilon_0}{1+z}$) the polarization properties are determined by the function (see (18)):

$$f(z) = g(z)/z^{3/2} \quad (26)$$

which is a smooth function z, note that f(0.5) = 0.31, f(1) = 0.39, f(2) = 0.46, $f_{max}$ = 0.52 at $z_m$ = 7.6. It should be also mentioned that z = 1 at $\varepsilon$ = 250 GeV and $X_0$ = 0.05 is equivalent to the range in a magnetic field of ~ 1.6 cm. A detailed analysis of highest approximations is given in ref.[2]. It is shown that this approach is applicable at $X_0/z \lesssim 0.1$.

In ref.[3] the effect is considered in an inhomogeneous magnetic field H(z) = B$\nu$(z). In the Focker-Plank approximation it has turned out to be possible to find the solution of the kinetic equation which has the form (14) but the sense of entering quantities is another, in particular the dispersions of the distributions are

$$\Delta^2 = b_1 \int_{-\infty}^{z}\nu^3(z')dz'$$

$$\delta^2 = 2b_1 \int_{-\infty}^{z}\nu(z_1)dz_1\int_{-\infty}^{z_1}\nu(z_2)dz_2\int_{-\infty}^{z_2}\nu^3(z_3)dz_3 \quad (27)$$

Here $b_1$ is given by formula (3), where $X_0 = B\varepsilon_0/H_0 m$

Two cases are analysed as examples of inhomogeneous field in ref.[3]:

$$\nu_1(z) = \frac{d_1}{ch\,K_1 z}, \quad d_1 = \frac{\sqrt{\pi}}{2}, \quad K_1 = \frac{\sqrt{\pi}^2}{2z_0}$$

$$\nu_2(z) = \frac{d_2}{1+(K_2 z)^2}, \quad d_2 = 2, \quad K_2 = \frac{2\pi}{z_0} \quad (28)$$

where the constants $d_{1,2}$ and $k_{1,2}$ are chosen so that classical values of energy and angle, after passing the electron through a field $H(z)$ coincide with the corresponding characteristics at the range $z_0$ for homogeneous field $B$ $(\varepsilon_c = \frac{\varepsilon_o}{1+z}, \varphi_c = \beta z_o (1 + \frac{z_o}{2}))$. Substituting the result obtained in formula (22), one can find a polarization degree. The result is presented in Fig.3, where the angles are measured again in units (23). It is seen that in

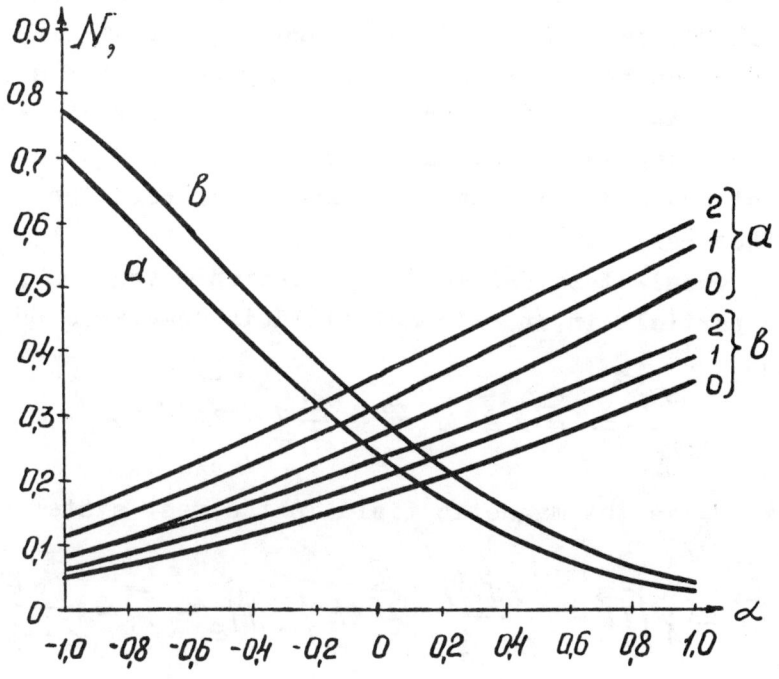

Fig.3 Polarization $\xi(\alpha, \infty)$ in fields $H(z) = B\nu(z)$, $\nu_1$ and $\nu_2$ is given by (28) (indices 1,2). For comparison the result for homogeneous field is presented (index 0). All the results for $z = 1$, $\chi_o = 0.1$ (curves (a)) and $\chi_o = 0.05$ (curves (b)). Number of particles with given polarization $N(\alpha,\infty)$ for $\nu = \nu_2$.

an inhomogeneous field the polarization is higher than

that the homogeneous one, the difference in the number of particles is unsignificant. This means that in each case, carrying out an optimal separation not only over angle but also over energy it is possible to obtain a quite high polarization degree.

## RADIATIVE POLARIZATION

The radiative polarization effect in an homogeneous field was calculated 15 years ago[8]. For that time the phenomenon being of outstanding importance for storage rings, has been received detailed and comprehensive theoretical analysis. This analysis covers both practical questions and understanding, and interpretation of this effect (see reviews[4-6]).

Quasi-classical motion of the spin vector $\vec{\zeta}$ of the particle in an external field is described by equation

$$\frac{d\vec{\zeta}}{dt} = \vec{F} \times \vec{\zeta} \, , \quad \vec{F} = \frac{e}{\varepsilon} \left[ a\vec{H}_R + \vec{H}_E \right] \tag{29}$$

Here $\vec{H}_R$ is the magnetic field in the rest system of the particle.

$$\vec{H}_R = \gamma \left[ \vec{H} - \frac{\vec{v}(\vec{v}\vec{H})}{1 + 1/\gamma} - \vec{v} \times \vec{E} \right], \quad \vec{H}_E = \vec{H} - \frac{\vec{v} \times \vec{E}}{1 + 1/\gamma} \tag{30}$$

$\vec{H}, \vec{E}$ are the fields in laboratory system, $\gamma = \varepsilon/m$, $a = \frac{g-2}{2}$ is the anomalous magnetic moment (in units $e\hbar/2mc$) for the electron $a = \alpha/2\pi + ..$ . Equation (29) was derived in the twenties by Frenkel and Thomas. Later on this equation was deduced by many methods, in particular, in [10,4] it was obtained in the framework of quantum electrodynamics.

Equation (29) evidently describes the rotation. For the latter it is natural to introduce the action-angle variables. If the action

$$\vec{\zeta}_n = \vec{\zeta}\vec{n}\,(\vec{p},\vec{z}) \tag{31}$$

where $\vec{n}(\vec{\rho},\vec{z})$ is the unit vector independent explicitly of time and being a solution of equation (29). This vector is completely determined by a particle trajectory on which the fields (30) are determined, then the variable conjugate to $\vec{\zeta}_n$ is a precession phase of the vector $\vec{\zeta}$ around the axis $\vec{n}$. If the evolution of the vector $\vec{n}$ is represented in the form of harmonic expansion, then its spectrum will contain only the frequencies of orbital motion. When the mean spin precession frequency $\langle \dot{\psi} \rangle$ is close to linear combinations from the orbital motion frequencies (synchrotron oscillations are omitted in our consideration)

$$\nu_o \equiv \frac{\langle \dot{\psi} \rangle}{\langle \dot{\vartheta} \rangle} \simeq K_\vartheta + K_z \nu_z + K_x \nu_x \qquad (32)$$

where $\langle \dot{\vartheta} \rangle$ is the mean revolution frequency, $K_{\vartheta,z,x}$ are integers, $\nu_z$ and $\nu_x$ are the numbers of vertical and radial betatron oscillations per turn, then the spin resonance takes place near which the vector $\vec{n}$ moves with large deviations at small variations of parameters. After phase mixing, because of the frequency of the precession spread $\langle \dot{\psi} \rangle$, the average spin of the particles moving near equilibrium orbit will be directed near the precession axis for the equilibrium trajectory: $\vec{n}_s(\vartheta) \equiv \vec{n}(\rho_s, z_s)$, namely:

$$\langle \vec{\zeta} \rangle = \langle \zeta_n \vec{n} \rangle + \langle \vec{\zeta}_\perp \rangle \simeq \vec{n}_s \langle \zeta_n \rangle \qquad (33)$$

A significant feature of the vector $\vec{n}_s$ introduced in [9] is the periodicity with respect to the generalized azimuth of particle $\vartheta$: $\vec{n}_s(\vartheta) = \vec{n}_s(\vartheta + 2\pi)$. Thus, stable polarization repeated on a given azimuth may be done in arbitrary external stationary fields.

In existing storage rings the direction $\vec{F}$ differs unessentially from the vertical one, therefore, the direction $\vec{n}_s$ is also close to the latter. Knowledge of small deviations of $\vec{n}$ from the vertical direction is needed for quantitative description of quantum depolarization.

If unpolarized elecrons are injected into a storage ring, then under the action of radiation the spin projection on the axis $\vec{n}$ will slowly (compared to typical periods of motion) varied approaching to some equilibrium value. This value is determined by a common action of polarizing and depolarizing factors. The mean rate of the variation can be written as follows:

$$\left\langle \frac{d\delta\vec{\zeta}_n}{dt}\right\rangle = \left\langle \vec{n}\frac{d\delta\vec{\zeta}}{dt}\right\rangle + \left\langle \vec{\zeta}\frac{d\delta\vec{n}}{dt}\right\rangle + \left\langle \delta\vec{\zeta}\frac{d\vec{n}}{dt}+\frac{d\vec{\zeta}}{dt}\delta\vec{n}\right\rangle \quad (34)$$

where $\delta\vec{\zeta}$ and $\delta\vec{n}$ are the increments $\vec{\zeta}$ and $\vec{n}$, respectively under the radiation action. The first term $\vec{n}\frac{d\delta\vec{\zeta}}{dt}$ describes the radiation action on a spin vector[10], the second - the radiation effect on a precession axis via perturbation of orbital motion. The correlative term $\left(\delta\vec{\zeta}\frac{d\vec{n}}{dt}+\frac{d\vec{\zeta}}{dt}\delta\vec{n}\right)$ is small for ultrarelativistic electrons and may be neglected. In the case when the direction of the vector $\vec{F}$ nearly coincides with the vertical one, the term $\left\langle \vec{\zeta}\frac{d\delta\vec{n}}{dt}\right\rangle$ describes depolarization under the action of quantum energy fluctuations (quantum depolarization). This mechanism was first described in ref.[11], and then it was analysed in [12] (see also [4]):

(35)
$$\left\langle \vec{\zeta}\frac{d\delta\vec{n}}{dt}\right\rangle = \left\langle \zeta_n\vec{n}\frac{d}{dt}\left(\frac{\partial\vec{n}}{\partial\gamma}\delta\gamma+\frac{1}{2}\frac{\partial^2\vec{n}}{\partial\gamma^2}(\delta\gamma)^2\right)\right\rangle = -\frac{\zeta_n}{2}\left\langle \left(\frac{\partial\vec{n}}{\partial\gamma}\right)^2\frac{d(\delta\gamma)^2}{dt}\right\rangle$$

Here it is taken into account that $\vec{n}^2= 1$, the r.m.s. energy fluctuation is

$$\frac{1}{\varepsilon^2}\frac{d(\delta\varepsilon)^2}{dt}=\frac{1}{2\varepsilon^2}\int(\hbar\omega)^2 dW(\vartheta,\varphi) = \frac{55}{48\sqrt{3}}\frac{z_0\hbar\gamma^5}{m\langle R^3\rangle} \quad (36)$$

where $dW(\vartheta,\varphi)$ is the photon radiation probability per unit time.

Substituting (35) and the explicit expression for $d\delta\vec{\zeta}/dt$ [10] into eq.(34). one gets the equation (in this form it was discussed in [4]):

$$\frac{d\vec{q}_n}{dt} = -A\vec{q}_n - B \qquad (37)$$

where

$$A = \frac{1}{T} + \frac{1}{\tau_d}, \quad \frac{1}{T} = \frac{5\sqrt{3}}{8} \frac{z_0 \lambda \gamma^5}{R^3} \langle |K|^3 \rangle,$$

$$\frac{1}{\tau_d} = \frac{55}{48\sqrt{3}} \frac{z_0 \lambda}{R^3} \gamma^7 \left\langle \left(\frac{d\vec{n}}{d\gamma}\right)^2 |K|^3 \right\rangle, \quad B = \frac{z_0 \lambda \gamma^5}{R^3} \left\langle \frac{\vec{n}(\vec{v} \times \vec{\dot{v}})}{|\vec{v}|} K^3 \right\rangle \qquad (38)$$

$K = \frac{H_z}{\langle H_z \rangle}$ is the dimensionless curvature of an ideal trajectory, $\langle H_z \rangle = \int_0^{2\pi} H_z d\vartheta/2\pi$, R is the radius of an ideal orbit, $z_0 = e^2/m$, $\lambda = \hbar/m$.

Thus, the particles are polarized in the direction opposite to $(\vec{v} \times \vec{\dot{v}})$, where the polarization degree

$$\vec{q}_n(t \to \infty) = -\frac{B}{A} = -\frac{8}{5\sqrt{3}} \frac{\langle K^3 \rangle}{\left\langle |K|^3 \left(1 + \frac{11}{18}\left(\gamma \frac{d\vec{n}}{d\gamma}\right)^2\right)\right\rangle} \qquad (39)$$

and the characteristic time of polarization is $\tau = T\tau_d/(T+\tau_d)$. Note, that eq.(37) and the general equation for $\vec{q}_n$ at an arbitrary vector $\vec{n}$ [13] can be derived by the technique used in [10], if it is applied to the combination $\vec{\sigma}\vec{n}$.

In the case of producing the polarized beams with the use of the mechanism described, it is necessary to choose an operation point beyond the resonances (32) (at least, the strongest from them). Quantum depolarization is also intensified near spin resonances. In a general case quantum depolarization is strongly connected with the imperfectness of the storage ring magnetic system. At superhigh energies $a\gamma \gg 1$ the most dangerous is the vertical distortion of closed orbits[11,12]. A comprehensive analysis of quantum depolarization at superhigh energies was carried in ref.[14] wherein both the case $\delta\nu_0 = a\delta\gamma \ll \Delta\nu$ (cf.(32)) and the case of $a\delta\gamma \gg \Delta\nu$ were considered (where $\Delta\nu$ is the distance between spin resonances), as well as the interaction between the beams. It was shown that the conservation of radiative polarization requires the fulfilment of strong limitations concerning the quality of the magnetic system of a storage ring. For example, permissible angular deviations of orientation of bending magnets $\alpha_M$ and lenses $\alpha_L$ in a plane normal to the orbit, and the permissible vertical displacement of lenses $\Delta z_L$ for a storage ring with a large number of uniform elements at $\alpha\delta_\gamma \ll \Delta \nu$ are:

$$\alpha_M \lesssim 0.1 \frac{\sqrt{\nu_z}}{\varepsilon^2}, \quad \alpha_L \lesssim 0.2 \frac{\sqrt{\nu_z}}{\varepsilon}, \quad \frac{\Delta z_L}{R} \lesssim \frac{5 \cdot 10^{-2}}{\nu_z^{3/2} \varepsilon^2} \quad (40)$$

where the energy $\varepsilon$ is taken in GeV. In [14] the critical analysis of the papers [15-18] was made whose authors are rather pessimistic relative to the possibility to produce the polarized particles in the biggest storage rings PEP, PETRA, LEP.

It may appear that the conditions of [14] do not satisfied. Then it is desirable to reduce the polarization time, this is possible, e.g., with magnetic "snakes" [5]. Such a "snake" consists of the sections with a strong sign-variable vertical magnetic field with an average value equal to zero and with a large $\langle H_z^3 \rangle$. In the simplest version there are three sections with lengths $\vartheta_-, \vartheta_+, \vartheta_-$ and fields $H_-, H_+, H_-$ in them. The zero average and symmetry with respect to the middle section ensure the conservation of a trajectory beyond the "snake" region. If the fields $H_+$ and $H_-$ are high enough, then both the polarizing and depolarizing processes will be mainly determined by the "snake", while the polarization degree will go to the value (it is assumed that in the "snake" section $\left(\gamma \frac{d\vec{n}}{d\gamma}\right)^2 \ll 1$):

$$\bar{\zeta}_n = -\frac{8}{5\sqrt{3}} \frac{\langle H_z^3 \rangle}{\langle |H_z|^3 \rangle} = -\frac{8}{5\sqrt{3}} \frac{H_+^2 - H_-^2}{H_+^2 + H_-^2} \quad (41)$$

and the polarization time to $T_f$:

$$\frac{1}{T_f} = \frac{5\sqrt{3}}{8} \frac{z_0 \lambda}{R^3} \gamma^5 \frac{|H_+|^3 g_+ + 2|H_-|^3 g_-}{2\pi |\langle H_z \rangle|^3} \quad (42)$$

so that the polarization time essentially decreases. There are other methods of forced polarization. Among them is the method based on using the interaction with a circularly-polarized electromagnetic wave (it has been recently proposed[19]).

There is also a proposal to increase the spin stability due to multiple spin rotations[20]. It is asserted that in an optimal case, one can decrease the effect of depolarizing factors by a factor of $M^2$ in the accelerator with 2M spin-rotating sections.

It is possible that at superhigh energies in a storage ring the versions of radiative separation may be realized such that on the small sections of the storage ring the particles with opposite-oriented spins move along different trajectories.

Thus, the radiative polarization seems to be important physical phenomenon with the help of which the polarized electrons and positrons were generated in the storage rings VEPP-2M, SPEAR, ACO. This turns out to be extremely useful in carrying out physical experiments. It is reached some level of theoretical understanding concerning the production of polarized particles in higher energy storage rings VEPP-4, PEP, PETRA, though this requires, as has noted above, special and laborious efforts. It should also bear in mind that in all the storage rings at the operation with polarized particles the luminosity is substantially, at least, a few times lower than the maximally attained. In connection with this, it is required an additional efforst to obtain polarized colliding beams with a high luminosity.

## ACCELERATION OF POLARIZED PARTICLES

The acceleration of polarized particles is a complicated problem. This is due to intersection of a large number of resonances (32) (if the field in the whole orbit is close to the vertical one, then in (32) $\nu_o = \alpha\gamma$). With an assumption that the phase mixing of the spin precession occurs, the vertical spin projections before and after the traversal of the resonance are connected by the relation [21,22]

$$\langle \vec{q}_z \rangle^{t \to +\infty} = (2e^{-2J_K} - 1) \langle \vec{q}_z \rangle^{t \to -\infty} \quad (43)$$

where the quantity $J_K = \pi |w_K|^2 [4(\dot{\nu}_o - \dot{\nu}_K)]^{-1}$ is determined by the rate of a passage of the resonance $\dot{\nu}_o - \dot{\nu}_K$ and its power $|w_K|$. Variation in the polarization degree may be small at a fast passage $J_K \ll 1$ or $J_K \gg 1$ at slow passage. When intersecting the set of resonances the summary variation in $\delta \langle \vec{q}_z \rangle$ must be small. Otherwise, the beam is depolarized.

In the case of fast passage the harmonics of integer resonances $\nu_o = K_\theta$ are compensated and the jumps of betatron frequencies are provided at the moment of passing through betatron resonances. All these means enable one to produce the polarized protons with an energy of 12 GeV in Argonne ZGS[23].

The proposal exists to improve the stability of polarization, introducing additional fields wherein the spin vector precesses around $\vec{n}(\theta)$ [20]. It is asserted that the rotation angles may be chosen so that the spin resonances will become impossible. This is an example of how the dynamical stability of spin motion is ensured. In order to solve this problem in an optimal way, a further study is probably required.

It is important to be able to obtain the required particle polarization in definite sections of the trajectory. This task is solved by a choice of electromagnetic fields in such a way that vector $\vec{n}(\vartheta)$ has a desirable orientation in the particle interaction region. In particular, the possibility to produce the longitudinally polarized particles at the interaction point in electron-positron storage ring was discussed in refs.[9,4]. The set of such tasks has been recently analysed in ref.[24]. In ref.[25] it has been discussed the concrete task of obtaining the longitudinal electron polarization in the interaction region for the designed facility with colliding e-p beams at CERN.

MEASUREMENT OF PARTICLE POLARIZATION IN A STORAGE RING

The following methods of measuring the particle beam polarization are known (see ref.[4]).

1) Due to dependence of the internal scattering effects in the beam particles, "Touchek effect", on polarization, the number of particles going out from the beam for this reason also depends on polarization[26]. Comparing this number for the polarized and unpolarized beam, one can measure the polarization. In ref.[26] it is considered the case of a plane beam with the assumption that $\eta = \frac{\Delta p}{\varepsilon} \ll 1$ ( $\Delta p$ is the maximum permissible deviation of the momentum from the equilibrium one). A general case has been quite recently analysed in ref.[27] where some uncorrectnesses of the early papers were eliminated and for the case $\eta \ll 1$ the distinction of elastic electron-electron scattering from the Born one was taken into account. This method is most effective at an energy of the order of a GeV and was used at Novosibirsk[28], Orsay[29], and Stanford[30].

2) At scattering of the circularly-polarized photons on the polarized electrons in the scattered photon distribution there is the azimuthal asymmetry, by which the measurement of particle polarization is possible[31]. The maximum asymmetry $P_{max} \simeq 1/3$ is achieved at a photon

energy of $\hbar\omega \simeq \frac{m}{2\gamma}$. For the particles with a few GeV energy, it is required to use an ultraviolet part of the spectrum to achieve this maximum. As possible sources of ultraviolet circularly-polarized radiation it should mention the synchrotron radiation of the beam thereof [32] and also the so-called ondulator radiation on a magnetic helical lattice, which is located in the rectilinear section of a storage ring [33]. With the use of a laser in an optical range the quantity $P$ is substantially lower than $P_{max}$ at an energy of a few GeV, nevertheless, it is quite measurable. This method has been recently used at Stanford [34].

3) At scattering of the polarized photons on a polarized electron target [31] the azimuthal asymmetry of scattered electrons takes place too, that may be used for measuring the polarization. For transversely polarized electrons the maximum anisotropy constitutes $\pm 10\%$ and does not depend on energy, whereas for longitudinally polarized electrons this effect is very essential (the ratio of cross sections is equal to 8).

It is important that the methods 2) and 3) do not require the beam depolarization and, hence, are useful for fast measurements of polarization.

At interaction of the polarized electrons and positrons the azimuthal asymmetry appears in the cross sections of many processes. Some of these processes have been discussed in ref.[4]. In addition, it should mention the inclusive cross section of hadron production, distribution of jets, production of three pions [35].

References

1. V.N.Baier, V.M.Katkov, V.M.Strakhovenko, Phys. Lett. 70B, 83 (1977).
2. V.N.Baier, V.M.Katkov, V.M.Strakhovenko, Yadern. Fiz. 27, 728 (1978).
3. V.N.Baier, V.M.Katkov, V.M.Strakhovenko, Dokl. Akad. Nauk 237, 548 (1977).
4. V.N.Baier, Sov. Phys. - Uspekhi 14, 695 (1972).
5. A.N.Skrinsky, Proceedings of the XVIII Intern. Conf. on High Energy Physics, Tbilisi (1976).
6. J.D.Jackson, Rev. Mod. Phys. 48, 417 (1976).
7. G.Loew, Proceedings of the X Intern. Conf. on High Energy Accelerators, Protvino (1977), v.1. p.58.
8. A.A.Sokolov, I.M.Ternov, Sov. Phys. - Doklady 8,1203 (1964).
9. Ya.S.Derbenev, A.M.Kondratenko, A.N.Skrinsky, Sov. Phys. - Doklady, 15, 583 (1970).
10. V.N.Baier, V.M.Katkov, V.M.Strakhovenko, Sov. Phys. - JETP 31, 908 (1970).
11. V.N.Baier, Yu.F.Orlov. Sov.Phys. - Doklady 10, 1145 (1966).
12. Ya.S.Derbenev, A.M.Kondratenko, Sov. Phys. - JETP 35, 230 (1972).
13. Ya.S.Derbenev, A.M.Kondratenko, Dokl. Akad. Nauk 217, 311 (1974).
14. Ya.S.Derbenev, A.M.Kondratenko, A.N.Skrinsky, Preprint INP 77-60 (1977).
15. R.F.Schwitters. Nucl.Inst. and Methods 137, 331 (1974).
16. A.W.Chao, R.F.Schwitters, Preprint SPEAR 197, PEP 21 (1976).
17. A.W.Chao, R.F.Schwitters, Preprint PEP -233 (1977).
18. D.Möll, B.W.Montague, Nucl. Inst. and Methods 137, 423 (1976).
19. Ya.S.Derbenev, A.M.Kondratenko, E.L.Saldin, Preprint INP 78-64(1978).

20. Ya.S.Derbenev, A.M.Kondratenko, Paper submitted to All-Union Conference on Accelerators, Dubna, 1978.
21. M.Froissart, R.Stora, Nucl.Instr. and Methods 7, 297 (1960).
22. Ya.S.Derbenev, A.M.Kondratenko, A.N.Skrinsky, Sov. Phys. - JETP 33, 658 (1971).
23. R.L.Martin, Proceedings of the X Intern. Conf. on High Energy Accelerators, Protvino (1977), v.2,p.64.
24. Ya.S.Derbenev, A.M.Kondratenko, A.N.Skrinsky, Preprint INP 76-62, 1976.
25. B.W.Montague, Preprint CERN-ISR-TH/77-34 (1977).
26. V.N.Baier, V.A.Knore. Atomnaya energiya 25, 440 (1967).
27. V.N.Baier, V.M.Katkov, V.M.Strakhovenko. Dokl. Akad. Nauk 241, 797 (1978).
28. S.I.Serednyakov, A.N.Skrinsky, G.M.Tumaikin et al. Sov.Phys. - JETP 44, 1063 (1976).
29. I.Le Duff et al. Preprint Orsay 4-73 (1973).
30. U.Camerini et.al. Phys.Rev. D12, 1885 (1975).
31. V.N.Baier, V.A.Khoze. Sov.Phys. - Nucl.Phys. 9, 238 (1969).
32. Ya.S.Derbenev et al. Proceedings of the X Intern. Conf. on High Energy Accelerators. Protvino (1977), v.2, p.55.
33. V.N.Baier, V.M.Katkov, V.M.Strakhovenko, Sov.Phys. JETP, 36, 1120 (1973).
34. R.F.Schwitters (private communication)(1977).
35. V.N.Baier, Preprint CERN TH-2081 (1975).

EXPERIMENTAL REVIEW OF
BEAM POLARIZATION IN HIGH ENERGY $e^+e^-$ STORAGE RINGS*

R. F. Schwitters
Stanford Linear Accelerator Center
Stanford University, Stanford, California 94305

## I. INTRODUCTION

In 1964, Sokolov and Ternov[1] showed that under certain conditions, synchrotron radiation with spin-flip should lead to a gradual build-up of spin polarization of electrons and positrons circulating in a storage ring. This observation opened up the possibility of having practical sources of polarized $e^+e^-$ beams which could be exploited for studying high energy physics.

In this paper, I review what information is provided by beam polarization in high energy $e^+e^-$ interactions, how polarized beams are produced, the experimental evidence for radiative beam polarization, and the results which have been obtained to date using polarized beams. I conclude with a discussion of prospects for beam polarization experiments in future generations of storage rings.

## II. POLARIZATION EFFECTS IN $e^+e^-$ ANNIHILATION

One of the principal advantages of studying high energy phenomena with $e^+e^-$ collisions is the simplicity of the $e^+e^-$ initial state. At currently accessible energies ($E_{c.m.} \lesssim 25$ GeV), reactions involving annihilation of an electron and positron are mediated predominantly by the one-photon intermediate state. Thus, the net quantum numbers of the final state particles are the same as those of the photon, namely $J^{PC} = 1^{--}$, and only two helicity states, $\lambda = \pm 1$, are allowed.

As discussed below, synchrotron radiation involving spin-flip yields transversely polarized beams where the spins of the electrons, positrons are predominantly parallel, antiparallel to the guide magnetic field of the storage ring. If the beams are so polarized, the $e^+e^-$ pair annihilate through a state having the properties of a linearly polarized photon, and the most general single particle inclusive angular distribution can be written as:[2]

$$\frac{d\sigma}{d\Omega} = \frac{1}{2}\left[(\sigma_t + \sigma_\ell) + (\sigma_t - \sigma_\ell)(\cos^2\theta + P^2\sin^2\theta\cos 2\phi)\right] \quad (1)$$

where $\sigma_t$, $\sigma_\ell$ are non-negative functions of particle type, particle energy, and center-of-mass energy; P is the degree of transverse

---

*Work supported by the Department of Energy under contract no. EY-76-C-03-0515.

ISSN: 0094-243X/79/510091-19$1.50 Copyright 1979 American Institute of Physics

polarization of the beams (assumed to be equal in magnitude and opposite in direction for the two beams); $\theta$, $\phi$ are the polar and azimuthal angles of the produced particle, as defined in Fig. 1.

The quantity of interest that can be determined by measurements of angular distributions is

$$\alpha = \frac{\sigma_t - \sigma_\ell}{\sigma_t + \sigma_\ell} ;$$

it gives information on the production dynamics of the particle being studied. For example, $\alpha = +1$ in the case of pair production of spin-½ particles such as muons,

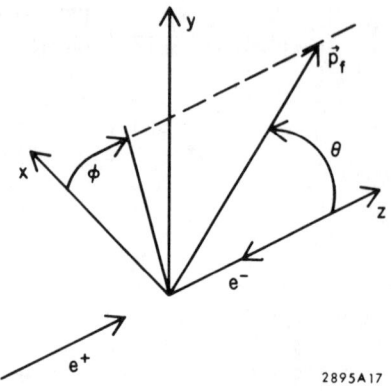

Fig. 1. Coordinate system and definition of angles.

while $\alpha = -1$ for pair production of pseudoscalars, such as pions. In multihadron production, $\alpha$ is bounded between these two extreme values and will, in general, depend on particle type, momentum, and center-of-mass energy.

It is evident from Eq. (1) that transverse beam polarization does not provide any new information that could not be determined by measurement of the polar angle dependence above. However, the azimuthal angle variation of inclusive cross sections is often technically easier to measure than the $\cos^2 \theta$ dependence because of the geometry of most detectors. This is because most detectors are nearly cylindrically symmetric with respect to the incident $e^+e^-$ direction and have essentially full acceptance in azimuthal angle. Most detectors have less than complete acceptance in polar angle with losses occurring at large values of $\cos^2 \theta$. Thus, transversely polarized beams provide a convenient, if not necessary, means for determining the parameter in processes dominated by the one-photon exchange approximation. The combined $\theta$-$\phi$ information with polarized beams allows a check of this approximation.

At the next generation of $e^+e^-$ storage rings ($E_{c.m.} \gtrsim 30$ GeV), weak neutral current effects are expected to begin to play a significant role in $e^+e^-$ annihilation. Various parity-violating phenomena could occur and some of these may be observed using longitudinally polarized beams. In what follows, the weak neutral current for spin-½ particles is assumed to be a mixture of vector and axial-vector parts. The interference between the electromagnetic current and the weak neutral current leads to the following $e^+e^-$ spin dependence of the cross section for producing spin-½, point-like fermion pairs $f\bar{f}$:[3]

$$\sigma_f(\lambda_+, \lambda_-) = [(1-\lambda_+\lambda_-) + (\lambda_- - \lambda_+)H_f]\sigma_f(0,0) \qquad (2)$$

where $\lambda_+$, $\lambda_-$ are the longitudinal polarization of the incident $e^+$, $e^-$ beams, measured with respect to their directions of motion. The "low" energy behavior of $H_f$ and $\sigma_f(0,0)$ can be described by:

$$\sigma_f(0,0) \simeq \frac{4\pi}{3} \frac{\alpha^2}{s} Q_f^2$$

$$H_f \simeq \frac{-Gs}{\sqrt{2}\, 4\pi\alpha} \frac{g_A^e\, g_V^f}{Q_f}$$

(3)

where $\alpha$ is the fine structure constant, $G$ is the Fermi constant, $s$ is the square of the center-of-mass energy, $Q_f$ is the electric charge of particle f measured in units of the charge on the positron, $g_A^e$ is the weak axial vector coupling constant of the electron, and $g_V^f$ is the weak vector coupling constant of the final state fermion $\bar{f}$. At $E_{c.m.} = 30$ GeV, the quantity $Gs/\sqrt{2}\, 4\pi\alpha$ is approximately 0.081.

From Eqs. (2) and (3), it is seen that longitudinally polarized beams would provide new information, namely a measurement of the coupling constant $g_V^f$ (assuming $g_A^e$ is known from other measurements). (It is possible to determine the weak axial-vector coupling constant of f, $g_A^f$, by measuring the front-back angular asymmetry in the production of $f\bar{f}$ pairs.) It is interesting to note that if the incident beams are fully longitudinally polarized in opposite directions, the total annihilation production rate will vanish!

## III. RADIATIVE BEAM POLARIZATION

All experimental work performed to date with polarization in high energy $e^+e^-$ storage rings has relied on the fact that under certain conditions, the beams become transversely polarized through the mechanism of synchrotron radiation with spin-flip. This is called radiative beam polarization and was first discussed by Ternov, Lokutov, and Korovina in 1961.[4] In 1963, Sokolov and Ternov[1] showed that the transverse polarization for particles circulating in a uniform magnetic field would build up in time according to:

$$P(t) = \frac{8\sqrt{3}}{15}\left(1 - e^{-t/T_{pol}}\right)$$

$$\frac{1}{T_{pol}} = \frac{5\sqrt{3}}{8} \frac{e^2 \hbar \gamma^5}{m^2 c^2 \rho^3}$$

(5)

where $\gamma$ is the Lorentz factor of the particle ($\equiv E/m$) and $\rho$ is the bending radius of the orbit. Positrons, electrons would become polarized parallel, antiparallel to the magnetic field. Baier and Katkov,[5] in 1967, generalized this result to include inhomogeneous magnetic fields and obtained the following general expression for the transition probability per unit time for spin flip:

$$W^{\uparrow\downarrow} = \frac{5\sqrt{3}}{16} \frac{e^2\hbar}{m^2c^5} \gamma^5 |\vec{\dot{\beta}}|^3 \left[1 - \frac{2}{9}(\vec{S}\cdot\hat{\beta})^2 + \frac{8\sqrt{3}}{15}\vec{S}\cdot(\hat{\beta}\times\hat{\dot{\beta}})\right] \quad (5)$$

where $\vec{S}$ is the initial spin direction in the electron rest frame, $\hat{\beta}$ and $\hat{\dot{\beta}}$ are unit vectors in the velocity, acceleration directions, respectively, and $\vec{\dot{\beta}}$ is the acceleration, measured in the laboratory frame. A complete discussion of this phenomenon is presented in the review article by Baier[6] and a detailed pedigodical derivation of Eq. (5) is given in the review article by Jackson.[7]

When combined with the usual Thomas-BMT equation of spin motion,[8] Eq. (5) leads to damping terms that give rise to a build-up of polarization described in Eq. (4), in the case of a conventional separated function storage ring. When the guide bending magnets all have the same value of magnetic field, the time constant for polarization build-up is

$$T_{pol}(\text{sec}) = \frac{98.7 \times |\rho(m)|^3}{|E(\text{GeV})|^5} \times \frac{R}{\rho} \quad (6)$$

where E is the beam energy in GeV, $\rho$ is the bending radius in units of meters and R is the average radius of the storage ring. The most distinctive feature of Eq. (6) is the very strong energy dependence. For example, the SPEAR storage ring, operating at 3.7 GeV per beam, has a build-up time of approximately 14 minutes.

A necessary condition for radiative polarization to occur is that spin motion along the polarization direction be stable over time scales on the order of or greater than $T_{pol}$. The Novosibirsk group has made the major contributions to the study of the stability of spin motion in $e^+e^-$ storage rings. The classic papers of Derbenev and Kondratenko[9,10] contain the general results; the review papers of Baier[6] and Derbenev, Kondratenko, and Skrinsky[11] are useful references on depolarization phenomena. The basic result of this work is that there exist depolarization mechanisms in conventional storage rings that will lead to a reduction in the asymptotic polarization $P_{max}$ from 92.4% to:

$$P_{max} = \frac{8\sqrt{3}}{15} \times \frac{1}{1 + \frac{T_{pol}}{T_{depol}}} \quad (7)$$

where $T_{depol}$ is the characteristic time for depolarization. (The results on depolarization presented in Refs. 9 and 10 are not restricted to the simple storage ring geometry considered here.)

Briefly, spin motion in a conventional storage ring is simply Thomas-Larmor precession about the vertical direction. The spin precession frequency ν represents the number of precessions in advance of the orbital motion that the spin experiences during each orbital period; it is given by

$$\nu = \gamma \frac{(g-2)}{2} \simeq \frac{E(GeV)}{0.44065} \qquad (8)$$

where g is the gyromagnetic ratio of the electron.

In addition to the main bending field, the particles exhibit betatron and synchrotron motion and experience various focusing and acceleration fields. These also affect the spin motion. The electromagnetic fields set up by the opposing beam are a further strong perturbation to both orbital and spin motion. These three general mechanisms that can lead to depolarization may be summarized by:

1. Resonance depolarization
2. Stochastic depolarization
3. Beam-beam effects.

Resonance depolarization occurs when the spin precession frequency is integrally related to characteristic frequencies of orbital motion according to:

$$\nu = n \pm i\nu_x \pm j\nu_y \pm k\nu_s \qquad (9)$$

where n, i, j, k are integers, $\nu_x$ and $\nu_y$ are the horizontal and vertical betatron tunes, and $\nu_s$ is the synchrotron frequency. The most prominent depolarization resonances are usually the integer resonances, ν=n, and the first order sidebands where i or j equals 1. The integer resonances repeat every 440 MeV in beam energy.

Stochastic depolarization arises primarily from the transverse focusing fields experienced by a particle during cycles consisting of the emission of synchrotron radiation followed by the build-up and damping of betatron and synchrotron motion. This gives rise to depolarization away from the important resonances. Stochastic depolarization rates are sensitive to beam sizes and orbit distortions. There now exist standard computer codes[12] for calculating resonance and stochastic depolarization effects.

The beam-beam interaction exerts strong, non-linear forces on the particles and is expected to play a role in depolarization through its effect on orbital motion as well as its direct influence on spin. Currently, beam-beam forces are poorly understood and their effect on polarization cannot be computed with reliability. A discussion of beam-beam interaction effects on polarization is contained in Ref. 11.

The first experimental indications of radiative polarization were obtained in the late 1960's and 1970 at Orsay and Novosibirsk. Unambiguous evidence for the radiative build-up of polarization was reported in 1971 by both the Novosibirsk group[6] and the Orsay group[13]

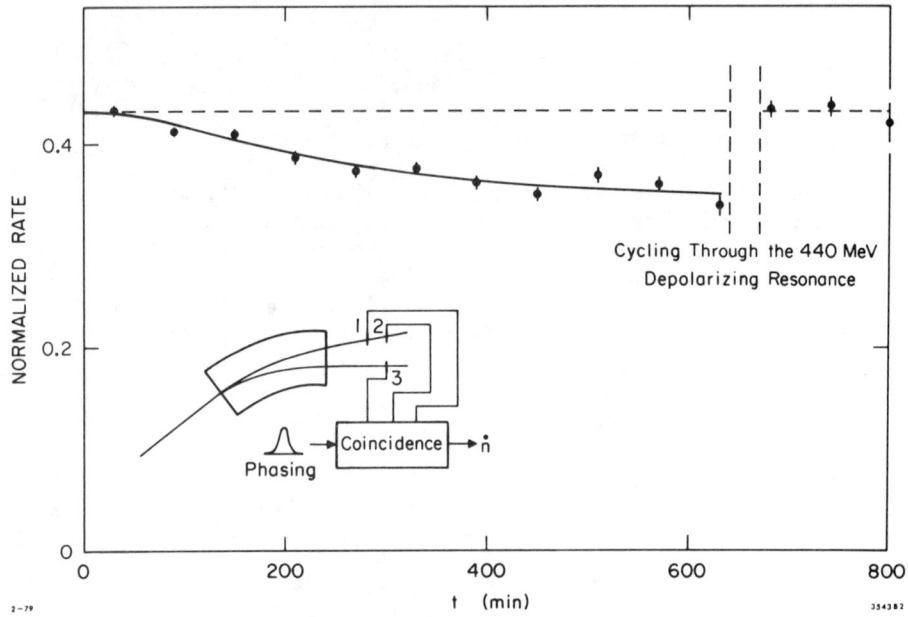

Fig. 2. Schematic diagram of Touschek scattering setup and experimental results of the Orsay group.[13] The data show the reduction in Touschek rate caused by the build-up of transverse polarization. Near the end of the data run, the beam was depolarized by changing beam energy through a spin resonance.

using the storage ring ACO. The Orsay group, in 1973, reported[14] on more detailed measurements performed at ACO. They found evidence for betatron-frequency-sideband depolarizing resonances and showed that polarization persists in the presence of colliding beams. One particularly interesting measurement made during this study was the polarization of one beam as a function of the current of the other colliding beam when the beam energy was fixed near the vertical betatron sideband resonance. At high currents, the polarization was large, at somewhat lower current it was reduced, and at still lower current, it became large again. This is interpreted as evidence for the linear betatron tune shift of one beam caused by focusing forces that are proportional to the intensity of the oncoming beam. At one value of current, the tune shift was the value necessary to satisfy the spin resonance condition and the beam depolarized. At higher or lower currents the tune shift was either too large or too small to satisfy the resonance condition. This is a good example of the accelerator diagnostic possibilities of polarized beams. In a completely nonperturbative way, spin frequency information can be used to probe betatron motion where conventional techniques may not be possible.

In 1975, the observation of radiative beam polarization of the expected level and build-up rate was reported from SPEAR.[15] At about this time, as will be discussed later, experiments making use of polarized beams began at Novisibirsk and at SPEAR.

In all of these early observations of radiative polarization, the one basic technique employed for measuring polarization was the measurement of the Touschek scattering rate. The details of the method can be found in the review article of Baier[6] and in a paper by Ford, Mann, and Ling.[16] Briefly, Touschek scattering is Møller scattering of electrons or positrons within a single rf bucket. When viewed from a reference frame moving with a bunch of electrons, the individual particles have typical momenta on the order of a few hundred keV and they scatter with other particles in the bunch. The scattering rate will depend on polarization. If two particles, which were originally moving toward each other in a direction perpendicular to the bunch velocity direction, scatter at a large angle such that they travel nearly parallel to the bunch direction after scattering then their Lorentz transformed laboratory energies will be significantly different after the scattering occurred and the particles may be lost from the beam. This can be an important loss mechanism in $e^+e^-$ storage rings. By measuring the rate for the correlated loss of pairs of particles from a single beam, one has a measure of the intra-beam Møller scattering which, in turn, depends on polarization. A typical experimental set-up and series of measurements is shown in Fig. 2.

The experiments performed at Novosibirsk and Orsay demonstrated the essential features of radiative beam polarization. First, the beams do indeed become transversely polarized with the time dependence given by Eq. (4). This can be seen quite well from the nice results of the Novosibirsk group[17] shown in Fig. 3. As previously mentioned, depolarizing effects can be important and their expected behavior seems to be born out by the Novosibirsk and Orsay experiments.

In order to perform more detailed measurements of depolarizing phenomena, our group[18] has recently developed a back-scattered laser polarimeter. As first pointed out by Baier and Khoze,[19] Compton scattering of circularly polarized optical photons by a high energy transversely polarized beam is a sensitive and direct method for

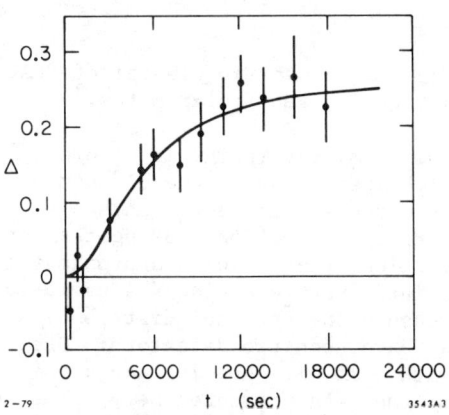

Fig. 3. Polarization build-up observed by Novosibirsk group[17] with Touschek scattering method. Build-up rate agrees with Eq. (4).

measuring the beam polarization. We chose this method rather than
Touschek scattering for several reasons. First, the analyzing power
and counting rate for a laser system can be accurately calculated,
while the Touschek rate depends sensitively on the beam size and intensity, introducing uncertainty in the analyzing power. Many possible systematic errors can be checked by varying the polarization of
the laser beam; such systematic checks are not available in the
Touschek method. At high energies, Touschek scattering losses represent a relatively small fraction of the total particle loss rate.
Thus, backgrounds became very important, whereas the backgrounds in
the laser case can be made quite small and can be accurately measured
by simply turning off the laser. The goal of our development project
was to design a monitor that could make polarization measurements to
the 10% level of accuracy in one or two minutes and to study depolarization effects with this device.

The experimental set-up is shown schematically in Fig. 4. An
Argon-ion laser supplies the photons which are alternately switched
between right and left circular polarization at a rate of approximately 25 Hz by means of an electro-optic device known as a Pockel's cell. The laser is operated in a cavity-dumped mode so that the beam of circularly polarized photons can be pulsed in synchronism with the 1.28 MHz resolution frequency of the single $e^+$ bunch normally stored in SPEAR. The peak laser intensity is approximately 80 watts; the photon energy is about 2.4 eV (green in color). The backscattered gamma rays are contained in a cone of characteristic angle $1/\gamma$ where $\gamma$ is the Lorentz factor of the $e^+$ and the maximum gamma-ray energy is of order 100 MeV depending on $e^+$ energy. The
light beam crosses the $e^+$ beam vertically at an angle of 8 mr; the
intersection point was chosen so that the $e^+$ beam converged in the
vertical direction causing the backscattered gamma rays to be focused
vertically at a point approximately 13 meters from the intersection
point. Here was placed a gamma-ray detector that could accurately
measure the vertical distribution of the backscattered gamma rays.
We have used two detectors: One, a multiwire proportional chamber
with 1 mm wire spacing and a gamma-ray converter, was used for most
of the data presented here. The second, consisting of a converter
and single cell drift chamber with approximately 0.2 mm resolution
vertically, is currently being used. Both detectors have performed
satisfactorily.

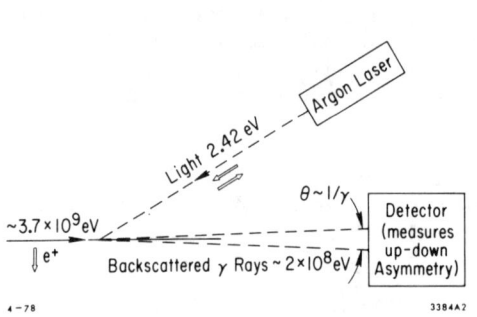

Fig. 4. Schematic diagram of SLAC-Wisconsin laser polarimeter.[18]

The basic measurement being made is an up-down asymmetry in the yield of backscattered gamma rays that is proportional to the transverse beam polarization. This asymmetry changes sign when the helicity of the incident photon beam is reversed. Thus, to minimize systematic errors, we rapidly alternate between right and left circular polarization of the laser beam and compute the average asymmetry, taking into account the change in sign, for the two helicities. A microcomputer system tallies asymmetries, gates the laser on and off for determining background, times the experimental runs, and prints out results. The analyzing power for the system is approximately 2.5% and a typical counting rate is 10 kHz; the usual data run requires two minutes and yields a value for the average up-down asymmetry accurate to ±0.1%.

Figure 5 shows the up-down asymmetry as a function of time for a beam energy of 3.7 GeV. The data are in excellent agreement with expected polarization build-up rate.

The first set of detailed measurements that we have made was to study the ratio of depolarization rate to build-up rate as a function of beam energy for single $e^+$ beams. This ratio is extracted from the polarization build-up time constant, without resort to using absolute asymmetry, according to:

$$\frac{1}{T_{obs}} = \frac{1}{T_{pol}} \left(1 + \frac{T_{pol}}{T_{depol}}\right) \qquad (10)$$

$$A_{obs}(t) = A\left(1 - e^{-t/T_{obs}}\right)$$

where $T_{obs}$ is the observed time constant for the measured asymmetry $A_{obs}$, which can be determined without independently knowing the value A. In such measurement, the polarization is allowed to reach nearly its asymptotic value, then the beam energy is changed by a few MeV and a new asymmetry is reached. This process is repeated several times until a given sweep in energy is completed, or the positron intensity drops to an unacceptably low level. The data from a typical scan are presented in Fig. 6. The curve is a fit to the data assuming a constant analyzing power and that the polarization at the beginning of each new energy point equals the

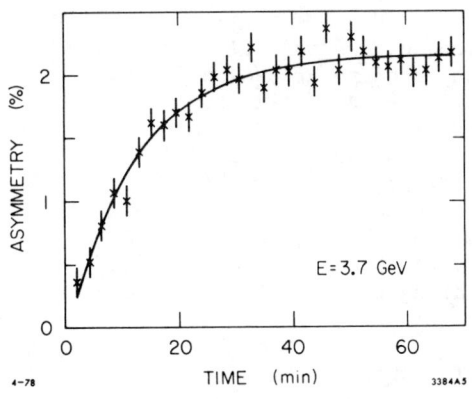

Fig. 5. Polarization build-up measured at SPEAR.[18] Solid line is a fit to the data using Eq. (10). At this energy, depolarization is negligible.

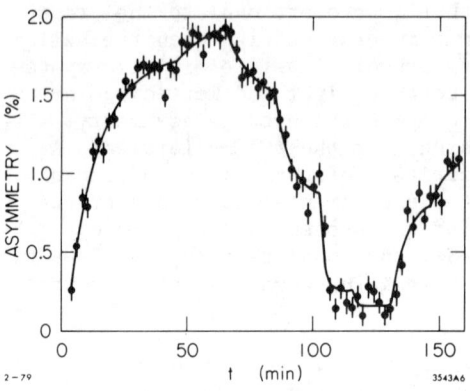

Fig. 6. Polarization scan measured at SPEAR.[18] Breaks in solid line indicate times where the beam energy was changed. The line is a fit to the data using Eq. (10), but with different depolarization rates for each energy setting.

polarization at the end of the previous energy setting. Breaks in the solid curve indicate new energy settings. The data show the usual build-up, then as the energy or spin frequency approaches a resonant value, the asymmetry quickly drops according to Eqs. (9) and (10), until it essentially vanishes. At slightly higher energies, the asymmetry re-emerges. The fitted values for the ratio of depolarization rate to build-up rate are given in Fig. 7 along with a theoretical calculation.[12] The theory and data are in excellent agreement and clearly show the presence of a strong spin resonance at a sideband due to horizontal betatron motion and a weaker resonance corresponding to the vertical betatron tune.

These measurements will continue with single beams and colliding beams with the goals of thoroughly checking the single beam theory and attempting to learn more about the beam-beam interaction. From the measurements we have already made, we can make the following preliminary conclusions:

1. The theoretical descriptions of the radiative build-up of transverse polarization and depolarization due to resonant and stochastic effects in single beams appear to be accurate. Depolarization resulting from forces that are non-linear in excursions of the particle motion from the equilibrium orbit appears to be rather weak.
2. A high degree of polarization can exist when the beams are colliding. However, various non-reproducible effects were observed indicating the need for much more study in this area.

## IV. EXPERIMENTS WITH POLARIZED BEAMS

The natural, transverse polarization acquired by stored beams of high energy electrons and positrons have been put to use in a number of experiments. These experiments fit into two general categories, those where the Thomas-Larmor precession of the $e^+$ or $e^-$ is exploited, and those where angular distributions of final state particles from $e^+e^-$ annihilation are measured.

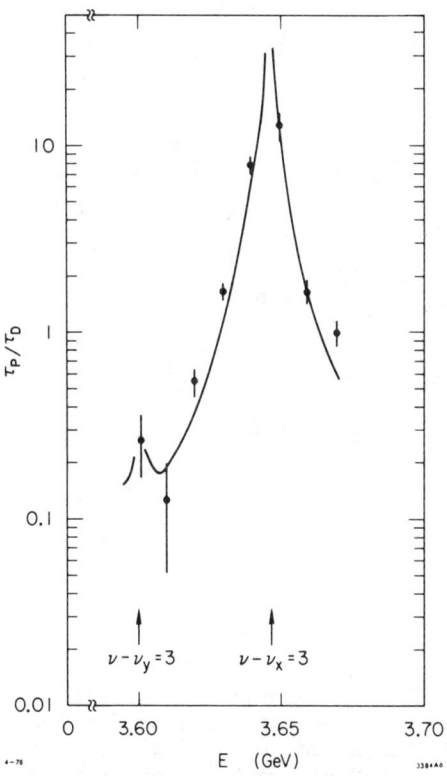

Fig. 7. Ratio of depolarization rate to polarization rate as a function of beam energy derived from the data shown in the previous figure. Positions of the two important betatron sideband spin resonances in this energy range are indicated.

The Novosibirsk group has reported results in both categories. S. I. Serednyokov et al.[20] performed a high precision comparison of the anomalous magnetic moments of the electron and positron in an experiment using the VEPP-2M storage ring. Their basic measurement was a comparison of the spin precession frequencies $\nu^+$, $\nu^-$ of $e^+$, $e^-$ beams simultaneously stored. The beams were allowed to polarize for two characteristic time intervals ($T_{pol} \approx 1$ hr, $E = 625$ Mev). Then an oscillating longitudinal magnetic field was applied to the beams and the driving frequency $f_D$ of this field was slowly swept. The polarization was monitored by two identical Touschek scattering detectors, one for each of the $e^+$ and $e^-$ beams. When the resonance condition

$$f_D^\pm = \frac{\Omega_0}{2\pi} (\nu^\pm - 1) \qquad (11)$$

is met ($\Omega_0$ is the revolution frequency), the $e^\pm$ beam depolarizes rapidly and will give a jump in the corresponding Touschek scattering rate at that frequency. The group found that $|f_D^+ - f_D^-| < 250$ Hz which, according to Eq. (8), provides the following limit on the difference between the anomalous magnetic moments ($a = (g-2)/2$) of electrons and positrons:

$$|a_{e^+} - a_{e^-}| < 1.0 \times 10^{-5} \quad (95\% \text{ confidence level}) \quad .$$

This is about two orders of magnitude more sensitive than previous measurements, which involved direct measurements of $a_{e^+}$ and $a_{e^-}$ in different experiments.

The resonant depolarization technique developed at Novosibirsk,[17] combined with the very accurately known value of $a_{e^-}$ (Ref. 21) provide a precise tool for calibrating the energy of $e^+e^-$ storage rings. This has been done for VEPP-2M (Ref. 22) and yielded the following value for the Φ, meson mass, which is produced directly as a resonance in $e^+e^-$ annihilation:

$$M_\Phi = 1019.48 \pm 0.13 \text{ MeV}/c^2 .$$

The accurate energy calibration of VEPP-2M and a measurement of the kinetic energy of $K^\pm$ mesons using nuclear emulsion, allowed the Novosibirsk group[23] to determine the charged kaon mass to

$$M_{K^\pm} = 493.670 \pm 0.029 \text{ MeV} .$$

Observations of effects due to transverse beam polarization on angular distributions of particles produced by $e^+e^-$ annihilation were first reported in 1975. The electrodynamic reactions $e^+e^- \rightarrow e^+e^-$ (Bhabha scattering) and $e^+e^- \rightarrow \mu^+\mu^-$ were studied at SPEAR with the SLAC/LBL magnetic detector. Learned, Resvanis, and Spencer[24] analyzed the results of these measurements and found significant azimuthal variation in both reactions at

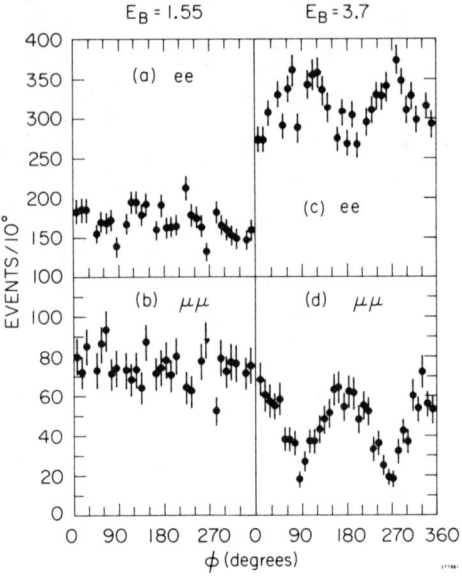

Fig. 8. Azimuthal angle distributions for muon pair production (μμ) and Bhabha scattering (ee) measured at SPEAR[24] at two beam energies. At $E_B$ = 1.55 GeV the beams were unpolarized; at $E_B$ = 3.7 GeV, they were over 70% polarized.

the center-of-mass energy $E_{c.m.} = 2E_B = 7.4$ GeV that was consistent with the theory of quantum electrodynamics and indicated that the beams were polarized within 80% of the expected maximum value. Their results are presented in Fig. 8. At $E_B$ = 1.55 GeV, the beams are expected to be unpolarized, while at $E_B$ = 3.7 GeV, large azimuthal variations are seen in both reactions. Muon pair production has the angular distribution of Eq. (1) with $\sigma_\ell = 0$. Bhabha scattering has a more complicated angular distribution[2] because it does not proceed entirely by single photon annihilation. Kurdadze et al.[25] have also reported observation of the azimuthal variation in muon pair production at VEPP-2M.

The single-photon exchange picture of hadron production by $e^+e^-$ annihilation has been confirmed through polarization studies. The Novosibirsk group[22] measured the azimuthal variation of K meson pair production at the Φ resonance and it is described by Eq. (1) with $\sigma_t = 0$, as expected.

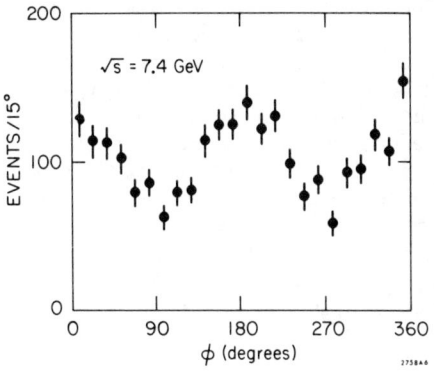

Fig. 9. Inclusive azimuthal angle distribution for hadrons with x > 0.3 measured at SPEAR.[26]

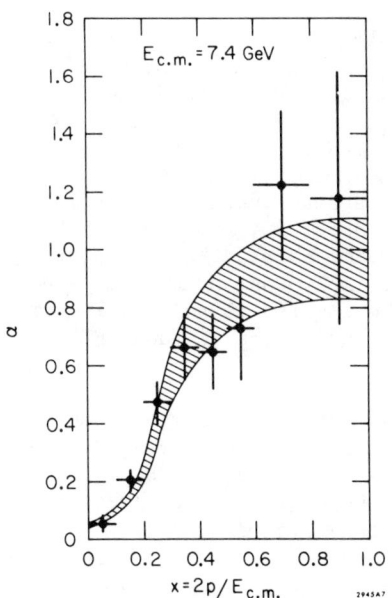

Fig. 10. The parameter α versus x for inclusive hadron production. The dashed region is the prediction of a jet model calculation.[27]

Studies of multihadron production with polarized beams at the center-of-mass energy 7.4 GeV have been made by the SLAC/LBL collaboration at SPEAR.[26] All events having three or more charged hadrons within the angular region $|\cos\theta| \leq 0.7$ were selected. The average beam polarization was about 70%. Figure 9 shows the inclusive azimuthal angular distribution for all prongs in this event sample having x > 0.3 where x is the scaling variable, $x = 2p/E_{c.m.}$, and p is the particle momentum. The strong φ dependence is evident. The combined θ, φ distributions were fitted to Eq. (1) and values of the parameter α were obtained as a function of the scaling variable x. These are shown in Fig. 10. At small values of x, α approaches zero, while at large values, it approaches unity. Most of the hadrons in these events are pions, so it is significant that at large x they display the value of α corresponding to pair production of spin-½ particles rather than the value α = -1 expected for meson pair production and observed in $K^+K^-$ production at the Φ mass. However, this is consistent with the quark-parton model where hadrons are created through the pair production of spin-½ quarks followed by their subsequent decay to ordinary hadrons. In this picture, the hadrons of higher momentum more closely follow the initial quark direction and retain the quark value of α, namely α = +1.

This view is strengthened by the observation of jets in these same data.[27] Jets are multiparticle correlations where hadronic events display a preferred axis where components of momentum perpendicular to this axis are limited to small values, the mean value of which is more or less independent of center-of-mass energy. Momenta parallel to the preferred or jet axis grow essentially linearly with increasing center-of-mass energy. Jets are exactly the kind of multiparticle correlations expected in the quark-parton model, where the direction of motion of the initial quark pair is the jet axis, and the fragmentation of the quarks into hadrons takes place along this axis with relatively small $p_t$. The polarized beam data show that the angular distribution of the jet axis corresponds to $\alpha_{jet} = 0.97 \pm 0.14$, in agreement with unity, the value expected for spin-½ quarks. The shaded region of Fig. 10 shows values of $\alpha$ vs. x one would expect for hadrons produced by fragmentation of quark pairs; the jet model calculations are in good agreement with these data.

To summarize, from experimental work performed to date, beam polarization has been shown to be a convenient energy calibration tool, the one-photon exchange approximation has been shown to be valid for electrodynamic and hadronic processes, and multihadron production fits well the simple spin-½ quark picture.

## V. FUTURE PROSPECTS

We can expect polarized beams to continue to play an important role in future generations of $e^+e^-$ colliding beam machines. In the present generation, PETRA and PEP, conventional radiative polarization should occur and it will provide convenient tools for precision energy calibration, non-destructive beam diagnostics, and probing one-photon exchange processes through the measurement of azimuthal asymmetries. Chao[28] has calculated the expected depolarization forces for PEP; these results are summarized in Fig. 11, where the asymptotic polarization is shown as a function of the beam energy. The polarization build-up time constant for PEP is about 20 minutes for a beam energy of 15 GeV. This relatively long time constant and the strong depolarizing resonances shown in Fig. 11 imply that practical experiments using radiative polarization may be limited to relatively narrow bands of energies at the high energy end of the PEP and PETRA energy ranges.

As we go to $e^+e^-$ storage rings of even higher energy, these two effects combine to make it more difficult to use the natural radiative polarization. As discussed above, $T_{pol}$ scales as $R^3 E^{-5}$. To optimize costs and performance for high energy storage rings,[29] it is necessary to increase the radius with energy according to:

$$R \propto E_{max}^p$$

where p lies between 2 and 3 for current technology. Therefore we can expect $T_{pol} \propto E_{max}^\epsilon$ where $1 \lesssim \epsilon \lesssim 4$. Thus, if nothing is done to control the polarization build-up rate, build-up times will become impractically long. The second deleterious effect present in storage

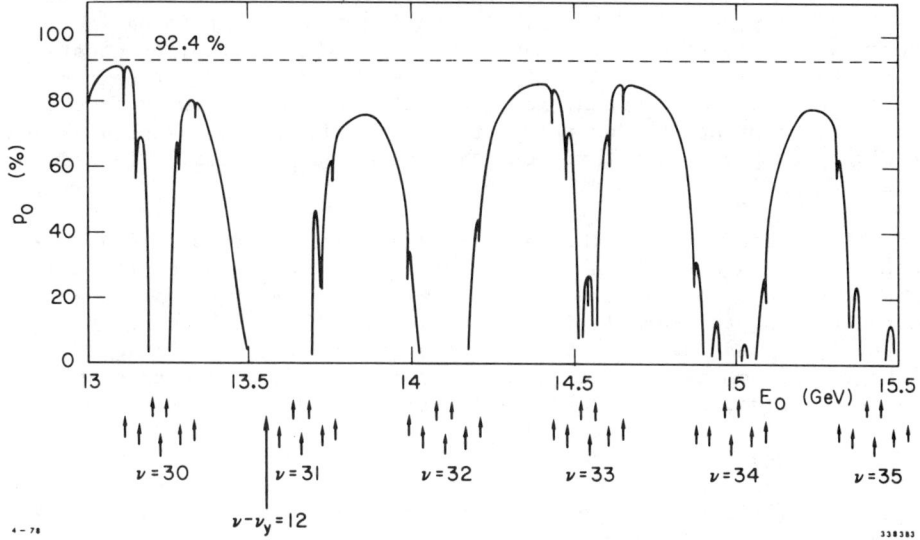

Fig. 11. Maximum transverse beam polarization expected at PEP.[28]

rings of higher energy is evident in Fig. 11. Depolarization resonances are spaced in bands every 440 MeV in beam energy and, therefore, these bands become relatively more dense as the energy is raised. Möhl and Montague[30] have estimated that at very high energies, above 100 GeV, the overlap of these resonance may overwhelm radiative build-up of polarization in conventional storage rings.

There appears to be at least a partial solution to the problem of prohibitively long time constants. Several workers[31] have suggested using "wiggler" magnets to increase the polarization rate. These are reversed guide magnets that increase the total synchrotron radiation emitted on each resolution. Since $T_{pol}$ is related to the strength B of the guide field by:

$$\frac{1}{T_{pol}} \propto \oint |B|^3 \, ds$$

where the integral is over a complete revolution, it is reduced by the addition of wiggler magnets. However, one must take care not to reduce the asymptotic polarization which is related to B by:

$$p_{max} = 92.4\% \times \frac{\oint B^3 \, ds}{\oint |B|^3 \, ds} \quad .$$

The reverse wiggler magnets should be as weak as possible, while those having the same field direction as the guide field should be as strong as possible.

Looking further ahead, it would be extremely desirable to have longitudinally polarized beams. As mentioned in the introduction, such beams could be used to measure the weak vector coupling constants of quarks and leptons. At energies above the threshold for pair production of charged W bosons, longitudinal polarization would provide a detailed test of the interplay of the various subprocesses involved in $W^+W^-$ pair production that are required by gauge theories in order to obtain finite cross sections.[32]

If longitudinally polarized particles were injected into a conventional storage ring, they would quickly depolarize unless special care is taken with the guide magnetic field.[33] The basic strategy for overcoming this depolarization relies on the well known fact (Refs. 6, 9, and 10) that at every point on the closed orbit of a storage ring composed of essentially arbitrary static electric and magnetic fields, one can define a polarization direction about which the spin of an electron precesses by a constant phase advance on successive revolutions about the storage ring. Particles initially polarized along this direction will remain polarized unless subjected to the depolarization forces discussed previously. In a conventional storage ring, the polarization direction coincides with the guide field direction. However, by suitable choice of magnetic elements, it is possible to arrange the polarization direction to be parallel to the beam direction at certain positions on the equilibrium orbit, in particular, at intersection regions.

Two basic schemes have been deviced for doing this. In one method, a special set of magnets is inserted adjacent to an intersection region. These magnets are arranged to rotate the spin from the transverse direction to longitudinal at the interaction point, then rotate it back to transverse before re-entering the normal lattice of the storage ring. In this way, normal radiative build-up of polarization can take place in the regular guide magnets and no additional source of polarized particles is needed. Many specific magnet arrangements have been proposed for this transformation;[11,31,33,34] a simple illustrative example is shown in Fig. 12. Here, a series of six vertical bending magnets, each having a bending power of approximately 2.3 t-m, will transform transverse spins to longitudinal and vice versa for electrons and positrons of all momenta. Chao[28] has calculated the maximum polarization that could be expected at PEP if such a scheme were employed; his results are plotted in Fig. 13. Beyond the obvious technical difficulties with these special spin rotating insertions, they

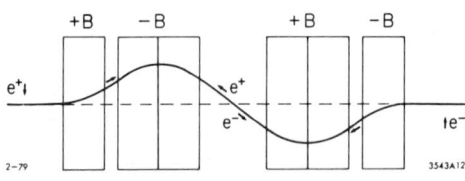

Fig. 12. Schematic diagram of one scheme for producing collisions of longitudinally polarized beams using the natural build-up of transverse polarization. Arrows indicate polarization directions. Boxes represent vertical bending magnets with horizontal fields ±B.

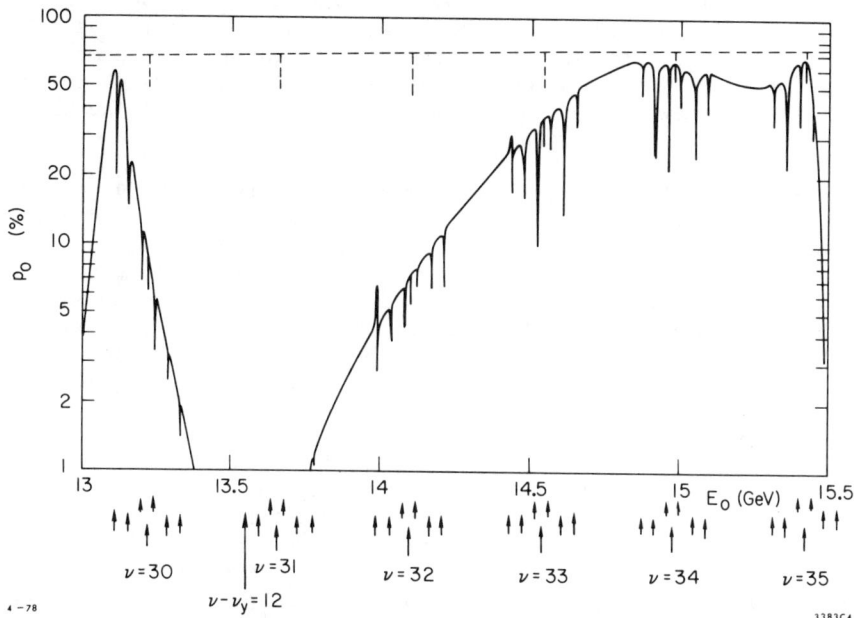

Fig. 13. Maximum longitudinal beam polarization expected at PEP[28] with a magnet system like the one shown in Fig. 12.

have the curious property that the electrons and positrons will have antiparallel spins. Therefore, if we are to avoid having the annihilation cross section vanish, it will be necessary to depolarize selectively one of the beams.

The second general scheme for providing longitudinally polarized beams has been christened the Siberian Snake in honor of its development at Novosibirsk through the work of Derbenev and Kondratenko[35] and others. Here one plays a topological trick on the spin motion. By placing a solenoid of suitable strength or other special set of magnetic elements at some point on a storage ring, it is possible to transform the spin orbit into a kind of Mobius strip such that, at a point on the storage ring opposite to the special magnets, the normal spin precession direction is longitudinal. The condition on the solenoid or set of magnets is that the spin rotates by exactly 180° about the beam direction in passing through the "Snake". When coupled to the usual precession about the vertical direction in the rest of the storage ring, the polarization direction will be in the plane of the orbit and be parallel to the beam diametrically opposite the Snake. The other curious feature of such a topology is that the spin advances by exactly π on successive orbital revolutions, not the usual 2πν. This means that the effective spin tune is ½ and the resonance condition Eq. (9) can, in general, be avoided at all energies. Thus, spin motion with a Siberian Snake is expected to be very stable.

In the simplest version of the Siberian Snake considered here, normal radiative polarization takes place perpendicular to the stable spin precession direction and it can only depolarize the beams. Thus, to use such a device, it will be necessary to inject already polarized beams. Somewhat more complicated arrangements should be able to avoid this effect if needed.[36]

Currently, there is much theoretical activity in studying new approaches for providing longitudinally polarized beams[37] and polarized beams at high energies. Much of this work has been discussed at this Conference, and I shall not cover it here, except to refer to the talk of Derbenev.[36] There is still much work to be done in order to find practical arrangements for polarized beams that also allow the storage ring to function! Nevertheless, we can be reasonably optimistic about the prospect for such beams in future storage rings.

In conclusion, we have seen that beam polarization in high energy $e^+e^-$ storage rings is a versatile tool for studying many phenomena. Probably the most significant result obtained to date is the very clear demonstration of the spin-½ quark character of the basic processes involved in hadron production by $e^+e^-$ annihilation. The same behavior is seen in muon pair production, as expected from the theory of QED. The well understood process of spin precession makes beam polarization useful for energy calibration and other diagnostic functions. In future generations of storage rings and experiments using $e^+e^-$ storage rings, there are many interesting measurements to be performed with polarized beams.

## REFERENCES

1. A. A. Sokolov and I. M. Ternov, Sov. Phys.-Dokl. 8, 1203 (1964).
2. Y. S. Tsai, Phys. Rev. D 12, 3533 (1975).
3. See, for example: R. Budny, Phys. Lett. 45B, 340 (1973), and A. McDonald, Nucl. Phys. B75, 343 (1974).
4. I. M. Ternov, Yu. M. Loskutov, and L. I. Korovina, Sov. Phys.-JETP 14, 921 (1962).
5. V. N. Baier and V. M. Katkov, Phys. Lett. 24A, 327 (1967), and Sov. Phys.-JETP 25, 944 (1967).
6. V. N. Baier, Sov. Phys.-Usp. 14, 695 (1972), and Proc. of the International School of Physics "Enrico Fermi," Varenna, 1969, Course XLVI (Academic Press, New York and London, 1972), pp. 1-49.
7. J. D. Jackson, Rev. Mod. Phys. 48, 417 (1976).
8. V. Bargmann, L. Michel, and V. L. Telegdi, Phys. Rev. Lett. 2, 435 (1959).
9. Ya. S. Derbenev and A. M. Kondratenko, Sov. Phys.-JETP 35, 230 (1972).
10. Ya. S. Derbenev and A. M. Kondratenko, Sov. Phys.-JETP 37, 968 (1973).
11. Ya. S. Derbenev, A. M. Kondratenko, and A. N. Skrimskii, Novosibirsk preprint #77-60 (1977).
12. A. W. Chao, SLAC technical note PEP-257/SPEAR-208 (1977).

13. D. Potaux, Proc. of the 8th International Conference on High-Energy Accelerators, CERN, 1971, edited by M. H. Blewett (European Organization for Nuclear Research, Geneva, 1971), p. 127.
14. J. LeDuff et al., Orsay technical report #4-73 (1973).
15. U. Camerini et al., Phys. Rev. D $\underline{12}$, 1855 (1975).
16. W. T. Ford, A. K. Mann, and T. Y. Ling, SLAC report SLAC-158 (1972).
17. S. I. Serednyakov et al., Sov. Phys.-JETP $\underline{44}$, 1063 (1976).
18. The members of the SPEAR polarization group are: G. E. Fischer, D. B. Gustavson, J. R. Johnson, J. J. Murray, T. J. Phillips, R. Prepost, R. F. Schwitters, C. K. Sinclair, and D. E. Wiser. A preliminary report of this work was presented to this conference by J. R. Johnson.
19. V. N. Baier and V. A. Khoze, Sov. J. Nucl. Phys. $\underline{9}$, 238 (1969).
20. S. I. Serednyakov et al., Phys. Lett. $\underline{66B}$, 102 (1977).
21. R. S. Van Dyck, Jr., P. B. Schwinberg, and H. G. Dehmelt, Phys. Rev. Lett. $\underline{38}$, 310 (1977).
22. Ya. S. Derbenev et al., Particle Accelerators $\underline{8}$, 115 (1978).
23. L. M. Barkov et al., Novosibirsk preprint #77-74 (1977).
24. J. G. Learned, L. K. Resvanis, and C. M. Spencer, Phys. Rev. Lett. $\underline{35}$, 1688 (1975).
25. L. M. Kurdadze et al., Novosibirsk preprint #75-66 (1975).
26. R. F. Schwitters et al., Phys. Rev. Lett. $\underline{35}$, 1320 (1975).
27. G. Hanson et al., Phys. Rev. Lett. $\underline{35}$, 1609 (1975).
28. A. W. Chao, SLAC technical note PEP-263 (1978).
29. B. Richter, Nucl. Instrum. Methods $\underline{136}$, 47 (1976).
30. D. Möhl and B. W. Montague, Nucl. Instrum. Methods $\underline{137}$, 423 (1976).
31. Ya. S. Derbenev, A. M. Kondratenko, and A. N. Skrinsky, Novosibirsk preprint #76-62 (1976).
32. See, for example: J. Ellis and M. K. Gaillard, Physics With Very High Energy $e^+e^-$ Colliding Beams, CERN report 76-18 (1976), p. 39.
33. N. Christ, F.J.M. Farley, and H. G. Hereward, Nucl. Instrum. Methods $\underline{115}$, 227 (1974).
34. R. Schwitters and B. Richter, SLAC technical note PEP-87/SPEAR-175 (1974).
35. Ya. S. Derbenev and A. M. Kondratenko, Novosibirsk preprint #76-84 (1976).
36. Ya. S. Derbenev, invited talk at this Conference.
37. Ya. S. Derbenev, A. M. Kondratenko, and E. L. Saldin, Novosibirsk preprint #78-61 (1978).

# MEASUREMENTS OF SPEAR BEAM POLARIZATION USING A BACK-SCATTERED LASER TECHNIQUE

J. R. Johnson, R. Prepost, D. E. Wiser
University of Wisconsin, Madison, WI 53706

G. E. Fischer, D. B. Gustavson, J. J. Murray,
T. J. Phillips, R. F. Schwitters, C. K. Sinclair
Stanford Linear Accelerator Center, Stanford, CA 94305

## ABSTRACT

Fast beam-polarization measurements have been made at the SPEAR $e^+e^-$ storage ring by observing the up-down asymmetry in the distribution of laser photons back-scattered from the circulating positrons. The technique is described, and sample preliminary results are presented and discussed.

## INTRODUCTION

Beam polarization in $e^+e^-$ storage rings has been observed in the past[1-5] through measurements such as angular distributions of hadron jets and muon pairs, or the Touschek effect. The polarization builds up[5,6] parallel to the guide field due to spin-flip terms in the synchrotron radiation mechanism, but can be destroyed by machine resonances and other effects. There is interest[7] in the possibility of using polarized beams at PEP or PETRA to measure certain weak/electromagnetic interference terms in the cross-section for $e^+e^- \to \mu^+\mu^-$.

If $e^+e^-$ beam polarization is to be useful for experiments, it must be possible to determine at what energies the polarization can be found and to monitor its value as a function of time and machine conditions. This requires a way of making polarization measurements on a time scale that is short compared with the polarization build-up time constant ($\gtrsim$ 14 minutes at SPEAR). To test such a technique, and to check theoretical calculations, an experiment has been built and operated at SPEAR which is capable of measuring the beam polarization to a few percent in times of the order of 1 minute. The method[8] depends on the fact that for the special case of circularly polarized photons incident on transversely polarized electrons, spin-dependent terms in the Compton scattering cross-section yield an up-down asymmetry in the vertical distribution of the back-scattered photons.

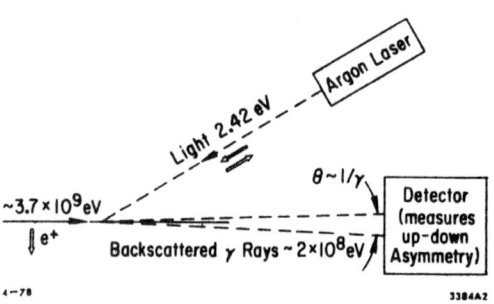

Fig. 1. Schematic layout of the experiment.

## METHOD

Figure 1 shows

schematically the layout (elevation view) of the experiment. Right or left circularly polarized photons of 2.4 eV are aimed at the circulating SPEAR positrons at a vertical angle of 8 mrad. The photons backscatter with energies up to 400 MeV, in a cone of characteristic opening angle $\theta \geq 0.14$ mrad. The vertical distribution is measured with a small detector package placed 12.9 m from the interaction point, where the distribution covers typically a few millimeters. Both the interaction point and detector distance were chosen from machine parameters to give a vertical focus of the projected beam directions at the position of the detector, so as to minimize the degradation of the vertical distribution due to finite beam emittance.

The polarized photons are supplied by a Spectra-Physics Model 166 Argon-ion laser equipped with a Model 365 acousto-optic coupler. The coupler allows the laser output to be pulsed at the SPEAR repetition rate (1.28 MHz) in 15 nsec wide bursts of up to 80W peak power. The linearly polarized light passes through a Pockels cell, which applies a voltage-controlled phase shift tuned to yield right or left circular polarization. In practice, the polarization is varied between right and left at a rate of about 25 Hz, to allow the measurement to average out possible systematic biases.

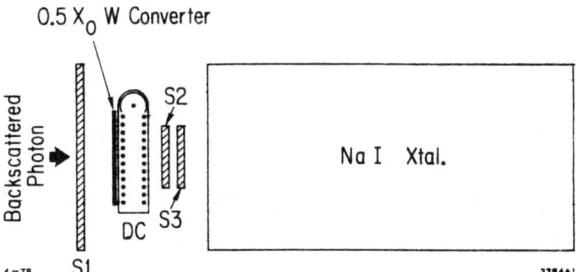

Fig. 2. Detector arrangement.

The detector arrangement is shown schematically in Fig. 2. A photon trigger is created with a telescope of three small scintillation counters: S1 vetoes charged particles, while S2 and S3 follow a tungsten converter plate and are put in coincidence. A single small drift cell (DC) is placed immediately behind the converter to measure the vertical position of each detected photon. This arrangement is followed by a 3-inch diameter by 6-inch long NaI (Tℓ) crystal to measure the photon energy. In the earlier stages of the experiment, a 1-mm wire spacing proportional chamber was used as the detector, with satisfactory results.

Figure 3 shows a diagram of the data acquisition system. The drift cell is digitized

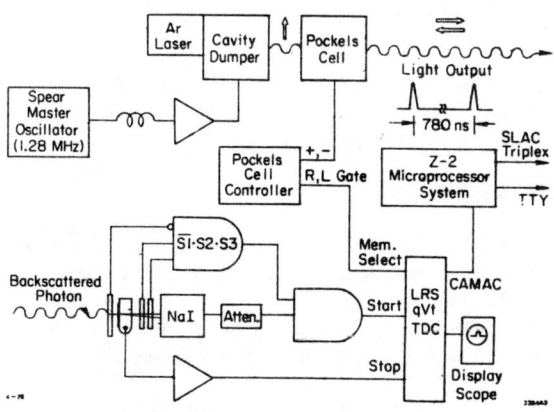

Fig. 3. Data acquisition system.

by the TDC feature of a LRS Model 3001 "qVt" analyzer. The "start" pulse is obtained from the basic $\overline{S1 \cdot S2 \cdot S3}$ photon trigger, perhaps including a cut on photon energy from the NaI counter; the "stop" pulse comes from the drift cell. The events are accumulated in two different regions of the analyzer memory, selected according to the photon polarization state determined by the Pockels cell controller. The system can accept data at typical back-scattered rates of ~ 10 kHz without serious dead-time losses.

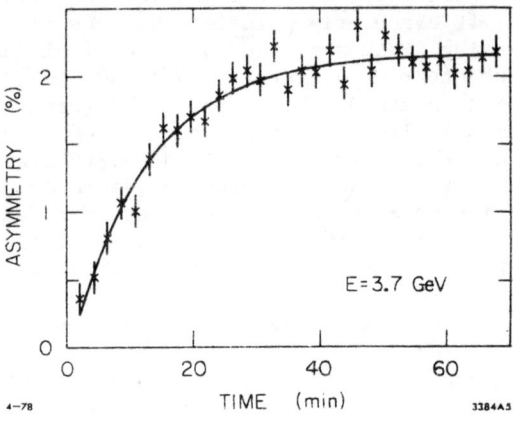

Fig. 4. Asymmetry measurements as a function of time, for a SPEAR energy of 3.7 GeV.

At the end of a run (typically 2-3 minutes), the accumulated vertical distributions corresponding to right (R) and left (L) circular photon polarization are read out through CAMAC by a Cromemco Z-2 microcomputer system.

For each distribution the microcomputer calculates a vertical asymmetry, defined as

$$A \equiv \frac{(up)-(down)}{(up)+(down)}$$

where (up) and (down) are specified regions on either side of center of the distribution, chosen to optimize the analyzing power and statistical precision of the measurement. The two asymmetries are then averaged,

$$A_{exp} \equiv \frac{A_R - A_L}{2},$$

to remove possible systematic errors or drifts. These results are printed out and sent over a phone line to the main SLAC Triplex computer system for storage and later analysis.

## PRELIMINARY RESULTS

Figure 4 shows a sample set of asymmetries as a function of time measured during a single SPEAR fill, starting (time $\equiv$ 0) when the storage ring reached its final energy of 3.7 GeV. As can be seen, typical maximum asymmetries of greater than 2% are measured with an absolute statistical precision of about 0.1% in times of about two minutes.

To study changes in beam polarization as a function of machine energy, the asymmetry can be measured while the SPEAR energy is raised in a series of small steps. Figure 5 shows the results of such a scan. Here the measured asymmetry first increases, then decreases to near zero as the beam energy moves into the region of a depolarizing machine resonance, then increases again as the resonance is passed by. This scan covered SPEAR energies around the region of 3.60 to 3.66 GeV.

To illustrate the resonance structure expected theoretically, Fig. 6 shows results of calculations by Chao.[9] The quantity plotted is the ratio of polarizing to depolarizing time constant; where this ratio is small, the polarizing effect will dominate. A scan such as shown in Fig. 5 may be analyzed to extract a measurement of this quantity for each step in machine energy. Such a fit appears as the solid line in Fig. 5, and the results are plotted in Fig. 7 along with the appropriate portion of the calculation of Fig. 6. The qualitative agreement is quite good.

The measurements presented above were obtained with only positrons in SPEAR. The general features of the single-beam results obtained to date are in reasonable agreement with theoretical expectations, including

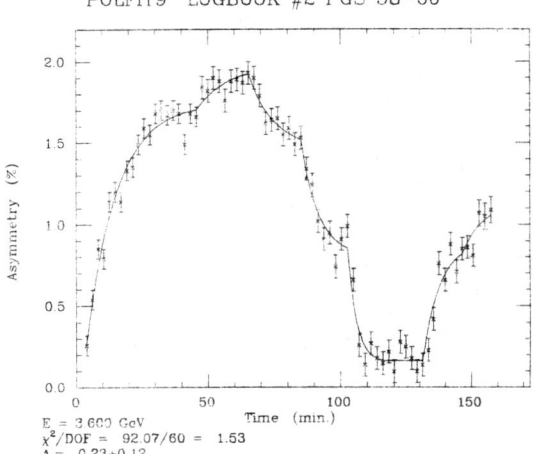

Fig. 5. Asymmetry measurements as a function of time, for 10 MeV steps of SPEAR energy from about 3.60 to 3.66 GeV. The solid line is a fit to a theoretical function.

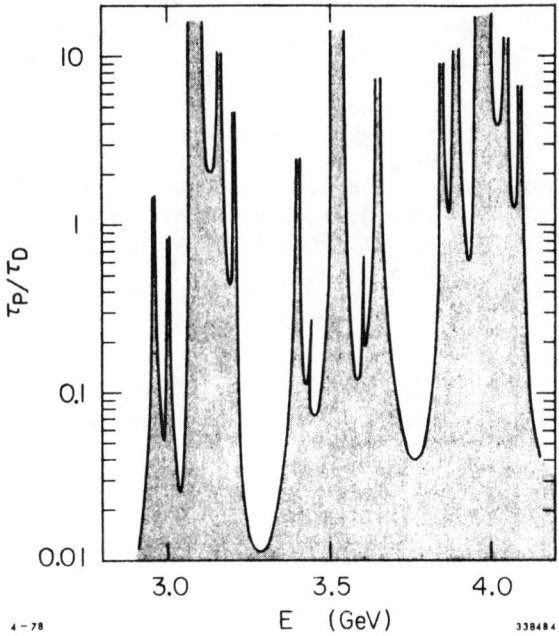

Fig. 6. Theoretical calculations from Ref. 9 showing some of the expected depolarizing resonances. The parameter plotted is the ratio of the polarizing to depolarizing time constant.

several narrow resonances that do not appear in Fig. 6. However, measurements with colliding beams have not been very reproducible and have shown that this situation is more complex. Further studies will be made to attempt to understand these effects.

The authors would like to acknowledge the important contributions of R. Eisele and J. Jurow in the design and implementation of the necessary changes to the SPEAR vacuum and cooling system, and the help of E. Taylor in the instrumentation and installation of the beam lines and experimental apparatus.

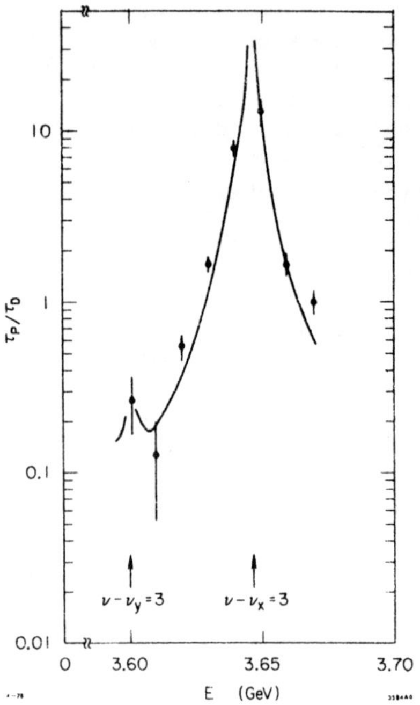

Fig. 7. Results of the fit shown in Fig. 5. The solid line is the theoretically expected behavior.

## REFERENCES

1. R. F. Schwitters et al., Phys. Rev. Letters 35, 1320 (1975).
2. J. G. Learned et al., Phys. Rev. Letters 35, 1688 (1975).
3. U. Camerini et al., Phys. Rev. D12, 1855 (1975).
4. D. Potaux, Proceedings of the Eighth International Conference on High Energy Accelerators, CERN, 1971.
5. V. N. Baier, Sov. Phys. Uspekhi 14, 695 (1972).
6. J. D. Jackson, Rev. Mod. Phys. 48, 417 (1976).
7. For example, Proceedings of the 1975 PEP Summer Study, SLAC-190, December 1975.
8. V. N. Baier and V. A. Khoze, Sov. J. Nucl. Phys. 9, 238 (1969).
9. A. W. Chao, PEP-257 (1977).

# MEASUREMENT OF BEAM-POLARIZATION IN THE STORAGE RING "PETRA"

H.C. Dehne, R. Rossmanith
Deutsches Elektronen-Synchrotron DESY, Hamburg, Germany

R. Schmidt
Universität Hamburg, Germany

## ABSTRACT

The degree of beam polarization in the electron-positron storage ring PETRA will be measured by backscattering circularly polarized photons from an $Ar^+$-laser. The existence of polarization manifests itself by an up-down asymmetry of the backscattered photons. These photons are emitted due to relativistics kinematics within an angle of ± 30μrad. Due to the small scattering angle of the photons, the detector must be 45 m away from the point where the laser beam and the particle beam interact. Detection is performed by the aid of a total absorbing shower counter. A movable converter in front of the shower counter converts the photon beam so that the counting rates in the upper and the lower plane can be measured in succession.

## INTRODUCTION

In electron or positron storage rings the beam is polarized by the emission of synchrotron-radiation. At the moment of injection the beam is assumed to be totally unpolarized, meaning that the number of particles with spin parallel to the magnetic guiding field and the number of particles with spin antiparallel to the guiding field are nearly equal. The emission of the synchrotron-radiation in the bending magnets generally does not flip the spin. The spin direction is only flipped at an amount of $10^{-11}$ of the total synchrotron radiation power. Spin-transition from ↑ to ↓ and ↓ to ↑ occur. Both these transitions are not equally probable and in the course of time the beam becomes polarized.[1]

The time dependence of polarization is given by the formula

$$P = P_o(1-e^{-t/\tau})$$

wherein P = degree of polarization

$$P_o \approx 92 \ \% \ ; \ \tau \ (sec) = 98 \ \frac{R^3 (m)}{E^3 (GeV)} \ \frac{<R>}{R}$$

&lt;R&gt; .. average radius of the storage ring,
R ... bending radius of the bending magnets,
E ... energy of the stored beam.

The polarization mechanism is depleted by depolarization effects. Hence, depolarization generally occurs when

$$(\frac{g-2}{2}) \cdot \frac{E}{E_o} = n + mQ_z$$

g-2/2 .. anomalous part of the electron magnetic moment
E ..... energy of the electron
$E_o$ ..... rest energy of the electron
$Q_z$ ..... vertical Q-value of the storage ring
n,m ... integers

## EXPECTED POLARIZATION AT PETRA

PETRA is a storage ring designed for about 7 to 15 GeV in the first stage. The bending radius is 197.15 m and the average radius is 367 m. According to the aforegoing formulas the polarization time is only less than one hour at energies greater than 10 to 11 GeV. Assuming the lifetime of the stored beam to be about one hour then reasonable amount of polarization is only to be expected at energies above 10 GeV.

The high Q-value of about 22, the high energy of the beam and the relatively high energy of the emitted quanta of synchrotron-radiation [2] will tend to depolarize the beam more than in storage rings working at lower energies than PETRA where polarization effects have already been detected. [3,4]

On the other hand, beam polarization is wanted and knowledge of the beam polarization is important for the study of the influence of the weak interaction in scattering processes between electrons and positrons.

## THE PRINCIPLE IDEAS OF THE PETRA-POLARIMETER

The only effect which can be used for the measurement of beam polarization seems to be the spin-dependent backscattering of circularly polarized photons by the polarized particle beam [5]. The system is sketched in fig. 1 and fig. 2. The cavity dumper is triggered by the particle beam bunch. The laser pulse is circularly polarized by a KDP crystal and enters the vacuum chamber via two mirrors. The cross section for scattering in the restframe of the electrons is described by two terms

$$\sigma = \sigma_o + \sigma_{spin}$$

where the first term is spin-independent (see fig. 3)

$$\sigma_o \approx (1+\cos^2\Theta) + \left(\frac{|\vec{K}_o|}{m_e} - \frac{|\vec{K}|}{m_e}\right)(1-\cos\Theta)$$

$K_o$ .. energy of the incoming photon
$K^o$ .. energy of the backscattered photo
$m_e$ .. restmass of the electron
$\Theta$ .. scattering angle

and the second is spin-dependent

$$\sigma_{spin} \approx \frac{1}{m_e}(1-\cos\Theta)\vec{S}\cdot(\vec{K}_o\cos\Theta+\vec{K})$$

$\vec{S}$ .. unity vector in the direction of spin

The term $\vec{S}\cdot\vec{K}$ leads to asymmetry of scattered particles meaning that more particles are scattered into the upper plane of fig. 3 than into the lower plane. The ratio $\sigma_{spin}/\sigma_o$ is measured:

$$\sigma_{spin}/\sigma_o \approx \frac{(1/m_e)(1-\cos\theta) \vec{S} \cdot (\vec{K}_o \cos\theta + \vec{K})}{1+\cos^2\theta}$$

The asymmetry ratio increases with $K_o$. $K_o$ is defined both by the relativistic kinematics and the energy of the incoming photon. The interaction must be head-on.

On the other hand, maximum asymmetry is achieved when $1+\cos^2\theta$ has a minimum, meaning that $\theta$ is 90 degrees. Transforming this back into the laboratory system, this anlge must be multiplied by a factor $1/\gamma$ so that maximum asymmetry occurs in PETRA (15 GeV) within an angle of the backscattered photons of 30μrad. The interaction between photons and electrons must occur on a position where the natural divergence of the electron beam is smaller than these 30μrad. This condition is only fulfilled in the interaction quadrupoles where the divergence is about 10μrad. The detector must be at a distance of 45 m from the interaction point for two reasons: firstly, at this distance electrons and photons are so far apart that the detector can be installed on the side of the vacuum chamber: secondly, the distance must be so great that the particles being scattered in the upper or in the lower plane can be distinguished. The backscattered γ quantas have energies between 3 and 6 GeV.

The detector is sketched in fig. 4. The energy of the photons is detected by a total absorbing shower counter and a converter. The counting rates in the upper and lower plane are measured in succession. Charged particles are suppressed by a Veto-counter and unwanted uncharged particles beyond the converter by a trigger counter.

## STATUS REPORT

PETRA began operating in the summer of 1978. Until present beams up to about 5 mA at energies between 5 and 7 GeV have been successfully stored and with these beams the first tests of the polarization monitors have been performed. Fig.5 demonstrates the horizontal beam profile of the electron beam measured by backscatterering the laser photons. When PETRA reaches reasonable currents at energies near 10 GeV, hopefully within the next weeks, the first attempts to measure beam-polarization will be made.

## REFERENCES

1. V.N. Baier, XLVI Corso Scuola International di Fisica "Enrico Fermi", 1969, Physics with intersecting storage rings, Academic Press, New York, 1971
2. B.W. Montague, Depolarising Effects in PETRA, Technical Note DESY PET-77/28
3. J. Le Duff, P.C. Marin, J.L. Masnou, M. Sommer, Measurement of Beam Polarisation in the Orsay Colliding Beam Ring A.C.O. Rapport Technique, 5-72, Institute National de Physique Nucleaire

et de Physique des Particules, Orsay
4. Report of the SPEAR polarization group at this meeting.
5. F.W. Lipps and H.A. Tolhoek, Polarization Phenomena of Electrons and Photons, Physica XX (1954) 85-98.

## FIGURE CAPTIONS

Fig. 1: Laser beam-electron beam interaction region. In this drawing the laser is seen to be in the paper plane and is in reality perpendicular to the paper plane.
Fig. 2: Path of the electron- and the gamma beam
Fig. 3: Definition of the scattering angle
Fig. 4: Detector arrangement
Fig. 5: Beam profile measured by backscattering the laser photons.

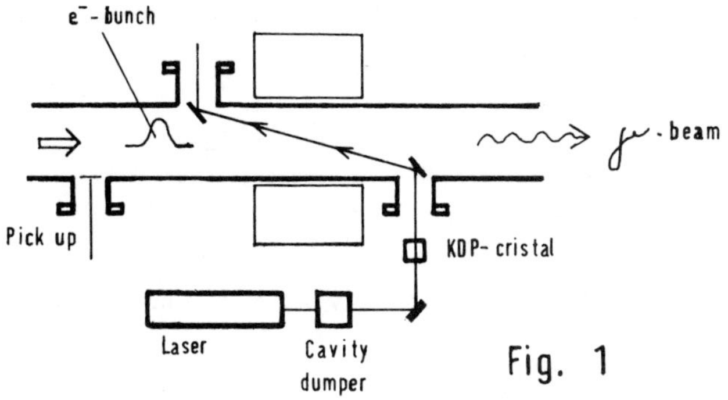

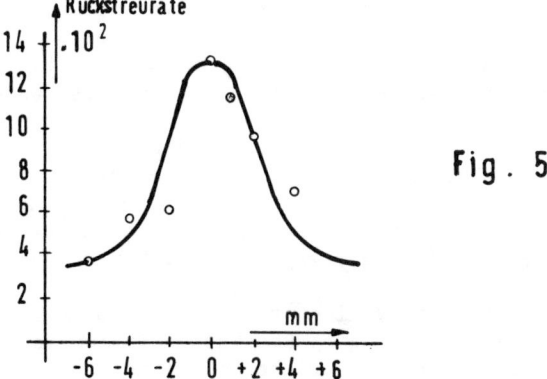

Fig. 1

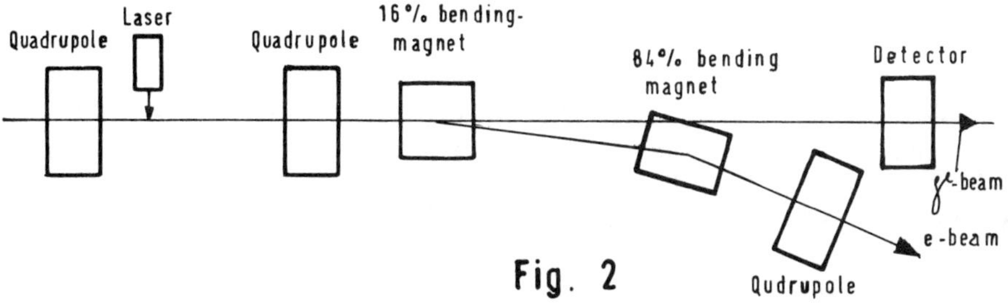

Fig. 2

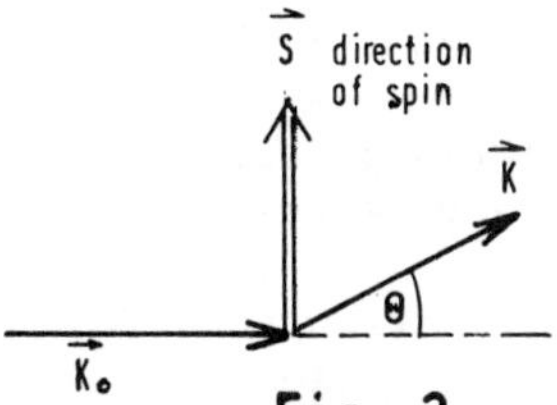

Fig. 3

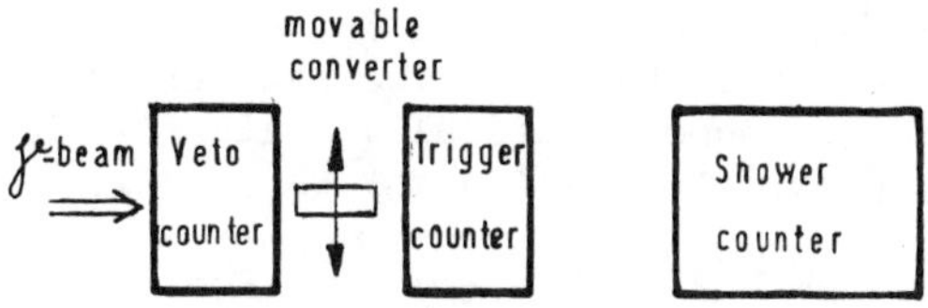

Fig. 4

Fig. 5

## ACCELERATION OF POLARIZED ELECTRONS IN THE 2,5 GEV SYNCHROTRON AT BONN

W. Brefeld, V. Burkert, K.P. Crone, W. von Drachenfels, E. Ehses,
M. Hofmann, D. Husmann, G. Knop, Th. M. Müller, W. Paul and
H.R. Schaefer

Physikalisches Institut der Universität Bonn

Federal Republic of Germany

### ABSTRACT

Polarized electrons have been accelerated for the first time in a synchrotron with the results that the polarization is maintained to a high degree up to 2.0 GeV. Strong depolarization only is observed in narrow energy bands due to imperfection and intrinsic resonances.

A pulsed source of polarized electrons[1] with a repetition rate of 50 Hz has been developed for use in the Bonn 2.5 GeV synchrotron[2]. The source is based on the Fano effect on rubidium i.e. photoionisation of unpolarized alkali atoms with circularly polarized light of appropriate wavelength. The quality of the source is given by

$$Q = P^2 \times N \quad (1) \quad (P,N = \text{degree of electron polarization and the number of electrons})$$

It depends on the photoionisation cross section, the available light pulse energy and the chosen wavelength.

As an optimized arrangement for the 50 Hz repetition rate of the synchrotron we used a light beam of 266 nm wavelength crossing an array of 20 atomic beams of rubidium. A Nd-YAG laser quadrupled in frequency served as the light source with pulse energies of 5-8 mJ at 266 nm. The linear polarized light of the laser is converted to circular by a quarter wave plate.

The number of photoelectrons extracted from the interaction region was up to $2 \times 10^9$ electrons per pulse within 20 nsec. The photoelectrons are extracted along the light direction. According to the Fano effect they are longitudinally polarized.

The Rb-oven is built as a recycling system. It permits a running time of two weeks with only 60 g of Rb. The operation is only interrupted for 40 min every 8 hours for the recycling procedure.

The extracted electrons are accelerated to 120 KeV. Their polariza-

tion vector is turned from longitudinal to transverse direction by
a Wien-filter. The electron polarization degree is determined by
Mott-scattering. The observed polarization was up to $(65 \pm 5)\%$ at
$2 \times 10^9$ electrons/pulse. During a two weeks run no decrease in polarization was observed.

The polarization can be reversed from pulse to pulse by changing
the circular polarization of the UV-light from left to right hand
by rotating the quarter-wave plate at 12.5 Hz, a quarter of the repetition rate of the synchrotron.

The general arrangement of the polarized electron source, the accelerator and the polarimeter is shown in fig.1. From the source the
polarized electron beam is bent to the linac by two 90° dipole magnets. To focus the beam on the 5 m long way to the linac four magnetic double lenses are used. In order to align the polarization vector parallel to the guiding field of the synchrotron, the beam
transport system must provide a rotation of the polarization vector.
Therefore each two lenses are connected in series but with opposite
field direction. By this means the polarization vector does not depend on the lens current. An additional long solenoid provides rotation of the polarization vector over a range of 270°.

The beam transfer line for the low momentum electrons is screened
against the influence of the magnetic earth-field by an iron-shield
(µ-metal). Position and mean current of the beam is registered by
movable Zinc-sulfid screens and Faraday-cups.

During acceleration in the synchrotron two different resonance effects may change the polarization of the electrons. Such resonances
occur whenever the Thomas-frequency due to the electron's anomalous
magnetic moment

$$\omega_T = (eBa)/(mc) \quad (2) \quad (a=(g-2)/2; \text{ g=electron g-factor})$$

is an integer multiple of either the cyclotron frequency

$$\omega_T = (eB)/(mc\gamma) \quad (3)$$

or the vertical betatron frequency

$$\omega_B = Q_z \omega_c \quad (4) \quad (Q_z = \text{number of vertical betatron oscillations per revolution})$$

In the first case, the so called imperfection resonances occur at
$\gamma$-values

$$\gamma = \frac{n}{a} \quad (5)$$

corresponding to electron energies $E = n \times 440$ MeV. These energies
are independent of the special magnet configuration of the accelerator. The energy of the second type of depolarization resonances (intrinsic resonances) depends on the $Q_z$ value of the accelerator. For

the Bonn 2.5 GeV synchrotron Q is 3.4. Therefore the lowest energy for such a depolarization resonance is 1.5 GeV.

The polarization of the ejected electron beam has been measured with a polarimeter[3] using the spin dependence of the elastic electron-electron scattering (Møller-scattering). For longitudinally polarized electrons this process shows a large asymmetry in the cross sections for polarization of the projectile electrons parallel and antiparallel to the polarization of the target electrons. The ejected electrons are transversely polarized. In order to change the polarization to longitudinal direction they are deflected by magnetic dipole fields in the plane given by momentum and polarization vector. The transformation is 100% at about 1.8 GeV but only 67% at 0.85 GeV.

As a polarized electron target a 40 μm thick foil of supermendur[*] was used. The degree of polarization was about 8%. The foil was inclined to the beam direction by an angle of 23°. Therefore the effective longitudinal polarization was 7.4%. The polarimeter detects in coincidence both elastic scattered electrons in the region of the maximum Møller-scattering asymmetry $A_{max}$ = 7/9. The expected asymmetry of the counting rates is about 4% at 100% beam polarization.

To reduce systematic errors the polarization of the beam was reversed from pulse to pulse. To detect apparativ asymmetries, the polarization of the target was reversed for succeeding runs. Several runs were also made with polarized target and unpolarized beam, and with unpolarized target and polarized beam.

At present only 2% of the electrons emerging from the source and passing the accelerator and ejection system are reaching the Møller-target. The number of the detected pairs of Møller-electrons are normalized to the number of electrons from nuclear scattering.

For the first test, the synchrotron has been run at an endpoint energy of 0.85 GeV, far above 0.44 GeV and just below 0.88 GeV. This allowed to pass fast through the 0.44 GeV resonance and thus limit depolarization.

To optimize the direction of the polarization vector of the electrons at injection the long solenoid in the injection line is used. The polarization of the beam has been measured as a function of the solenoid current. The result is shown in fig.2. The data are in agreement with the expected sine dependence. In the maximum the primary polarization is maintained for about 80%.

With this setting the degree of polarization at several endpoint energies of the synchrotron up to 2.0 GeV has been measured. The experimental results are shown in fig.3. It is to be seen that the degree of polarization of the accelerated electrons up to 2.0 GeV is

---

[*]Alloy of 49% Fe, 49% Co, 2% V with special magnetic treatment.

still a considerable fraction of the polarization degree of the injected electrons. This behaviour was expected from calculations[4] based on the observed small misalignments of the synchrotron magnets (see fig.4).

To study the influence of the imperfection resonances careful measurements have been done in a narrow energy band around 880 MeV and 1320 MeV. The polarization was measured changing the endpoint energy of the synchrotron in steps down to 0.25 MeV. The stability of the power supplies permits such a procedure. At the resonance energy the sign of the polarization changes between two steps - indicating a very small resonance width - followed by a smooth rise to higher energies. This is due to a decreasing resonance crossing time with increasing endpoint energy.

The measurements allow to calculate the zero crossing with high accuracy and make possible a very precise absolute energy calibration of the synchrotron in the order of $10^{-4}$.

The paper presents only preliminary results.

### REFERENCES

1) W. von Drachenfels, U.T. Koch, Th.M. Müller, W. Paul, H.R. Schaefer; Nucl.Inst.a.Meth. 140, 47(1977)

2) K.H. Althoff, K. Bätzner, J. Drees, A. Febel, O. Gildemeister, G. von Holtey, G. Knop, P. Lüttner, H. Netter, W. Paul, F.J. Schittko, A. Schultz von Dratzig, H.E. Stier and E. Weisse; Nucl.Inst.a.Meth. 61, 1-30(1968)

3) W. Brefeld, V. Burkert, E. Ehses, M. Hofmann, G. Knop, E. Pohlen; submitted to Nucl.Inst.a.Meth.

4) M. Hofmann; Internal Report

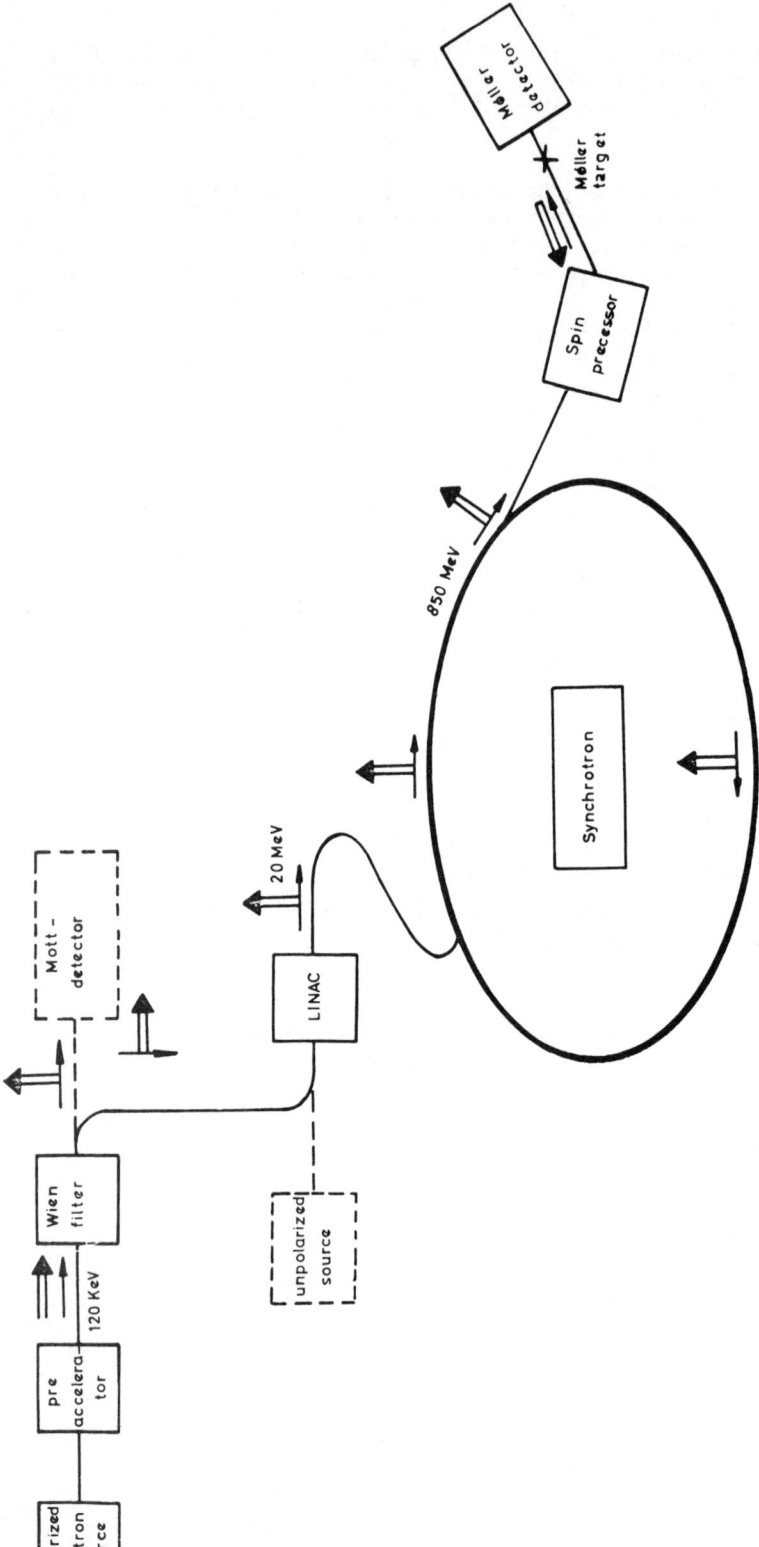

Fig.1: Schematic diagram of the experimental set up : production, acceleration and detection of the polarized electrons. The arrows indicate the direction of the momentum vector (→) and the polarization vector (⇒).

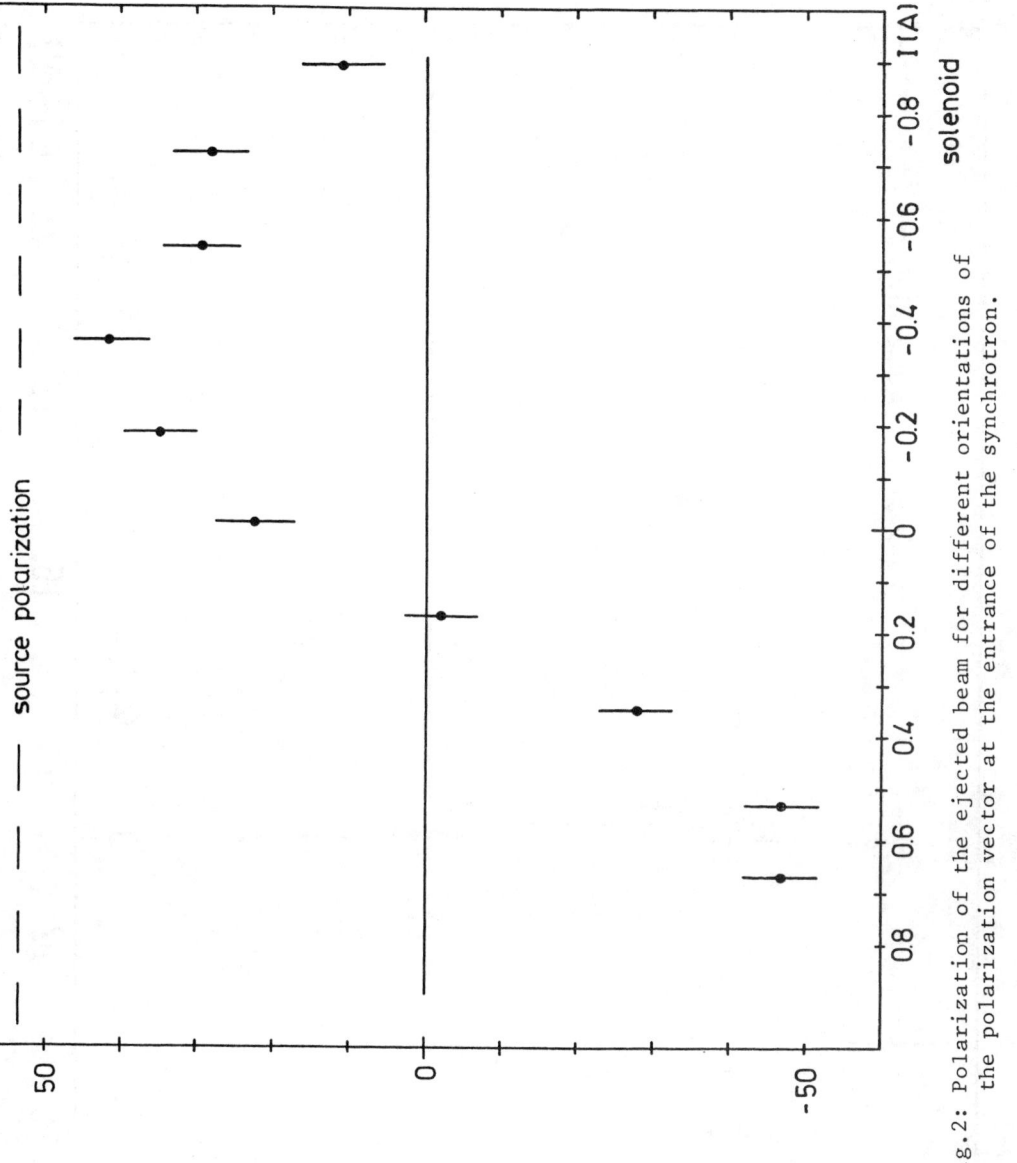

Fig.2: Polarization of the ejected beam for different orientations of the polarization vector at the entrance of the synchrotron.

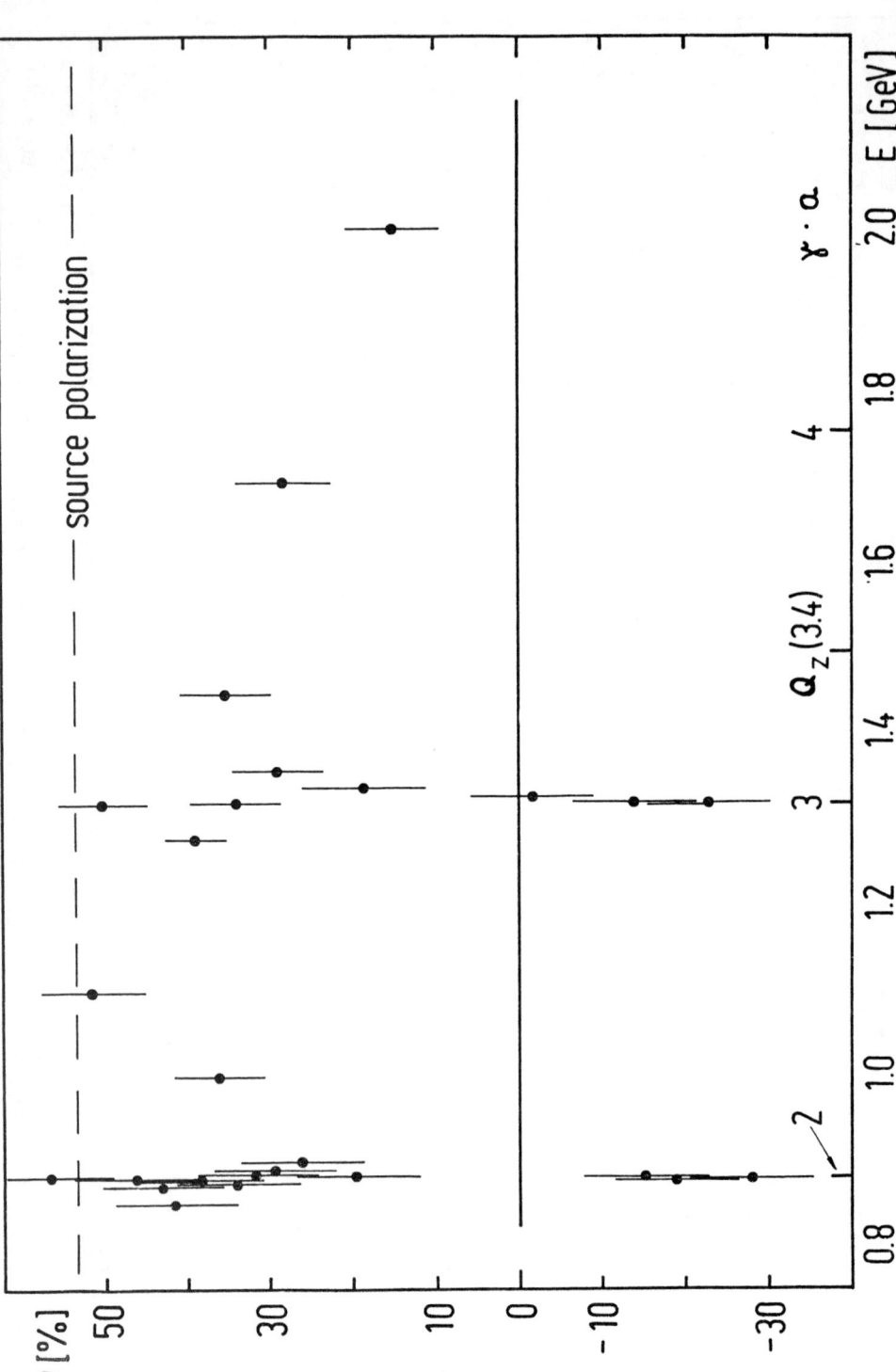

Fig.3: Polarization of the ejected beam at endpoint energies of the accelerator between 0.85 and 2.0 GeV.

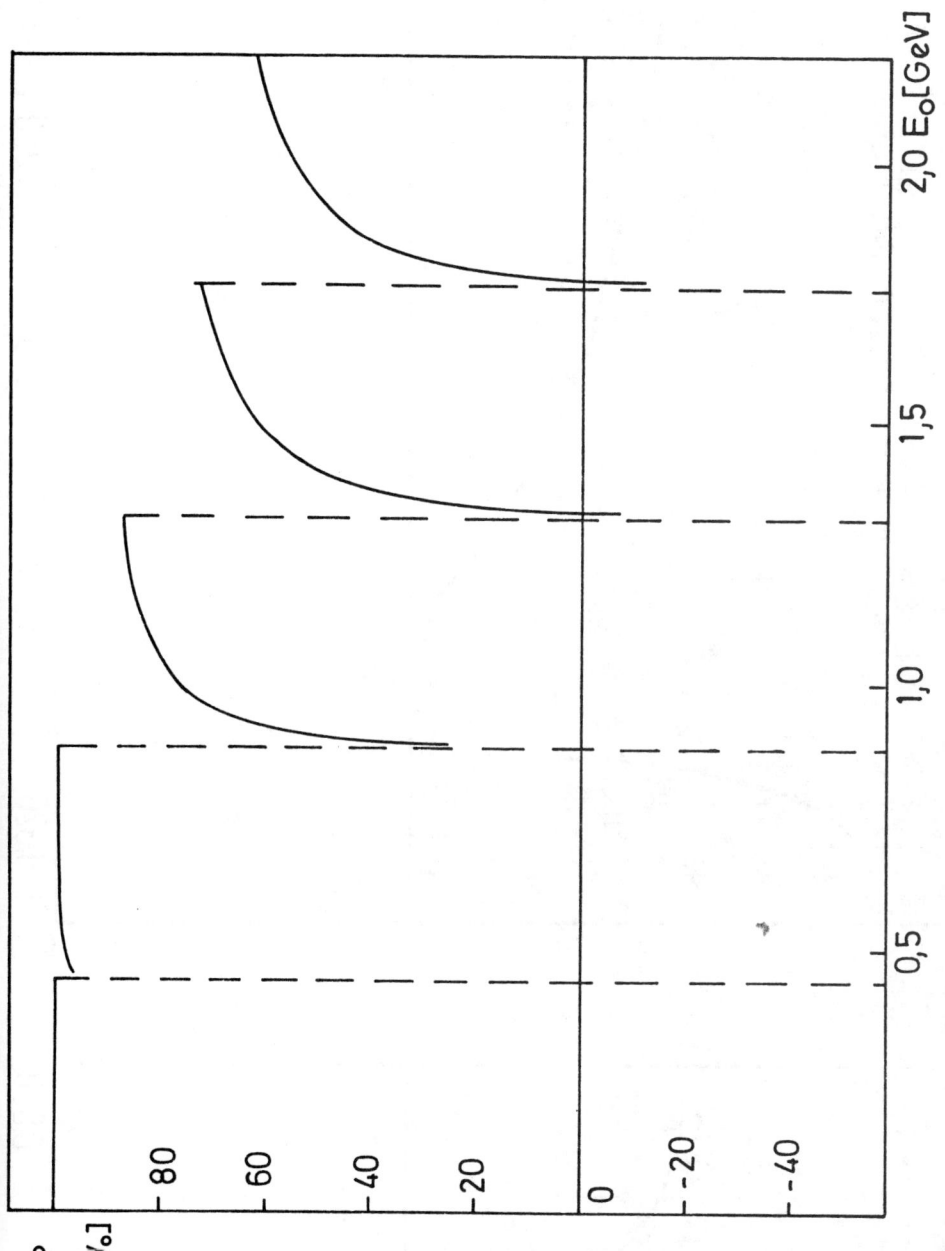

Fig.4: Calculated behaviour of the imperfection resonances for different endpoint energies. The calculations are based on measurements of small misalignments of the synchrotron magnets.

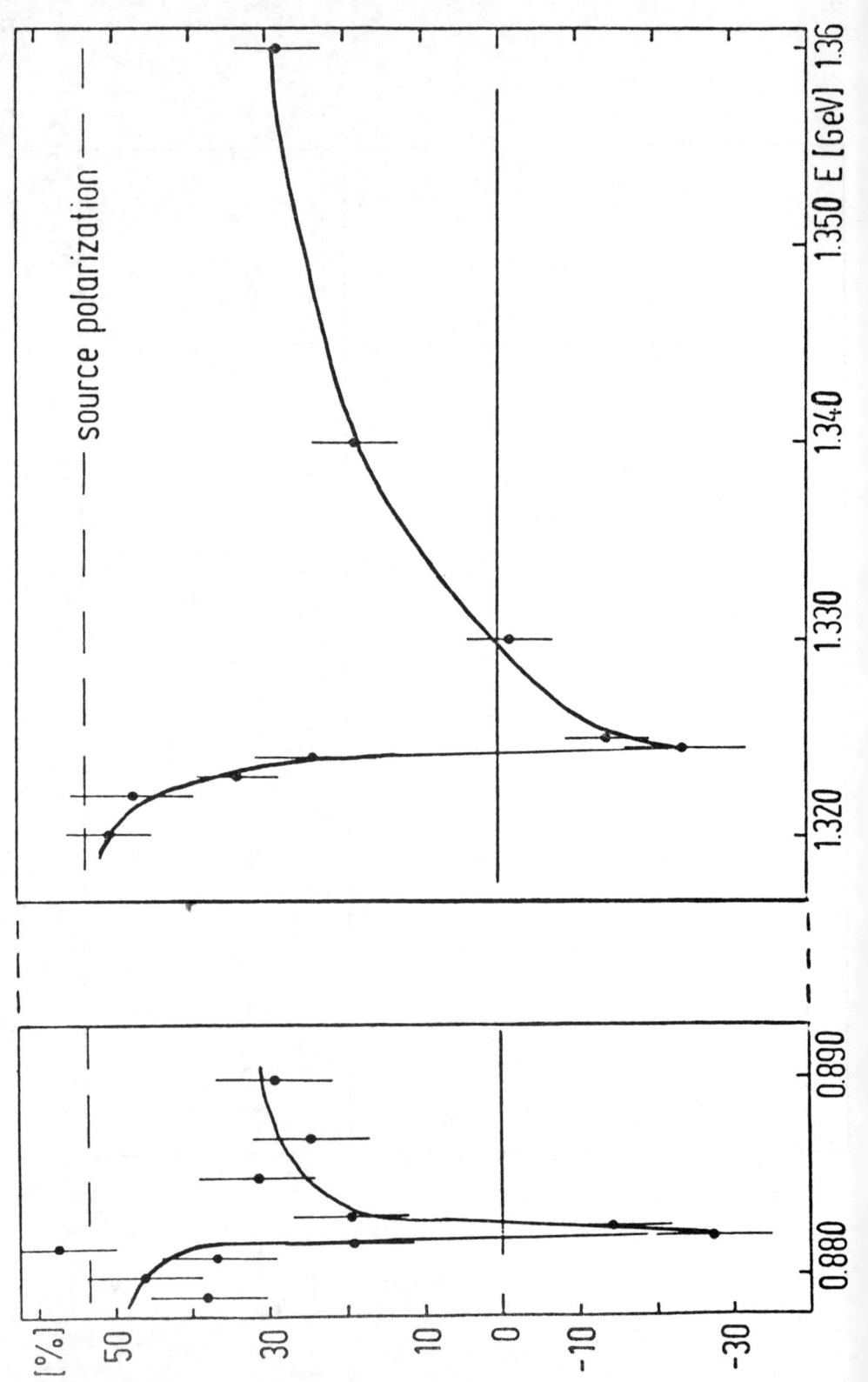

POLARIZED $e^{\pm}$ BEAMS IN LEP?

B.W. Montague
CERN, CH-1211 Geneva 23

ABSTRACT

Physics with the Large Electron Positron Storage Ring under study at CERN would derive much benefit if polarized beams could be stored. Hitherto, the depolarizing effects of spin resonances seemed to exclude this possibility. The "Siberian Snake" proposed by the Novosibirsk group indicates a way of avoiding these resonances and offers some hope of maintaining polarized $e^{\pm}$ beams over at least part of the interesting energy range of LEP. The implementation of such a scheme has important consequences on the theoretical, technological and financial aspects of the LEP design.

POLARIZATION IN NORMAL $e^{\pm}$ STORAGE RINGS

The radiative polarization of electron (and positron) beams, predicted theoretically by Sokolov and Ternov[1], has been observed experimentally in several storage rings[2]. The effect arises from an asymmetry in the spin-flip probability on emission of a synchrotron-radiation photon. Since spin flip occurs typically at only about $10^{-11}$ of the quantum emission rate, the polarization time $\tau_p$ is rather long, in the range of minutes to hours; it is given by:

$$\tau_p = \frac{8}{5\sqrt{3}} \frac{m_o R \rho^2}{\hbar r_o \gamma^5} \qquad (1)$$

where $m_o$, $r_o$ are the mass and classical radius of the particle, $\hbar$ is the reduced Planck's constant, $\gamma$ is the Lorentz energy factor, $\rho$ is the bending radius and R is the average radius of the storage ring. The limiting degree of polarization that can result from the spin-flip asymmetry is 0.924.

The polarization state of a beam can be described by classical equations, the projection of the spins parallel ($e^+$) or anti-parallel ($e^-$) to the magnetic guide field giving the degree of beam polarization. Spin motion then follows the precession equation:

$$\frac{d\vec{P}}{dt} = \vec{\Omega} \times \vec{P} \qquad (2)$$

where the axial vector $\vec{\Omega}$ is given by the BMT equation[3]. In a co-ordinate system following the orbit of the particle, and in the absence of electric fields, the precession frequency $\Omega_\perp$ in a uniform transverse magnetic field $B_\perp$ is

ISSN: 0094-243X/79?510129-09$1.50 Copyright 1979 American Institute of Physics

$$\Omega_\perp = -\frac{eB}{m_o\gamma} \cdot \gamma\left(\frac{g-2}{2}\right) = (\gamma a)\Omega_c \qquad (3)$$

where $a = \frac{g-2}{2}$ is the anomalous part of the gyromagnetic ratio g and $\Omega_c$ is the relativistic cyclotron frequency. In contrast, the precession frequency $\Omega_\parallel$ in the longitudinal field $B_\parallel$ of a solenoid is given by

$$\Omega_\parallel = -\frac{eB_\parallel}{m_o\gamma} \cdot \frac{g}{2} , \qquad (4)$$

a factor of about $(\gamma a)$ smaller for a given field strength.

The ideal precession motion around the vertical guide field $B_z$ is perturbed by magnetic-field components perpendicular to $B_z$ arising, for example, from closed-orbit deviations, betatron oscillations and magnet-end fields. In electron machines above a few GeV $(\gamma a \gg 1)$ one sees from equations (3) and (4) that perturbations are dominated by radial field components. At energies where the main precession frequency $(\gamma a)\Omega_c$ is close to a Fourier harmonic of the perturbing fields, spin resonances occur, leading to depolarization of the beam. The general resonance condition is:

$$\upsilon = \gamma a = k_o + k_z Q_z + k_x Q_x + k_s Q_s \qquad (5)$$

where the k are positive or negative integers and $Q_z$, $Q_x$ and $Q_s$ are the vertical and horizontal betatron and the synchrotron oscillation wave numbers.

Of special concern in high-energy storage rings are the integer (or imperfection) resonances $\upsilon = k_o$ and the vertical betatron (intrinsic) resonances $\upsilon = k_o + k_z Q_z$. The former, excited by orbit imperfections, occur for electrons (positrons) at intervals of $(m_o c^2)/a = 440$ MeV throughout the energy spectrum. These integer resonances constitute a basic problem for very high-energy $e^+-e^-$ storage rings[4] because of the large energy spread $\sigma_E$ in such machines, given by

$$\sigma_E = \left[\frac{55\hbar c\, m_o c^2}{64\sqrt{3}}\right]^{1/2} \frac{\gamma^2}{\sqrt{\rho}} \qquad (6)$$

For the LEP 70 GeV design[5] the standard deviation $\sigma_E$ is 87 MeV and the 440 MeV spacing corresponds to $\pm 2.5\,\sigma_E$ for a beam energy centred between two adjacent resonances. Particles sweep across these resonances in the course of synchrotron (energy) oscillations and, if the resonances cannot be made sufficiently weak, the beam depolarization will be faster than the radiative polarization rate $\tau_p^{-1}$ given by equation (1). With the criteria used in reference 4 it appears practically impossible to reduce the relevant harmonic ($\nu = k_o = 159$ at 70 GeV) of the closed-orbit deviation to a sufficiently low level to maintain a beam polarized. Even if these criteria could be considerably relaxed, it seems excluded to compensate the many harmonics in the energy range between injection (15 GeV) and 70 GeV. Although, in a given machine, the energy spread is reduced at lower energies, comparison of equations (6) and (1) shows that the increase in $\tau_p$ goes much faster with energy.

## THE SIBERIAN SNAKE

It is evident that the main problem in avoiding spin resonances arises from the direct energy dependence of precession frequency $\nu = \gamma a$. A radical and ingenious proposal of Derbenev and Kondratenko[5] indicates a way of making the precession frequency essentially energy-independent by introducing a special magnet arrangement into one section of a storage ring. To illustrate this we consider the racetrack storage ring of Fig. 1. The straight section BC contains a magnet configuration which rotates spin vectors around the y (velocity) axis by an angle $\pi$; for the purpose of discussion we can suppose this to be a solenoid magnet. The other straight section A is field-free.

We follow separately the evolution of the three orthogonal components of an arbitrary spin vector for a particle moving on an ideal closed orbit. In Fig. 1 (a) the z-component starts at A in position 1 and remains along the z-axis parallel to the bending field up to B. Between B and C the field of the solenoid rotates the vector by $\pi$ around y so that it goes through the second bending arc anti-parallel to the z-axis. Back at A in position 2 it has simply been rotated by $\pi$ around the y-axis relative to its starting position 1. In Fig. 1 (b) the x-component precesses in the x,y plane through an angle $2k\pi + \theta$ around the z-axis between A and B. Between B and C, rotation by $\pi$ around the y-axis is equivalent to a rotation by $\pi - 2\theta$ around the z-axis. Finally, between C and A a further precession by $2k\pi + \theta$ about the z-axis brings the total rotation to $(4k+1)\pi$ between positions 1 and 2. The y-component of the spin vector, Fig. 1 (c), rotates $2k\pi + \theta$ between A and B; the rotation of $\pi$ around the y-axis from B to C corresponds to a z-rotation of $-2\theta$ and a further $2k\pi + \theta$ along CA brings the total to $4k\pi$, making positions 1 and 2 identical.

From the above we see that the y-component at A corresponds to a periodic solution; this is the real eigenvector $\vec{n}$ of the rotation matrix around one revolution at A. Any spin solution in this system is a linear combination of the three components discussed above, and any component transverse to $\vec{n}$ rotates around $\vec{n}$ by $\pi$ in each revolution of the machine, <u>independently of the basic precession angle</u> $2k\pi + \theta$ in each main arc. Although $2k\pi + \theta$ is a function of energy, the effective precession wave number $\nu$ is exactly one-half and is independent of energy. The topology of the $\vec{n}$-motion is evidently that of a Möbius strip.

Since spin motion couples only to dipole-like fields there are no half-integer resonances and $\upsilon = \frac{1}{2}$ is as far from adjacent integers as it can be. Thus, with some reservations to be discussed later, a polarized beam can be accelerated over a wide energy range without crossing any integer spin resonances. Furthermore, the energy spread of the beam no longer gives rise to a proportional spread in $\upsilon$ and the major source of depolarization in normal high-energy electron machines is absent with this configuration.

In the Siberian Snake scheme the normal radiative polarization mechanism of Sokolov and Ternov is of no help, since the periodic solution $\vec{n}$ is perpendicular to the direction z of the magnetic field in the main arcs. (In fact this becomes a depolarizing effect[7] and may limit the upper energy for polarized $e^{\pm}$ beams.) In the absence of self-polarization, the beams must be injected already polarized, requiring the use of a storage-ring injector to hold the beams long enough for radiative polarization to develop.

It has been shown by Derbenev and Kondratenko[6,8] that another polarization mechanism may exist, depending on the properties of the machine. It is too early as yet to determine whether this process is applicable to the LEP storage ring, but the possibility is under active study.

## TRANSVERSE-FIELD MAGNETS

The use of a solenoid in the "snake" section is impracticable at high energies because of the low precession rate given by equation (4); a rotation of $\pi$ at 70 GeV would require a strength of 733 Tm! Fortunately, the rotation around the velocity axis can be simulated by a succession of transverse-field magnets[5] which, however, have the undesirable consequence of modifying the local geometry in both vertical and horizontal planes. Also, since transverse fields of a given magnitude produce precession rates which are independent of energy, as can be seen from equation (3), it follows that one cannot have both fixed geometry and fixed precession conditions over a range of energies. In a high-energy $e^+$-$e^-$ storage ring such as LEP the problems of introducing a variable snake geometry look sufficiently imposing that it is of interest first to examine the limitations of a fixed-geometry scheme, in which the precession kinematics vary with energy.

A transverse-field magnet configuration which simulates the solenoid of Fig. 1 is shown schematically in Fig. 2, together with the geometrical projections of the orbit in both planes. The arrows show the directions of the field in the three successive bending magnets and the arguments in brackets indicate the corresponding precession angles. This arrangement is shorter than that proposed in reference 5 and seems more favourable for LEP; both require a further pair of magnets to restore the orbit displacement in one plane.

## VARIATION OF CHARACTERISTICS WITH ENERGY

The precession kinematics are conveniently calculated using two-component spinor algebra. The matrix over one revolution for the system of Fig. 2 can then be written

$$M = IC_o - i\sigma_x C_x - i\sigma_y C_y - i\sigma_z C_z \tag{7}$$

where $\sigma_x$, $\sigma_y$, $\sigma_z$ are the Pauli matrices, I is the unit matrix and the coefficients are given by:

$$\left.\begin{aligned}
C_o &= \cos\tfrac{\phi}{2}\left\{\cos\tfrac{\chi}{2} - \sin\tfrac{\phi}{2}\sin\tfrac{\chi}{2}\right\} \\
C_x &= \sin^2\tfrac{\phi}{2}\sin(2\lambda-1)\tfrac{\chi}{2} \\
C_y &= \sin^2\tfrac{\phi}{2}\cos(2\lambda-1)\tfrac{\chi}{2} \\
C_z &= \cos\tfrac{\phi}{2}\left\{\sin\tfrac{\chi}{2} + \sin\tfrac{\phi}{2}\cos\tfrac{\chi}{2}\right\}
\end{aligned}\right\} \tag{8}$$

in which:

$\chi = 2\pi\gamma a$    is the total precession angle in the two main bending arcs and

$\phi = \pi\dfrac{E}{E_o}$    is the rotation angle of the snake section at energy E for (fixed geometry) variable magnetic fields corresponding to $\phi = \pi$ at some reference energy $E_o$.

The parameter $\lambda$ ($0 \leq \lambda \leq 1$) defines the azimuth in the ring at which the matrix M is calculated, excluding the snake section itself.

The effective precession frequency $\upsilon$ is given by:

$$\cos(\pi\upsilon) = \tfrac{1}{2}\text{Tr } M = C_o \tag{9}$$

The behaviour of this is shown in Fig. 3 as a function of $E/E_o$ for fixed geometry; the further E deviates from the reference energy $E_o$ the closer the excursions of $\upsilon$ approach the integer resonances. The presence of betatron resonances and synchro-betatron sidebands may further restrict the available region and suggest that an energy range of about three to one may be the largest practicable with fixed geometry.

The components of the periodic solution $\vec{n}$ can also be calculated from equations (7) and (8). Quite generally:

$$n_u = \frac{\pm C_u}{\sqrt{1-C_o^2}} \quad (u = x,y,z) \tag{10}$$

both + and - solutions being equally valid. The variation with

$E/E_o$ of the longitudinal component $n_y$ in the "privileged" interaction region, ($\lambda = 0.5$) opposite the snake section, is shown in Fig. 4. At interaction regions elsewhere on the machine circumference the longitudinal projection $\pm n_y$ of the polarization approaches unity at more widely spaced energy steps.

## ARE POLARIZED $e^{\pm}$ BEAMS FEASIBLE IN LEP?

At the Ann Arbor workshop[9] in 1977, the use of a Siberian Snake for high-energy polarized protons aroused much interest. The absence of synchrotron radiation in such an application makes the evaluation of the problems more straightforward than for high-energy $e^+$-$e^-$ machines.

In the LEP energy range the feasibility of polarized $e^{\pm}$ beams is mainly governed by the effects of synchrotron radiation, even with the beneficial precession kinematics of the Siberian Snake. Spin flip in the normal bending arcs is likely to restrict the energy range over which the beams remain polarized to below the maximum machine energy[7]. The transverse-field magnets of the snake itself exert some depolarizing influence which can be kept low only by operating with weak fields and therefore a long snake insertion. This in turn increases the difficulty in designing a practical variable-geometry scheme for a wide energy range. The need to inject $e^{\pm}$ already polarized into LEP would require a storage-ring injector, probably of higher energy than the 15 GeV synchrotron injector at present foreseen, with a consequent increase in the project cost.

In addition to the problems of spin kinematics, the normal orbit dynamics of the storage ring are perturbed, introducing a unit superperiodicity which could have important consequences on the closed orbit excursions, the betatron motion and the various chromatic effects. These aspects have not yet been studied for LEP and, whilst we have no reason to expect unsurmountable difficulties, detailed work will be necessary to ensure that the beam dynamics can be adequately controlled with such a scheme.

Assuming it could be made to work, is it worth it? At the LEP Summer Study, Les Houches and CERN, 1978[10], the utility of polarized $e^{\pm}$ beams for LEP physics was widely discussed. The general conclusion was that, in the framework of conventional gauge theories, polarized beams do not appear to be an essential tool, but could nevertheless be useful for cross-checking in some types of experiment. On the other hand, polarization could be a valuable revelator of unorthodox phenomena, in particular non-gauge theories. Consequently it seems very desirable to keep open the option in the LEP design and to avoid taking any step which might preclude implementation of a polarized-beam facility. It was further recommended that studies of polarization and depolarization mechanisms be continued and that other possibilities for Siberian-Snake or equivalent schemes be explored.

There exists an extensive literature on the basic kinematics of spin motion in storage rings, much of which is cited in the review papers of references 6 and 11. From this we can expect that, with careful machine adjustment, useful polarized beams can be obtained at PETRA and PEP, just as they have been in the lower energy range. At LEP energies, where new tricks appear to be necessary, the situation is more uncertain and much work will have to be done before we can predict with confidence whether polarized $e^{\pm}$ beams are a practical proposition for such a machine.

## ACKNOWLEDGEMENTS

I wish to thank Miss Monica Hanney for the computations and for help in preparing the figures.

## REFERENCES

1. A.A. Sokolov and I.M. Ternov, Sov. Phys. Doklady, $\underline{8}$, 1203 (1964).
2. The Orsay Storage Ring Group, Proc. VIIIth Int. Conf. on High Energy Accelerators, CERN (1971), p. 127.
   V.N. Baier, Sov. Phys. Uspekhi, $\underline{14}$, 695 (1972).
   U. Camerini et al., Phys. Rev. $\underline{D12}$, 1885 (1975).
   L.M. Kurdadze et al.; A.D. Bukin et al.; Proc. Vth Int. Conf. on High Energy Physics, Warsaw (1975).
   R.F. Schwitters, see ref. 10.
3. V. Bargmann, L. Michel and V.L. Telegdi, Phys. Rev. Letters, $\underline{2}$, 435 (1959).
4. D. Möhl and B.W. Montague, Nucl. Instr. and Methods, $\underline{137}$, 423 (1976).
5. The LEP Study Group, Report CERN/ISR-LEP/78-17 (1978).
6. Ya.S. Derbenev et al., Particle Accelerators, $\underline{8}$, 115 (1978).
   Ya.S. Derbenev and A.M. Kondratenko, Novosibirsk Preprint 76-84 (1976).
7. R.F. Schwitters, Note ECFA/LEP 40, private communication (1978).
8. Ya.S. Derberev and A.M. Kondratenko, Sov. Phys. JETP $\underline{37}$, 968 (1973).
9. Workshop on Higher Energy Polarized Beams (Ann Arbor, 1977), AIP Conf. Proceedings no. 42 (1978).
10. Proceedings of the LEP Summer Study (Les Houches and CERN, 1978), to be published.
11. V.N. Baier, Proceedings of the 46th Varenna Summer School on Storage Rings (1971).
    J.D. Jackson, Rev. Mod. Phys. $\underline{48}$, 417 (1976).

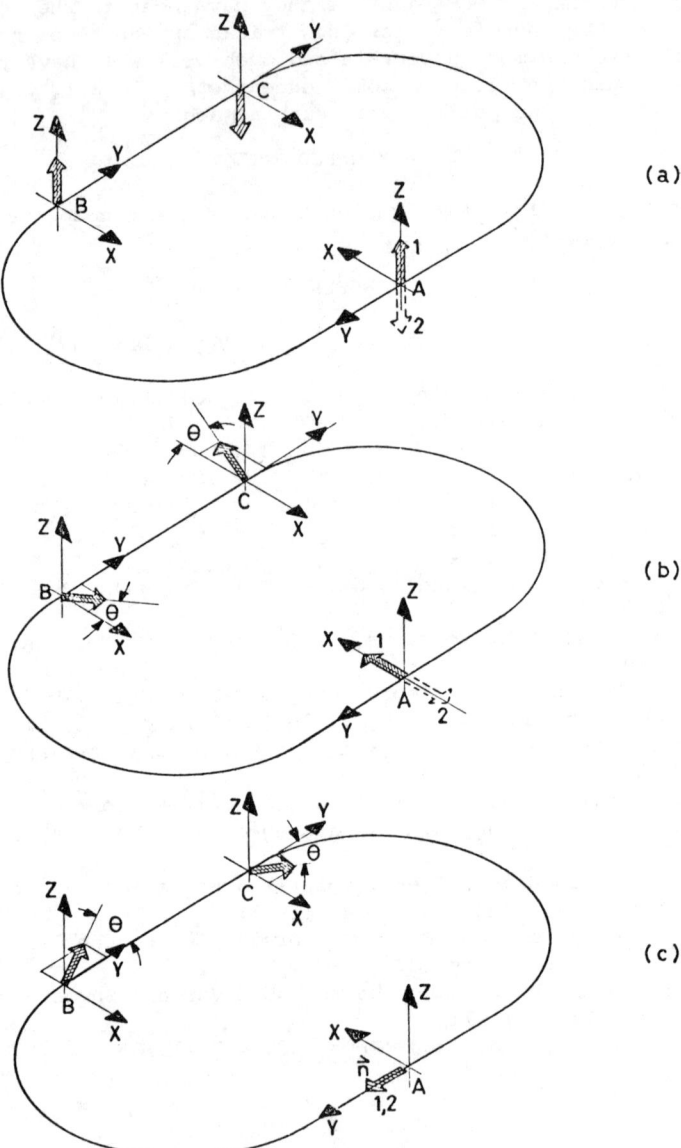

Fig. 1. Principle of the Siberian Snake; evolution of spin-vector components between positions 1 and 2 in one revolution. Section BC contains a solenoid field, section A is field free. The component $\vec{n}$ in (c) is the periodic solution.

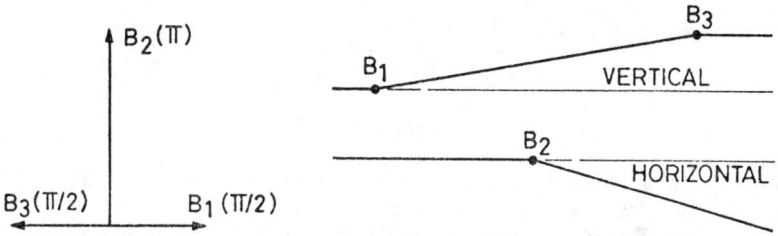

Fig. 2. Snake configuration for LEP; sequence of bending-field vectors and resultant orbit projections. The arguments in brackets are the corresponding precession angles.

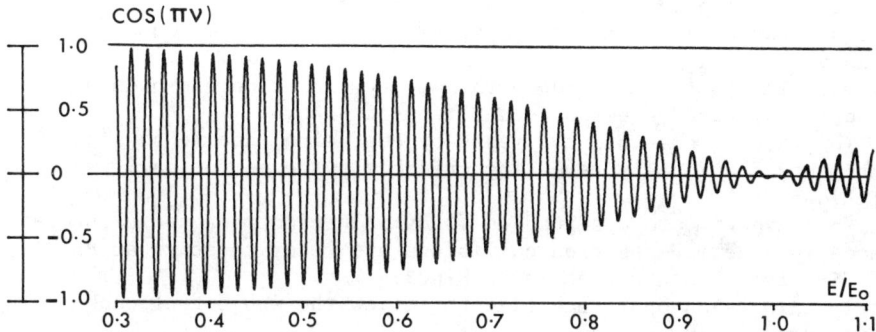

Fig. 3. Variation of precession frequency $\nu$ with normalized energy $E/E_o$.

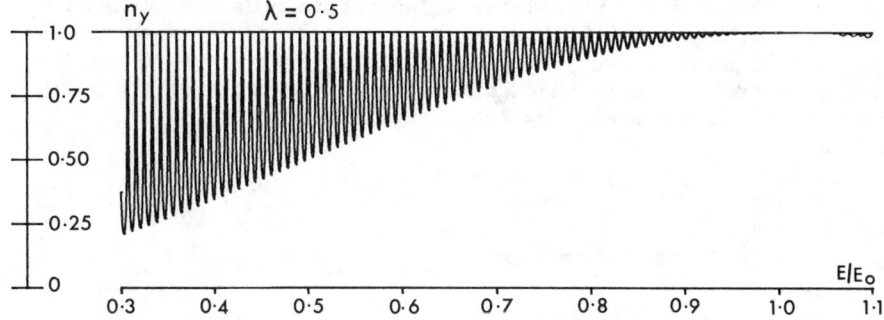

Fig. 4. Variation of longitudinal polarization $n_y$ opposite the snake as a function of normalized energy $E/E_o$.

WEAK-ELECTROMAGNETIC INTERFERENCE*

J. J. Sakurai
University of California, Los Angeles, CA 90024

I. INTRODUCTION

Until the spring of this year the weak interactions had been detected in only three classes of phenomena:

(i) The decays of metastable states--nuclei, leptons and hadrons--and closely related reactions such as muon capture by nuclei.

(ii) Neutrino-induced reactions at reactor and accelerator energies.

(iii) Parity violation in nuclear physics.

In recent months the situation has changed dramatically. We now have a fourth class of weak interaction phenomena:

(iv) Weak-electromagnetic interference.

There are three important developments relevant to the fourth category.

(a) Indisputable evidence for parity violation has been presented in polarized-electron scattering at SLAC.

(b) After more than two years of confusion, atomic physics experiments are finally revealing credible evidence for parity violation.

(c) Experiments designed to study weak-electromagnetic interference in electron-positron collisions are ready to start at PETRA. (If PETRA is on, can PEP be far behind?)

Clearly this is an opportune time to review the subject of weak-electromagnetic interference.

I hasten to add that even though this topic is only a few months old to many of us, the possibility of observing weak-electromagnetic interference has been contemplated by some theorists for nearly twenty years. As early as 1959, Zel'dovich[1] quite correctly pointed out that an asymmetry as large as $10^{-4}$ to $10^{-3}$ may be observable in the scattering of polarized electrons by protons at $q^2 \simeq m_p^2$ provided the weak interactions contain parity-violating neutral currents with strength similar to that of the charged currents. His argument is very simple; a weak asymmetry due to parity-violating neutral currents must go roughly as the ratio of the Fermi constant G to the one-photon exchange amplitude $e^2/q^2$,

$$A_{em} A_{weak} / [|A_{em}|^2 + |A_{weak}|^2] \simeq A_{weak}/A_{em} \sim G/(e^2/q^2) \ . \quad (1.1)$$

If we put the numbers in, we obtain

*Supported in part by the National Science Foundation (U.S.).

$$G/(e^2/q^2) = (10^{-5} \times 137/4\pi) \, q^2/m_p^2$$
$$\simeq 10^{-4} q^2/m_p^2 \,. \qquad (1.2)$$

See Fig. 1. Zel'dovich's expectation was based on a model of weak

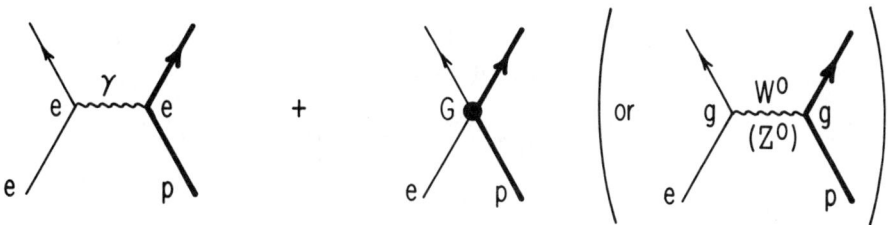

Fig. 1. Weak-electromagnetic interference in electron-proton scattering.

interactions where we make the V-A charged-current Lagrangian SU(2) symmetric by adding a V-A neutral-current piece. Such a model, considered somewhat earlier by Bludman,[2] gives essentially the same predictions at low energies as the now popular Weinberg-Salam model[3] with $\sin^2\theta_W$ set equal to zero. What is even more remarkable, Zel'dovich also discussed in the same paper the possibility of observing parity violation in atoms through optical rotation, an experiment which became feasible only with the advent of laser technology.

The plan of my talk is as follows. I start with a quick review of the rudiments of neutral-current physics; if you are already familiar with this subject you are advised to take a nap. This is followed by a detailed discussion on parity violation in atoms. I then discuss polarized-electron scattering with emphasis on phenomenological implications. Comparisons with gauge models are presented to show that the standard (Weinberg-Salam) model is just about the sole survivor. After that I discuss what we are likely to learn by studying weak interaction effects in electron-positron collisions at PETRA/PEP energies and beyond.

## II. RUDIMENTS OF NEUTRAL CURRENT PHYSICS

I would like to present a very elementary review of the basic theoretical framework needed to understand neutral-current phenomena. This part of my talk is intended for those who have been too busy to follow an extremely exciting chapter in particle physics which began with the Gargamelle discovery of neutral currents in 1973.[4]

First of all, why do we usually ignore the charged currents in discussing weak-electromagnetic interference? We consider only neutral currents because to obtain a sizable effect, namely of order G in the interference, we must have diagonal weak interactions like $(\bar{e}e)(\bar{u}u)$ where u stands for the isospin-up quark. In contrast, the charged-current interactions are off-diagonal, e.g. $(\bar{e}\nu)(\bar{u}d)$, so they do not interfere to lowest order with the diagonal electromagnetic interactions. In the weak boson language, we would need at least two $W^{\pm}$ exchange to get weak-electromagnetic interference if we just had charged currents.

Until the spring of this year weak neutral currents had been detected only in neutrino-induced reactions

$$\overset{(-)}{\nu} + N \rightarrow \overset{(-)}{\nu} + \text{hadron(s)} \;.$$
$$\overset{(-)}{\nu} + e^- \rightarrow \overset{(-)}{\nu} + e^- \;. \qquad (2.1)$$

These reactions probe the couplings of $(\bar{\nu}\nu)$ to $(\bar{q}q)$ with $q = u,d$ -- for all practical purposes s and c escape -- and of $(\bar{\nu}\nu)$ to $(\bar{e}e)$. In contrast neutral-current couplings not involving neutrinos -- $(\bar{e}e)(\bar{q}q)$ and $(\bar{e}e)(\bar{\mu}\mu)$ -- are my main concern today.

Traditionally there have been two approaches to neutral-current phenomenology -- model-independent and model-dependent. In the first model-independent approach we analyze various neutral-current data using the most general neutral-current interactions compatible with V,A structure. In doing so we do not assume that the neutral-current interactions are mediated by one or more $Z^o(W^o)$ bosons, nor do we start with any particular gauge group. Instead we simply write down completely phenomenological Lagrangians of the four-fermion type.

Let us count how many independent coupling parameters we have in this model-independent approach. Take $(\bar{\nu}\nu)(\bar{q}q)$; the quark current can be isovector $(\bar{u}u - \bar{d}d)$ or isoscalar $(\bar{u}u + \bar{d}d)$. In addition, its space-time (Lorentz) structure can be vector or axial vector. So ignoring $\bar{s}s$, $\bar{c}c$ etc. we have four constants to be determined:

$$\alpha: \text{ I=1 vector,} \quad \beta: \text{ I=1 axial-vector,}$$
$$\gamma: \text{ I=0 vector,} \quad \delta: \text{ I=0 axial-vector.} \qquad (2.2)$$

The $(\bar{\nu}\nu)(\bar{e}e)$ case is simpler; we just have two constants $g_V$ and $g_A$. Turning now to $(\bar{e}e)(\bar{q}q)$, we notice that only the parity-violating part is accessible to atomic parity experiments and electron scattering experiments at present energies. So, again there are four constants to be determined:

$$\tilde{\alpha}: \quad A_{lept}V_{quark}^{I=1}, \quad \tilde{\beta}: \quad V_{lept}A_{quark}^{I=1}$$
$$\tilde{\gamma}: \quad A_{lept}V_{quark}^{I=0}, \quad \tilde{\delta}: \quad V_{lept}A_{quark}^{I=0}, \quad (2.3)$$

where the subscript "lept" stands for e (or $\mu, \tau ...$). Finally $(\overline{ee})(\overline{\mu\mu})$ is characterized by three constants $h_{VV}$, $h_{AA}$ and $h_{VA}$ where the notation should be obvious.[5] So, omitting the couplings of $(\overline{ss})$, $(\overline{cc})$ etc. and also the parity-conserving part of $(\overline{ee})(\overline{qq})$, there are altogether 13 measurable constants in low $q^2$ neutral-current interactions.[6] If any of you are interested in the precise manner in which these constants are normalized, just consult Appendix.

From a purely phenomenological point of view the goal of low $q^2$ neutral-current physics is to determine uniquely and completely all these 13 constants. In Section V I'll show that we have made non-trivial advances in the past year towards achieving this goal.

Let us now turn to the model-dependent approach. Here we go to the opposite extreme and proceed with theoretically elegant models that make very specific predictions. As is well known, the most popular model along this line is the standard $SU(2) \otimes U(1)$ model à la Weinberg and Salam,[3] which has its origin in the 1961 paper of Glashow.[7,8] At the present moment this model is the embodiment of the QFD Orthodoxy.

The fundamental creed of the Orthodoxy goes as follows. The basic group of "electroweak" interactions is $SU(2) \otimes U(1)$. For the $U(1)$ part we associate a singlet (Abelian) gauge boson $B^0$ coupled to weak hypercharge $Y_W$ in much the same way as we have the photon coupled to electric charge in the "old," non-unified theory. For the $SU(2)$ part we associate a triplet of (non-Abelian) gauge bosons, $W^{\pm,0}$, coupled to weak isospin $\underline{T}$. The meaning of weak isospin and weak hypercharge satisfying a Gell-Mann-Nishijima-like relation

$$Q = T_3 + Y_W \quad (2.4)$$

should be apparent from Table 1 where the multiplet assignment of $\nu$ and e is given. Notice that $\nu_L$ and $e_L^-$ form a left-handed weak-isospin doublet while the right-handed electron appears as a singlet.

The gauge bosons are initially massless but when the gauge symmetry is broken spontaneously by the so-called "Higgs mechanism,"[9] which I won't explain here, we get massive $W^{\pm}$ that remain coupled to the charged currents corresponding to the weak-isospin lowering and raising operators $T_{\mp}$. As for the $W^0$-$B^0$ complex, one linear combination of $W^0$ and $B^0$, called $Z^0$,

$$Z_\lambda^0 = \cos\theta_W W_\lambda^0 - \sin\theta_W B_\lambda^0, \quad (2.5)$$

Table I. Multiplet assignment for the electron and neutrino

|  | Q | $T_3$ | $Y_W$ | |
|---|---|---|---|---|
| $\nu_L$ | 0 | $\frac{1}{2}$ | $-\frac{1}{2}$ | LH doublet |
| $e_L^-$ | $-1$ | $-\frac{1}{2}$ | $-\frac{1}{2}$ | |
| $e_R^-$ | $-1$ | 0 | $-1$ | RH singlet |

now also massive, is coupled to the current that corresponds to

$$T_3 - Q \sin^2\theta_W , \qquad (2.6)$$

while the other orthogonal linear combination

$$A_\lambda = \cos\theta_W B_\lambda^o + \sin\theta_W W_\lambda^o \qquad (2.7)$$

can be shown to remain massless and is, in fact, to be identified with the photon coupled to the usual electric charge.

The effective four-fermion-like charged-current interactions at low $q^2$ ($q^2 \ll m_W^2$) mediated by $W^\pm$ exchange are characterized by

$$G/\sqrt{2} = g^2/8m_W^2 \qquad (2.8)$$

where G is Fermi's G ($\simeq 10^{-5}/m_p^2$) and g stands for the dimensionless coupling constant of the W triplet to the weak isospin. The neutral-current interactions are mediated by the $Z^o$ boson which becomes more massive than the $W^\pm$ boson due to $W^o$-$B^o$ mixing; however, the dimensionless coupling constant of the Z to $T_3 - Q \sin^2\theta_W$ turns out to be also larger by a factor of $1/\cos\theta_W$. If the symmetry breakdown à la Higgs is due to a scalar field that transforms like a weak isospin doublet, we can derive

$$m_W^2/m_Z^2 \cos^2\theta_W = 1 , \qquad (2.9)$$

so that we have effective neutral-current couplings of the four-fermion type with strength again given by

$$(g^2/\cos^2\theta_W)/8m_Z^2 = g^2/8m_W^2 = G/\sqrt{2} . \qquad (2.10)$$

The parameter $\sin\theta_W$ that appears in (2.5) and (2.6) is also related to the ratio of the residues at the photon pole and the $W^\pm$ pole as follows:

$$e^2/g^2 = \sin^2\theta_W . \qquad (2.11)$$

Using this relation together with (2.8) and (2.9), we obtain the celebrated mass formulas of Weinberg[3]

$$m_W = \sqrt{\sqrt{2}\, e^2/8G}\Big/\sin\theta_W \simeq 37.4 \text{ GeV}/\sin\theta_W ,$$
$$m_Z = m_W/\cos\theta_W \simeq 75 \text{ GeV}/\sin 2\theta_W . \qquad (2.12)$$

A realistic model of the world must also specify how fermions other than $e^-$ and $\nu$ transform under $SU(2) \otimes U(1)$. The simplest (hence least imaginative) possibility is to repeat the pattern shown in Table I.[10] We then have left-handed doublets

$$\begin{pmatrix}\nu_e \\ e^-\end{pmatrix}_L, \begin{pmatrix}\nu_\mu \\ \mu^-\end{pmatrix}_L, \begin{pmatrix}\nu_\tau \\ \tau^-\end{pmatrix}_L ; \begin{pmatrix}u \\ d'\end{pmatrix}_L, \begin{pmatrix}c \\ s'\end{pmatrix}_L, \begin{pmatrix}t \\ b\end{pmatrix}_L \qquad (2.13)$$

and right-handed singlets

$$e_R^-, \mu_R^-, \tau_R^- ; u_R, d_R, c_R, s_R, b_R, t_R . \qquad (2.14)$$

With this multiplet assignment there is very little arbitrariness in the model. In particular the 13 phenomenological coupling parameters I defined earlier are completely determined once $\sin^2\theta_W$, the single parameter in the model, is given. More on this in Section V.

In the literature there have been many attempts to build models which are essentially variations on the theme by Glashow, Weinberg and Salam.[11] We may stay within the same group $SU(2) \otimes U(1)$ and introduce more complicated multiplets such as

$$\begin{pmatrix}E^0 \\ e^-\end{pmatrix}_R, \begin{pmatrix}u \\ b\end{pmatrix}_R , \qquad (2.15)$$

or, alternatively, we may consider more complicated groups such as $SU(3) \otimes U(1)$ and $SU(2) \otimes SU(2) \otimes U(1)$ which require, among other things, more than one $Z^0$ boson. It is fair to say that such attempts were at least partially motivated by experiments which have subsequently been discredited.

## III. PARITY VIOLATION IN ATOMS

How large an effect do we expect for parity violation in atoms? At first it may appear that parity-violation experiments are hopeless because the momentum transfer involved is so tiny. From my earlier discussion weak-electromagnetic interference effects are of order $10^{-4}q^2$ where $q^2$ is in GeV$^2$ but in atomic physics $\sqrt{q^2}$ is like the inverse of atomic dimension $R_{atom} \sim 137/m_e$; substituting $q^2 \sim \alpha^2 m_e^2$, we get $\sim 10^{-15}$, indeed a very small number. So, to be able to detect anything, we have to be extraordinarily clever.

Let us examine how parity-violation actually arises in atomic transitions. If there are parity-violating interactions between the nucleon and the electron, atomic levels are expected to contain small opposite-parity components. To estimate the admixture we first derive an effective parity-violating potential between the atomic electrons and the nucleus. Parity violation due to interactions among the atomic electrons can be shown to be negligible in comparison.

Nuclei are made up of nucleons, and nucleons are believed to be made up of quarks. So we construct an effective potential based on the fundamental interaction $(\bar{e}e)(\bar{q}q)$. As already mentioned, the parity-violating part of this interaction is characterized by four coupling parameters, $\tilde{\alpha}$, $\tilde{\beta}$, $\tilde{\gamma}$ and $\tilde{\delta}$.

The fact that the nucleons are slow implies that for the $A_{lept} V_{quark}^{I=1,0}$ interaction (the $\tilde{\alpha}$ and $\tilde{\gamma}$ terms) only the time component of $V_{quark}$ is important; likewise, for the $V_{lept} A_{quark}^{I=1,0}$ interaction (the $\tilde{\beta}$ and $\tilde{\delta}$ terms), only the space components are important. All this is familiar from nuclear beta decay. However, unlike nuclear beta decay there is coherence for the vector part because the "charges" are additive. The matrix element of the time component of $V_{quark}^{I=1,0}$ goes like the number of nucleons; more precisely it is given by the "weak charge" $Q_W$:

$$Q_W(Z,N) = -[\tilde{\alpha}(Z-N) + 3\tilde{\gamma}(Z+N)] \qquad (3.1)$$

where Z and N stand for the numbers of protons and neutrons, respectively. In the standard model (3.1) can be written as

$$Q_W(Z,N)\Big|_{W-S} = -[Z(4\sin^2\theta_W - 1) + N] . \qquad (3.2)$$

In contrast, because the nuclear spins tend to cancel, only the nucleons outside the closed shell contribute to the matrix elements of $A_{quark}^{I=1,0}$; so for heavy atoms the $V_{lept} A_{quark}^{I=1,0}$ interactions are relatively less important. As for the lepton side, we note, for example, that the time component of the axial-vector current gives

$$\bar{e}\gamma_4\gamma_5 e \to -(\vec{p}+\vec{p}')\cdot\vec{\sigma}_e/2m_e \tag{3.3}$$

in the plane-wave representation. Putting everything together we obtain a short-ranged parity-violating potential between the electron and the nucleus, first derived by Bouchiat and Bouchiat:[12]

$$\begin{aligned}H_{PV} &= (G/4\sqrt{2})\, Q_W\, \vec{\sigma}_e \cdot \{\delta^{(3)}(\vec{x}), \vec{P}_e/m_e\} \\ &+ (G/2\sqrt{2})[\{\vec{P}_e/m_e, \delta^{(3)}(\vec{x})\} - i[\delta^{(3)}(\vec{x}), (\vec{\sigma}_e \times \vec{P}_e)/m_e]] \\ &\times [(1.25\tilde{\beta} + 0.75\tilde{\delta})\vec{S}_p + (-1.25\tilde{\beta} + 0.75\tilde{\delta})\vec{S}_n]\end{aligned} \tag{3.4}$$

with $Q_W$ given as in (3.2).[13] I may mention that the nonrelativistic reduction I made is for illustration only. In actual calculations we must use potentials appropriate for the Dirac electron.

Because of $H_{PV}$, an atomic level is no longer a pure eigenstate of parity. Consider for definiteness a one-valence-electron atom in a $P_{1/2}$ state with principal quantum number n. Due to parity mixing this $P_{1/2}$ state actually contains $S_{1/2}$ components as follows:

$$|n\ ^2P_{1/2}\rangle' = |n\ ^2P_{1/2}\rangle + \sum_{n'} \varepsilon(nP, n'S)|n'\ ^2S_{1/2}\rangle \tag{3.5}$$

where the parity-mixing coefficients $\varepsilon(nP, n'S)$ are given by

$$\varepsilon(nP, n'S) = \langle n'\ ^2S_{1/2}|H_{PV}|n\ ^2P_{1/2}\rangle/(E_{nP_{1/2}} - E_{n'S_{1/2}}). \tag{3.6}$$

Consequently, a radiative transition between two atomic levels, for example, the M1 transition between $7P_{1/2}$ and $6P_{1/2}$ of the thallium atom---acquires a very small amount of an opposite-parity E1 component. To detect parity nonconservation in such a transition we look for the dependence of observable quantities on some pseudoscalars, e.g. the photon helicity (circular polarization) $\vec{S}_{photon} \cdot \vec{P}_{photon}$.

Having sketched the basic physics involved, let me now discuss how to enhance possible parity-violation effects. There are essentially three different ways:

(i) Work with heavy atoms. This helps because the matrix element of $H_{PV}$ due to $A_{lept}V_{quark}$ is proportional to $Z^3$; one factor of Z from the electron velocity [$\vec{p}_e/m_e$ in (3.4)], another factor of Z from $|\psi(0)|^2$, and yet another factor due to the coherence of $Q_W$ [see (3.1)]. Note $Z^3 \simeq 5.3 \times 10^5$ for thallium.

(ii) Study a transition where the main (parity-conserving) matrix element is highly suppressed. For instance, we may work with an M1 transition with a change in n, for which the matrix element vanishes in the nonrelativistic approximation due to orthogonality of the initial and final wave functions. This is helpful because parity-violating asymmetries go like $E_{PV}/M$ where $E_{PV}$ and $M$ stand for the parity-violating E1 and the parity-conserving M1 matrix element, respectively; obviously it pays to make the denominator small.

(iii) Make the energy denominator that appears in the expression for the mixing parameter (3.6) small. This can be accomplished by choosing, to start with, a level with an almost degenerate opposite-parity level nearby and adjusting the hyperfine splittings by an external magnetic field until a pair of the opposite parity levels actually cross.

The importance of (i) and (ii) was first emphasized by Bouchiat and Bouchiat.[12] The famous bismuth (Bi) experiments and other heavy atoms--cesium (Cs) and thallium (Tℓ)--take advantage of (i); the Cs and Tℓ experiments carried out at the Ecole Normale Supérieure and Berkeley, respectively, rely, in addition, on (ii). The key to Z = 1 hydrogen and deuterium experiments proposed by Lewis and Williams and others[14] is precisely (iii).

Conceptually the simplest atomic-parity experiment is of the kind conceived by Bouchiat and Bouchiat schematically shown in Fig. 2. Atoms--in practice metallic vapor--are irradiated by a

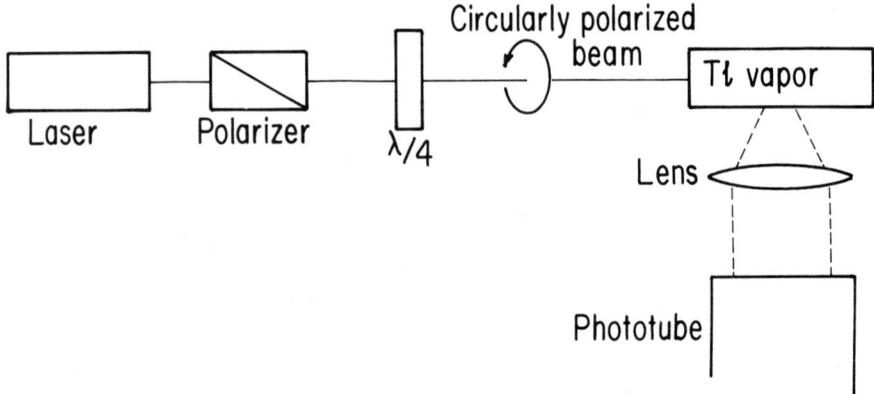

Fig. 2. Circular dichroism experiment.

circularly polarized laser beam tuned to some particular radiative transition satisfying (i) and (ii) above. We then measure the photon absorption cross section by detecting resonance fluorence due to the decay of the level excited by the laser beam. A nonvanishing value of "circular dichroism"

$$A = [\sigma(+) - \sigma(-)]/[\sigma(+) + \sigma(-)] \tag{3.7}$$

where + and – stands for photon helicity positive and negative (or right and left circular polarization) would signal conclusive evidence for parity violation in atoms, independently of any detailed theory. For an absorption line which would be pure M1 in the absence of parity violation, the circular dichroism A directly measures $\text{Im}(E_{PV})/M$:

$$A = 2\,\text{Im}(E_{PV}M)/(|M|^2 + |E_{PV}|^2)$$

$$\simeq 2\,\text{Im}(E_{PV})/M \ . \tag{3.8}$$

Of course, it is a nontrivial task to relate the parity-violating matrix element $E_{PV}$ to $Q_W$. I'll come back to this point later but here I simply mention that A is predicted on the basis of the standard model to be of the order of $10^{-4}$ to $10^{-3}$ in the transitions involved in the Cs and Tℓ experiments.

The Tℓ experiment just completed by Commins and coworkers[15] at Berkeley is a slight variation on the original Bouchiat-Bouchiat proposal. Let us look at the energy level diagram (Fig. 3) of the Tℓ atom which has one (6p) electron outside a core made up of 80 electrons. The original scheme, as applied to this atom, would be to detect fluorescence at the 535 nm line ($7\,^2S_{1/2} \to 6\,^2P_{3/2}$) following the absorption of 292.7 nm photons ($6\,^2P_{1/2} \to 7\,^2P_{1/2}$) provided by a circularly polarized laser beam. According to Commins that did not work; before the experiment was successfully completed, several modifications had to be made. In the final arrangement the Tℓ vapor is placed in a region with a static electric field which gives, on top of parity mixing due to genuine parity violation, a <u>controlled</u> parity mixing à la Stark; the Tℓ vapor is irradiated by another (ultra-violet) laser beam that takes the upper $7\,^2P_{1/2}$ state to yet another higher state, $8\,^2S_{1/2}$; the fluorescence actually detected

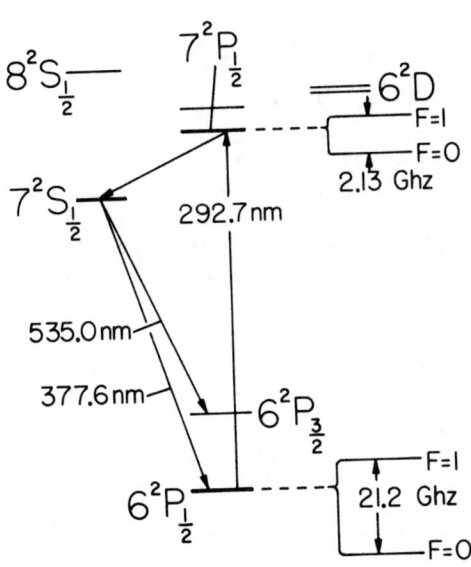

Fig. 3. The level diagram of the Tℓ atom.

is $8\,{}^2S_{1/2} \to 6\,{}^2P_{3/2}$ rather than $7\,{}^2S_{1/2} \to 6\,{}^2P_{3/2}$. Yet the basic principle is still the same; one measures $\text{Im}(E_{PV})/M$ by detecting the dependence of resonance fluorescence on the circular polarization of laser beam tuned to $6\,{}^2P_{1/2} \to 7\,{}^2P_{1/2}$.

As of this week, the experimental result Commins wishes to be quoted is[15]

$$Q_W = -280 \pm 140 \qquad (3.9)$$

where we expect $Q_W = -123$ on the basis of the standard model with $\sin^2\theta_W = 1/4$. The error shown includes theoretical errors due to atomic physics uncertainties; the actual parity-violation effect is established to more than 2 standard deviations.[16]

To sum up there is now credible evidence for parity violation in atoms. When compared to the standard-model prediction, the observed effect has the correct sign with about the correct order of magnitude. The quoted result, if anything, appears to be a little too large.

A few words on the atomic physics calculations used to interpret the results. Let us first recall that the Tℓ atom has 81 electrons as follows:

$$(1s)^2\,(2s)^2\,(2p)^6\,(3s)^2\,(3p)^6\,(4s)^2\,(3d)^{10}\,(4p)^6$$
$$(5s)^2\,(4d)^{10}\,(5p)^6\,(6s)^2\,(4f)^{14}\,(5d)^{10}\,(6p)^1\,. \qquad (3.10)$$

In other words it is made up of a spherically symmetric core consisting of 80 electrons plus one valence electron. So this atom is somewhat simpler than Bi which has three (6p) electrons outside its core. The circular dichroism A for the 292.7 nm line was calculated by Neufer and Commins[17] using the wave functions based on a "Tietz potential" of the form

$$V(r) = \{-e^2(Z-1)/[\gamma(1+\eta r)^2]\}\exp(-\gamma r) - e^2/r \qquad (3.11)$$

with $\eta$ and $\gamma$ determined from the observed 6P and 7P levels. This potential is known to account for other energy levels, hyperfine splittings, $g_J$ factors (Zeeman splittings), E1 rates, and Stark mixing parameters. We may therefore have some confidence in the atomic physics calculations needed to interpret the experimental results.

Let us now turn to the "other" atomic experiments. By this time almost everybody has heard about various Bi experiments where attempts are made to detect rotation of the plane polarization as a linearly polarized laser beam goes through Bi vapor. In the Berkeley Tℓ experiment, we measure the dependence of the absorption cross section on the incident photon helicity; in the langauge of optics we are studying the imaginary part of the index of refraction $n_\pm$, related to the forward elastic photon scattering amplitude $f_\pm(\omega)$ via the Lorentz relation

$$n_\pm = 1 + (2\pi/\omega^2) \, N f_\pm(\omega) \qquad (3.12)$$

where $\pm$ stands for the photon helicity. In contrast, in the Bi (optical rotation) experiments, conceived by Khriplovich and others,[18] we study the real part of the index of refraction; more quantitatively, the rotation angle is related to $n_\pm$ as follows:

$$\delta\theta = \frac{1}{2} \, \mathrm{Re}(n_+ - n_-) \, k\ell \, , \qquad (3.13)$$

which is expected to have a dispersive shape near the resonance peak. Perhaps an analogy with neutral K meson physics may be helpful here; recall how a $K_L$ beam going through matter with $\sigma_{tot}(K^0 N) \neq \sigma_{tot}(\bar{K}^0 N)$ acquires a regenerated $K_S$ component in the forward direction. The circular polarization states are analogous to the strangeness eigenstates $K^0$ and $\bar{K}^0$ while the linear polarization states are analogous to $K_L$ and $K_S$.[19]

I won't describe the details of the various Bi experiments performed so far. I simply present here a table of theoretical and experimental values of $\mathrm{Im}(E_{PV})/M$ (which can be shown to be twice the rotation angle at a dispersion peak for density corresponding to one unit of absorption length), as summarized at the Tokyo Conference:[20]

Table II: Summary of the Bi (optical rotation) experiments

| Experiment | Transition | $\mathrm{Im}(E_{PV})/M$ $(10^{-8})$ | Theoretical prediction $(10^{-8})$ |
|---|---|---|---|
| Seattle | 876 nm | $-0.5 \pm 1.7$ | $-10$ to $-18$ |
| Oxford | 648 nm | $-5 \pm 1.6 \pm ?$ | $-13$ to $-23$ |
| Novosibirsk | 648 nm | $-19 \pm 5$ | $-13$ to $-23$ |

The most obvious conclusion here is: "Draw your own conclusion." It appears that we have to be prepared to accept one of the following:

(i) The $T\ell$ experiment and the Novosibirsk Bi experiment are wrong.

(ii) The null Bi results are wrong.

(iii) The atomic physics calculations are much less reliable for Bi than for $T\ell$.

I have no time to discuss the hydrogen and deuterium experiments now in progress in various parts of the world--by Williams (Michigan), Hughes (Yale), Fortson (Seattle), Sandars (Oxford), Telegdi (ETH) ... . These experiments appear even more difficult but they are of extreme importance because, in principle, they could separately determine all four constants $\alpha$, $\beta$, $\gamma$ and $\delta$, not just one linear combination of $\alpha$ and $\tilde{\gamma}$.[21] Furthermore the atomic physics calculations are absolutely clean here. So good luck to all those who are involved in these formidable experiments![22]

## IV. PARITY VIOLATION IN POLARIZED ELECTRON SCATTERING

To high-brow theorists a transformation that mixes integer-spin bosons and half-integer-spin fermions is known as a "supersymmetry" transformation. To go from the T$\ell$ circular dichroism experiment to the now famous SLAC polarized-electron experiment we resort to such a transformation:

circularly polarized photons → longitudinally polarized electrons

$$\lambda = \pm 1 \rightarrow \lambda = \pm 1/2 ,$$

T$\ell$ vapor → liquid deuterium,

Commins ... → Prescott ..., (4.1)

Berkeley → SLAC.

More importantly, we must also perform scale transformations:

$$4.2 \text{ eV} \rightarrow 19.4 \text{ GeV} ,$$

$$3 \text{ meters} \rightarrow 2 \text{ miles}, \quad (4.2)$$

$$\$? \rightarrow \$?$$

Just as the Berkeley T$\ell$ experiment studies how the intensity of inelastically scattered photons depends on the circular polarization of the incident laser beam, the SLAC electron experiment studies how the intensity of inelastically scattered electrons depends on the helicity of the incident electrons. In other words this is a "single-arm" experiment

$$e^-(\lambda = \pm 1/2) + D \rightarrow e^- + \text{any} \quad (4.3)$$

where the final electron is detected at some definite angle $\theta$ and energy $E'$. The polarization asymmetry

$$A \equiv \frac{d^2\sigma(+)/d\Omega dE' - d^2\sigma(-)/d\Omega dE'}{d^2\sigma(+)/d\Omega dE' + d^2\sigma(-)/d\Omega dE'}, \quad (4.4)$$

measured is completely analogous to the circular dichroism (3.7).

About four months ago a dramatic announcement was made at a SLAC colloquium given by C. Prescott in the Taylor Group:[23]

$$A/q^2 = -(9.5 \pm 1.6) \times 10^{-5} \text{GeV}^{-2} \quad (4.5)$$

where the bulk of the data are taken under the kinematical conditions

$$E = 19.4 \text{ GeV}, \quad \theta = 4°, \quad q^2 \simeq 1.6 \text{ GeV}^2,$$
$$x \equiv q^2/[2m(E-E')] \simeq 0.2, \quad y \equiv (E-E')/E \simeq 0.21. \quad (4.6)$$

The experiment, as discussed elsewhere at this Symposium,[24] is absolutely beautiful; parity violation in the electron-nucleon interactions is established beyond any shadow of doubt. Furthermore the sign and magnitude of A are precisely what we expect on the basis of the standard model, as I'll show shortly.

What does the experiment tell us about the phenomenological parameters $\tilde{\alpha}$, $\tilde{\beta}$, $\tilde{\gamma}$ and $\tilde{\delta}$? Using Bjorken scaling, $m_Z^2 \gg q^2$ and $\sigma_S/\sigma_T \simeq 0$, we can write the asymmetry A as

$$A(x,y,q^2) = [a_1(x) + a_2(x)\{[1 - (1-y)^2]/[1 + (1-y)^2]\}]q^2. \quad (4.7)$$

The two independent asymmetry functions $a_1(x)$ and $a_2(x)$ receive contributions from $A_{lept}V_{quark}$ and $V_{lept}A_{quark}$, respectively, and can be expressed in terms of the coupling parameters as follows:

$$a_1 = (G/\sqrt{2}\ e^2)(9\tilde{\alpha} + 3\tilde{\gamma})/5,$$
$$a_2 = (G/\sqrt{2}\ e^2)(9\tilde{\beta} + 3\tilde{\delta})/5, \quad (4.8)$$

independent of x for the deuteron target. These formulas were originally derived within the framework of quark parton models; they are gauge-model-independent ways of writing formulas found in the papers of Cahn, Gilman, Yoshimura,[25] etc. who expressed $a_1$ and $a_2$ in terms of the weak isospin components of quarks and leptons. It has subsequently been pointed out that they are relatively insensitive to the details of quark parton models.[26] Notice that by studying the y dependence it is possible to separate out the $V_{lept}A_{quark}$ contribution ($\tilde{\beta}$ and $\tilde{\delta}$) from the $A_{lept}V_{quark}$ contribution ($\tilde{\alpha}$ and $\tilde{\gamma}$). Numerically the basic scale is set by

$$G/\sqrt{2}\ e^2 = G/4\pi\sqrt{2}\ \alpha \simeq 8.94 \times 10^{-5} \text{GeV}^{-2} \ . \tag{4.9}$$

The published SLAC experiment measured A for just one value of y; so only one linear combination of the four constants can be extracted from the data:

$$\left(\tilde{\alpha} + \frac{1}{3}\tilde{\gamma}\right) + 0.23\left(\tilde{\beta} + \frac{1}{3}\tilde{\delta}\right) = -0.59 \pm 0.10 \ . \tag{4.10}$$

A new run, starting just this week (?), attempts to separate out $\tilde{\alpha} + \frac{1}{3}\tilde{\gamma}$ from $\tilde{\beta} + \frac{1}{3}\tilde{\delta}$ by measuring A at different values of y, from $y \simeq 0.15$ to $y \simeq 0.45$.

The standard (Weinberg-Salam) model makes very specific predictions on the coupling parameters:

$$\tilde{\alpha} = -(1 - 2\sin^2\theta_W), \quad \tilde{\gamma} = \frac{2}{3}\sin^2\theta_W \ ,$$
$$\tilde{\beta} = -(1 - 4\sin^2\theta_W), \quad \tilde{\delta} = 0 \ . \tag{4.11}$$

Using (4.10) and (4.11), we can determine $\sin^2\theta_W$, the only adjustable parameter of the model; we obtain

$$\sin^2\theta_W = 0.20 \pm 0.03 \tag{4.12}$$

in agreement with determination based on neutrino-induced reactions; more on this in Section V. This value of $\sin^2\theta_W$ makes a very specific prediction for the y dependence; in terms of $a_2/a_1$, which is the same as the slope-to-intercept ratio at y = 0, we must have

$$a_2/a_1 \simeq 0.36 \pm 0.23 \ , \tag{4.13}$$

which implies a gentle y dependence with $a_1$ and $a_2$ both negative. This is also apparent from Fig. 4, taken from the SLAC paper.[24] We'll soon find out whether this prediction is fulfilled.

I now discuss information obtained by combining the polarized-electron results with the heavy-atom results. Let us first note that the SLAC eD experiment and the Bi and Tℓ experiments measure different linear combinations of the coupling parameters. Once the y dependence is determined, we can deduce from the SLAC data $3\tilde{\alpha} + \tilde{\gamma}$ while a measurement of the weak charge $Q_W$ in the heavy-atom experiments determines $-0.07\tilde{\alpha} + \tilde{\gamma}$. In other words the two types of experiments study almost orthogonal linear combinations in an $\tilde{\alpha}$-$\tilde{\gamma}$ plane. This should be apparent from Fig. 5 where I plot the constraints from the SLAC eD and Berkeley Tℓ results together with strictly null Bi results.[27] In the same figure I indicate which region of the $\tilde{\alpha}$-$\tilde{\gamma}$ plane is allowed by the single Z boson

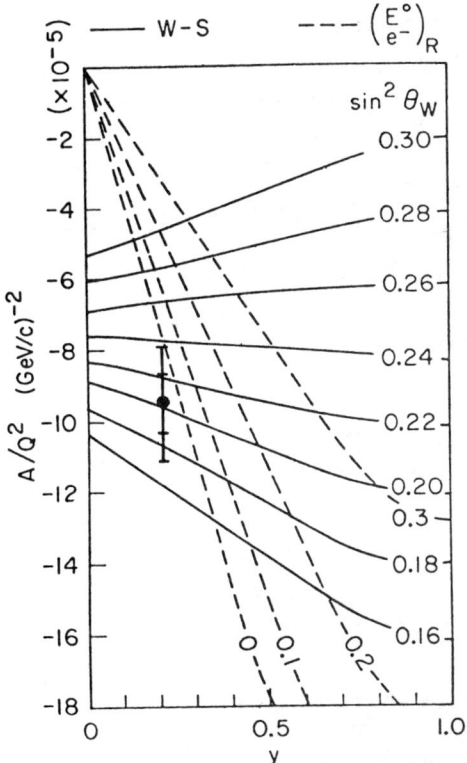

Fig. 4. The polarization asymmetry in electron-deuteron scattering as a function of y.

(or factorization) hypothesis where the basic input is phenomological analyses of various neutrino-induced reactions;[28] the indicated region is obtained by assuming that the isoscalar-to-isovector ratio in the vector part of the quark current measured in the electron-hadron interactions (polarized eD, Bi, Tℓ) is the same as the corresponding ratio measured in the neutrino-hadron interactions:

$$\tilde{\gamma}/\tilde{\alpha} = \gamma/\alpha , \qquad (4.14)$$

which follows from a rather weak form of factorization.[6] Clearly, as long as the y dependence in the SLAC experiment is not too steep, say $|a_2/a_1| \lesssim 5$, the null Bi results cannot be reconciled with the SLAC eD experiment within the framework of <u>any</u> model with just a single Z boson. This far-reaching conclusion is completely independent of gauge-model considerations.

It is in principle possible to determine $\alpha$ and $\gamma$ separately from polarized electron scattering <u>alone</u>. Just use proton targets where the ratio of the u quark distribution to the d quark distribution are believed to be very different for large x.[25,27] With ep we can determine a different linear combination of $\tilde{\alpha}$ and $\tilde{\gamma}$; for instance,

$$[2a_1^{(ep)}\sigma_{ep} - a_1^{(eD)}\sigma_{eD}]/[2\sigma_{ep} - \sigma_{eD}] = \tilde{\alpha} + 3\tilde{\gamma} , \qquad (4.15)$$

again independent of x.[29] However, to get nontrivial results it is essential to measure the left-hand side of (4.15) for x close to 1 where the u quark and d quark distributions for the proton are substantially different. Otherwise we are liable to obtain 0/0. The ep asymmetry reported in the same SLAC paper[24] unfortunately does not add any new information because both the eD and ep measurements were taken at $x \simeq 0.2$.

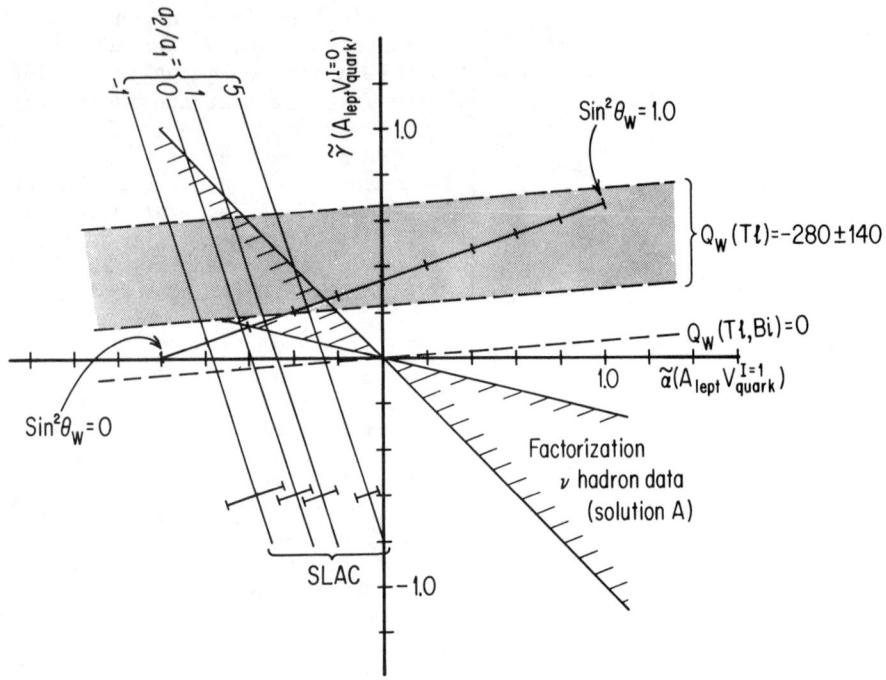

Fig. 5. Constraints from the polarized eD and heavy-atom experiments.

I now would like to relate the eD asymmetry to yet another process, $\overset{(-)}{\nu}e$ scattering. With the single Z hypothesis (factorization) it is possible to express the coupling constant ratio $g_V/g_A$ in $\overset{(-)}{\nu}e$ scattering in terms of the slope-to-intercept ratio in the eD asymmetry experiment and the $\nu$ quark coupling constants:

$$g_V/g_A = [(3\alpha+\gamma)/(3\beta+\delta)] \, a_2/a_1 \simeq (0.5 \pm 0.2) \, a_2/a_1 \, . \quad (4.16)$$

Phenomenological studies of $\nu_\mu e$, $\bar{\nu}_\mu e$ and $\bar{\nu}_e e$ scattering have left us with a two-fold ambiguity in the coupling constants $g_V$ and $g_A$:[30]

axial-vector dominant solution: $|g_V| \lesssim 0.15$, $g_A \simeq -0.55\pm0.10$ ,

vector dominant solution: $g_V \simeq -0.55\pm0.10$, $|g_A| \lesssim 0.15$ .

(4.17)

We see that unless $a_2/a_1$ turns out to be rather large, say $a_2/a_1 \gtrsim 5$, we must reject the vector dominant solution, at least within the framework of single Z boson models.[31]

## V. COMPARISONS WITH GAUGE MODELS

Prior to the atomic parity and polarized electron experiments there were many gauge models all alleged to be viable as far as the neutrino-induced reactions are concerned. With the null Bi results the following classes of models became very popular--a class of models with parity-conserving neutral current couplings and a class of models with no couplings of the form $A_{lept}V_{quark}$. I describe here just one model each from the two classes above.

Even though inelastic and elastic neutrino-nucleon scattering have taught us that both V and A are present for the hadronic part of the neutral current,[32] this does not necessarily imply parity violation; for one thing, no pseudoscalar quantities that would define "handedness" are measured in such experiments. Indeed, with a large electroweak interaction group such as $SU(2) \otimes SU(2) \otimes U(1)$ which accommodates two $Z^0$ bosons, parity conservation can be quite compatible with the coexistence of V and A. All we have to do is let one of the Z bosons couple to a pure vector current, the other to a pure axial-vector current; the effective interaction would then be of the parity-conserving VV + AA form. The parameters of the VV + AA model can be arranged so that, as far as neutrino-induced reactions at present energies are concerned, its predictions agree exactly with the standard $SU(2) \otimes U(1)$ model;[33] the model, at the same time, "explained" the null Bi experiments. However, the VV + AA model is now convincingly ruled out by the SLAC polarized-electron experiment and also by the Berkeley Tℓ and the Novosibirsk Bi experiment.

Another popular model prior to the SLAC experiment was the so-called "hybrid model" based on $(E^0, e^-)_R$.[34] Here the basic electroweak interaction group is taken to be $SU(2) \otimes U(1)$, just as in the standard model, but the right-handed electron forms a weak-isospin doublet together with a yet-to-be-discovered neutral heavy lepton $E^0$; the $(E^0, e^-)_R$ doublet is assumed to enter symmetrically with the $(\nu_e, e^-)_L$ doublet so that the electron part of the neutral current is pure vector. [Recall $\frac{1}{2}\gamma_\lambda(1+\gamma_5) + \frac{1}{2}\gamma_\lambda(1-\gamma_5) = \gamma_\lambda$.] In the ν-quark sector the usual (standard) multiplet assignment (no right-handed doublets) may be assumed; that is why the model is called the hybrid model. It is easy to show that the hybrid-model predictions for the neutrino-hadron interactions are completely identical to the standard-model predictions with the same value of $\sin^2\theta_W$. The predictions for $\overset{(-)}{\nu_\mu}e$ and $\overset{(-)}{\nu_{e_2}}e$ scattering with $m_e$ ignored can also be shown to be the same for $\sin^2\theta_W = 1/4$; phenomenologically this model chooses the vector dominant rather than the axial-vector dominant solution in (4.17). In the electron-quark sector there is no $A_{lept}V_{quark}$ interaction, so essentially null results are predicted for <u>heavy</u> atom experiments. The cleanest way to test this model, which requires $a_1$ in (4.7) to vanish, is to study the y dependence of the SLAC asymmetry to see whether it extrapolates to 0 as $y \to 0$. Such a test is being carried out in the new SLAC run. Actually the published measurement at y ≃ 0.21 already provides strong evidence against this hybrid model. If we try to determine $\sin^2\theta_W$ using

$$\tilde{\alpha} = \tilde{\gamma} = \tilde{\delta} = 0 ,$$
$$\tilde{\beta} = -2 + 4\sin^2\theta_W \quad (5.1)$$

predicted by the model, we obtain an unphysical value

$$\sin^2\theta_W = -0.14 \pm 0.11 , \quad (5.2)$$

totally in disagreement with $\sin^2\theta_W$ determined from the neutrino-induced reactions. This is also apparent from Fig. 4.

There are many other models ruled out by the SLAC experiment. In fact, it is fair to say that all the models that used to explain the null Bi experiments prior to June, 1978, have been brutally killed by Prescott et al.[24] For this reason the polarized-electron experiment is known as the 1978 SLAC Massacre. It essentially put an end to "model building industry." One could still argue that the Bi experiments and the SLAC experiment measure different quantities and try constructing a complicated (artificial?) gauge scheme with more than one Z boson to explain both the null Bi results and the SLAC data but most theorists now believe that with the Berkeley Tℓ result, even the last remaining cloud has been cleared away in favor of simplicity.

It appears from the foregoing discussion that out of ~50 models that have been contemplated by model builders, the standard (Weinberg-Salam) model--or, more precisely, models that give the same low-energy predictions as the standard model--is the sole survivor. So, let us scrutinize its predictions a little more closely. This time we look also at the results of phenomenological analyses of various neutrino-induced reactions to examine whether the standard model gives a unified and coherent picture of all neutral-current phenomena, not just weak-electromagnetic interference.

The most traditional way to test the Weinberg-Salam model is to determine $\sin^2\theta_W$, the only parameter of the model, in many different ways. We can then ask whether we always obtain the same number within errors. This was most recently attempted at the Tokyo Conference two months ago by Baltay[35] who came up with Fig. 6. You see that Baltay's 9 ways to determine $\sin^2\theta_W$ indeed gives a consistent value around $\sin^2\theta_W = 0.23$.

Impressive as Baltay's figure may appear, a sceptic may still raise the following objection. When we determine $\sin^2\theta_W$ from a given experiment, one is implicitly assuming much of the basic structure of the model. For instance, in $\overset{(-)}{\nu}e$ scattering we determine $\sin^2\theta_W$ under the assumption that $g_A$ has a pre-assigned value, viz. -1/2; in deducing $\sin\theta_W$ from $\nu$-induced hadron reactions we are working

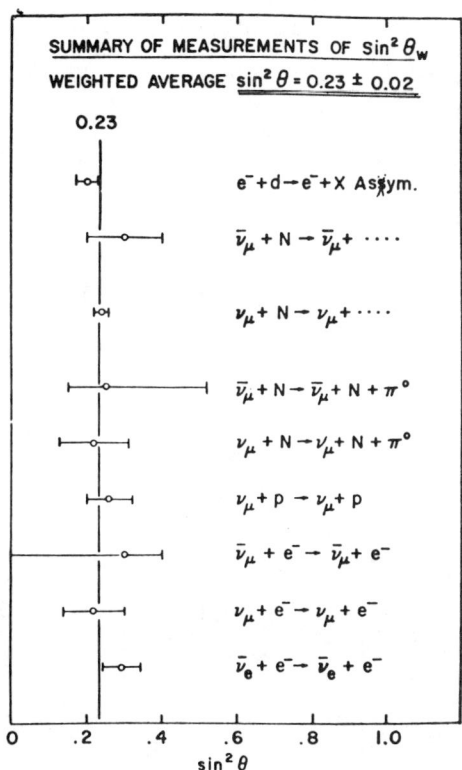

Fig. 6. Determination of $\sin^2\theta_W$ à la Baltay.

within the constraint that the isoscalar axial-vector current made up of ordinary quarks is absent etc. Even a more powerful and objective way of confronting the standard model (or, for that matter, any model) with the experimental data goes as follows. We first attempt to obtain each and every one of the 13 coupling parameters discussed in Section II in a model-independent way--i.e. without gauge-model constraints--and then and only then do we attempt to compare the phenomenologically determined parameters with the standard model predictions. An attempt along this line is summarized in Table 3, taken from my Oxford Conference talk[27] a few months ago. You see that 7 out of the 13 phenomenological parameters are now shown to agree within errors with the one-parameter standard-model predictions.

The agreement achieved here is impressive even to the eyes of a diehard sceptic. If all these parameters were just random numbers of order plus or minus one, the probability that all 7 of the parameters agree within the standard-model predictions with the kind

TABLE III. Objective tests of the standard model

| | | | $\sin^2\theta_W = 0.22$ | |
|---|---|---|---|---|
| $\nu$ quark | $\alpha$ | $1-2\sin^2\theta_W$ | 0.56 | $(\pm\atop-)$ 0.58 ± 0.14 |
| | $\beta$ | 1 | 1 | $(\pm\atop-)$ 0.92 ± 0.14 |
| | $\gamma$ | $-\frac{2}{3}\sin^2\theta_W$ | $-0.147$ | $(\mp)$ 0.28 ± 0.14 |
| | $\delta$ | 0 | 0 | $(\pm\atop-)$ 0.06 ± 0.14 |
| $\nu e$ | $g_V$ | $-\frac{1}{2}(1-4\sin^2\theta_W)$ | $-0.06$ | $-0.03 \pm 0.12$ (or $-0.52 \pm 0.15$) |
| | $g_A$ | $-\frac{1}{2}$ | $-0.5$ | $-0.52 \pm 0.15$ (or $-0.03 \pm 0.12$) |
| e quark | $\tilde{\alpha}$ | $-(1-2\sin^2\theta_W)$ | $-0.56$ | $\left(\tilde{\alpha}+\frac{\tilde{\gamma}}{3}\right) + 0.23\left(\tilde{\beta}+\frac{\tilde{\delta}}{3}\right)$ |
| | $\tilde{\beta}$ | $-(1-4\sin^2\theta_W)$ | $-0.12$ | $= -0.59 \pm 0.10$ |
| | $\tilde{\gamma}$ | $\frac{2}{3}\sin^2\theta_W$ | 0.147 | ($-0.54$ for |
| | $\tilde{\delta}$ | 0 | 0 | $\sin^2\theta_W = 0.22$) |
| (ee), ($\bar{\mu}\mu$) | $h_{VV}$ | $\frac{1}{4}(1-4\sin^2\theta_W)^2$ | 0.004 | |
| | $h_{AA}$ | $\frac{1}{4}$ | 0.25 | Wait for PETRA and PEP |
| | $h_{VA}$ | $\frac{1}{4}(1-4\sin^2\theta_W)$ | 0.03 | |

of errors indicated would be essentially negligible.

## VI. WEAK INTERACTION EFFECTS IN ELECTRON-POSITRON COLLISIONS

The last subject I would like to discuss is weak interaction effects that can be studied at electron-positron colliding-beam facilities like PETRA and PEP. Most of the time I'll be concerned with

$$e^+ + e^- \to \mu^+ + \mu^- . \tag{6.1}$$

The invariant four-momentum transfer squared $q^2$ is now time-like, negative in my metric; we use

$$s \equiv -q^2 = (\text{c.m. energy of } e^+e^-)^2 \ . \qquad (6.2)$$

As already mentioned in Section II, with $\mu e$ universality there are three low-energy parameters to be determined, $h_{VV}$, $h_{AA}$ and $h_{VA}$.
The history of this subject is again old. As early as 1961, when electron-positron colliding beam facilities were first contemplated at Orsay and Frascati, Cabibbo and Gatto[36] worked out the phenomenology of the muon pair reaction (6.1) within the framework of a weak-interaction model that contains a pure V-A neutral current.[2] Now, seven machine generations later—ACO, VEPP, ADONE, CEA-Bypass, SPEAR, DORIS, DCI—a study of weak-electromagnetic interference is finally becoming a reality. It is universally recognized that PETRA and PEP are excellent places to detect weak interaction effects. This is clear from our $10^{-4}q^2$ rule; with $s \simeq 1000$ GeV$^2$ typically at PETRA and PEP we should find ~10% effect.

If neither the electron, nor the positron beam is polarized, there are two types of experiments we can perform. First, we may examine whether the observed total cross section for (6.1) deviates from the QED prediction; for $s \gg m_Z^2$, this measures $h_{VV}$ as follows:[37]

$$(\sigma - \sigma_{QED})/\sigma = -4(G/\sqrt{2}\, e^2)\, h_{VV}\, s \ . \qquad (6.3)$$

Second, we may examine whether there is any forward-backward asymmetry in the angular distribution for (6.1); this measures $h_{AA}$:[37]

$$A_{FB} \equiv [(d\sigma/d\Omega)_\theta - (d\sigma/d\Omega)_{\pi-\theta}]/[(d\sigma/d\Omega)_\theta + (d\sigma/d\Omega)_{\pi-\theta}]$$

$$= -8(G/\sqrt{2}\, e^2)\, h_{AA}\, s\, \cos\theta/(1 + \cos^2\theta) \ , \qquad (6.4)$$

again for $s \ll m_Z^2$.

Now this Symposium is about polarized beams; we have heard a lot about how electrons get transversely polarized (or depolarized) in storage rings. So let us see what we can learn when the electron beam is polarized, the positron beam, unpolarized. If the polarization is transverse, there will be a $\phi$ dependence but it turns out that we won't learn anything new; the "right-left" asymmetry we observe in such an experiment conserves parity, so it can depend only on the parity-conserving parameters $h_{VV}$ and $h_{AA}$ already studied in (6.3) and (6.4). The situation is quite different with a longitudinally polarized electron beam. Here there is a possibility of measuring a genuine parity violation effect analogous to the $T\ell$ circular dichroism or the SLAC polarized-electron asymmetry:[37]

$$\frac{d\sigma(+)/d\Omega - d\sigma(-)/d\Omega}{d\sigma(+)/d\Omega + d\sigma(-)/d\Omega} = -4(G/\sqrt{2}\ e^2)\ h_{VA}\left[1 + \frac{2\cos\theta}{1+\cos^2\theta}\right] \quad (6.5)$$

for $s \ll m_Z^2$, where $\pm$ refer to the electron helicity $\lambda = \pm 1/2$. Similar information can be obtained by a measurement of the final muon polarization starting with unpolarized electron and positron beams.

Let us now look at the standard-model predictions. For $\sin^2\theta_W = 1/4$, the leptonic current becomes pure axial-vector, so $h_{VV}$ and $h_{VA}$ vanish. Because $\sin^2\theta_W$ is now believed to be close to $1/4$, prospects are rather bleak for measuring nonvanishing values for (6.3) and (6.5). On the other hand, the forward-backward asymmetry that depends on $h_{AA}$ is expected to be comfortably large to be measurable because $h_{AA}$ in the standard model is $1/4$ (independent of $\sin^2\theta_W$). At $\sqrt{s}$ = 30 GeV, $8(G/\sqrt{2}\ e^2)\ h_{AA}s$ is predicted to be as large as 0.16. For completeness I may mention that the upper limit on $h_{AA}$ from SPEAR is[38]

$$h_{AA} < 2.0\ (90\%\ \text{CL}) \quad (6.6)$$

At the Tokyo Conference two months ago, Barbiellini of Frascati asked after Weinberg's rapporteur talk what we might learn by studying weak interaction effects in electron-positron annihilations into muons; Weinberg confidently replied that such an experiment merely checks the parameters already determined from other experiments. In a certain sense, Weinberg is correct; as long as $\sqrt{s}$ is much less than $m_Z$, we are unlikely to encounter any surprise. Given the successes of the standard model in the $\nu e$, $\nu q$ and $eq$ sectors, the expectations based on the standard model for this process are essentially identical to the predictions of any model with a single Z boson and $\mu e$ universality. Conversely, by measuring $h_{VV}$, $h_{AA}$ and $h_{VA}$, we test the single Z (or factorization) hypothesis and/or $\mu e$ universality in a nontrivial, model-independent way.[6]

It is, of course, evident that we do expect deviations from the predictions above if we sit right on heavy quarkoniums, e.g. $(t\bar{t})$ bound states with $J^P = 1^-$. A careful study of muon pair production via the vacuum polarization mechanism

$$e^+ + e^- \to \gamma, Z \to (t\bar{t}) \to \gamma, Z \to \mu^+ + \mu^- \quad (6.7)$$

could reveal important information on the neutral-current couplings of the t quark. For a $(t\bar{t})$ bound state mass of ~35 GeV, the polarization asymmetry [the left-hand side of (6.5)] is expected to be as large as ~7% for the standard t quark assignment.[39]

The formulas (6.3)-(6.5) must be modified if the Z mass is not completely negligible compared to $\sqrt{s}$. To first order in $s/m_Z^2$ all we have to do is multiply (6.3)-(6.5) by $1 + m_Z^2/s$. The basic question is whether we can perform an experiment precise enough to see the effect of a finite Z boson mass.

According to the currently accepted Orthodoxy the Z boson, the basic mediator of the weak neutral-current interactions, must be in the 85-95 GeV range. So even the PETRA/PEP energy is not high enough for direct formation of the Z boson. Yet a very precise experiment performed at the highest PETRA/PEP energy, $\sqrt{s} \simeq 40$ GeV, may be able to distinguish between the orthodox boson theory with $m_Z \simeq 90$ GeV from the four-fermion theory. This is shown in Fig. 7 where the

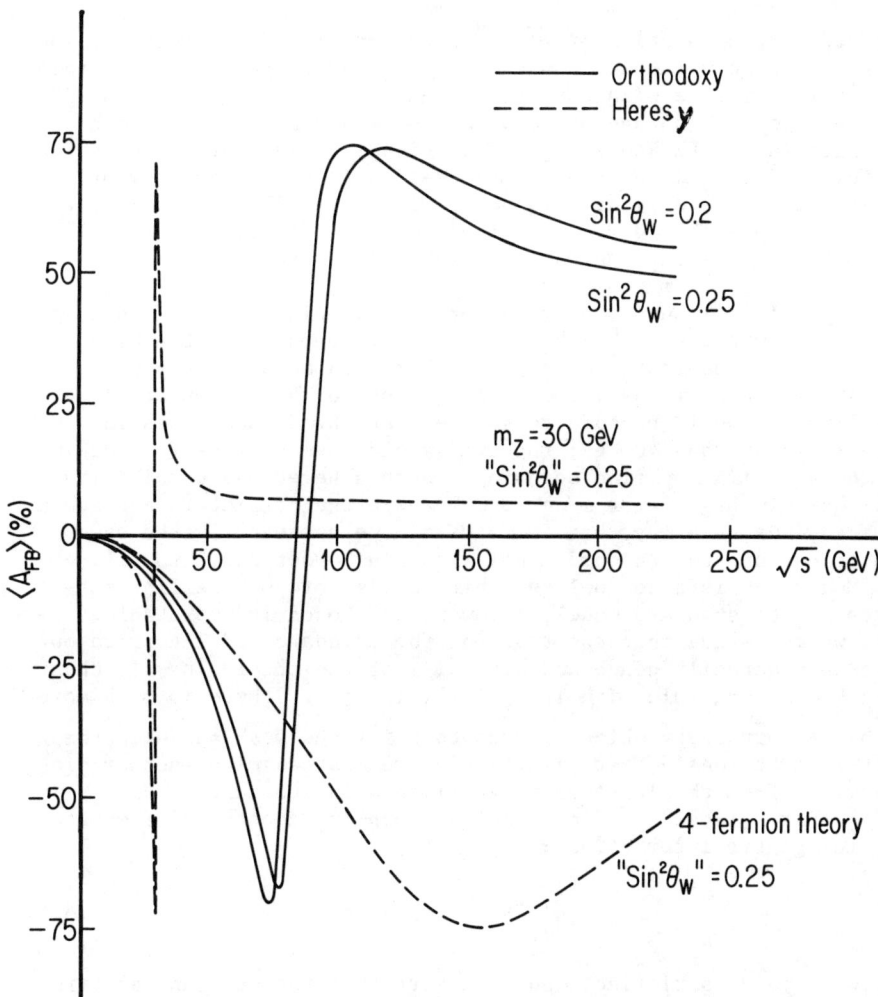

Fig. 7. Integrated angular asymmetry in electron-positron annihilations into muon pairs.

integrated forward-backward asymmetry, defined by

$$\langle A_{FB}\rangle \equiv \frac{\int_{\theta=0}^{\theta=\pi/2}(d\sigma/d\Omega)d\Omega - \int_{\theta=\pi/2}^{\theta=\pi}(d\sigma/d\Omega)d\Omega}{\int_{\theta=0}^{\theta=\pi/2}(d\sigma/d\Omega)d\Omega + \int_{\theta=\pi/2}^{\theta=\pi}(d\sigma/d\Omega)d\Omega}, \qquad (6.8)$$

is plotted as a function of $\sqrt{s}$. The parameters of the four-fermion theory are chosen in such a way that its predictions at low energies completely coincide with the standard-model predictions with $\sin^2\theta_W$ set equal to 1/4 (no vector coupling for e and μ). If the Mark J Collaboration at PETRA--a big international collaboration including physicists from Peking--really measures $\langle A_{FB}\rangle$ to an accuracy of ~2%, as claimed in the original proposal, we may be able to see, for the first time, possible deviations from the four-fermion theory. The possibility of seeing the "dispersive tail" of the Z boson peak in this manner is, in my opinion, quite exciting.

It will, of course, be even more spectacular if the Z boson peak itself lies within the reach of PETRA/PEP, contrary to the Orthodoxy predictions. Should the Mark J Collaboration measure, say at $\sqrt{s}$ = 40 GeV, a positive value of $A_{FB}$, short of breakdown of μe universality, it would be safe to assume that the Z boson mass is actually <u>lower</u> than 40 GeV; the time would then be ripe for looking for the gold mine by fine scanning. Such a heretical possibility is also shown in Fig. 7 for $m_Z$ = 30 GeV where the parameters are again adjusted in such a way that for $s \ll m_Z^2$ we get exactly the same predictions as the standard model with $\sin^2\theta_W$ set equal to 1/4.

Most theorists now believe that in view of the dazzling successes of the standard model, a low mass Z boson is improbable. However, we may argue that the tests of the standard model carried out so far are actually concerned with ~1/2 of the theory, namely the $T_{3L,R} - Q\sin^2\theta_W$ rule with $T_{3L} = \pm 1/2$, $T_{3R} = 0$. There is much more to the QFD orthodoxy which alleges to unify the weak and electromagnetic interactions. Most importantly, we must examine whether $\sin^2\theta_W$ determined from the low-energy experiments is indeed the same $\sin^2\theta_W$ that relates the dimensionless coupling constants of weak and electromagnetic interactions:

$$\sin^2\theta_W = (e/g)^2. \qquad (6.9)$$

Only with (6.9) satisfied, can we assert that the weak and electromagnetic interactions are truly unified. The most practical way to test this is to verify the Weinberg mass relations because the dimensionless coupling constant $g^2$ is related to Fermi's G via (2.8).

I emphasize all this because all the successes of the standard model in accounting for neutral-current phenomena at present energies can be reproduced just as well in a "non-unified" model where[40]

(i) The basic weak symmetry is SU(2) à la Bludman.[2]

(ii) There is a large electromagnetic correction phenomenologically described by a gauge-invariant $\gamma$-$W^0$ mixing:

$$L_{\gamma W} = -\frac{1}{2} \lambda F_{\mu\nu} W_{\mu\nu}^o . \qquad (6.10)$$

In this heretical model we can derive

$$\text{"}\sin^2\theta_W\text{"} = e\lambda/g , \qquad (6.11)$$

where "$\sin^2\theta_W$" is what experimentalists like Baltay call $\sin^2\theta_W$, but the unification condition (6.9) does not follow; as a result the Weinberg mass relations need not be satisfied. For this reason it is of utmost importance to have an open mind on the W and Z masses despite the brilliant successes of the Weinberg-Salam model evidenced in Fig. 6 and Table III.

It may, of course, turn out that the Orthodoxy predictions are completely correct, in which case the W and Z bosons are to be found at 78 GeV and 90 GeV, respectively, with $\sin^2\theta_W \simeq 0.23$. The bosons will then be first discovered at proton colliding-beam facilities--the $\bar{p}$ cooler at CERN, the main ring collider at Fermilab, or Isabelle at Brookhaven--rather than at electron-positron colliding-beam facilities. Even so, it will be spectacular to see Z boson formation in $e^+e^-$ collisions--at LEP?--; the famous R at the Z peak, defined in the usual way, can be computed on the basis of the 3 lepton-doublet, 6 colored-quark model to be[41]

$$R \equiv \sigma(e^+e^- \to Z \to \text{any})/(4\pi\alpha^2/3s) \simeq 5000 , \qquad (6.12)$$

not 5, not 50 but 5000.

After the discoveries of W and Z, our next major task will be to examine whether the weak interactions, or better the "electroweak" interactions, are indeed damped at high energies, i.e. at energies far beyond the weak boson resonances, as demanded by renormalizable theories. It is right here that weak-electromagnetic interference plays an absolutely essential role in insuring decent high energy behavior, as can be seen from the following example. Consider

$$e^+ + e^- \to W^+ + W^- . \qquad (6.13)$$

The cross section for this process is supposed to go down at sufficiently high energies in renormalizable models, but how does this come about explicitly? If we calculate the cross section using $\gamma$ exchange, Z exchange or $\nu$ exchange, as shown in Fig. 8, each exchange mechanism separately gives a cross section that rises with energy. See the upper part of Fig. 9 taken from the calculations of Alles, Boyer and Buras.[42] Yet when we also look at the various

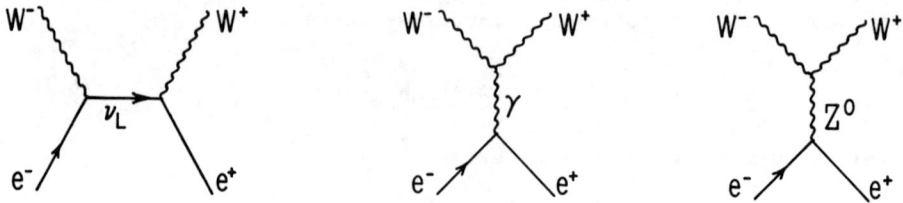

Fig. 8. Mechanism for electron-positron annihilations into $W^+W^-$ pairs.

interference terms using the Orthodoxy couplings,[43] a miracle takes place. The interference contributions sum up to just the right

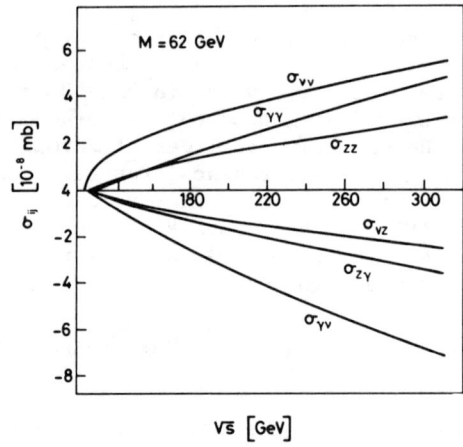

Fig. 9. Individual contributions to W pair production.

amount to cancel the rising cross sections of the individual exchange mechanisms; in fact, the total cross section for $W^+W^-$ production is predicted to start decreasing around 40 GeV above the threshold, as shown in Fig. 10, also taken from Alles et al.[42] These cancellations are the hallmark of renormalizable gauge theories;[44] to quote Weinberg,[45]

"Indeed the best way to convince oneself that gauge theories may have something to do with nature is to carry out some specific calculation and watch the cancellations before one's very eyes."

Does all this sound convincing? In any case it would be fantastic to see how the predicted cancellations take place experimentally at colliding beam facilities--LEP II?--in the

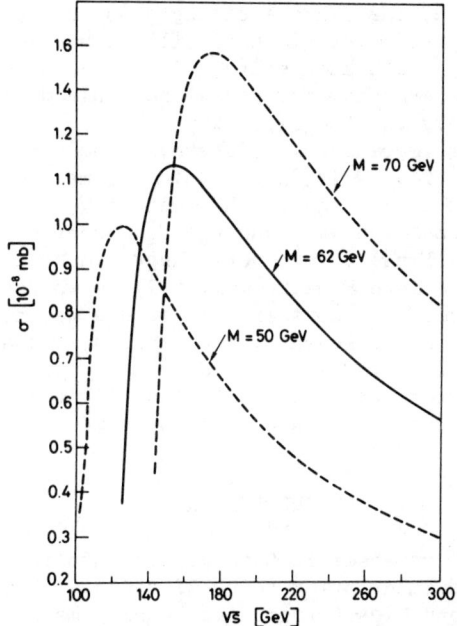

Fig. 10. The total cross section for W pair production.

200-300 GeV range.

The cancellation mechanism discussed above must, of course, be operative separately for each helicity amplitude. This is of some interest because, due to the pure V-A nature of the charged-current interactions, the $\nu$ exchange diagram can be turned off in collisions of left-handed ($\lambda = -1/2$) positrons and right-handed ($\lambda = +1/2$) electrons;[46] the gauge-theory cancellation must then be complete just between $\gamma$ exchange and Z exchange. Perhaps Dr. Schwitters will tell us how such an experiment may become feasible in the late '80s or early '90s.

## VII. CONCLUSION

With the recent experiments showing parity violation in polarized electron scattering and in atomic transitions, we have clearly entered a new era in weak-interaction physics. Thanks to those who successfully executed experimental tasks which, in the beginning, appeared impossible, nontrivial advances have been made in our understanding of weak-interaction phenomena. The SLAC eD experiment literally massacred a large number of gauge models, and even the last remaining doubt on atomic parity violation appears to have been cleared away by the Berkeley T$\ell$ experiment. The standard (Weinberg-Salam) model--the simplest gauge model that incorporates weak neutral currents and weak-electromagnetic unification--has emerged as the clear winner with flying colors.

As for the future, there are exciting prospects at PETRA/PEP where weak-electromagnetic interference will be studied in electron-positron annihilations into muon pairs. A precision measurement of the forward-backward asymmetry may reveal possible deviations from the four-fermion theory for the first time.

It is of obvious importance to determine the W and Z masses-- from a certain point of view it is the Weinberg mass relations, not so much the successes in low-energy phenomenology, that critically test the hypothesis that the weak and electromagnetic interactions are truly unified.[47] Finally the energy dependence of electron-positron annihilations into W pairs will tell us whether weak-electromagnetic interference conspires in such a way to insure decent high-energy behavior, as demanded by renormalizable theories.

It took nearly 20 years since Zel'dovich's paper to detect weak-electromagnetic interference at all. Fortunately there are some indications that the rate of progress in the next 20 years will be considerably faster.

## ACKNOWLEDGEMENT

I wish to thank Professor E. Commins for informing me of the status of the T$\ell$ experiment prior to publication. Thanks are also due to Minh D. Tran and Nigel Wright for helping me with the preparation of figures.

## APPENDIX: DEFINITION OF COUPLING PARAMETERS

$\overset{(-)}{\nu} + N \to \overset{(-)}{\nu} + $ hadrons:

$$L = -\frac{G}{\sqrt{2}} \bar{\nu} \gamma_\lambda (1+\gamma_5) \nu \Big\{ \frac{1}{2} [\bar{u} \gamma_\lambda (\alpha+\beta\gamma_5)u - \bar{d}\gamma_\lambda(\alpha+\beta\gamma_5)d] \\ + \frac{1}{2} [\bar{u}\gamma_\lambda(\gamma+\delta\gamma_5)u + \bar{d}\gamma_\lambda(\gamma+\delta\gamma_5)d] \Big\} . \quad (A.1)$$

Chiral constants:

$$\varepsilon_L(u) = \frac{1}{4}(\alpha+\beta+\gamma+\delta), \quad \varepsilon_L(d) = \frac{1}{4}(-\alpha-\beta+\gamma+\delta),$$

$$\varepsilon_R(u) = \frac{1}{4}(\alpha-\beta+\gamma-\delta), \quad \varepsilon_R(d) = \frac{1}{4}(-\alpha+\beta+\gamma-\delta) . \quad (A.2)$$

$\overset{(-)}{\nu}_\mu + e^- \to \overset{(-)}{\nu}_\mu + e^-$:

$$L = -\frac{G}{\sqrt{2}} [\bar{\nu}_\mu \gamma_\lambda (1+\gamma_5) \nu_\mu][\bar{e}(g_V \gamma_\lambda + g_A \gamma_\lambda \gamma_5) e] \,. \tag{A.3}$$

$e^+ + e^- \to \mu^+ + \mu^-$:

$$\begin{aligned} L = -\frac{G}{\sqrt{2}} [ &h_{VV}(\bar{e}\gamma_\lambda e + \bar{\mu}\gamma_\lambda \mu)(\bar{e}\gamma_\lambda e + \bar{\mu}\gamma_\lambda \mu) \\ &+ 2h_{VA}(\bar{e}\gamma_\lambda e + \bar{\mu}\gamma_\lambda \mu)(\bar{e}\gamma_\lambda \gamma_5 e + \bar{\mu}\gamma_\lambda \gamma_5 \mu) \\ &+ h_{AA}(\bar{e}\gamma_\lambda \gamma_5 e + \bar{\mu}\gamma_\lambda \gamma_5 \mu)(\bar{e}\gamma_\lambda \gamma_5 e + \bar{\mu}\gamma_\lambda \gamma_5 \mu)] \,. \end{aligned} \tag{A.4}$$

Parity violation in atoms and in polarized electron scattering:

$$\begin{aligned} L = -\frac{G}{\sqrt{2}} \Big\{ &\bar{e}\gamma_\lambda \gamma_5 e \Big[ \frac{\tilde{\alpha}}{2}(\bar{u}\gamma_\lambda u - \bar{d}\gamma_\lambda d) + \frac{\tilde{\gamma}}{2}(\bar{u}\gamma_\lambda u + \bar{d}\gamma_\lambda d) \Big] \\ &+ \bar{e}\gamma_\lambda e \Big[ \frac{\tilde{\beta}}{2}(\bar{u}\gamma_\lambda \gamma_5 u - \bar{d}\gamma_\lambda \gamma_5 d) + \frac{\tilde{\delta}}{2}(\bar{u}\gamma_\lambda \gamma_5 u + \bar{d}\gamma_\lambda \gamma_5 d) \Big] \Big\} \,. \end{aligned} \tag{A.5}$$

Electron-nucleon constants:

$$\tilde{\alpha} = -(C_{1p} - C_{1n}), \quad 1.25\tilde{\beta} = -(C_{2p} - C_{2n}),$$

$$\tilde{\gamma} = -\frac{1}{3}(C_{1p} + C_{1n}), \quad 0.75\tilde{\delta} = -(C_{2p} + C_{2n}) \,. \tag{A.6}$$

## REFERENCES AND FOOTNOTES

1. Ya B. Zel'dovich, JETP **9**, 682 (1959).
2. S. A. Bludman, Nuovo Cimento **9**, 443 (1958).
3. S. Weinberg, Phys. Rev. Lett. **19**, 1264 (1967); A. Salam, Elementary Particle Theory, ed. N. Svartholm (Almquist and Wiksell, Stockholm, 1968), p. 367.
4. F. J. Hasert et al., Phys. Lett. **46B**, 138 (1973). See also A. Benvenuti et al., Phys. Rev. Lett. **32**, 800 (1974).
5. If we relax μe universality, $h_{AV}$ becomes another independent parameter.
6. If we assume that the neutral-current interactions are mediated by a single Z boson, then there are factorization relations which make only 7 of them independent. See P. Q. Hung, and J. J. Sakurai, Phys. Lett. **69B**, 323 (1977).

7.  S. L. Glashow, Nucl. Phys. $\underline{22}$, 579 (1961). See also A. Salam and J. C. Ward, Phys. Lett. $\underline{13}$, 168 (1964).
8.  The history of this subject is traced in: M. Veltman, <u>Proceedings of the VIth International Symposium on Electron and Photon Interactions at High Energies</u>, Bonn, August, 1973, ed. H. Rollnik and W. Pfeil (North-Holland, Amsterdam, 1974), p. 429; J. J. Sakurai, UCLA/78/TEP/9 (to be published in AIP Proceedings).
9.  P. W. Higgs, Phys. Lett. $\underline{12}$, 132 (1964); Phys. Rev. Lett. $\underline{13}$, 508 (1964). See also F. Englert and R. Brout, Phys. Rev. Lett. $\underline{13}$, 321 (1964); G. S. Guralnik, C. R. Hagen and T. W. B. Kibble, Phys. Rev. Lett. $\underline{13}$, 585 (1964).
10. Because the quarks are fractionally charged, their $Y_W$ values are 1/6 for left-handed quarks in doublets, and 2/3 and -1/3 for right-handed quarks in singlets. In (2.13) d' and s' stand for Cabibbo-rotated quarks; when d and s enter in this manner, the absence of strangeness-changing neutral currents is guaranteed, as is well known from work of S. L. Glashow, J. Iliopoulos and L. Maiani, Phys. Rev. D$\underline{2}$, 1285 (1970).
11. I cite no references because a complete list would be too long to be reproduced here.
12. M. A. Bouchiat and C. Bouchiat, Phys. Lett. $\underline{48B}$, 111 (1974); Journ. de Phys. $\underline{35}$, 899 (1974); $\underline{36}$, 493 (1975). See also F. C. Michel, Phys. Rev. $\underline{138B}$, 408 (1965).
13. In writing down the contribution from the $V_{lept} A^{I=0}_{quark}$ ($\tilde{\delta}$ term) interaction, the isoscalar axial-vector matrix element is assumed to have the value predicted by the non-relativistic quark model. This may be reasonable because the non-relativistic quark model successfully accounts for the magnetic moment ratio of the nucleon. See e.g. P. Q. Hung, Phys. Rev. D$\underline{17}$, 1893 (1978), which discusses various approaches to this subject.
14. R. R. Lewis and W. L. Williams, Phys. Lett. $\underline{59B}$, 70 (1975); E. A. Hinds and V. W. Hughes, Phys. Lett. $\underline{67B}$, 487 (1977).
15. R. Conti et al. (Berkeley preprint, submitted to Phys. Rev. Lett.)
16. In terms of A defined in (3.8) the Berkeley Group finds $(5.2 \pm 2.4) \times 10^{-3}$.
17. D. V. Neuffer and E. Commins, Phys. Rev. A$\underline{16}$, 844 (1977). See also O. P. Sushkov, V. V. Flambaum and I. B. Khriplovich, JETP Lett. $\underline{24}$, 502 (1976); E. M. Henley and L. Wilets, Phys. Rev. A$\underline{14}$, 1411 (1976).
18. I. B. Khriplovich, JETP Lett. $\underline{20}$, 315 (1974); P. G. H. Sandars, <u>Atomic Physics</u> IV, ed. Z. Putlitz et al. (Plenum Press, New York 1975); E. N. Fortson, Bull. Am. Phys. Soc. $\underline{20}$, 491 (1975).
19. Remember the old days when an analogy with optics was invoked to explain neutral K meson phenomena? The times have changed!

20. This table is taken from C. Baltay's rapporteur talk at the Tokyo Conference (Columbia preprint, to be published in the Proceedings of the XIXth International Conference on High-Energy Physics, Tokyo, August 1978). The experimental data are from N. Fortson (to be published in the Proceedings of Neutrino '78, Purdue University, May 1978); P. G. H. Sandars (results presented at the Riga Conference, August 1978); L. M. Barkov and M. S. Zolotorev, JETP Lett. $\underline{27}$, 379 (1978). For earlier data see R. L. Lewis et al., Phys. Rev. Lett. $\underline{39}$, 795 (1977); P. E. Baird et al., Phys. Rev. Lett. $\underline{39}$, 798 (1977).
21. G. Feinberg and M. Y. Chen, Phys. Rev. $\underline{D10}$, 190 (1974); R. N. Cahn and G. L. Kane, Phys. Lett. $\underline{71B}$, 348 (1977). These authors use $C_{1p}$, $C_{1n}$, $C_{2p}$, $C_{2n}$ defined by (A.6).
22. For further discussion of various atomic-parity experiments see e.g. G. Feinberg (Columbia preprint CU-TP-111, to be published in the Proceedings of the Ben Lee Memorial International Con- on Parity Non-conservation, Weak Neutral Currents and Gauge Theories, Fermilab, October 1977).
23. C. Y. Prescott et al., Phys. Lett. $\underline{77B}$, 347 (1978).
24. C. Y. Prescott, these Proceedings.
25. R. N. Cahn and F. J. Gilman, Phys. Rev. $\underline{D17}$, 1313 (1978); M. Yoshimura, Progr. Theoret. Phys. $\underline{59}$, 231 (1978). For earlier discussion see e.g. A. Love, G. G. Ross and D. V. Nanopoulos, Nucl. Phys. $\underline{B49}$, 513 (1972); E. Derman, Phys. Rev. $\underline{D7}$, 2755 (1973).
26. J. D. Bjorken, SLAC-PUB-2146 (1978). See also L. Wolfenstein (unpublished Carnegie Mellon report); E. Derman, Rockefeller preprint COO-2332B-158 (1978).
27. This figure is an update of a figure I presented at the Oxford Conference: J. J. Sakurai, UCLA/78/TEP/18, to be published in the Proceedings of the Topical Conference on "Neutrino Physics at Accelerators," Oxford, July 1978. See also J. D. Bjorken (Reference 26).
28. Model-independent analyses of the neutrino-quark couplings have been attempted by a number of authors. A list of recent papers may include: L. M. Sehgal, Phys. Lett. $\underline{71B}$, 99 (1977); Aachen preprint PITHA-102 (to be published in the Proceedings of Neutrino '78), P. Q. Hung and J. J. Sakurai, Phys. Lett. $\underline{72B}$, 208 (1977); G. Ecker, Phys. Lett. $\underline{72B}$, 450 (1978); L. F. Abbott and R. M. Barnett, Phys. Rev. Lett. $\underline{40}$, 1303 (1978); D. P. Sidhu and P. Langacker, Phys. Rev. Lett. $\underline{41}$, 732 (1978); E. A. Paschos, BNL-24619 (1978); M. Gourdin and X. Y. Pham, Université Pierre et Marie Curie preprint, PAR-LPTHE 78/14. For recent review see e.g. J. J. Sakurai, Reference 27; C. Baltay, Reference 20.
29. This combination is advocated particularly by L. Wolfenstein, Ref. 26.
30. See e.g. J. J. Sakurai (Reference 27) based largely on the $\bar{\nu}_e e$ date of F. Reines et al., Phys. Rev. Lett. $\underline{37}$, 315 (1976) and the $\nu_\mu e$ data of A. M. Cnops et al., Phys. Rev. Lett. $\underline{41}$, 357 (1978).

31. For other approaches to this subject see L. F. Abbott and R. M. Barnett, SLAC-PUB-2136 (1978); M. Konuma and T. Oka, Kyoto preprint, RIFP-338 (1978). However, these authors implicitly assume, in addition to the single Z boson hypothesis, that the $\nu\nu$ scattering amplitude has the "normal" strength [SU(2) ⊗ U(1) with the simplest Higgs mechanism].
32. For review see e.g. A. Mann (to be published in the <u>Proceedings of the Ben Lee Memorial International Conference on Parity Nonconservation, Weak Neutral Currents and Gauge Theories</u>, Fermilab, October 1977).
33. See e.g. H. Fritzsch and P. Minkowski, Nucl. Phys. <u>B103</u>, 61 (1976).
34. See e.g. P. Fayet, Nucl. Phys. <u>B78</u>, 14 (1974); T. P. Cheng and L. F. Li, Phys. Rev. Lett. <u>38</u>, 381 (1977).
35. C. Baltay, Reference 20.
36. N. Cabibbo and R. Gatto, Phys. Rev. <u>124</u>, 1577 (1961).
37. The phenomenology of this reaction has been discussed by a number of authors. For a model-independent treatment see e.g. T. Kinoshita et al., Phys. Rev. <u>D2</u>, 910 (1970); L. Wolfenstein, AIP Proceedings <u>23</u>, 84 (1974). For predictions based on the Weinberg-Salam model see e.g. J. Godine and A. Hankey, Phys. Rev. <u>D6</u>, 3301 (1972); R. Budny, Phys. Lett. <u>45B</u>, 340 (1973).
38. T. Himel et al., Phys. Rev. Lett. <u>41</u>, 449 (1978). In this reference the experimental result is presented as $em_Z/g_a > 53$ GeV. To the extent that no attempt has been made to study the energy dependence of $A_{FB}$, (6.6) is a more objective way of presenting the experimental result.
39. I. I. Y. Bigi, J. H. Kühn and H. Schneider, Max-Planck preprint, MPI-PAE/PTh 28/78; T. H. Nieh and Y. S. Teh, Stony Brook preprint (1978).
40. P. Q. Hung and J. J. Sakurai, UCLA/78/TEP/8, Nucl. Phys. (to be published). See also J. D. Bjorken SLAC-PUB-2062 (1977); SLAC-PUB-2133 (1978).
41. Because the Z boson width is predicted to be of order 2 GeV, we need not worry here about the beam resolution. However, Bjorken has warned me that the radiative corrections can reduce the peak value of R by a factor of 2.
42. W. Alles, Ch. Boyer and A. J. Buras, Nucl. Phys. <u>B119</u>, 125 (1977).
43. The SU(2) ⊗ U(1) value of the W boson gyromagnetic ratio $[g(W^{\pm}) = 2]$ is quite essential here.
44. Also to be mentioned in this connection is a theorem due to C. H. Llewellyn Smith [Phys. Lett. <u>46B</u>, 233 (1973)] and J. M. Cornwall, D. Levin and G. Tiktopoulos [Phys. Rev. Lett. <u>30</u>, 1268 (1973)], which states that tree unitarity demands gauge-theory couplings.
45. S. Weinberg, Revs. Mod. Phys. 46, 255 (1974).
46. K. J. F. Gaemers and G. J. Gounaris, CERN-TH-2548 (1978).
47. Even if the W and Z bosons show up at the correct masses, some theorists may still reserve judgement on the Weinberg-Salam model until we discover the Higgs boson with the right coupling properties.

# POLARIZED LEPTON-HADRON SCATTERING*

Vernon W. Hughes
Gibbs Laboratory, Physics Department, Yale University
New Haven, Connecticut 06520

## ABSTRACT

Within the past 10 years high energy polarized lepton-hadron scattering experiments have contributed important tests of invariance principles and a new type of information about the internal spin structure of the nucleon. Time reversal invariance in the electromagnetic interaction has been established at a significant level, and very recently the important discovery of parity nonconservation has been made. The new information on the proton spin structure agrees with the general predictions of the quark-parton model of the nucleon and with a basic current algebra sum rule. These experiments have been made possible by the recent development of polarized electron sources for the SLAC linear accelerator and by some improvements in polarized proton targets.

## INTRODUCTION

This paper is intended to provide a brief review and survey of experiments on high energy polarized lepton-hadron scattering, including their theoretical implications. The leptons involved are $e^{\pm}$ and $\mu^{\pm}$, and the hadrons are p, n and certain nuclei.

The principal physics motivations for these experiments have been, firstly, the study of the invariance principles of space inversion P and time reversal T in the electromagnetic interaction, and, secondly, the study of the internal spin structure of the nucleon.

The first high energy polarization experiment involving lepton-hadron scattering was a test of time reversal invariance in the scattering of unpolarized electrons by polarized protons done at the Cambridge Electron Accelerator (CEA).[1] The first high energy polarized electron beam was produced at the 20 GeV Stanford Linear Accelerator Center (SLAC),[2] and its polarization was measured at high energy by electron-electron (Møller) scattering.[3] The experiments done at CEA[1] and at SLAC[4,5] to test T invariance have established an upper limit to a possible violation. Following initial experiments at SLAC with polarized electrons[6-9] and at Serpukhov with polarized muons[10] which set an upper limit to parity nonconservation, a more sensitive experiment at SLAC[11] has revealed a breakdown of P invariance. Experiments have been done at SLAC to study the scattering of polarized electrons by polarized protons in the elastic,[12] deep inelastic,[6,8] and resonance[13] regions, which have provided information about the internal spin structure of the proton.

*Research supported in part by the Department of Energy under Contract No. EY-76-C-02-3075.

High energy experiments with polarized electrons have been done thus far only at SLAC and have used two different sources of polarized electrons - one based on photoionization of a polarized $^6$Li atomic beam,[2,14] and the second based on photoemission from GaAs using polarized laser light.[11,15,16,17] The parity nonconservation experiment with muons[10] used polarized muons arising naturally from pion decays. The polarized proton targets used in the electron scattering experiments were of the standard type based on the method of dynamic nuclear orientation with hydrocarbon samples, but special attention was required for the high radiation damage from the electron beam.[12,18,19]

## POLARIZED BEAMS AND TARGETS

Two types of polarized electron sources have been used thus far in high energy scattering experiments, both at the 20 GeV Stanford Linear Accelerator. The first source, designated PEGGY I, is based on photoionization of a polarized $^6$Li atomic beam, and the second, PEGGY II, is based on photoemission of electrons from negative electron affinity GaAs using polarized laser light.

Figure 1 shows the energy levels and magnetic moments of the hyperfine structure magnetic substates of the ground $^2S_{1/2}$ state of $^6$Li as a function of magnetic field H, and Figure 2 is a schematic drawing of PEGGY I. An intense atomic beam of $^6$Li in its ground state is formed by heating $^6$Li in an oven with an orifice and by collimating the resulting flux of atoms. The sixpole magnet with its strong inhomogeneous magnetic field transmits only atoms with

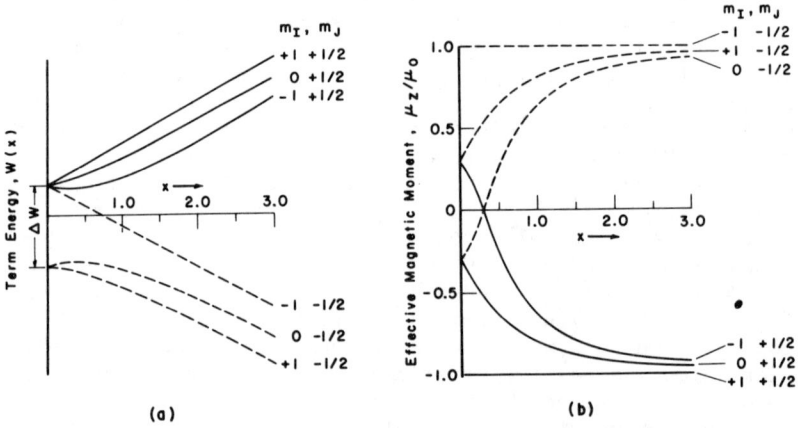

Fig. 1. Energy levels and magnetic moments of $^6$Li in the ground $^2S_{1/2}$ state and with nuclear spin I=1 as a function of magnetic field H, where $m_J$, $m_I$ are the electronic and nuclear magnetic quantum numbers, and $x=(g_J-g_I)\mu_0 H/\Delta W$ in which $g_J$ and $g_I$ are the nuclear and electronic g values, $\mu_0$ is the Bohr magneton, and $\Delta W = h\Delta\nu$ is the hyperfine structure interval.

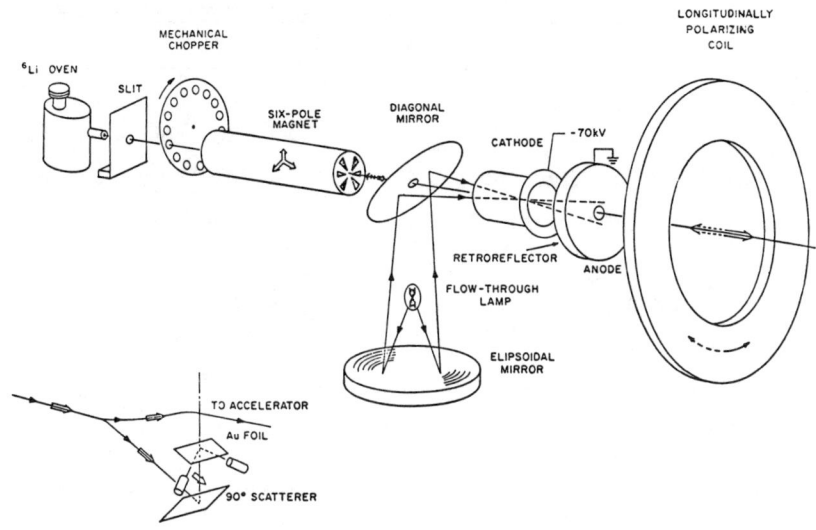

Fig. 2. Schematic diagram of PEGGY I showing the principal components of the lithium atomic beam, the uv optics, the ionization region electron optics, and the double Mott scattering polarization analysis.

electronic magnetic quantum number $m_J = +1/2$, which then pass adiabatically into the ionization region where there is a longitudinal magnetic field of about 200 G provided by the indicated coil. Intense light in the uv range from 1700 Å to 2300 Å is produced by a vortex-stabilized argon flash lamp and is focussed onto the polarized $^6$Li atomic beam. The resulting photoelectrons are extracted from the photoionization region with a kinetic energy of about 70 keV, and then transported either into the accelerator or into the polarization analyzer which employs double Mott scattering. The PEGGY beam is longitudinally polarized either parallel or antiparallel to the beam direction depending on the direction of the current in the polarizing coil. The operating characteristics of the resulting high energy polarized electron beam from PEGGY I are given in Table I. A complete report on PEGGY I, together with earlier references, has been written.[20]

With reference to PEGGY II, Figure 3 shows an energy band diagram for the nonmagnetic semi-conductor GaAs, and possible electric dipole transitions from the valence band to the conduction band. Optical pumping from the valence $P_{3/2}$ magnetic substates to the conduction $^2S_{1/2}$ magnetic substates with pulsed circularly polarized laser light of about 7100 Å wavelength produces polarized electrons in the conduction band with a magnitude of polarization up to about 0.50 and with a sign of polarization depending upon whether the laser light is right or left circularly polarized. If the clean surface of the GaAs crystal is treated with cesium and oxygen to obtain a negative electron affinity, the spin polarized electrons can be

Table I. Operating Characteristics of PEGGY I

| Characteristic | Value |
|---|---|
| Pulse length | 1.6 μs |
| Repetition rate | 180 pps |
| Electron intensity (at high energy) | ~$10^9$ $e^-$/pulse |
| Pulse to pulse intensity variation | <5% |
| Electron polarization | 0.85±0.07 |
| Polarization reversal time | <1 s |
| Time between reversals | 2 min |
| Intensity difference upon reversal | <5% |
| Lifetime of Li oven load | 175 h |
| Time to reload Li | 43 h |

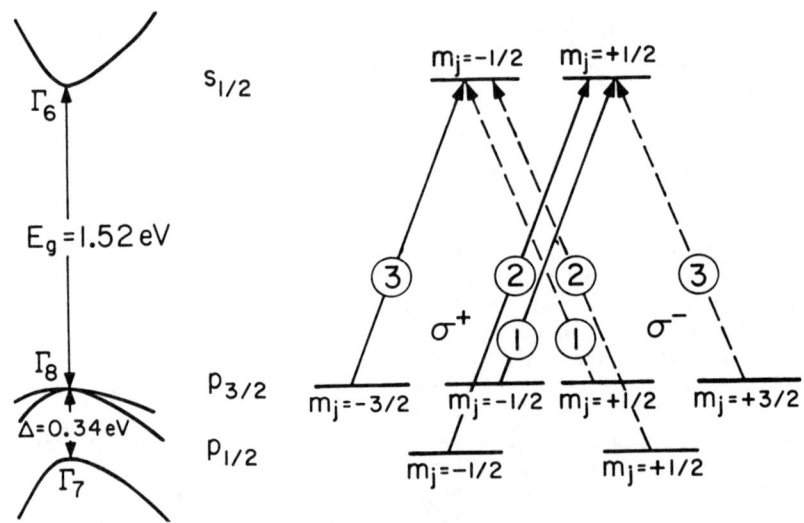

Fig. 3. Energy bands of GaAs at k=0 (left) and transitions from the $P_{1/2}$ and $P_{3/2}$ levels to the $S_{1/2}$ states. Solid and broken lines are for $\sigma^+$ and $\sigma^-$ light, respectively. The circled numbers give the relative transition intensities.

emitted from the conduction band. A Pockels cell was used to reverse the circular polarization of the light and hence the helicity of the photoemitted electrons. A schematic diagram of the PEGGY II source is shown in Figure 4. The operating characteristics of PEGGY II for the experiment in which parity nonconservation was

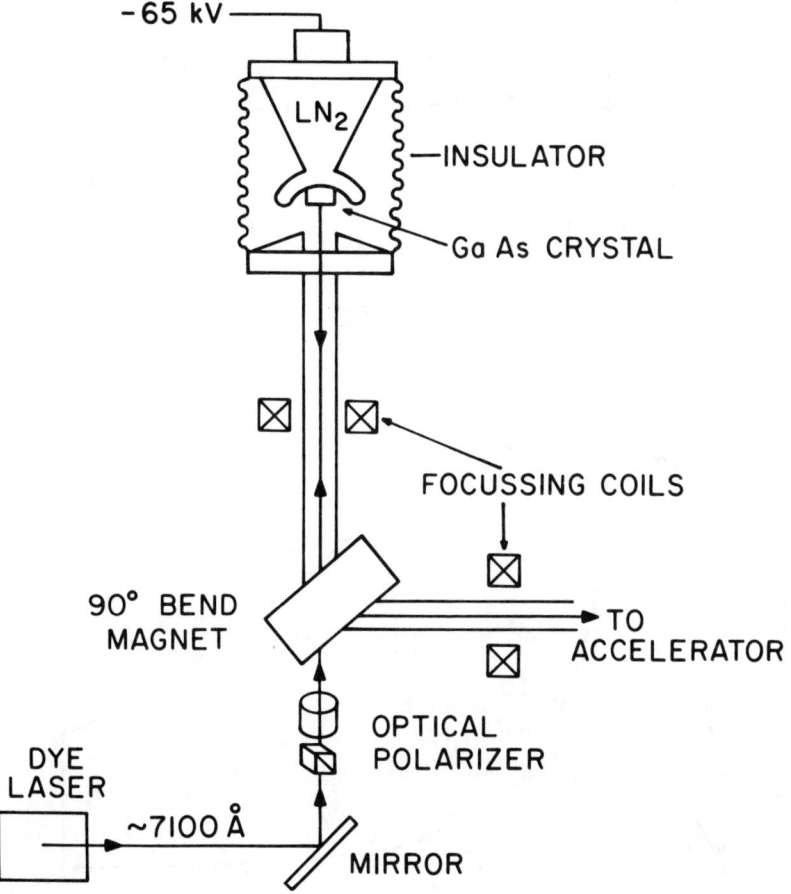

Fig. 4. Schematic diagram of PEGGY II showing the principal components of the GaAs crystal mount, the laser optics, and the electron optics.

discovered are given in Table II.[11]

The longitudinal polarization of the high energy electron beam has been measured by elastic electron-electron scattering (Møller scattering) from a magnetized iron foil.[3] Møller scattering has been chosen for all measurements done thus far because the cross section and analyzing power are large and the process is purely quantum electrodynamic. Figure 5 shows the Møller asymmetry and laboratory cross section at a representative incident beam energy of 9.71 GeV. Since the laboratory scattering angle is small, provision must be made in the measurement to separate physically the scattered electrons from the primary beam. Rather clean elastic

Table II. Operating Characteristics of PEGGY II

| Characteristic | Value |
|---|---|
| Pulse length | 1.5 μs |
| Repetition rate | 120 pps |
| Electron intensity (at high energy) | (1 to 4)x10$^{11}$ e$^-$/pulse |
| Pulse to pulse intensity variation | ~3% |
| Electron polarization | 0.37, average |
| Polarization reversal time | pulse to pulse |

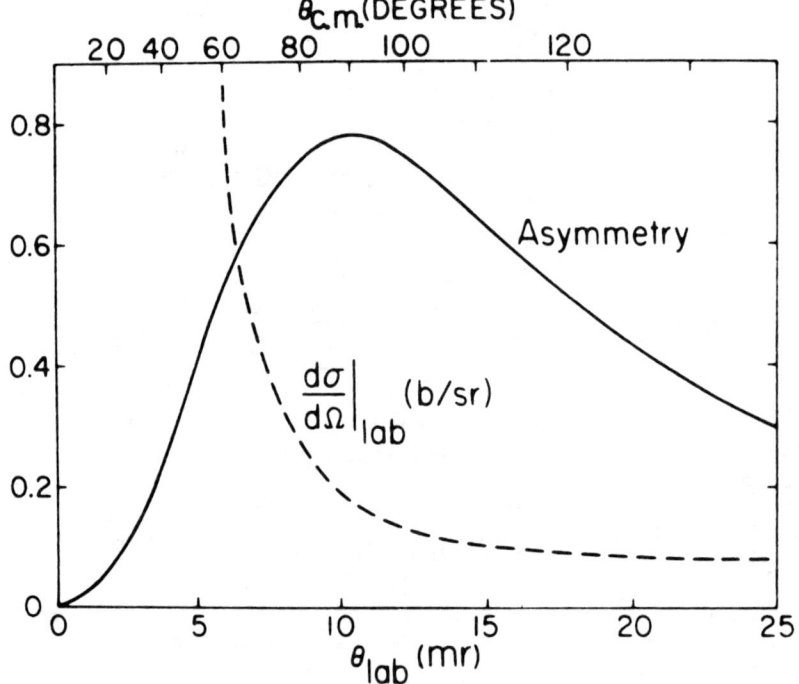

Fig. 5. The Møller asymmetry and laboratory cross section plotted versus laboratory angle for the representative incident energy of 9.712 GeV.

peaks have been observed with a background believed due to radiative Coulomb scattering of about 10% of the signal. The longitudinal beam polarization P measured as a function of beam energy E is shown in Fig. 6. The variation of P with E is caused by the g-2 precession of the spin relative to the momentum in the beam switchyard, where the beam from the accelerator is bent into the experimental

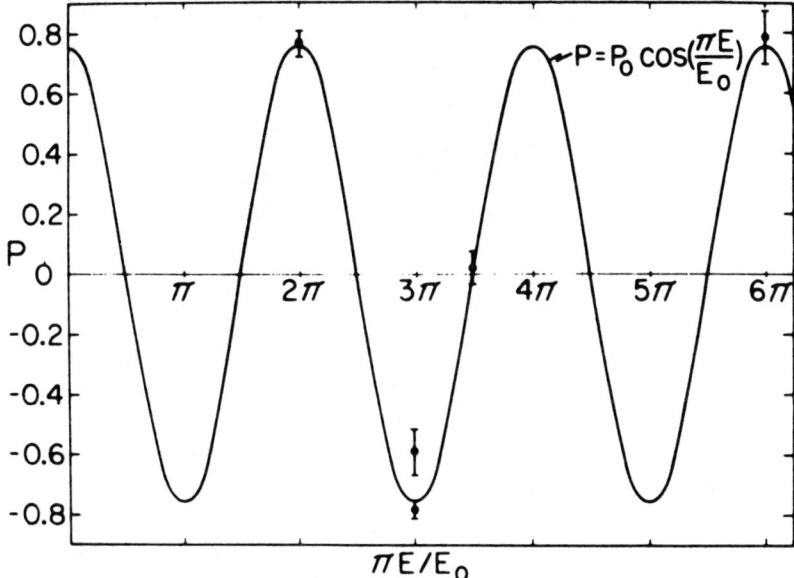

Fig. 6. The longitudinal component, P, of the beam polarization plotted versus $\pi E/E_o$, the angle through which the spin precesses relative to the momentum during the 24.5° bend into the experimental area. E is the beam energy and $E_o$=3.237 GeV. The curve shown is a best fit to the data and has an amplitude $P_o$=0.76±0.03. $P_o$ is the only free parameter.

area. The accuracy with which the longitudinal electron polarization has been determined in the experiments with polarized electrons is about 5% to 10%.[11,12] The error is due to the statistical counting error in the measured asymmetry, the uncertainty in the background subtraction, and the uncertainty about the spin magnetism of the magnetized iron foils.[3]

The polarized proton targets used in the lepton-hadron scattering experiments have employed the well known method of dynamic nuclear orientation using paramagnetically-doped hydrocarbon samples. The principal special feature has been the large energy dissipation in the target due to the relatively intense electron beam and the associated radiation damage to the target. Techniques for rastering the beam over the target area in order to minimize the effect of radiation damage and provide uniform target polarization, for annealing radiation damage, and for rapid changing of target material have been developed.[12,18,19] The target of most recent design which has been used at SLAC in experiments to study the scattering of polarized electrons by polarized protons operates at 1°K and uses a 50 kG magnetic field. A drawing of the target assembly is shown in Figure 7, and the operating characteristics of the target are given in Table III.

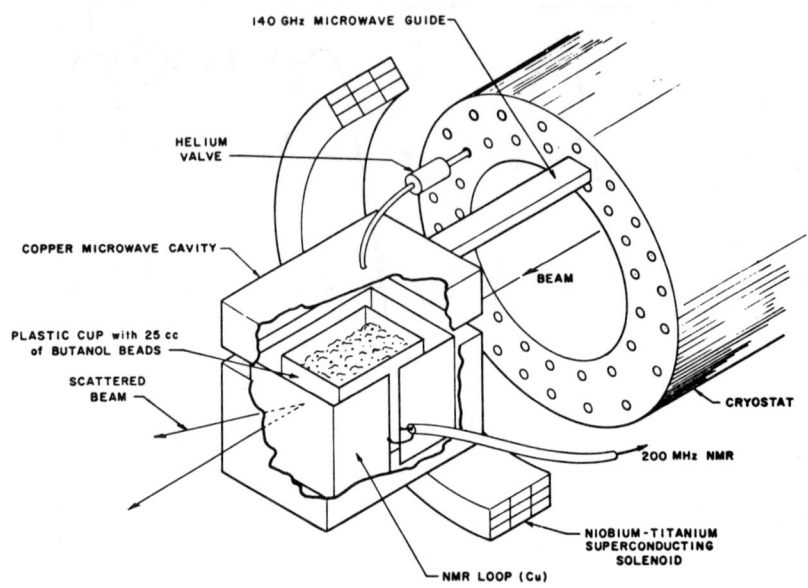

Fig. 7. Schematic diagram of the SLAC-Yale polarized proton target which operates at $1°K$ and 50 kG.

Table III. Operating Characteristics of the SLAC-Yale Polarized Proton Target

| Characteristic | Value |
|---|---|
| Magnetic field (longitudinal field of superconducting magnet) | 50 kG |
| Temperature | $1.05°K$ |
| Target material | 25 cm$^3$ of butanol-porphyrexide beads (~1.7 mm diam) |
| Initial polarization of free protons[a] | 0.50 to 0.65 |
| Depolarizing dose (1/e) | ~$3 \times 10^{14}$ $e^-/cm^2$ |
| Polarizing time (1/e) | ~4 min |
| Anneal or target change time (including polarizing) | ~45 min |

[a] Improvements in target operation gave the larger polarization values in the later parts of the experiment.

## TESTS OF INVARIANCE PRINCIPLES

Spin polarization, of course, has played a central role in the study of invariance principles. Sensitive tests for time reversal invariance and the discovery of parity nonconservation in the electromagnetic interactions have been made through polarized lepton-hadron scattering.

A test of T invariance in the electromagnetic interaction of hadrons has been made in experiments involving the inelastic scattering of electrons from polarized protons.[1,4] The test involved measurement of the asymmetry A where

$$A = \left(\frac{\sigma^\uparrow - \sigma^\downarrow}{\sigma^\uparrow + \sigma^\downarrow}\right) \frac{1}{\vec{P}\cdot\hat{n}}$$

The quantity $\sigma^\uparrow(\sigma^\downarrow)$ is the differential cross section for the inclusive inelastic reaction ep→eX with only the outgoing electron being observed and with the target polarization $\vec{P}$ parallel (antiparallel) to the normal $\hat{n}$ to the scattering plane; $\hat{n}=(\vec{p}_{in}\times\vec{p}_{out})/|\vec{p}_{in}\times\vec{p}_{out}|$ where $\vec{p}_{in}$ ($\vec{p}_{out}$) is the incident (scattered) electron momentum. In the approximation of single photon exchange a nonzero value for A would establish a violation of T invariance. The principal motivation for these experiments on T invariance was the suggestion that the violation of CP invariance discovered in the decay of the $K_L^0$ meson[21] might result from a violation of T invariance in the electromagnetic interactions of hadrons[22] which could be tested in such experiments.[23]

The experimental arrangement of the CEA experiment[1] is shown in Fig. 8. With an incident electron beam intensity of about $2\times10^{10}$ e-/sec and an average proton polarization of about 0.2 for free protons in an alcohol-water target, asymmetries were measured for resonance region inelastic scattering with missing mass W between 1170 and 1688 MeV and 4-momentum transfer squared $Q^2$ from 0.2 to 0.7 $(GeV/c)^2$. The measured asymmetries are shown in Figure 9 and are seen to be consistent with zero within the experimental errors of 4 to 12% which were dominantly statistical counting errors. Results of the similar experiment at SLAC,[4] which measured asymmetries for W up to 2.2 GeV and for $Q^2$ from 0.4 to 1.0 $(GeV/c)^2$ are shown in Figure 10. Again the authors conclude that within the 2 to 5% experimental errors no violation of T invariance has been observed, although the asymmetry of $(4.5\pm1.4)\%$ measured at W=1200 MeV, $Q^2=0.6$ $(GeV/c)^2$ is statistically unlikely. The positron data shown in Figure 10 are consistent with A=0 also, and, together with similar asymmetry measurements for elastic e-p scattering,[5] indicate as expected that two-photon exchange processes are unimportant. Although the violation of T invariance in the electromagnetic interaction of hadrons is not currently regarded as a likely explanation of CP violation in $K_L^0$ decay,[24] it would be interesting to do more sensitive experiments of the type discussed above, which should be possible with modern polarized proton targets and other improved instrumentation.

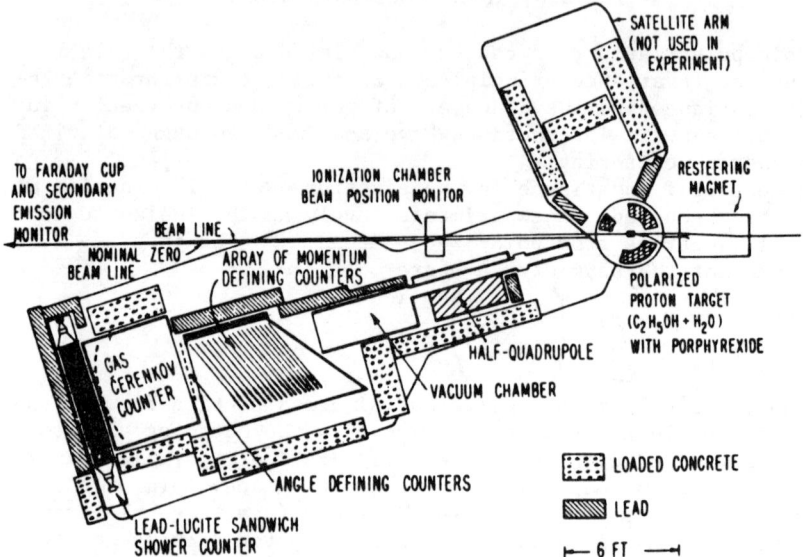

Fig. 8. Schematic diagram of inelastic electron scattering experiment[1] done at CEA to test T invariance.

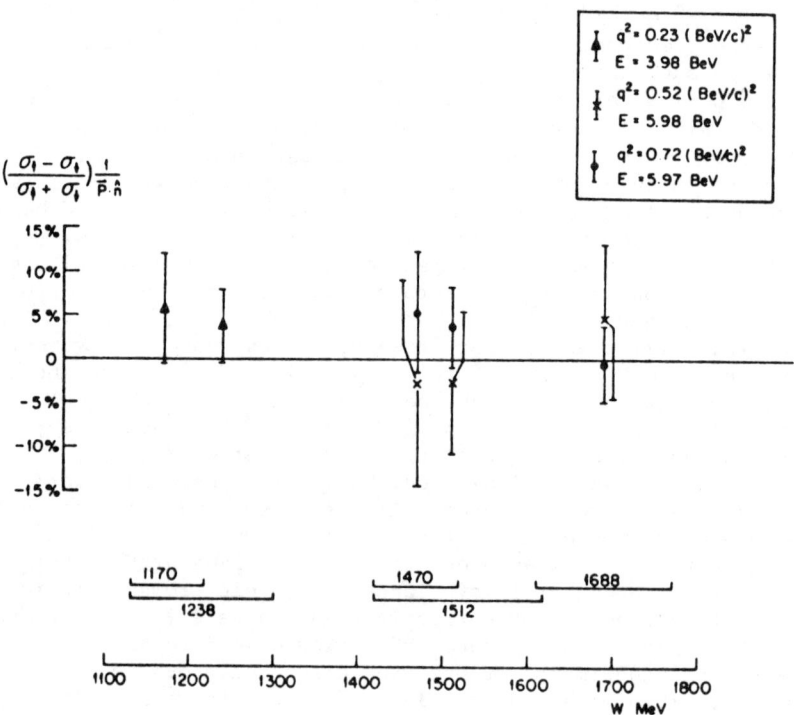

Fig. 9. Experimental asymmetries for different π-N center-of-mass system energies.[1]

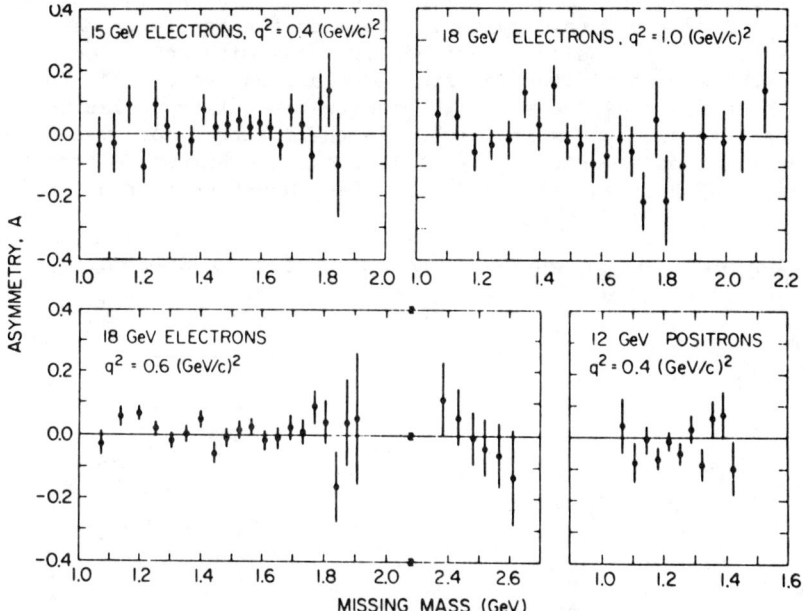

Fig. 10. The asymmetry values A are shown as a function of missing mass, where the errors are standard deviations calculated from counting statistics.[4] On each graph we indicate the incident beam (electrons or positrons), the incident energy, and the four-momentum transfer squared ($q^2$).

The topic of parity (or space inversion) nonconservation (PNC) in the electromagnetic interactions has seen important experimental and theoretical advances recently. In particular, the recent experimental discovery of parity nonconservation in the deep inelastic scattering of longitudinally polarized electrons by unpolarized deuterons and protons[11] is discussed in the article in these Proceedings by C.Y. Prescott. Here we give only a brief survey, including some historical comments.

The question of parity nonconservation in the electromagnetic interactions is an old one. Until recently P invariance in the electromagnetic interactions had received its principal tests in atomic physics.[25-27] The principal modern stimulus for the study of this topic has been the theoretical development of the unified gauge theories of the weak and electromagnetic interactions,[28,29] and the experimental discovery of neutral current interactions in neutrino scattering.[30,31,32] These theoretical and experimental discoveries suggest there exists a weakly interacting heavy neutral vector boson, called $Z^0$, which can be exchanged between a lepton and a hadron thus resulting in a weak interaction of first order in the Fermi coupling constant $G_F$. The theories further suggest that this interaction might violate parity conservation.[33]

A test for parity nonconservation in the interaction between an electron and a proton in high energy scattering can be made by measuring the difference (or asymmetry) between the scattering cross sections of longitudinally polarized electrons with spin along or opposite to the momentum direction from unpolarized protons.[34] Such an experiment effectively measures the manifestly parity violating pseudoscalar quantity $\vec{\sigma}\cdot\vec{p}$ in which $\vec{\sigma}$ is the electron spin vector and $\vec{p}$ is the electron momentum. Figure 11 shows the Feynman diagrams for single photon $\gamma$ and single $Z^0$ exchange in e-p scattering, and

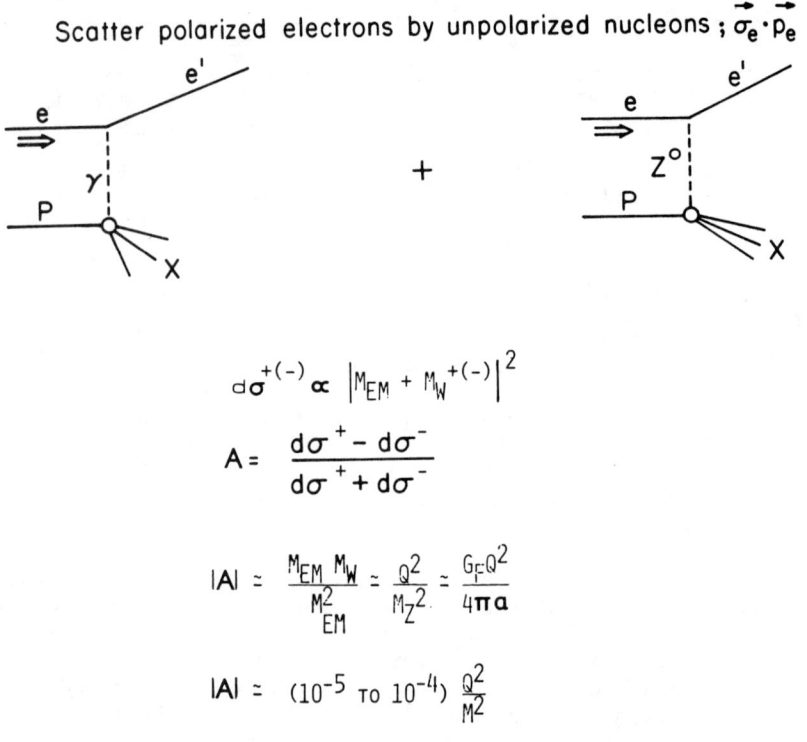

Fig. 11. Feynman diagrams indicating a cross section asymmetry A between + and - helicity polarized electrons scattered from unpolarized protons. The asymmetry arises from interference between the electromagnetic amplitude $M_{EM}$ associated with $\gamma$ exchange and the weak amplitude $M_W$ associated with $Z_0$ exchange. M is the proton mass in GeV and $Q^2$ is the 4-momentum transfer squared in $(GeV/c)^2$.

indicates how the interference between these diagrams can lead to an asymmetry in the scattering cross sections for + and - helicity incident electron beams. The numerical estimate for A is based on the standard (Weinberg-Salam) unified gauge theory.[35,36,37]

The first experimental test of parity nonconservation in high energy electron-nucleon scattering came as a byproduct of measurements at SLAC of asymmetries in elastic and deep inelastic scattering of longitudinally polarized electrons from PEGGY I by longitudinally polarized protons.[6,8] Asymmetries associated with the polarized protons can be eliminated by combining data for opposite orientations of proton polarization. Subsequently an experiment dedicated to the search for parity nonconservation with PEGGY I (and proposed[38] before the discovery of neutral current interactions) was also done at SLAC.[7,9] These experiments set an upper limit to parity nonconservation in electron-nucleon scattering of several parts in $10^3$ for $Q^2$ of about 1 $(GeV/c)^2$. The accuracies were limited by counting statistics and by possible asymmetries in the PEGGY I beam.

An experimental test of parity nonconservation in deep inelastic muon-nucleon scattering has been done at Serpukhov[10] using the polarized muons from pion decays, where $\mu^-$ of 21 GeV/c momentum and positive (negative) helicity are obtained from forward (backward) decays of $\pi^-$ of momentum 28(40) GeV/c. The experimental arrangement is shown in Figure 12 and the measured asymmetries in Figure 13. As seen in Figure 13 the observed asymmetries for incident $\mu^-$ momenta of about 21 GeV/c scattering from Fe with $Q^2$ from 1 to 11 $(GeV/c)^2$ are consistent with zero within the 5% to 40% statistical counting errors.

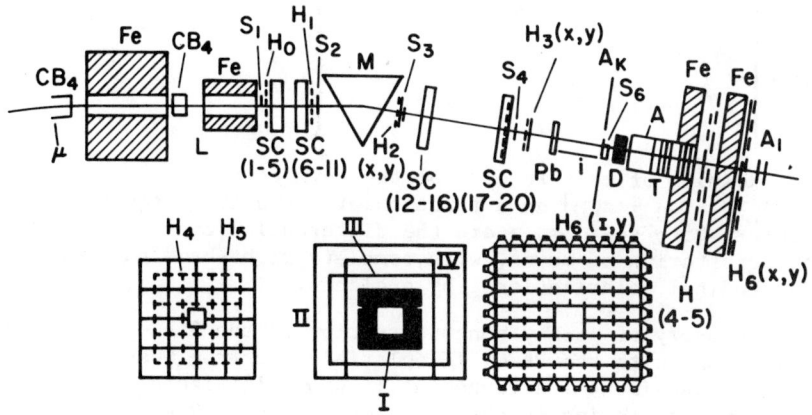

Fig. 12. Diagram of experiment to search for parity violation effects in deep inelastic μN interactions. The main elements of the apparatus are as follows: a) -muon-beam magnetic spectrometer (magnet M, counters $S_1$-$S_5$, hodoscopes $H_0$-$H_3$, wire spark chambers $SC_{1-20}$); b) -target-calorimeter T (total thickness 1130 g/cm$^2$ of iron, 12 counters); c) -hodoscopic muon detector for detection of scattered muons (hodoscopes $H_4$-$H_6$); d) -shower detector D (20 radiation lengths); e) -guard counters $A_k$, which removed the halo, anticoincidence counters $A_1$ and $A_2$ which identified muon interactions in the target T; auxiliary counter A.

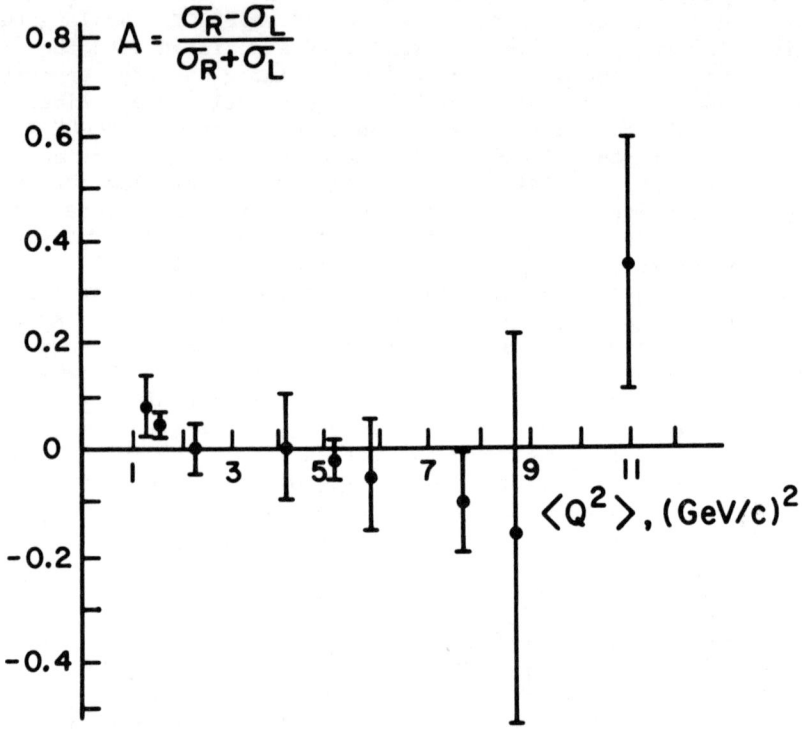

Fig. 13. Results of measurements of the asymmetry
$A=[\langle\sigma_R\rangle-\langle\sigma_L\rangle]/[\langle\sigma_R\rangle+\langle\sigma_L\rangle]$ as a function of the average square momentum transfer $\langle Q^2 \rangle$; $\langle\sigma_R\rangle$, $\langle\sigma_L\rangle$ are the differential cross sections for inelastic scattering of muons, with complete right-hand and left-hand longitudinal polarizations, by nucleons.

The discovery of parity nonconservation in the SLAC-Yale experiment[11] was done with the PEGGY II polarized electron beam and a $LD_2$ target. The experimental arrangement is shown in Figure 14, where deep inelastic scattering is observed for $Q^2 \approx 1.6$ $(GeV/c)^2$. The high intensity of the PEGGY II beam and its rapid polarization reversibility by optical means, together with careful monitoring and control of the accelerator output beam, made it possible to measure asymmetries at the $10^{-5}$ level. We note that this level of sensitivity is not unprecedented, and, indeed, in a similar experiment testing parity conservation in proton-proton scattering using a longitudinally polarized proton beam, asymmetries have been measured[39] at the $10^{-7}$ level at MeV energies[40] and at the $10^{-5}$ level at GeV energies.[41]

Results of the asymmetry measurements are plotted in Figure 15. Nonzero values for the asymmetry are clearly found, thus establishing the violation of parity conservation. The variation of asymmetry with the energy $E_o$ of the incident beam is due to the g-2 spin precession discussed in connection with Figure 6 and provides important confirmation that the observed asymmetry is associated

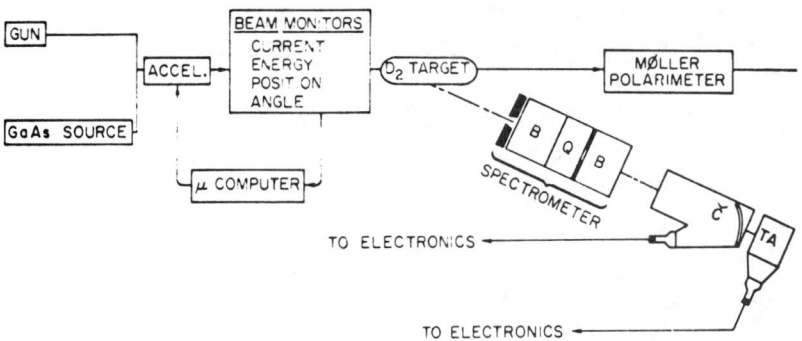

Fig. 14. Schematic layout of the experiment in which parity nonconservation was discovered. Electrons from the GaAs source or the regular gun are accelerated by the linac. After momentum analysis in the beam transport system the beam passes through a liquid deuterium target. Particles scattered at 4° are analyzed in the spectrometer (bend-quad-bend) and detected in two separate counters (a gas Cerenkov counter, and a lead-glass shower counter). A beam monitoring system and a polarization analyzer are only indicated, but they provide important information in the experiment.

with the longitudinal polarization of the beam. The result quoted[11] for the asymmetry A in deep inelastic e-D scattering is

$$\frac{A}{Q^2} = (-9.5 \pm 1.6) \times 10^{-5} \, (\text{GeV}/c)^2$$

for $<Q^2> = 1.6 \, (\text{GeV}/c)^2$ and $<y> = (E_o - E')/E_o = 0.21$, in which $E'$ is the energy of the scattered electron. The error is contributed about equally by the statistical counting error and by the systematic errors associated with asymmetries in the beam (principally its energy) and uncertainty in the beam polarization. For deep inelastic e-p scattering for which fewer data were obtained the quoted result is

$$\frac{A}{Q^2} = (-9.7 \pm 2.7) \times 10^{-5} \, (\text{GeV}/c)^2$$

for $<Q^2> = 1.6 \, (\text{GeV}/c)^2$ and $<y> = 0.21$.

Comparison of the measured asymmetry for deep inelastic e-D scattering with the theoretical predictions[37] of two gauge theory models - the standard model of Weinberg[28] and Salam[29] and a so-called hybrid model[42] - for several values of the parameter $\sin^2\theta_W$ in the theories is shown in Figure 16. There is good agreement of the measured point with the standard theory for $\sin^2\theta_W \approx 0.20 \pm 0.03$ which is consistent with the values obtained for $\sin^2\theta_W$ in neutrino experiments.[43] Agreement with the hybrid model is not good and requires a very low value of $\sin^2\theta_W$.

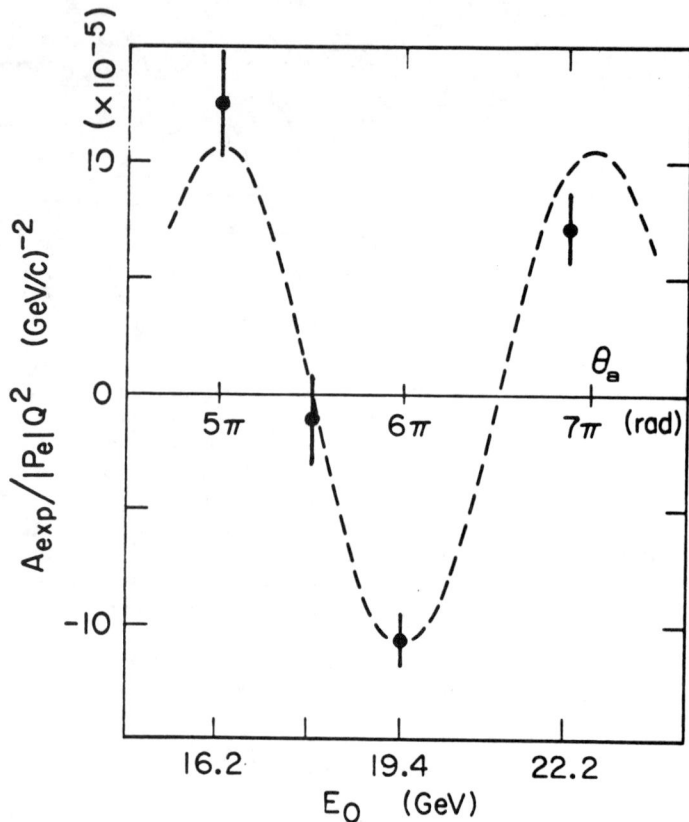

Fig. 15. The experimental asymmetry shows the expected variation (dashed line) as the beam helicity changes as a function of beam energy due to the g-2 precession in the beam transport system. The data are for the shower counter and the deuterium target. No systimatic errors are shown. No corrections have been made for helicity dependent differences in beam parameters.

The discovery of parity nonconservation in deep inelastic electron-nucleon scattering has revealed a new significant interaction between electrons and nucleons and has thus opened up a new field of study of weak-electromagnetic interference effects in lepton-nucleon interactions. According to present phenomenological theory[44,45] there should be four independent coupling constants characterizing the e-P and e-N vector and axial vector interactions, which could be determined by four suitable independent measurements. Thus far only one measurement has been made in deep inelastic scattering. Measurements at other values of y, and measurements at high $Q^2$ and $x(x=Q^2/2M\nu)$ where $\nu = (E_0-E')$, which are of adequate precision to distinguish the values of the coupling constants for neutron and proton, could determine these coupling constants.

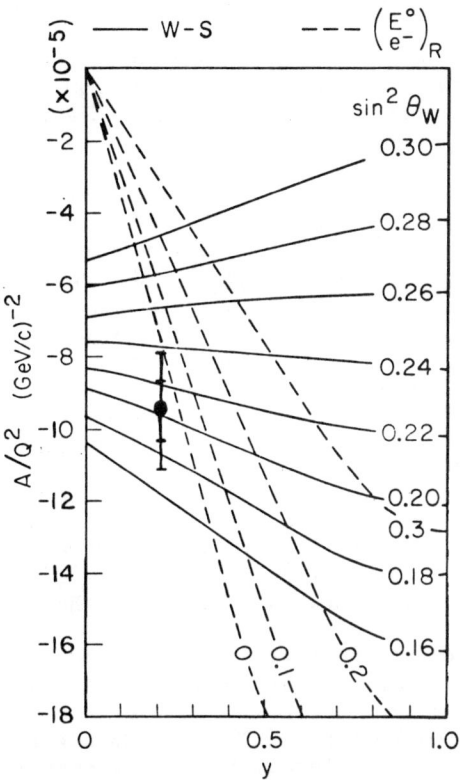

Fig. 16. Comparison of our result for deuterium with two SU(2)xU(1) predictions using the simple quark-parton model for nucleons. The outer error bars correspond to the error quoted in the text. The inner error bars correspond to the statistical error. The y-dependence of $A/Q^2$ for various values of $\sin^2\theta_W$ is shown for two models: Weinberg-Salam (solid lines) and the hybrid model (dashed line).

Parity nonconservation can also be studied in electron-nucleus scattering at lower energies and lower $Q^2$ values.[46,47] Elastic scattering from the ground state of carbon which has spin and isospin equal to zero is a very clean case to study. The predicted asymmetry is

$$A = -10^{-2} \frac{Q^2}{M^2} (\varepsilon_{1e}^{VA} + \varepsilon_{2e}^{VA})$$

in which $\varepsilon_{1e}^{VA}(\varepsilon_{2e}^{VA})$ are the vector quark 1(2) - axial vector electron coupling constants and M is the proton mass. In the standard theory this becomes

$$A = -4 \times 10^{-4} \frac{Q^2}{M^2} \sin^2\theta_W$$

For $Q^2 \simeq 0.01$ (GeV/c)$^2$ and with $\sin^2\theta_W \simeq 0.25$, the predicted asymmetry A is about $10^{-6}$. Elastic scattering from the proton, particularly in the backward direction, or quasi-elastic scattering from the deuteron are also interesting possibilities for lower $E_o$, $Q^2$ scattering.[48,49,50]

The modern gauge theories predict that parity nonconservation should also be present in the electron-electron interaction. From the initial high energy Møller scattering experiment[3] to measure the electron polarization, it can be concluded[51] that $A<10^{-3}$ for $Q^2 \simeq 0.01$ (GeV/c)$^2$. The theoretically predicted value for A based on the standard theory is at least three orders of magnitude less than this. In view of the large cross section for Møller scattering a much more sensitive experiment could certainly be done.

The European Muon Collaboration at CERN has proposed[52] to test parity conservation in the high energy scattering of polarized muons by nucleons for incident muon energies up to about 200 GeV and for $Q^2$ in the scattering up to 100 (GeV/c)$^2$. In view of the fact that and asymmetry of about $10^{-4}$ has been observed[11] in e-D deep inelastic scattering at $Q^2 \simeq 1.6$ (GeV/c)$^2$ and that the theory predicts the asymmetry should vary as $Q^2$, large asymmetries of the order of $10^{-2}$ should be observed. Such new data involving the muon would be most interesting.

As a final comment on invariance principles we note that a test of Lorentz invariance or special relativity is provided by comparing the value of the electron anomalous g-factor a as measured at high energy with the PEGGY beam at SLAC in the Møller scattering experiment with the value of a determined from low energy experiments.[53]

## POLARIZED ELECTROPRODUCTION

A new type of information on proton structure, its internal spin structure, has recently become available from a new type of experiment, polarized electroproduction. The scattering of longitudinally polarized electrons (i.e. polarized along beam direction) by longitudinally polarized protons is studied (Figure 17). Only the scattered electron is observed in an inclusive scattering experiment. With longitudinally polarized electrons the virtual photon will have a circular polarization with a component along the direction of the proton spin. The quantity measured is the asymmetry A, the normalized difference between the differential scattering cross sections for the antiparallel and parallel spin configurations. Data have been obtained for elastic,[12] deep inelastic,[6,8] and resonance[13] region scattering.

In Figure 18 the expression for the differential cross section for deep inelastic scattering is given and now includes, in addition to the familiar spin-averaged proton structure functions $W_1$ and $W_2$, two new spin-dependent structure functions $G_1$ and $G_2$, which can only

## POLARIZED ELECTRON-PROTON SCATTERING

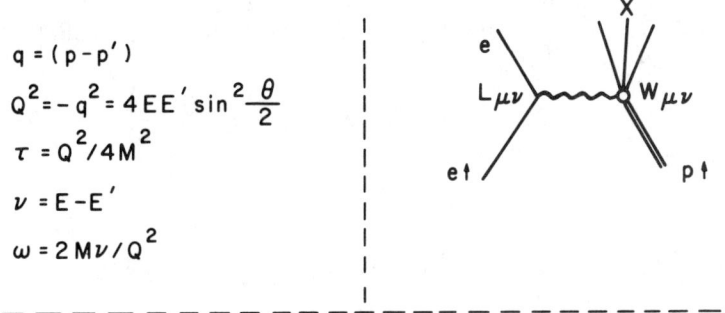

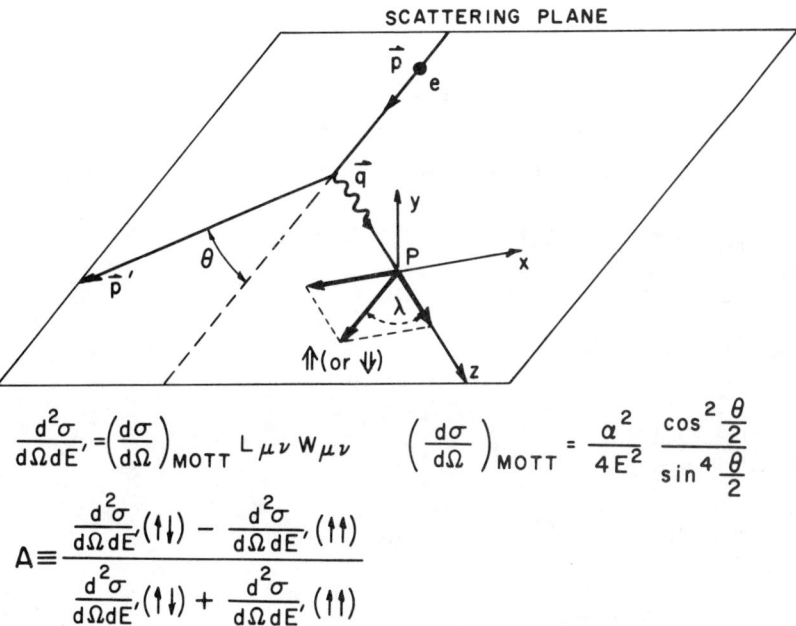

$$\frac{d^2\sigma}{d\Omega dE'} = \left(\frac{d\sigma}{d\Omega}\right)_{MOTT} L^{\mu\nu} W_{\mu\nu} \qquad \left(\frac{d\sigma}{d\Omega}\right)_{MOTT} = \frac{\alpha^2}{4E^2} \frac{\cos^2\frac{\theta}{2}}{\sin^4\frac{\theta}{2}}$$

$$A \equiv \frac{\frac{d^2\sigma}{d\Omega dE'}(\uparrow\downarrow) - \frac{d^2\sigma}{d\Omega dE'}(\uparrow\uparrow)}{\frac{d^2\sigma}{d\Omega dE'}(\uparrow\downarrow) + \frac{d^2\sigma}{d\Omega dE'}(\uparrow\uparrow)}$$

Fig. 17. The kinematics of polarized electron-proton scattering for which the asymmetry A is measured. In the expressions, E = energy of incident electron, E' = energy of scattered electron, M = proton mass, and α = fine structure constant.

be determined from polarized electroproduction.[54,55] Alternatively, we can consider $A_1$ and $A_2$ which refer to the virtual photon-proton interaction as the spin dependent quantities. The measured asymmetry A is related to $A_1$ and $A_2$ as indicated. D is a kinematic depolarization factor of the virtual photon, and η is a small kinematic factor. $A_1$ is the normalized difference between $\sigma_{1/2}$ and $\sigma_{3/2}$, where $\sigma_{1/2}$ is the total absorption cross of the virtual photon by the proton when the z component (z is the direction of the vir-

## CROSS SECTION AND ASYMMETRY

$$\frac{d^2\sigma}{d\Omega dE'} = \left(\frac{\alpha^2 \cos^2 \frac{\theta}{2}}{4E^2 \sin^4 \frac{\theta}{2}}\right)\left[W_2 + 2\tan^2\frac{\theta}{2} W_1 \pm 2\tan^2\frac{\theta}{2}(\varepsilon + E'\cos\theta)MG\right.$$
$$\left. \pm 8 EE'\tan^2\frac{\theta}{2}\sin^2\frac{\theta}{2} G_2\right]$$

+(A)
−(P)

$$\frac{d^2\sigma}{d\Omega dE'} = \left(\frac{d\sigma}{d\Omega}\right)_M \left(\frac{1}{\varepsilon(1+\nu^2/Q^2)}\right) W_1 \left\{1 + \varepsilon R \pm (1-\varepsilon^2)^{1/2}\cos\psi A_1 \right.$$
$$\left. \pm \left[2\varepsilon(1-\varepsilon)\right]^{1/2} \sin\psi A_2\right\}$$

$$\varepsilon = \left[1 + (1+\nu^2/Q^2)\tan^2\frac{\theta}{2}\right]^{-1}$$

$$R = \sigma_L/\sigma_T; \quad \sigma_T = (\sigma_{1/2} + \sigma_{3/2})/2$$

$$A = \frac{d\sigma(\uparrow\downarrow) - d\sigma(\uparrow\uparrow)}{d\sigma(\uparrow\downarrow) + d\sigma(\uparrow\uparrow)}$$

$$A = D(A_1 + \eta A_2)$$

$$D = \frac{E - E'\varepsilon}{E(1+\varepsilon R)} = \frac{(1-\varepsilon^2)^{1/2}\cos\psi}{(1+\varepsilon R)}$$

$$\eta = \frac{\varepsilon(Q^2)^{1/2}}{E-E'\varepsilon} = \left(\frac{2\varepsilon}{1+\varepsilon}\right)^{1/2} \tan\psi \simeq \tan\psi$$

$$A_1 = \frac{\sigma_{1/2} - \sigma_{3/2}}{\sigma_{1/2} + \sigma_{3/2}}$$

$$A_2 = \frac{2\sigma_{TL}}{\sigma_{1/2} + \sigma_{3/2}}$$

$$|A_1| \leq 1; \quad |A_2| \leq \sqrt{R}$$

Fig. 18. Formulae for cross section and asymmetry relevant to polarized electron-proton scattering.

tual photon momentum) of angular momentum of the virtual photon plus proton is 1/2, and for $\sigma_{3/2}$ it is 3/2. $\sigma_{TL}$ arises from the interference between transverse and longitudinal photon-nucleon amplitudes. Positivity limits on $A_1$ and $A_2$ are $|A_1| \leq 1$, $|A_2| \leq \sqrt{R}$. To a good approximation our experiment determines $A_1$.

For the polarized electroproduction experiment two new elements must be added to the usual inclusive e-p scattering experiment: a polarized electron source and a polarized proton target. PEGGY I was used as the polarized electron source and has the characteristics listed in Table I. The polarized proton target assembly is shown in Figure 7 and its characteristics are listed in Table III.

The method of the experiment was checked by measuring[12] the asymmetry A in e-p elastic scattering where the theoretical value is predicted from the measured proton form factors $G_E$ and $G_M$ as indicated in Figure 19. The measured quantity is the counting rate asymmetry $\Delta$ between the anti-parallel and parallel spin configurations. It is related to the intrinsic electron-proton scattering asymmetry A by the factors $P_e$, the electron polarization, $P_p$, the free proton polarization, and F, the fraction of the scattered electrons originating from free protons. At our kinematic point E=6.47 GeV, $\theta=8°$, and $Q^2=0.76$ (GeV/c)$^2$, $A_{theor} = 0.112 \pm 0.001$ in excellent agreement with the experimental value of $A_{expt} = 0.103 \pm 0.015$ within the 15% experimental error. Our measurement determined the sign of $G_M/G_E$ to be positive, which had not been measured previously.

Figure 20 indicates our measured data points.[6,8,12,13] The solid dots are seven deep inelastic points with missing mass W between 2 and 4 GeV, $Q^2$ between 1 and 4 (GeV/c)$^2$ and $\omega$ between 2 and 10. The unifilled squares are seven resonance region points. The crossed point is the elastic point. The open circles are data points planned in our upcoming SLAC experiment.

Figure 21 shows our measured asymmetry values[8] A/D ~ $A_1$, the virtual photon-proton asymmetry, for the deep inelastic data, in which the errors (vertical bars) are due principally to counting statistics and are typically about 25% of the measured values, whereas systematic errors in $P_e$, $P_p$ and F are 5% to 10%. Radiative corrections, which are relatively small, are included in the plotted points; the horizontal bars give the range in x associated with the radiative corrections. Note that intrinsically the spin dependent effect is large with $A_1$ being a large fraction of its positivity limit of 1, and hence implies that there is a large difference between the total absorption cross sections of virtual photons by protons for the antiparallel and parallel spin configurations. On the other hand, our measured counting rate asymmetries $\Delta$ are small, 0.5% to 1%, due to the small value of the product $P_e P_p F$ (~0.05, with F alone ~0.1) and the depolarization factor $D \simeq 1/3$, which relate $\Delta$ and $A_1$. However, our asymmetry measurement is rather free of systematic errors associated with electron beam helicity and hence the error has been limited principally to counting statistics which are about 0.1%.

There are several implications of these data: 1) test of

## ASYMMETRY IN ELASTIC SCATTERING

**Theory**

$$A = \frac{\left.\frac{d^2\sigma}{dpd\Omega}\right|_A - \left.\frac{d^2\sigma}{dpd\Omega}\right|_P}{\left.\frac{d^2\sigma}{dpd\Omega}\right|_A + \left.\frac{d^2\sigma}{dpd\Omega}\right|_P}$$

$$A(\text{THEOR}) = \frac{G_M}{G_E} \frac{\tau\{2\frac{M}{E} + \frac{G_M}{G_E}[2\tau\frac{M}{E} + 2(1+\tau)\tan^2\frac{\theta}{2}]\}}{1 + \tau\left(\frac{G_M}{G_E}\right)^2[1 + 2(1+\tau)\tan^2\frac{\theta}{2}]}$$

$E$ = initial electron energy; $\tau = \frac{Q^2}{4M^2}$

$G_E(G_M)$=electric (magnetic) form factors of proton: $G_E(0)$=1; $G_M(0)$=2.79.

**For kinematic point**

$E = 6.47$ GeV, $\theta = 8.0°$, $Q^2 = 0.76$ (GeV/c)$^2$

$A(\text{THEOR}) = +0.112 \pm 0.001$  ($\left|\frac{MG_E}{G_M}\right| = +0.98 \pm 0.04$).

**Experiment**

| E80(75)  $\Delta = P_e P_p FA$ | E80(76)  $(\Delta = \frac{N_{\uparrow\downarrow} - N_{\uparrow\uparrow}}{N_{\uparrow\downarrow} + N_{\uparrow\uparrow}})$ |
|---|---|
| $P_e = 0.51 \pm 0.06$ | $P_e = 0.85 \pm 0.08$ |
| $P_p = 0.34 \pm 0.03$ | $P_p = 0.51 \pm 0.04$ |
| $F = 0.27 \pm 0.02$ | $F = 0.33 \pm 0.03$ |
| $\Delta = 0.0063 \pm 0.0010$ | $\Delta = 0.0127 \pm 0.0015$ (899<W<999 MeV) |
| $A = 0.138 \pm 0.031(0.019)$ | $A = 0.092 \pm 0.017(0.010)$ |

$A_{\text{EXPT}} = 0.103 \pm 0.015$

$A_{\text{EXPT}} - A_{\text{THEOR}} = -0.009 \pm 0.015$

Fig. 19. Theoretical expression for and experimental measurement of asymmetry in elastic electron-proton scattering.

Bjorken sum rule, 2) scaling, 3) models of proton structure. Note that the values of $A_1$ are all positive, which is a firm prediction of the quark-parton model. Scaling is predicted[56] for spin dependent structure functions; in particular

$$A_1(\nu, Q^2) \to A_1(\omega) \text{ as } \nu, Q^2 \to \infty, \text{ with } \omega \text{ held constant}$$

Note that our data are consistent with scaling within their rather large errors, i.e. for fixed x, $A_1$ is independent of $Q^2$.

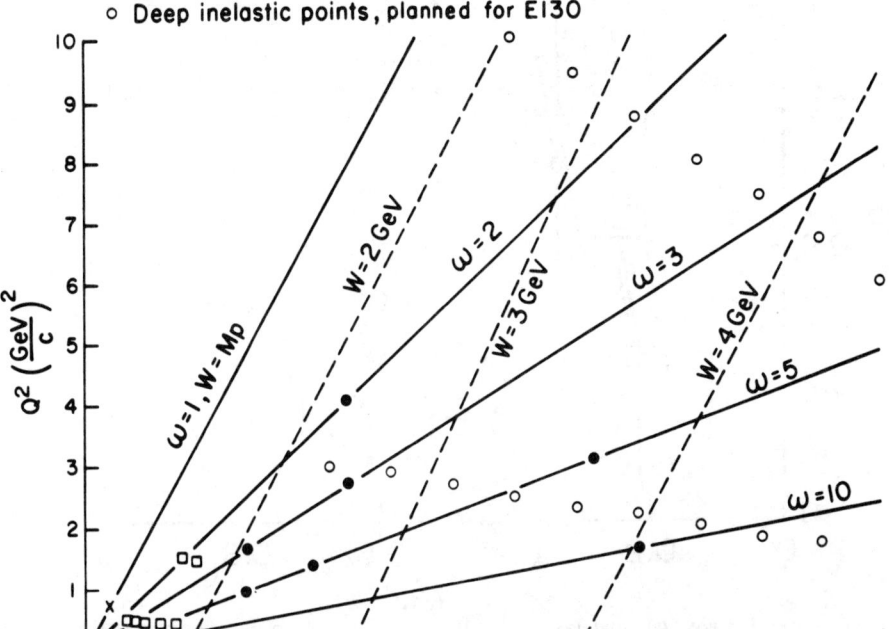

Fig. 20. Data points measured in SLAC experiment E80 and planned for SLAC experiment E130.

The Bjorken sum rule[57] as indicated in Figure 22 predicts equality between an integral over $\omega$ of a product of the spin-averaged nucleon structure function $\nu W_2$ and the spin dependent function $A_1$ in the scaling limit to the ratio of axial vector to vector weak coupling constants of beta decay. The difference of the proton and neutron structure factor products appears on the left hand side. This remarkable relation is based on quark current algebra and incorporates the general quark model of the nucleon and the view that the same weak current applies for quarks as for leptons. In the absence of experimental information on $A_1^n$ for the neutron we approximate $A_1^n = 0$ since quark-parton models of the neutron predict that $A_1^n$ is small. Using known measured values of $W_2^p$ and our measured values of $A_1$ [Assuming scaling we take the weighted average of the $A_1(\nu,Q^2)$ to give $A_1(\omega)$], we obtain the plotted points. Over our measur-

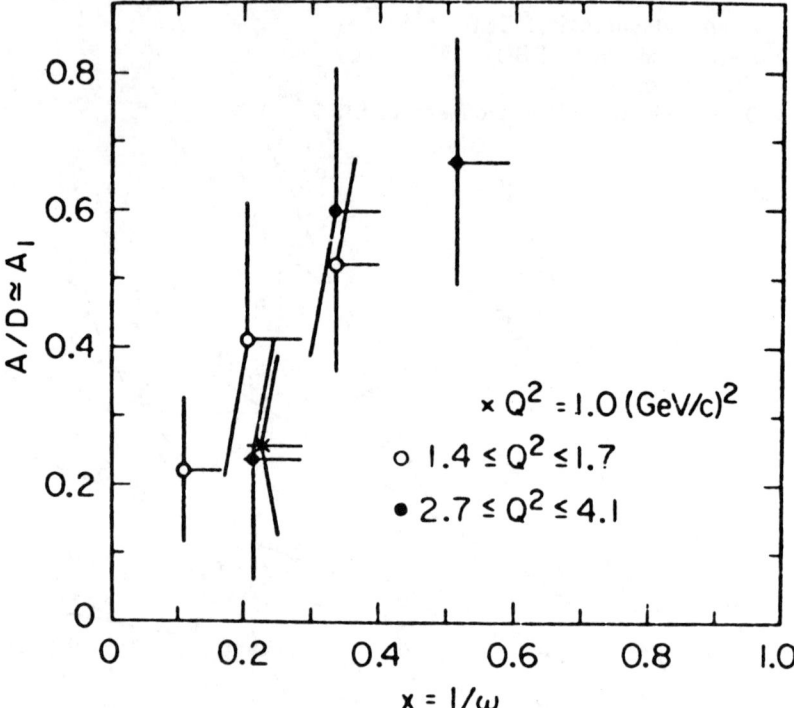

Fig. 21. Experimental values of $A/D \simeq A_1$ vs x. The horizontal bars give the range in x associated with the radiative corrections. (See K.P. Schüler in these Proceedings for a discussion of radiative corrections.)

ed interval from $\omega=2$ to $\omega=10$ we obtain the value of $0.16\pm0.03$ for the integral which saturates 40% of the sum rule. We fit our data to the form $A_1 = c/\sqrt{\omega}$ with $c=0.78$ which represents a satisfactory fit to our data and is suggested by Regge theory at large $\omega$. If we then extrapolate to small and large values of $\omega$, we obtain the value for the full integral of $0.34\pm0.05$, where the error includes only our errors in $A_1$. This result is very consistent with the sum rule and indeed saturates 82% of the predicted value.[8]

With regard to proton structure our asymmetry measurements in deep inelastic scattering probe the internal structure associated with both spin and momentum distributions.

Figure 23 indicates the basic quark-parton model of the proton and the nature of the impulse approximation which leads to the prediction of a positive value for the virtual photon-proton asymmetry $A_1$. We note that a virtual photon can be absorbed by a quark only when their spins are antiparallel, and then the absorption probability will be proportional to the square of the quark charge.

Figure 24 gives the spin-unitary spin parts of the nucleon wavefunction in the simple symmetrical quark-parton model of the nucleon,[5,8] together with the predicted values of $A_1^p = +5/9$ for the proton and of $A_1^n = 0$ for the neutron. Consideration of the momentum

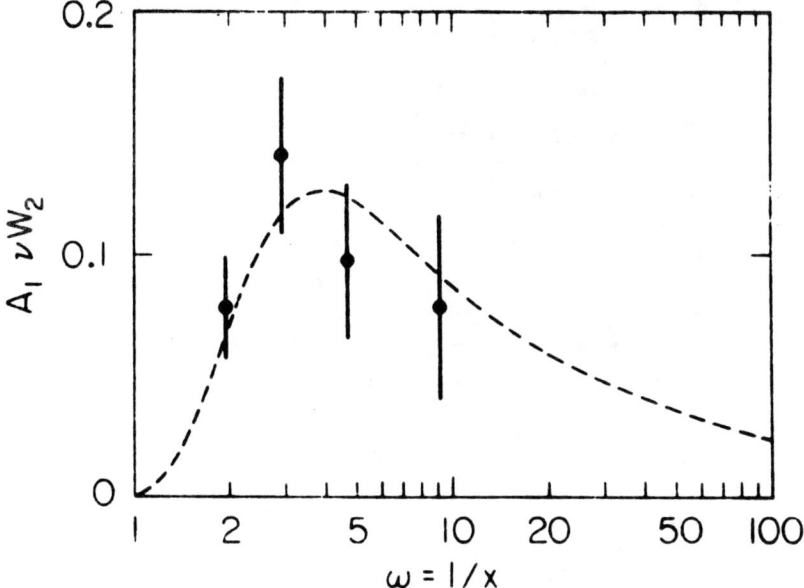

Fig. 22. Comparison of theory and experiment for the Bjorken sum rule. Bjorken sum rule for polarized electroproduction in the scaling limit:

$$\int_1^\infty \frac{d\omega}{\omega} [A_1^p \nu W_2^p - A_1^n \nu W_2^n] = \frac{1}{3}\left|\frac{g_A}{g_V}\right| = 0.417(3)$$

Experimental values of $A_1^p \nu W_2$ vs $\omega$ are indicated by the plotted points. The dashed curve uses $A_1 \dot= 0.78\ \omega^{-1/2}$.

distribution of the quarks and other assumptions about the spin wavefunction alter the above simple prediction to give an x dependence for $A_1^p$ and $A_1^n$.

Figure 25 shows our earliest asymmetry data compared with several nucleon models. (a) relativistic symmetrical valence quark model of Kuti and Weisskopf[54] (b) a quark model incorporating the Melosh transformation[59] (c) a resonance model[60] (d) a bare nucleon-bare meson model.[61] Our data strongly favor the quark-parton models.

Figure 26 shows our four deep inelastic points $A_1(x)$ obtained with the scaling assumption compared with some modern quark-parton models of the nucleon. 1) relativistic symmetrical valence quark model of Kuti and Weisskopf,[54,62] 2) quark model incorporating Melosh transformation,[59] 3) model incorporating nonvanishing quark orbital angular momentum,[63] 4) unsymmetrical model[64] in which the entire spin of the nucleon is carried by a single quark in the limit of x=1, 5) model incorporating the MIT bag model of quark confinement[65] 6) Schwinger's source theory[66]—not a quark model. There is general agreement of the data with the trend of the model predictions, but with the present experimental errors it is not possible to distinguish well among the

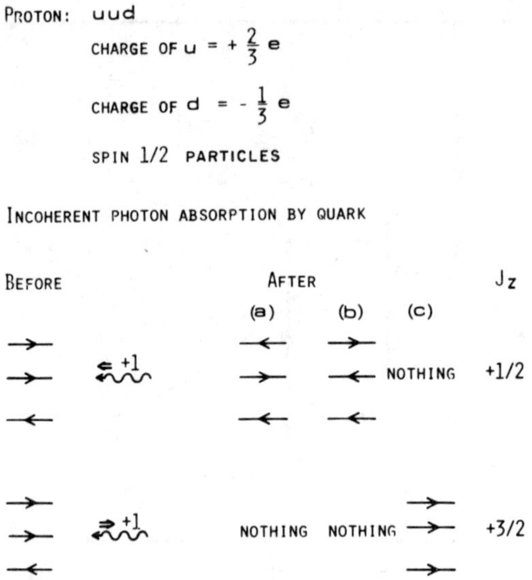

PROTON: uud

CHARGE OF u $= +\frac{2}{3}$ e

CHARGE OF d $= -\frac{1}{3}$ e

SPIN 1/2 PARTICLES

INCOHERENT PHOTON ABSORPTION BY QUARK

| BEFORE | AFTER | | | $J_z$ |
|---|---|---|---|---|
| | (a) | (b) | (c) | |
| | | | NOTHING | +1/2 |
| | NOTHING | NOTHING | | +3/2 |

$$A_1 = \frac{\sigma_{1/2} - \sigma_{3/2}}{\sigma_{1/2} + \sigma_{3/2}} > 0$$

$A_1 = \frac{5}{9}$, SIMPLEST QUARK-PARTON MODEL

Fig. 23. Virtual photon-proton asymmetry in the quark-parton model with the impulse approximation. $J_z$ is the component of angular momentum along the propagation direction of the virtual photon.

different models. In our next experiment, the errors are expected to be about 1/3 as great and hence tests of the models should be much improved.

Asymmetries have also been measured in the resonance region (see paper in these Proceedings by K.P. Schüler).

As to the future, an experiment E130[67] is planned soon at SLAC with a large acceptance spectrometer which should reduce the errors in determining the asymmetries $A_1^p$ in the deep inelastic region by about a factor of 3 and also extend the kinematic range as indicated in Figure 20. Also in this experiment $A_1$ for the neutron will be determined by measuring deuteron asymmetries. Although the experiment is not yet designed in detail, it should be possible to measure the other virtual photon-nucleon asymmetry factor $A_2$ by use of a target with transverse polarization.[55] The $A_2$ term, which involves an interference between amplitudes for longitudinal and transverse photons, is somewhat similar to $R \equiv \sigma_L/\sigma_T$ for the spin-averaged structure functions; measurement of $A_2$ in deep inelastic scattering should provide an interesting test of quantum chromodynamics calculations. In the future it will also be most interesting to compare our electron results with the anticipated high energy results from CERN[52] on polarized muon-polarized proton scattering with regard to scaling, the

$$|\text{Proton} \uparrow\rangle = \frac{1}{\sqrt{18}}\left[2|u^\uparrow d^\downarrow u^\uparrow\rangle + 2|u^\uparrow u^\uparrow d^\downarrow\rangle + 2|d^\downarrow u^\uparrow u^\uparrow\rangle\right.$$
$$-|u^\uparrow u^\downarrow d^\uparrow\rangle - |u^\uparrow d^\uparrow u^\downarrow\rangle - |u^\downarrow d^\uparrow u^\uparrow\rangle$$
$$\left.-|d^\uparrow u^\downarrow u^\uparrow\rangle - |d^\uparrow u^\uparrow u^\downarrow\rangle - |u^\downarrow u^\uparrow d^\uparrow\rangle\right]$$

PROBABILITIES

$u(\uparrow) = 5/9$  $\qquad e_u = +\frac{2}{3}e$

$u(\downarrow) = 1/9$  $\qquad e_d = -\frac{1}{3}e$

$d(\uparrow) = 1/9$

$d(\downarrow) = 2/9$

PROBABILITY OF ABSORPTION OF VIRTUAL PHOTON BY QUARK

$q\,\gamma$

$\uparrow\uparrow = 0$

$\uparrow\downarrow \propto e_i^2$

$$A_1^p = \frac{\sigma_{1/2} - \sigma_{3/2}}{\sigma_{1/2} + \sigma_{3/2}} = 5/9$$

$$|\text{Neutron} \uparrow\rangle = \frac{1}{\sqrt{18}}\left[2|d^\uparrow u^\downarrow d^\uparrow\rangle + 2|d^\uparrow d^\uparrow u^\downarrow\rangle + 2|u^\downarrow d^\uparrow d^\uparrow\rangle\right.$$
$$-|d^\uparrow d^\downarrow u^\uparrow\rangle - |d^\uparrow u^\uparrow d^\downarrow\rangle - |d^\downarrow u^\uparrow d^\uparrow\rangle$$
$$\left.-|u^\uparrow d^\downarrow d^\uparrow\rangle - |u^\uparrow d^\uparrow d^\downarrow\rangle - |d^\downarrow d^\uparrow u^\uparrow\rangle\right]$$

PROBABILITIES

$d(\uparrow) = 5/9$
$d(\downarrow) = 1/9$
$u(\uparrow) = 1/9$
$u(\downarrow) = 2/9$

$$A_1^n = \frac{\sigma_{1/2} - \sigma_{3/2}}{\sigma_{1/2} + \sigma_{3/2}} = 0$$

Fig. 24. Virtual photon-nucleon asymmetries in the simple symmetrical quark-parton model of the nucleon.

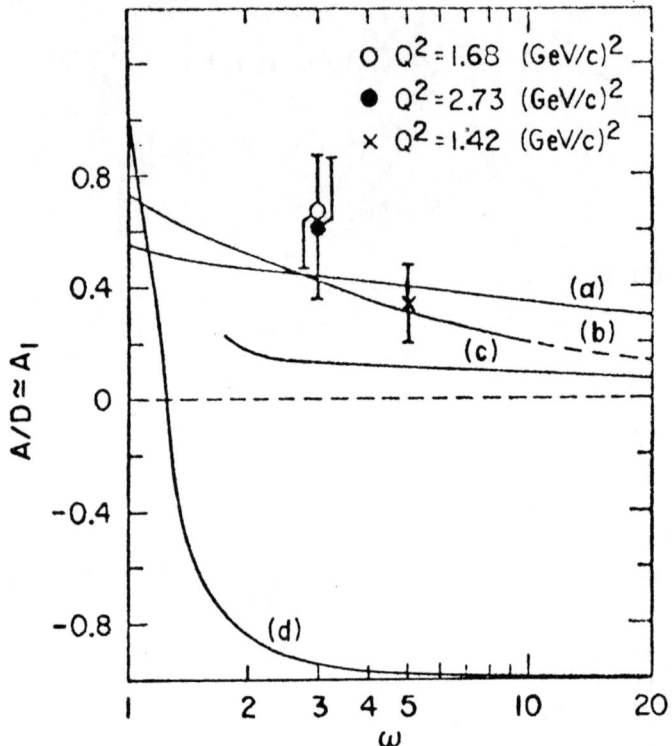

Fig. 25. Early asymmetry data compared with some early theoretical predictions.

Bjorken sum rule, and nucleon models.

Finally, as an interdisciplinary note, we remark that the asymmetry measurements that have now been made in the deep inelastic[6,8] and resonance[13] regions for e-p scattering should allow a quantitative estimate to be made of the famous old problem of the effect of proton polarizability on the hyperfine structure interval of hydrogen.[68,69,70,71]

## REFERENCES

1. J.R. Chen et al., Phys. Rev. Lett. 21, 1279 (1968).
2. M.J. Alguard et al., in Proceedings of the Ninth International Conference on High Energy Accelerators, Stanford Linear Accelerator Center, Stanford, California, 1974, Conf-740 522 (National Technical Information Service, Springfield, Va., 1974), p. 309.
3. P.S. Cooper, et al., Phys. Rev. Lett. 34, 1589 (1975).
4. S. Rock et al., Phys. Rev. Lett. 24, 748 (1970).
5. T. Powell, et al., Phys. Rev. Lett. 24, 753 (1970).
6. M.J. Alguard, et al., Phys. Rev. Lett. 37, 1261 (1976).
7. C.Y. Prescott et al., in High Energy Physics with Polarized Beams and Targets, Proceedings of the Argonne Symposium, ed. by M.L.

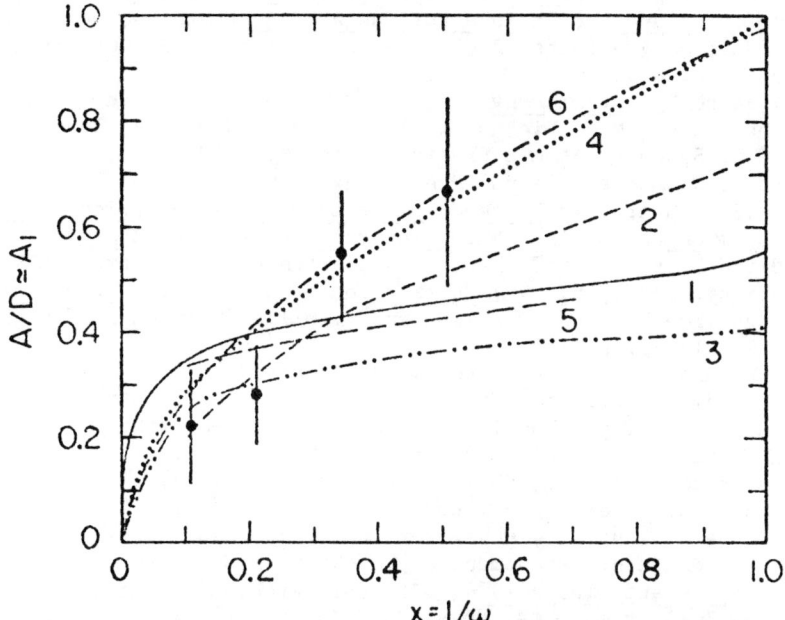

Fig. 26. Comparison of data on asymmetries in deep inelastic scattering with several theoretical predictions.

Marshak (AIP, New York, 1976), p. 315.
8. M.J. Alguard, et al., Phys. Rev. Lett. 41, 70 (1978).
9. W. Atwood, et al., Phys. Rev. D 18, 2223 (1978).
10. Y.B. Bushnin, et al., Soviet J. Nucl. Phys. 24, 279 (1976).
11. C.Y. Prescott, et al., Phys. Lett. 77B, 347 (1978).
12. M.J. Alguard, et al., Phys. Rev. Lett. 37, 1258 (1976).
13. M. Bergstrom, et al., Bull. Am. Phys. Soc. 23, 529 (1978).
14. V.W. Hughes, et al., Phys. Rev. A 5, 195 (1972).
15. D.T. Pierce, F. Meier, and P. Zürcher, Appl. Phys. Lett. 26, 670 (1975).
16. D.T. Pierce and F. Meier, Phys. Rev. B 13, 5484 (1976).
17. C.K. Sinclair et al., in High Energy Physics with Polarized Beams and Targets, Proceedings of the Argonne Symposium, edited by M.L. Marshak (AIP, New York, 1976), p. 424.
18. M. Borghini, et al., Nucl. Instrum. Methods 84, 168 (1970).
19. W.W. Ash, in High Energy Physics with Polarized Beams and Targets, Proceedings of the Argonne Symposium, edited by M.L. Marshak (AIP, New York, 1976), p. 485.
20. M.J. Alguard, et al., "A Source of Highly Polarized Electrons at the Stanford Linear Accelerator Center," SLAC PUB-2244. Submitted to Nuclear Instruments and Methods, January 1979.
21. J.H. Christenson, et al., Phys. Rev. Lett. 13, 138 (1964).
22. J. Bernstein, G. Feinberg, and T.D. Lee, Phys. Rev. 139, B1650 (1965).

23. N. Christ and T.D. Lee, Phys. Rev. <u>143</u>, 1310 (1966).
24. S. Weinberg, in <u>Proceedings of the 1977 International Symposium on Lepton and Photon Interactions at High Energies</u>, edited by F. Gutbrod (Deutsches Elektronen-Synchrotron DESY, Germany, 1977), p. 619.
25. G. Feinberg, <u>Atomic Physics</u>, ed. V.W. Hughes, B. Bederson, V.W. Cohen, and F.M.J. Pichanick (Plenum Press, New York, 1969), p. 1.
26. P.G.H. Sandars, <u>Atomic Physics 4</u>, ed. G. zu Putlitz, E.W. Weber, and A. Winnacker (Plenum Press, New York, 1975), p. 71.
27. P.G.H. Sandars, in <u>Proceedings of the 1977 International Symposium on Lepton and Photon Interactions at High Energies,</u> edited by F. Gutbrod (Deutsches Elektronen-Synchrotron DESY, Germany, 1977), p. 599; <u>Unification of Elementary Forces and Gauge Theories,</u> ed. D. Cline and F. Mills (Harwood Academic Publ., 1977), p. 153.
28. S. Weinberg, Phys. Rev. Lett. <u>19</u>, 1264 (1967).
29. A. Salam in "Elementary Particle Theory; Relativistic Groups and Analyticity," Nobel Symp. No. 8, ed. N. Svartholm (Almqvist and Wiksell, Stockholm, 1968), p. 367.
30. F.J. Hasert, et al., Phys. Lett. <u>46B</u>, 138 (1973).
31. B. Aubert, et al., Phys. Rev. Lett. <u>32</u>, 1454 (1974).
32. S.J. Barish, et al., Phys. Rev. Lett. <u>33</u>, 448 (1974).
33. L. Wolfenstein, "Phenomenology of Neutral Currents," in <u>Particles and Fields</u>-1974, AIP Conf. Proc. 23, p. 84.
34. Y. B. Zel'dovich, Sov. Phys.-JETP <u>36</u>, 682 (1959).
35. S.M. Berman and J.R. Primack, Phys. Rev. D <u>9</u>, 2171 (1974); D <u>10</u>, 3895 (1974).
36. W.W. Wilson, Phys. Rev. D <u>10</u>, 218 (1974).
37. R.H. Cahn and F.J. Gilman, Phys. Rev. D <u>17</u>, 1313 (1978).
38. C.Y. Prescott, et al., "Experimental Test for an Electromagnetic Axial-Vector Current of Hadrons in Inelastic Scattering of Polarized Electrons," SLAC Proposal #95, June 1972.
39. D.E. Nagle, <u>High-Energy Physics and Nuclear Structure-1975</u>, ed. D.E. Nagle, R.L. Burman, G.B. Storms, A.S. Goldhaber and C.K. Hargrøve (AIP, New York, 1975), p. 497.
40. J.M. Potter et al., Phys. Rev. Lett. <u>33</u>, 1307 (1974).
41. J.D. Bowman et al., Phys. Rev. Lett. <u>34</u>, 1184 (1975).
42. A. DeRujula, H. Georgi, S.L. Glashow, and H. Quinn, Rev. Mod. Phys. <u>46</u>, 391 (1974).
43. C. Baltay, <u>Proceedings of the 19th Intl. Conf. on High Energy Physics,</u> Tokyo, 1978.
44. J.D. Bjorken, Phys. Rev. D <u>18</u>, 3239 (1978).
45. L.F. Abbott and R.M. Barnett, Phys. Rev. D <u>18</u>, 3214 (1978).
46. G. Feinberg, Phys. Rev. D <u>12</u>, 3575 (1975).
47. G. Feinberg, <u>High Energy Physics and Nuclear Structure-1975</u>, ed. D.E. Nagle, R.L. Burman, G.B. Storms, A.S. Goldhaber and C.K. Hargrøve (AIP, New York, 1975), p. 468.
48. F.J. Gilman and T. Tsao, "Polarized Electron Elastic Scattering Asymmetries in SU(2)xU(1)," SLAC-PUB-2139, June 1978.
49. E.W. Otten, et al. "Proposal for Searching for Parity Violating Neutral Currents in Elastic e-d Scattering," 1976 (Univ. of Mainz).
50. P.A. Souder, et al., "Search for Parity Violation in the Elastic

Scattering of Polarized Electrons from Nuclei," Bates Proposal, September 1977.
51. P.S. Cooper, Ph.D. Thesis, Yale University (1975) (unpublished).
52. E. Gabathuler et al., "Proposed Experiments and Equipment for a Program of Muon Physics at the SPS," by European Muon Collaborators, CERN, July 1974.
53. P.S. Cooper, et al., Bull. Am. Phys. Soc. $\underline{24}$, 72 (1979).
54. J. Kuti and V.F. Weisskopf, Phys. Rev. D $\underline{4}$, 3418 (1971).
55. F.J. Gilman, Proceedings of Summer Institute on Particle Physics SLAC-167, Vol. I, November 1973, p. 71.
56. L. Galfi et al., Phys. Lett. $\underline{31B}$, 465 (1970).
57. J.D. Bjorken, Phys. Rev. D $\underline{1}$, 1376 (1970).
58. J.J.J. Kookkedee, The Quark Model, (W.A. Benjamin, Inc., New York, 1969).
59. F.E. Close, Nucl. Phys. $\underline{B80}$, 269 (1974).
60. G. Domokos et al., Phys. Rev. D $\underline{3}$, 1191 (1971).
61. S.D. Drell and T.D. Lee, Phys. Rev. D $\underline{5}$, 1738 (1972).
62. R. McElhaney and S.F. Tuan, Nucl. Phys. $\underline{B72}$, 487 (1974).
63. G.W. Look and E. Fischbach, Phys. Rev. D $\underline{16}$, 211 (1977); L.M. Sehgal, Phys. Rev. D $\underline{10}$, 1663 (1974).
64. R. Carlitz and J. Kaur, Phys. Rev. Lett. $\underline{38}$, 673, 1102(E) (1977); J. Kaur, Nucl. Phys. $\underline{B128}$, 219 (1977).
65. R.J. Hughes, Phys. Rev. D $\underline{16}$, 622 (1977); R.L. Jaffe, Phys. Rev. D $\underline{11}$, 1953 (1975).
66. J. Schwinger, Nucl. Phys. $\underline{B123}$, 223 (1977).
67. V.W. Hughes, et al. "Precise Measurements of Asymmetries in Deep Inelastic Scattering of Polarized Electrons by Polarized Protons and by Polarized Deuterons," SLAC Proposal E-130, May, 1977.
68. C.K. Iddings, Phys. Rev. $\underline{138}$, 446 (1965).
69. S.J. Brodsky and S.D. Drell, Ann. Rev. Nucl. Sci. $\underline{20}$, 147 (1970).
70. B.E. Lautrup et al., Phys. Rep. Phys. Lett. $\underline{3C}$, 193 (1972).
71. R.L. Heimann, Nucl. Phys. $\underline{B64}$, 429 (1973).

# PARITY VIOLATION IN INELASTIC SCATTERING OF POLARIZED ELECTRONS*

Charles Y. Prescott
Stanford Linear Accelerator Center, Stanford, Calif. 94305

## ABSTRACT

Parity non-conservation has been observed in the inelastic scattering of longitudinally polarized electrons from an unpolarized deuterium target at 19.4 and 22.2 GeV. We find an asymmetry $A = (\sigma_R - \sigma_L)/(\sigma_R + \sigma_L) = (-9.5 \pm 1.6) \times 10^{-5} Q^2$, $Q^2$ in $(GeV/c)^2$, for values of $Q^2$ near 1.4. The statistical and systematic errors are each about 9 percent of the measured asymmetry. This result is consistent with predictions from the standard Weinberg-Salam $SU(2) \times U(1)$ model. Using the simple quark-parton model of the nucleon, we obtain $\sin^2\theta_W = 0.20 \pm 0.03$.

## I. INTRODUCTION

The interest in parity non-conservation has been with us since the fifties, when those effects were first demonstrated in weak interaction charged current processes. The emergence of the Weinberg-Salam $SU(2) \times U(1)$ gauge theory of weak and electromagnetic interactions and more recently a wide variety of alternative gauge theories has again raised the issue of parity non-conservation, this time in connection with the weak neutral currents. The measurement of parity violating effects in electromagnetic interactions and in neutrino induced reactions serves to discriminate between the gauge theory models that have come into existence. Several years ago at SLAC, motivated primarily by the implications of the gauge theory ideas, a program was undertaken to develop the necessary experimental tools for investigating parity violating effects in electromagnetic interactions. Inelastic electron scattering, a process thoroughly investigated experimentally and presumably quite well understood phenomonologically, was a natural place to look for parity violation. Parity violation, observed as a helicity dependent term in the cross section, does not arise from electromagnetic processes even in higher order, and thus is a unique signature of non-electromagnetic processes, presumably the weak interactions. To measure helicity dependent effects in the cross section required development of an intense polarized electron source. The performance parameters of such a source and the sensitivity demanded of the experiments were dictated by the smallness of the anticipated parity violating effects. These effects arise from interference between weak and electromagnetic amplitudes, and are expected from quite general arguments to be in the order of

---

*Work supported by the Dept. of Energy

$G_F Q^2/(2\sqrt{2}\,\pi\alpha)$ where $G_F$ is the Fermi coupling constant, $\alpha$ is the fine structure constant, and $Q^2$ is the four-momentum-transfer-squared. For inelastic scattering at SLAC, the expected parity violating effects in the cross section are around $10^{-4} Q^2$, $Q^2$ in $(\text{GeV/c})^2$. Gauge theory predictions included the possibility that the effects could be even smaller, or zero. In order to achieve a significant test of gauge theories, the parity violating asymmetry

$$A = (\sigma_R - \sigma_L)/(\sigma_R + \sigma_L) \qquad (1)$$

where $\sigma_R(\sigma_L)$ is the cross section for $+(-)$ helicity of the incident beam, had to be measured to an accuracy of

$$\Delta A \leq 10^{-5} \qquad (2)$$

The then-existing techniques used in inelastic scattering experiments did not provide monitoring or control of experimental parameters at the level of sensitivity we required. The experimental program undertaken several years ago had as its objective the development of beam monitoring and counting techniques capable of achieving sensitivity sufficient for these measurements.

## II. FLUX COUNTING

Before discussing the technique of flux counting, a general description of the experiment is useful. Figure 1 shows the various components in a highly schematic form. Inelastic scattering of polarized electrons from unpolarized deuterium was detected at 4° in a spectrometer. This process had been studied earlier at this angle using unpolarized electrons, and a considerable amount of experimental information was available[1]. For the majority of the data the beam energy was 19.4 GeV and the secondary scattered electron energy was 14.5 GeV. The spectrometer consisted of two bending elements and a quadrupole, constructed from magnets taken from the 8 GeV/c and 20 GeV/c end-station spectrometers. These were arranged downstream and below the target to detect electrons which scattered at 4° down from the target. The angular acceptances were $\Delta\theta = \pm 7.5$ mrad and $\Delta\phi = \pm 16.6$ mrad. Two counters were placed behind the spectrometer to detect the scattered electrons. The first was an atmospheric gaseous $N_2$ Cerenkov counter (C). The counter was 3.35 meters long with horizontal and vertical apertures of 70cm each. At the end of the counter, a spherical mirror collected Cerenkov light and focussed it onto the photocathode of a photomultiplier tube placed off axis. Electrons passed out of the Cerenkov counter and into a second electron counter consisting of $9X_0$ thick lead glass (TA) with an aperture of 88cm (horizontal) by 52cm (vertical). Cerenkov light produced in the lead glass was viewed by an array of 4 photomultipliers.

Flux counting is a technique developed for this experiment to permit measurements of inelastically scattered electrons at the at the high counting rates in this experiment. For the gas Cerenkov counter, electrons which enter its aperture, produce Cerenkov light that was collected by a spherical mirror and focussed onto a photomultiplier photocathode. The anode current is sent to electronic circuitry, where for each beam pulse it is integrated, digitized and read by the computer. This signal (FLUX) is proportional to the flux of electrons through the counter. We generate a quantity proportional to the scattered cross section (averaged over the acceptance) by normalizing to the incident charge, Q, delivered to the target. For each beam pulse i, we form a cross section (units are arbitrary)

$$\sigma_i^\pm = FLUX/Q_i \qquad (3)$$

where + and - refer to the beam helicity.

Using unpolarized electrons from the conventional SLAC guns, we performed a series of runs from one to three hours in length. Individual beam pulses were arbitrarily assigned + or - helicity. For these runs we formed an experimental asymmetry

$$A_{exp} = \frac{\sigma^+ - \sigma^-}{\sigma^+ + \sigma^-} \qquad (4)$$

using the means of the distributions $\sigma^\pm$. The error on $A_{exp}$ was calculated from the errors on the means

$$\Delta\sigma^\pm = \Delta\sigma^\pm(rms\ width)/\sqrt{n^\pm} \qquad (5)$$

where the number of beam pulses, n±, was for these runs quite large.

The asymmetry averaged for 26 runs is $(-2.5 \pm 2.2) \times 10^{-5}$. If we calculate for each run the deviation of $A_{exp}/P_e$ from zero, divided by the error, we obtain a distribution consistent with the expected shape, a gaussian curve of unit width. The 26 runs have a standard deviation of $1.02 \pm 0.13$. This result leads us to believe that the errors are properly calculated, and that the techniques used for flux counting can be applied to measuring asymmetries at the level of $10^{-5}$.

### III. POLARIZED ELECTRON SOURCE

The polarized electron source is shown schematically in Figure 2. The principle by which this source operates is photoemission of longitudinally polarized electrons from a gallium arsenide crystal using circularly polarized light [2]. The possibility that this process could provide intense beams of polarized electrons was first suggested in 1974 by Ed Garwin, of SLAC, and D. Pierce [3] and H.C. Siegmann of ETH Zürich. The development of such a source to be an injector for the SLAC linac was undertaken at that time by Ed Garwin, C. Sinclair, R. Miller and myself [4].

The source was driven by a flash lamp pumped dye laser operating near 710nm wavelength. The flash lamp was pulsed at 120 pulses per second, synchronized to the SLAC linac running at the same rate. Pulses of photoemitted electrons, approximately 1.5μsec in length and up to 300 milliamperes in intensity, were accelerated from the -65 keV potential to ground and were transported to the SLAC linac by a series of lenses and d.c. magnets. The longitudinal polarization of electrons leaving the GaAs cathode are negligibly depolarized in the transport system. Beam intensities of $1 \times 10^{11}$ to $4 \times 10^{11}$ electrons per pulse were accelerated and delivered to the target. Loss of polarization in the linac was likewise negligible, as demonstrated in earlier tests [5].

Circular polarization of the laser pump light is achieved in optical polarizers consisting of two elements. These are shown in Figure 2, and again in detail in Figure 3. Laser light was first polarized linearly in a calcite prism. Circularly polarization was achieved in a Pockels cell, a uniaxial crystal which exhibits a birefringence linear in the applied electric field. It is cylindrical in shape with ring electrodes around its circumference. A high voltage pulse, either positive or negative, is applied prior to each beam pulse. Reversing the sign of the high voltage reversed the helicity of the photons, which in turn reversed the helicity of the photoemitted electrons. The reversals were done randomly on a pulse to pulse basis to minimize the effects of drifts on the experiment, and randomization avoided changing the helicity synchronously with any possible periodic changes of beam parameters.

The sign of the Pockels cell voltage was sent to the experimental computer prior to each beam pulse. This allowed us to form our basic experimental quantity:

$$A_{exp} = \frac{\sigma(V+) - \sigma(V-)}{\sigma(V+) + \sigma(V-)} \quad (6)$$

Throughout the experiment the experimental asymmetry, $A_{exp}$, is measured <u>relative to the Pockels cell voltage</u>.

The calcite prism was mounted in a rotatable holder; the plane of linear polarization could be rotated by 45° or 90° relative to its 0° position. Rotation of the linear polarization by 90° <u>reversed</u> the helicity of the photons. In general rotation by an angle $\phi_p$ causes the net helicity to vary by $\cos 2\phi_p$. Since the experimental asymmetry, eq.6, is measured relative to the Pockels cell voltage, we expect to find the relation

$$A_{exp} = \frac{\sigma(V+) - \sigma(V-)}{\sigma(V+) + \sigma(V-)} = |P_e| A \cos(2\phi_p) \quad (7)$$

where $|P_e|$ is the measured beam polarization (around 0.40), and A is the physics asymmetry arising from parity violating effects, defined by eq. 1.

The rotation of the plane of linear polarization provided a technique which could separate parity violating effects related to helicity of the beam from systematic effects which could arise due to perturbations of other beam parameters when the Pockels cell voltage was reversed. In section VI the asymmetries measured at $0°$, $45°$ and $90°$ orientations of the calcite prism are discussed.

## IV. BEAM MONITORING

A primary objective of the experimental techniques was to eliminate sources of systematic errors to the extent that corrections to the data were unnecessary. To determine the size of the systematic errors, monitoring of beam parameters which could affect measured cross section values was necessary. The parameters measured were average polarization (sign and magnitude), energy, beam current, beam position and angle at the target.

The heart of the beam monitoring system was a resonant microwave cavity with a node which was placed on the beam axis [6]. Beam pulses passing through the node induced no signals in the cavity. Beam pulses displaced from the node would induce signals proportional to the produce of the beam current times the displacement. Using phase-sensitive microwave electronics, both the sign and the amplitude of the signals were measured. By normalizing to the beam current, measured independently, the average displacement pulse by pulse was digitized and stored on tape along with other information for each pulse.

Two resonant microwave position monitors were placed 2 meters before the liquid deuterium target, one sensitive to horizontal displacements, the other sensitive to vertical displacements. These devices provided pulse to pulse measurements of beam position, averaged over the pulse, with resolutions better than 10μm. The pulse to pulse jitter, arising from instabilities in accelerator pulsed components, was typically 50μm to 150μm, varying somewhat with conditions from time to time.

Using additional microwave position cavities placed upstream, angle of the beam at the target was measured. Pulse to pulse jitter of 1 to 3 μradians was typical. Beam energy was measured with a microwave position cavity placed at a location in the beam transport where momentum was dispersed. Pulse to pulse jitter of 0.05% in energy was typical. Beam current was monitored in two separate non-intercepting beam toroids[7]. Absolute accuracy of better than 1% is achieved in the devices, and linearity of response is quite good.

The important role the beam monitoring played in the experiment was the determination of the equality (or inequality) of beam parameters between + and - helicities. Differences in + and - beam position can lead to measured asymmetries through geometric effects, energy differences could enter through the cross section dependence, and current imbalances could generate systematic asymmetries through electronic non-linearities. Quantitative estimates of the sensitivity to these imbalances were made through several techniques; calculation of geometric effects through Monte Carlo work, and estimation of position angle and energy dependence from known cross section formulae. Sensitivity to current imbalances were measured. These results are summarized in Table I. The parameters quoted are the differences between + and − helicity beams. The first column summarizes the imbalances measured. The second column shows the asymmetries which result from these imbalances. We treat these results as corrections to the asymmetries, but in addition <u>increase</u> the systematic errors by the same amount. This procedure reflects the preliminary nature of our errors; the understanding of systematic errors should improve with further study of the present and future data.

## V. POLARIZATION MONITORING

Polarization was measured every few hours during the experiment. The process by which polarization was measured was the elastic scattering of polarized beam electrons by polarized target electrons. The target was a thin foil of a highly permeable alloy of iron (Supermendur). An externally applied magnetic field saturated the material, providing a target with a known fraction (7.8%) of the electrons polarized along the beam direction. Elastic scattering of electrons by electrons (Moller scattering) is a simple QED process and the determination of the spin-dependent part is a straight forward first order QED calculation. We measured the asymmetry

$$A_M = \frac{\sigma_p - \sigma_a}{\sigma_p + \sigma_a} \qquad (8)$$

where $\sigma_p (\sigma_a)$ is the cross section for Moller scattering with the beam electron spin parallel (anti-parallel) to the target electron spin. For relativistic scattering, at 90° in the e-e center-of-mass, we expected a 100% incident polarization to give a value of $A_M = -0.057$. This value is the result of the QED calculation (−7/9) multiplied by the average target polarization (0.078) and the alignment at 20° to the beam direction (cos 20°).

Scattered electrons were detected in a proportional wire chamber hodoscope shower detector, constructed of brass and tungsten. The bins were separated by 4mm, dividing the lab scattering angle horizontally from 3mrad to 10mrad into 24 bins. Momentum was dispersed vertically; elastically scattered electrons fall in a nearly vertical stripe which crossed the center of the hodoscope. At the high rates we encountered, single electron counting was not possible. We measured instead the flux of electrons, using ideas described in section II. The current for each hodoscope wire was integrated and digitized for each beam pulse. We divided by the incident beam charge $Q_i$ to obtain <u>for each bin</u> and <u>for each beam pulse</u> a cross section $\sigma_{ij}$ ($i^{th}$ beam pulse, $j^{th}$ bin, units arbitrary). We formed the bin-by-bin asymmetry $A_j$, using all the beam pulses for a run.

The asymmetry for the bins below the elastic peak is non-zero because of the radiative tail for e-e scattering contributes counts in this region. Extraction of beam polarization values requires a subtraction of background signals, and uncertainties in the procedures leads to uncertainties in the measured polarization values, $P_e$. We have assigned a ± 0.002 uncertainty on $P_e$ which comes from the uncertainty of the background subtraction. This uncertainty contributed the largest part of the systematic error on the parity violation asymmetry, ± 5% of $A/Q^2$.

## V. THE DATA

This part will describe the data in some detail. The techniques of flux counting, described in Section I, and polarization reversals, described in Section III, give us the basic experimental asymmetry, $A_{exp}$, defined by equation 6. This experimental quantity, formed in the computer using the rapid random reversals of helicity, is stable even in the presence of drifts in the experimental apparatus. In addition to the rapid polarization reversals using the Pockels cell, two other methods for reversing beam helicity were available to us. They were: 1. Rotation of the plane of linear polarization, before the Pockels cell, by rotating the calcite prism in its mount. 2. Precession of the electron spin relative to its direction, due to the electron anomolous magnetic moment ("g-2 precession"), by taking data at different beam energies.

These two methods serve as consistency checks on our procedures; the experimental asymmetries should follow the expected changes if sources of systematic errors are negligible. In particular, each method contains null points where asymmetries are expected to vanish, and any non-zero measurements can arise only from systematic effects. Thus the measurements can place limits on systematic errors, independent of the results obtained from beam monitoring data.

We take data with the calcite prism set at three discrete orientations; $0°$, $45°$, and $90°$. We expect to obtain a $\cos(2\phi_p)$ dependence of equation (7).

Figure 4 shows the results at 19.4 GeV for $A_{exp}/|P_e|$ obtained in the shower counter. For the $45°$ point, we used a value 0.37 for $|P_e|$. The dashed curve is the expected $\cos(2\phi_p)$ form, fit to the $0°$ and $90°$ points. The $0°$ and $90°$ asymmetries are equal and opposite within errors, and the $45°$ point is consistent with zero, as expected. The errors shown are statistical only. No systematic corrections have been applied to these data.

These results, summarized by Figure 4, contain two null measurements, which are satisfied within statistical accuracy, and two consistency checks which are satisfactory. The null measurements are the zero asymmetries obtained with unpolarized electrons from the SLAC gun and from the (unpolarized) GaAs source, and the consistency checks are the two data points at $0°$ and $90°$, which are equal and opposite.

An additional check on the validity of the data is seen in Figure 4, where results for the gas Cerenkov counter are shown. The asymmetries are independently determined from flux counting measurements made on this counter. This counter is not statistically independent, since the same electrons detected in this counter pass through and are counted in the shower counter, but this counter has independent electronics, and responds quite differently to possible sources of backgrounds. Thus, we conclude that the measured asymmetries do not arise from an artifact of electronics or from unknown background counts of some kind.

Reversals of beam helicity can be achieved by the g-2 precession of the polarized electrons. At 19.4 GeV the helicity of the beam at the target was positive for positive Pockels cell voltage. However, the helicity depended on beam energy, owing to the anomolous magnetic moment of the electron, and the 24.5° deflection of the beam passing through the transport magnets. The spin precesses relative to the beam direction by

$$\theta_{prec} = \frac{E_o}{m_e} \frac{g-2}{2} \theta_{bend} = \frac{E_o (GeV)}{3.237} \pi \text{ radians} \quad (9)$$

where $m_e$ is the mass and g is the gyromagnetic ratio of the electron. For the experimental asymmetry we expect

$$A_{exp} = |P_e| A_{phys} \cos \frac{E_o (GeV)}{3.237} \pi \quad (10)$$

The asymmetries $A_{exp}/|P_e| Q^2$ are shown in Figure 5 for the shower counter and for the Cerenkov counter. In this figure we plot $A_{exp}/|P_e| Q^2$ because at different beam energies, the average $Q^2$ for the scattering varies. We expect A to grow linearly in $Q^2$ (8), and the quantity plotted to show the $\cos \frac{E_o \pi}{3.237}$ dependence. The point at $E_o$ = 17.8 GeV provides a third null test for thie experiment. At this energy, electrons are transversely polarized as they pass through the target, and asymmetries from highly relativistic transversely polarized electrons are expected to vanish. The orientation of the spectrometer is such that the transverse polarization is normal to the scattering plane of the electrons for $E_o$=17.8 GeV. The data point measured here, consistent with zero asymmetry, gives experimental evidence that the measured asymmetries do not arise from transverse components of the spin.

Figure 5 constitutes the ultimate experimental evidence that the observed asymmetries are associated with electron spin. No systematic effects or influence the source may have on the beam parameters can mimic the g-2 precession which arise from the anomolous magnetic moment of the electron and beam transport geometry.

Based on the results shown in Figures 4 and 5 we conclude that parity violation exists in this process. To determine the magnitude, we exclude the lowest energy point, because it contains fairly strong elastic and resonance contributions ans is the lowest $Q^2$

point, where the nucleon model, used for deep inelastic scattering, is lease likely to apply. The data point at 19.4 GeV and 22.2 GeV (with its sign changed) are combined to give

$$A/Q^2 = (-9.5 \pm 1.6) \times 10^{-5} \text{ (GeV/c)}^2 \quad (11)$$
$$\text{deuterium}$$

The sign implies negative helicity electrons have a greater probability for scattering than do positive helicity electrons. The quoted error is derived from a statistical error of $\pm 0.86 \times 10^{-5}$ added linearly to estimated systematic uncertainties of 5% in the value of $|P_e|$ and of e.e% from asymmetries in beam parameters. The $\pi^-$ background contributed less than $0.1 \times 10^{-5}$ to $A/Q^2$, but normalization corrections of 2% for $\pi^-$ background and 3% for radiative corrections were made.

## VI. IMPLICATIONS FOR GAUGE THEORIES

The high degree of interest in parity violation in the weak neutral currents arises from the ability of such effects to discriminate between different gauge theory models. The experimental observation of existence of neutral current processes, originally reported in 1973 by the Gargamelle collaboration at CERN[9], has to be considered a major success for the general ideas of gauge theories which proposed to unify the weak and electromagnetic interactions into one force of nature. Experimental information on the detailed nature of neutral currents was meager in the following several years. In the absence of experimental information, a number of models were proposed. Parity non-conservation in the neutral currents is a central issue because experimentally measurable effects arise that can be related to neutral current coupling constants. Neutrino cross section measurements are in principle unable to distinguish between parity violating models and more complex gauge theories have arisen which preserve parity invariance in the underlying dynamics while explaining the neutrino-anti-neutrino cross section differences as a consequence of comparing left handed neutrinos with right handed anti-neutrinos. The popularity of these models has been enhanced, of course, by the reported absence of parity violation in atomic bismuth[10,11]. The results of this experiment appear to contradict the atomic bismuth results and almost certainly exclude those left-right symmetric models which have parity conserved. The issue of parity violation in atoms is unresolved, and is being actively pursued at present.

The simplest gauge theories are those based on the SU(2)xU(1) gauge group. The original form, the Weinberg-Salam(W-S) model of the weak interactions, has the left handed electron and quarks assigned to weak isospin doublets, while the right handed quarks and electrons are singlets. Prediction of parity violation in inelastic scattering of electrons (and muons) can be made, but requires the knowledge of the hadronic vertex. Predictions have been made by a number of authors.[12] The usual assumption is to treat the nucleon in the simple quark-parton model $^{(12)}$. The scattering is taken as an incoherent process off 3 valence quarks only (for deuterium, six quarks). The neutral current couplings are specified in this model by the mixing parameter, $\sin^2\theta_W$, where $\theta_W$ is the Weinberg angle, and by the weak isospin assignment for the u,d quarks and for the electrons. The predicted asymmetries have the form

$$A/Q^2 = -\frac{G_F}{2\sqrt{2}\pi\alpha} \cdot \frac{9}{10} \left\{ \left(1+2T_{3R}^e\right)\left(1-\frac{20}{9}\sin^2\theta_W + \frac{4}{3}T_{3R}^u - \frac{2}{3}T_{3R}^d\right) \right.$$
$$\left. + \left(1-4\sin^2\theta_W - 2T_{3R}^e\right)\left(1-\frac{4}{3}T_{3R}^u + \frac{2}{3}T_{3R}^d\right)\left[\frac{1-(1-y)^2}{1+(1-y)^2}\right] \right\} \quad (12)$$

where $y = \nu/E_0$ is the fractional energy transfer to the hadrons, and $T_{3R}^{e,u,d}$ has values $\pm\frac{1}{2}$ or 0, depending on the weak isospin assignments of the right handed e,u and d. Following the arguments of Cahn and Gilman$^{(12)}$, if we take the left handed weak isospin assignments as determined to be doublets, and take all possible right handed combinations (either doublets or singlets) there are 8 possible SU(2)xU(1) "models" or predicted asymmetries.

Three of the models, those with the electron in a doublet along with one or both quarks assigned to right handed doublets, are immediately excluded. Two more models $\binom{u}{b}_R$ and $\binom{u}{b}_R \binom{t}{d}_R$, can be also excluded because they require large values of $\sin^2\theta_W$, inconsistent with neutrino data. A sixth model, $\binom{u}{b}_R$ is in good agreement with our data, but fails a more detailed study of neutrino results$^{(13)}$. Only two models remain of the original eight. One of these has the electron assigned to a doublet, $\binom{E^o}{e}_R$, while the other is the original Weinberg-Salam model. Figure 6 shows these two models in detail. The y-dependence is shown for the W-S model (solid lines) and for the "mixed" or "hybrid" model (dashed lines) for values of $\sin^2\theta_W$ shown. The measured asymmetry has inner error bars which are statistical only, and outer errors which are obtained by adding the systematic errors linearly. The agreement with the W-S model is satisfactory, provided

$$\sin^2\theta_W = 0.20 \pm 0.03 \quad (14)$$

while agreement with the hybrid model is questionable, requiring a value of $\sin^2\theta_W < 0$. The hybrid model gives a predicted asymmetry which is 3 standard deviations from the measured value of $\sin^2\theta \approx \frac{1}{4}$.

Near the end of the available running time, we took a limited amount of data at 19.4 GeV using a liquid hydrogen target. The result was

$$A/Q^2 = (-9.7 \pm 2.7) \times 10^{-5} \text{ (GeV/c)}^{-2} \quad \text{(hydrogen)} \quad (15)$$

The proton target is treated as two u and one d quark compared to the equal mix for deuterium, and asymmetries are expected to be slightly smaller. Our result is consistent within errors with this expectation and with the result obtained from deuterium.

## REFERENCES

1. S.Stein, et al., Phys. Rev. D12, 1884 (1975).
2. D.T. Pierce, et al., Phys. Lett. 51A, 465 (1975), and Appl. Phys. Lett. 26 (1975).
3. E.L. Garwin, D.T. Pierce and H.C. Siegmann, Swiss Physical Society Meeting, April 26, 1974; Helv. Phys. Acta 47, 393 (1974) (abstract only). The full paper is available as SLAC-PUB-1576 (1975) (unpublished).
4. C.K. Sinclair, et al., in High Energy Physics with Polarized Beams and Targets, proceedings of the Argonne Symposium, edited by M.L. Marshak (AIP, New York, 1976), p. 424.
5. P.S. Cooper, et al., Phys. Rev. Lett. 34, 1589 (1975). This experiment used the same target as Ref. 5, but used a different spectrometer and detectors.
6. Z.D. Farkas, et al., SLAC-PUB-1823, 1976 (unpublished).
7. R.S. Larsen and D. Horelick in Proc. of Symposium on Beam Intensity Measurement, DNPL/R1, Davesburg Nuclear Physics Laboratory, April 1968 and SLAC-PUB-389 (1968).
8. This fact arises from the electromagnetic amplitude, which has a $1/Q^2$ dependence, giving an asymmetry proportional to $Q^2$.
9. F.J. Hasert, et al., Phys. Lett. 46B, 191 (1973) and F.J.Hasert et al., Phys. Lett. 46B, 138 (1973).
10. L.L. Lewis, et al., Phys. Rev. Lett. 39, 795 (1977), and
11. P.E. Baird, et al., Phys. Rev. Lett, 39, 798 (1977).
12. R.N. Cahn and F.J. Gilman, Phys. Rev. D17, 1313 (1978) and references therin.
13. L.F. Abbott and R.M. Barnett, Phys. Rev. Lett. 40, 1303 (1978).

## TABLE I

### SYSTEMATIC ERRORS SUMMARIZED

a. Beam Monitoring-Deuterium target runs

| Parameter | Units | Measured Difference (+) (−) | Correction to $A/Q^2$ |
|---|---|---|---|
| $E_o$ | o/o | $(1.5 \pm .28) \times 10^{-4}$ | $-.37 \times 10^{-5}$ |
| $Q$ | ma | $(-2.2 \pm .4) \times 10^{-3}$ | $-.03$ |
| $X_T$ | µm | $(-8.9 \pm 3.3) \times 10^{-2}$ | $+.04$ |
| $Y_T$ | µm | $(-.65 \pm 1.8) \times 10^{-2}$ | $-.02$ |
| $\theta_x$ | µrad | $(-.37 \pm .7) \times 10^{-3}$ | $.00$ |
| $\theta_y$ | µrad | $(1.5 \pm .9) \times 10^{-3}$ | $+.01$ |
| | | | $-.37 \times 10^{-5}$ |

b. Beam Monitoring-Hydrogen target runs

| Parameter | Units | Measured Difference (+) (−) | Correction to $A/Q^2$ |
|---|---|---|---|
| $E_o$ | o/o | $(-2.4 \pm .44) \times 10^{-4}$ | $+.50 \times 10^{-5}$ |
| $Q$ | ma | $(1.9 \pm 1.3) \times 10^{-3}$ | $+.04$ |
| $X_T$ | µm | $(12.7 \pm 2.5) \times 10^{-2}$ | $-.05$ |
| $Y_T$ | µm | $(2.8 \pm 1.8) \times 10^{-2}$ | $-.02$ |
| $\theta_x$ | µrad | $(-1.5 \pm .59) \times 10^{-3}$ | $.00$ |
| $\theta_y$ | µrad | $(-3.0 \pm .75) \times 10^{-3}$ | $+.03$ |
| | | | $+.50 \times 10^{-5}$ |

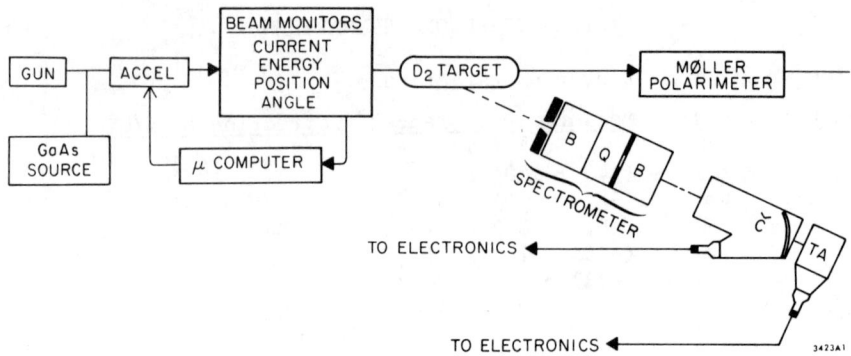

Figure 1: A schematic overview of the experiment. The two-mile accelerator (LINAC) is fed electrons from either the unpolarized SLAC gun or polarized electrons from the gallium arsenide source. Beam monitoring of position, angle, current, energy and polarization measure systematic errors. Detection of electrons scattered by a 30 cm liquid deuterium target is done in a spectrometer instrumented with a gas Cerenkov counter and a lead glass shower counter.

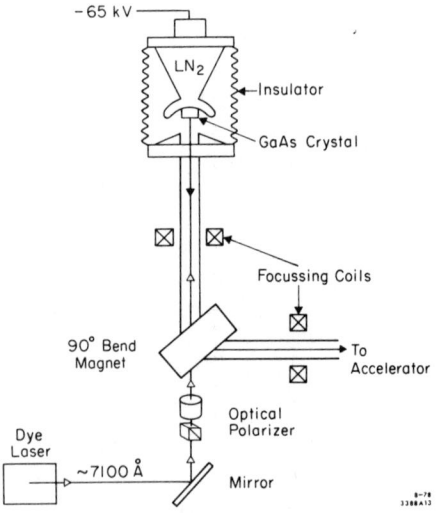

Figure 2: A schematic view of the gallium arsenide polarized electron source. Polarized electrons are photoemitted from the GaAs crystal using circularly polarized laser light. Electrons accelerate from -65 KV potential, are deflected out of the laser beam and transported to the injector.

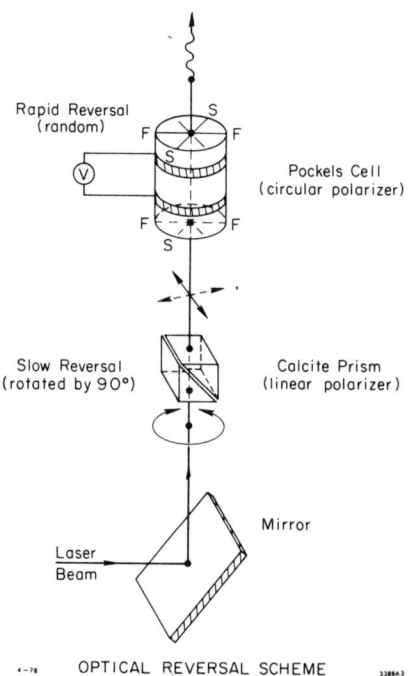

Figure 3: A detailed view of the laser polarizing optics. A calcite prism first linearly polarizes the light. A Pockels cell circularly polarizes the light to ± 100% for ± V applied to its ring electrodes. The experimental asymmetries are formed from beam pulses of + and − V. Rotation of the calcite prism through an angle $\phi_p$ about the axis of the beam should modulate the experimental asymmetry by $\cos 2\phi_p$.

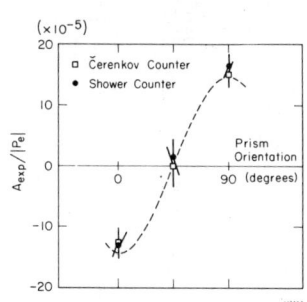

Figure 4: The experimental asymmetry shows the expected variation (dashed line) as the beam helicity changes due to the change in orientation of the calcite prism. The data are for 19.4 GeV and deuterium. Since the same scattered particles strike both counters, they are not statistically independent. No systematic errors are shown. No corrections have been made for helicity dependent differences in beam parameters.

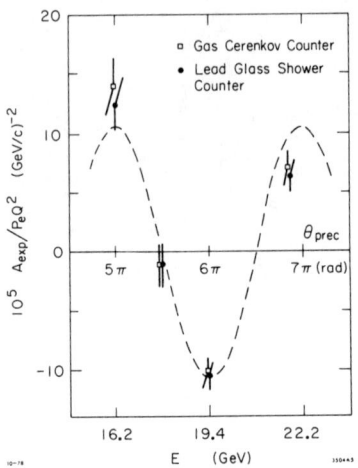

Figure 5: Experimental asymmetries, divided by $|P_e|Q^2$, for lead glass shower counter and gas Cerenkov counter, at four beam energies. The dotted curve shows the modulation of the asymmetry expected from the g-2 precession of the electron spin in the $24\frac{1}{2}°$ bend of the beam transport.

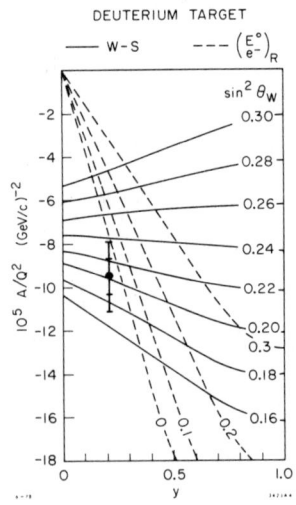

Figure 6: The y-dependence for the deuterium target asymmetry $A/Q^2$ two $SU(2) \times U(1)$ models and for values of $\sin^2\theta_W$ shown. The two models considered are the standard Weinberg-Salam model (solid curves) and the hybrid model which assigns the right handed electron to a doublet with an hypothesized heavy neutral lepton (dashed curves). Our data point is shown at y=0.21 with inner error bars (statistical only) and outer error bars (systematic errors added to statistical errors).

## POLARIZED ELECTRON - POLARIZED PROTON SCATTERING EXPERIMENTS AT SLAC*

Presented by K. P. Schüler
Yale University, New Haven, CT 06520

### ABSTRACT

We have measured the asymmetry in the deep inelastic scattering of longitudinally polarized electrons by longitudinally polarized protons in the kinematic range $2 \leq \omega \leq 10$, $1 \leq Q^2 \leq 4$ $(GeV/c)^2$, and $2 \leq W \leq 4$ GeV.[1,2] A comparison of our results with the Bjorken sum rule and with models of the internal spin structure of the proton is given elsewhere in these proceedings.[3] We have also measured the asymmetry in the region of the first three resonances at $Q^2 = 0.5$ and $1.5$ $(GeV/c)^2$. The radiative corrections to our experimental asymmetries will be discussed. Our continuing experimental program aims at better statistics over a larger kinematic range and will also include polarization data of the neutron.

### INTRODUCTION

Since the subject of polarized electron - polarized proton scattering[1,2] has also been covered by Vernon Hughes[3] I will try to minimize overlap. I will focus somewhat on radiation corrections. I will make a few remarks on what kind of asymmetry behavior we might expect in the resonance region. I will also show you some exploratory data in the resonance region, and I will say a few words about future experiments.

The following people have participated in the experiment: G. Baum and W. Raith from Bielefeld; W. Ash, D. Coward, R. Miller, and D. Sherden from SLAC; K. Kondo and S. Miyashita from Tsukuba; and M. Alguard, M. Bergstrom, J. Clendenin, P. Cooper, R. Ehrlich, V. Hughes, M. Lubell, D. Palmer, N. Sasao, P. Schüler, P. Souder, and M. Zeller from Yale.

For our experiments we used the polarized electron source PEGGY[4] which produces a highly polarized (85%) beam of modest intensity ($10^9$ e/pulse) at up to 180 pps. The high electron polarization makes this source particularly suitable for this type of experiment since the maximum tolerable beam intensity is limited by the radiation damage of the polarized proton target. (We have another type of polarized electron source at SLAC which has been used for the latest parity experiments and which produces much higher current at a polarization of 40%.[5]) Our polarized proton target (which uses butanol at 50 kG/1K) had an average free proton polarization of 50%. The 1/e depolarizing dose was $3 \times 10^{14}$ e/cm$^2$. The electron beam was rastered over the target cross section as usual in such applications.

ISSN: 0094-243X/79/510217-07$1.50 Copyright 1979 American Institute of Physics

The basic quantity which we determine is the electron-proton asymmetry $A = [d\sigma(\uparrow\downarrow) - d\sigma(\uparrow\uparrow)]/[d\sigma(\uparrow\downarrow) + d\sigma(\uparrow\uparrow)]$ which is related to the virtual photon-proton asymmetries $A_1 = (\sigma_{1/2} - \sigma_{3/2})/(\sigma_{1/2} + \sigma_{3/2})$, which is of main interest here, and $A_2 = \sigma_{TL}/\sigma_T$, which is a small interference term.[1,2,3]

## DEEP INELASTIC SCATTERING

Altogether we have amassed 1500 hours at 120 pps and $10^9$ e/pulse. Approximately 85% of this time was spent on deep inelastic scattering, 10% in the resonance region, and 5% for elastic scattering. Table I shows our results on deep inelastic scattering, Table II gives additional information on the kinematics and radiative corrections. The virtual photon asymmetry $A_1$ has been plotted as a function of the scaling variable x in Fig. 1. What have we learned from these data?

(i) The asymmetry $A_1$ is large and positive as predicted by Bjorken.

(ii) The data is consistent with scaling but more accurate data over a wide range in $Q^2$ would be desirable.

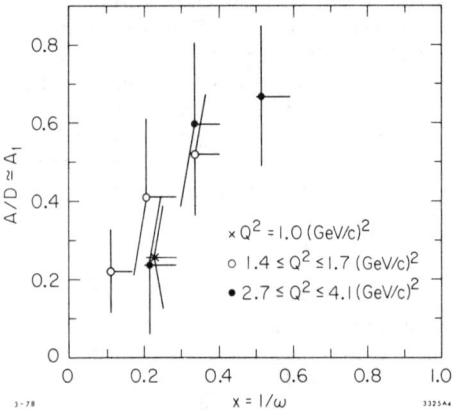

Fig. 1. Experimental values of the asymmetry $A/D \simeq A_1$ vs x. The horizontal bars give the range in x associated with the radiative corrections.

Table I: Results of Asymmetry Measurements

| ω | $Q^2$ (GeV/c)$^2$ | ν (GeV) | W (GeV) | Δ (%) | $A$[b] | D | $A$[c] | $A_1 + nA_2$[c],[d] | $\|nA_2\|$[e] |
|---|---|---|---|---|---|---|---|---|---|
| 2.0 | 4.09 | 4.35 | 2.22 | 1.28±0.26 | 0.211±0.051 (0.042) | 0.33 | 0.213±0.057 | 0.67±0.18 | <0.18 |
| 3.0 | 1.68 | 2.69 | 2.06 | 0.51±0.18 | 0.099±0.037 (0.034) | 0.26 | 0.131±0.039 | 0.52±0.15 | <0.20 |
| 3.0[a] | 1.68 | 2.69 | 2.06 | 0.44±0.11 | 0.191±0.057 (0.044) | | | | |
| 3.0 | 2.74 | 4.37 | 2.52 | 0.95±0.35 | 0.166±0.065 (0.060) | 0.32 | 0.188±0.066 | 0.60±0.21 | <0.15 |
| 3.0[a] | 2.74 | 4.37 | 2.52 | 0.50±0.17 | 0.215±0.089 (0.080) | | | | |
| 4.9 | 1.02 | 2.65 | 2.20 | 0.29±0.11 | 0.058±0.023 (0.022) | 0.25 | 0.062±0.031 | 0.25±0.13 | <0.16 |
| 5.0[a] | 1.42 | 3.78 | 2.56 | 0.28±0.11 | 0.141±0.058 (0.051) | 0.38 | 0.148±0.073 | 0.41±0.20 | <0.12 |
| 5.0 | 2.95 | 7.87 | 3.56 | 0.63±0.33 | 0.099±0.062 (0.061) | 0.49 | 0.109±0.081 | 0.23±0.17 | <0.07 |
| 10.0 | 1.70 | 9.12 | 4.04 | 0.42±0.19 | 0.100±0.038 (0.035) | 0.58 | 0.117±0.057 | 0.22±0.11 | <0.04 |

(a) Values previously reported in Ref. 2. (b) Measured values without radiative corrections. The total errors are statistical counting errors added in quadrature to the systematic errors in $P_e$, $P_p$ and F; the numbers in parentheses are the one standard deviation counting errors. (c) Radiatively corrected values (See Table II). (d) Calculated using weighted average of D. (e) Calculated upper limits using R=0.25.

Table II: Kinematics and Radiative Corrections

| ω | $Q^2$ (GeV/c)$^2$ | E (GeV) | θ (Deg) | Range of ω | Weighted average of ω | Fraction of events inside cuts | Radiative correction[a] | $(A_1+\eta A_2)$ Corrected |
|---|---|---|---|---|---|---|---|---|
| 2.0 | 4.09 | 12.95 | 11.0 | 1.7-2.0 | 1.95 | 0.86 | +0.02 | 0.67±0.18 |
| 3.0 | 1.68 | 9.71 | 9.0 | 2.5-3.0 | 2.93 | 0.75 | +0.02 | 0.52±0.15 |
| 3.0 | 2.74 | 12.95 | 9.0 | 2.5-3.0 | 2.92 | 0.77 | +0.03 | 0.60±0.21 |
| 4.9 | 1.02 | 9.71 | 7.0 | 3.5-4.9 | 4.63 | 0.73 | +0.03 | 0.25±0.13 |
| 5.0 | 1.42 | 9.71 | 9.0 | 3.5-5.0 | 4.73 | 0.75 | +0.04 | 0.41±0.20 |
| 5.0 | 2.95 | 16.18 | 8.5 | 3.5-5.0 | 4.72 | 0.77 | +0.03 | 0.23±0.17 |
| 10.0 | 1.70 | 16.18 | 7.0 | 6.0-10.0 | 9.21 | 0.66 | +0.05 | 0.22±0.11 |

(a) $(A_1+\eta A_2)_{corrected} - (A_1+\eta A_2)_{uncorrected}$

(iii) The prediction of $A_1$ = 5/9 by the naive quark model is in fact naive. The momentum distribution of the quarks has to be taken into account.

## RADIATIVE CORRECTIONS

(This section is based on work by Noboru Sasao.[6]) The incoming (outgoing) electron can radiate an unobserved brems-photon of energy $\omega_s(\omega_p)$ and this will alter the $Q^2$ and ν of the relevant virtual photon (Fig. 2). The angular distribution of the brems-photons is highly peaked in the directions of the incoming and outgoing electron. The two cases are referred to as s-peak and p-peak, respectively. The $Q^2$ and ν-values are constrained to two straight lines which merge at the nominal datapoint in the ν-$Q^2$ plane and which reach all the way towards the elastic limit. This is demonstrated in Fig. 3 for two of our datapoints. For the s-peak (p-peak) branch the effective $Q^2$ will be lowered (raised) with respect to the nominal datapoint. The effective ν, W, and ω will always be lowered.

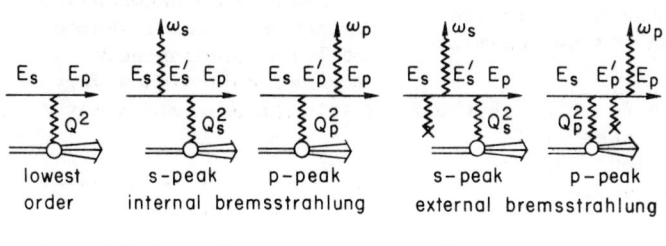

Fig. 2. Lowest order graph and radiation corrections in inelastic electron scattering.

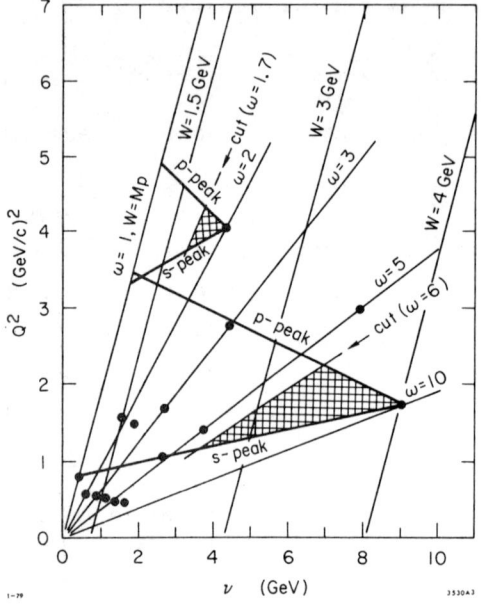

Fig. 3. $\nu$-$Q^2$ plane showing dots where asymmetry data have been obtained. Radiation effects cause contributions from lines labeled as s-peak and p-peak which merge at the nominal kinematic points as shown for two examples.

Extensive, radiatively corrected data are available on the spin averaged cross sections.[7] We have applied the equivalent radiator method of Mo and Tsai[8] to a parametrization of these cross sections. Fig. 4 shows a calculated spectrum and Fig. 5 gives a comparison with our cross section measurements. The non-zero slope shows the effect of external bremsstrahlung. Our measurements (dots) agree well with the calculated straight line.

In order to unfold an asymmetry measurement at any kinematic point from the radiative contaminations we have to know the cross section and the asymmetry on the s-peak and p-peak branches. In principle, this requires measurements everywhere in between the branches as well. Fortunately, our uncorrected asymmetry data show very little variation as a function of the kinematic variables and may therefore, for the purpose of the radiation corrections, be approximated by very simple functional forms.

The following procedure was carried out for each kinematic point:

(o) Choose a cut in $\omega$, which divides the contributing kinematic regions into a "signal" region near the nominal datapoint ($\omega > \omega_{cut}$) and a background region ($\omega < \omega_{cut}$).

(i) Construct an asymmetry model $A(\nu,Q^2)$, which approximates our uncorrected data.

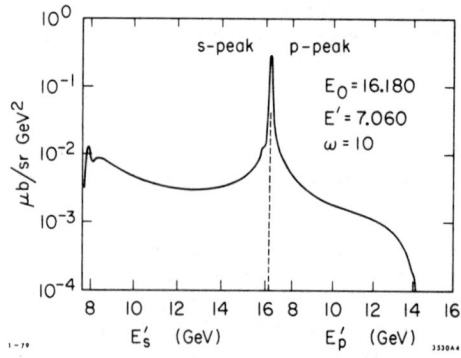

Fig. 4. Example of radiative spectrum. Shows contributions to the cross section from electrons with effective incoming (outgoing) energy $E'_s$ ($E'_p$).

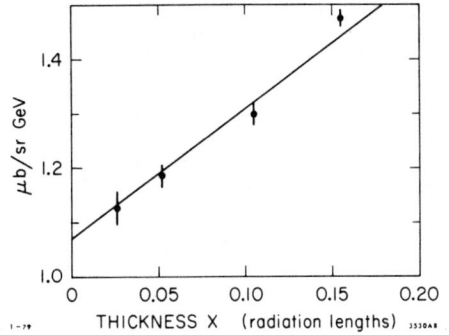

Fig. 5. Electron-carbon cross section/atom as a function of target thickness for the kinematic point $\omega = 10$.

(ii) Compute the radiated asymmetry for the signal and background region using the asymmetry model and the spin averaged cross section data.

(iii) Compute the difference between the asymmetries within and without the cut, apply this as a correction to our experimental asymmetry.

(iv) Go back to Step (i) and check for self-consistency.

The radiation corrections of the asymmetry were quite insensitive to the choice of model for $A(\nu, Q^2)$. The corrected values of the virtual photon asymmetry $A_1 + \eta A_2$ in Table II are larger than the uncorrected values by no more than 0.05.

## ASYMMETRY IN THE RESONANCE REGION

Our data show large and positive asymmetries in the deep inelastic region. However, there is reason to expect that this does not hold in the resonance region as well, because in the photoproduction limit ($Q^2 = 0$) the asymmetry $A_1$ of the resonance part can be shown to be negative throughout the first three resonances.[9] The first resonance is expected to maintain its magnetic dipole character out to all $Q^2$ and therefore the resonance contribution to $A_1$ is expected to remain negative. However, at large $Q^2$ the resonance disappears altogether and the non-resonant background dominates, which is expected to be predominantly s-wave ($A_1 = 1$). At the second and third resonance quark models predict a fast changeover from negative to positive $A_1$ for the resonance contributions themselves.[10] A multipole analysis of pion electroproduction data predicts a much slower changeover.[11] The behavior of the asymmetry in the resonance regions is also important with regard to duality.[12]

Fig. 6 shows the exploratory data which we have obtained in this interesting but difficult kinematic region. Radiation corrections were much larger than for the deep inelastic region. The analysis of these data was carried out by Mark Bergstrom.

## FUTURE EXPERIMENTS

We are currently preparing an experiment with a spectrometer of ten times larger acceptance and which will allow us to obtain ep-asymmetry data up to $Q^2 = 10$ $(GeV/c)^2$ and $x = 0.67$ (Fig. 7). This will improve our ability to test scaling and models. At smaller $Q^2$ we also plan to use a polarized deuterium target.

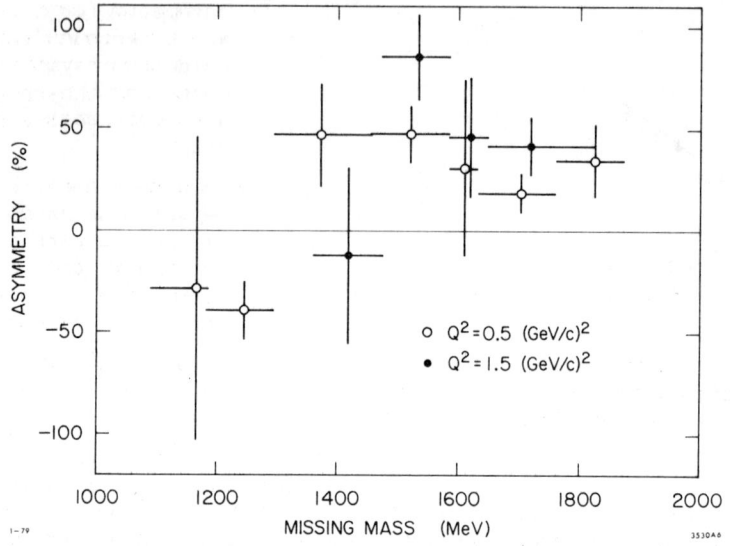

Fig. 6. Radiatively corrected asymmetry $A_1 + \eta A_2$ in the resonance region (preliminary results), $R = 0$ has been assumed.

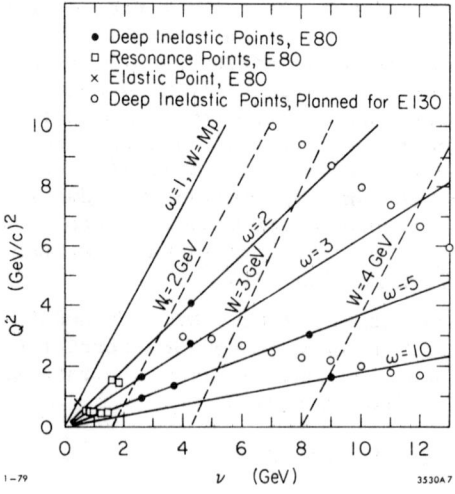

Fig. 7. Kinematic domain where ep-asymmetry measurements have been made (SLAC experiment E-80) or where such measurements are planned (SLAC experiment E-130). Polarized electron-polarized deuteron scattering is also planned at the low $Q^2$-points in E-130.

## REFERENCES

1. M. J. Alguard et al., Phys. Rev. Lett. $\underline{41}$, 70 (1978).

2. M. J. Alguard et al., Phys. Rev. Lett. $\underline{37}$, 1261 (1976).

3. Vernon W. Hughes, these proceedings.

4. M. J. Alguard et al., SLAC-PUB-2244 (1979), submitted to Nucl. Instr. Meth.

5. Charles Y. Prescott, these proceedings.

6. N. Sasao, "E-80 Radiative Correction" (1977), unpublished.

7. A. Bodek et al., Phys. Lett. $\underline{51B}$, 417 (1974), and references quoted therein; A. Bodek et al., to be published in Phys. Rev. D.

8. L. W. Mo and Y. S. Tsai, Rev. Mod. Phys. $\underline{41}$, 205 (1969).

9. S. D. Drell and A. C. Hearn, Phys. Rev. Lett. $\underline{16}$, 908 (1966); S. Gerasimov, Sov. J. Nucl. Phys. $\underline{2}$, 930 (1966); I. Karliner, Phys. Rev. $\underline{D7}$, 2717 (1973).

10. F. Ravndahl, Phys. Rev. $\underline{D4}$, 1466 (1971); S. Ono, Preprint, Technische Hochschule Aachen (1976).

11. R. C. E. Devenish and D. H. Lyth, Nucl. Phys. $\underline{B93}$, 109 (1975).

12. E. D. Bloom and F. J. Gilman, Phys. Rev. Lett. $\underline{25}$, 1140 (1970) and Phys. Rev. $\underline{D4}$, 2901 (1971); V. Rittenberg and H. V. Rubinstein, Phys. Lett. $\underline{35B}$, 50 (1971).

---

\* Research supported in part by the Department of Energy under Contract No. EY-76-C-02-3075 (Yale) and Contract No. EY-76-C-03-0515 (Stanford Linear Accelerator Center), the German Federal Ministry of Research and Technology and the University of Bielefeld, and the Japan Society for the Promotion of Science.

PARITY VIOLATION IN THE SCATTERING
OF 15 MEV PROTONS BY HYDROGEN*

D. E. Nagle, J. D. Bowman, C. Hoffman
J. McKibben, R. Mischke, J. M. Potter
Los Alamos Scientific Laboratory, Los Alamos, N. Mex. 87545

H. Frauenfelder, L. Sorenson
Univ. of Illinois, Urbana, Ill. 61801

ABSTRACT

A transmission experiment to measure the longitudinal asymmetry for polarized protons on hydrogen. The present result is $(-1.7\pm0.8)\times10^{-7}$).

## I. INTRODUCTION

Parity violating (PV) effects in the nucleon-nucleon interaction are at present the only practical way to study the strangeness-conserving nonleptonic part of the weak interaction. Although the expected effects in pp scattering are small ($\sim 10^{-7}$) it is important to study this simple nucleon-nucleon system. This is a status report of our 15 MeV experiment.

## II. PLAN OF THE EXPERIMENT

The idea is to measure the scattered beam current integrated over a solid angle of nearly $4\pi$, and divide it electronically by the transmitted beam. The quotient depends on the integrated cross section, and a change with beam helicity would indicate a PV piece in the scattering amplitude. The PV effect is estimated at $10^{-7}$ to $10^{-6}$, which requires five orders of magnitude increase in sensitivity over experiments of twenty years ago. At present, we have attained a sensitivity of seven parts in $10^8$.

To attain this level we use: integral counting, a synchronous detection detector design to minimize systematic errors, design of the source and use of servo control to minimize spin-correlated modulations of the beam, optical isolation, and a program of measuring and correcting for residual systematic errors.

Figure 1 shows layout of the experiment. The beam spin direction from the source is rapidly reversed at a 1 kHz frequency controlled by a reference oscillator. The scattering occurs when the beam passes along the axis of a cylinder containing hydrogen gas at about 100 psi (Fig. 2). The outer part of the vessel contains four slabs of scintillating plastic, symmetrically arranged above, below, to the right, and to the left of the beam. Each slab is viewed by three photomultipliers. The sum of the twelve scattering detectors is called the S signal. About 99% of the beam is transmitted through the scattering chamber to a set of five gold electrodes, arranged as center, up, down, and left and right with respect to the beam direction. The sum of these five currents is called B. (S/B is formed

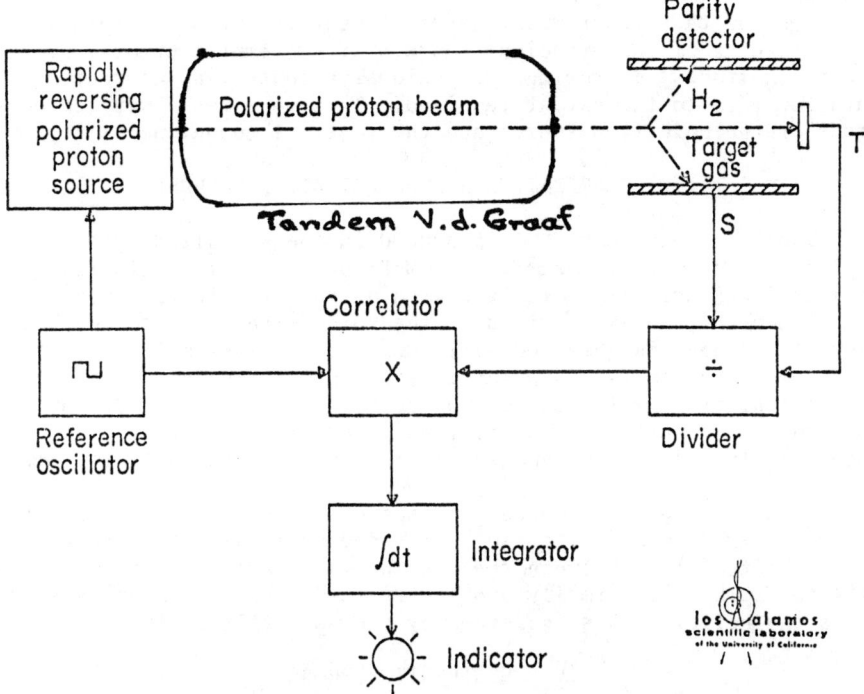

Fig. 1. Scheme for the parity violation experiment.

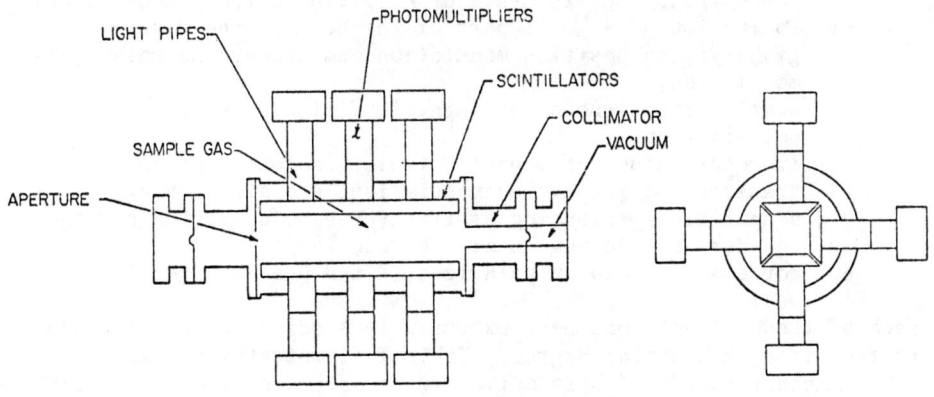

PNC SCATTERING CHAMBER

Fig. 2. Scattering chamber and detectors.

with an analog divider, and the synchronous component (S/B)(1 kHz) stored in an on-line computer as the nominal PV signal.

Most of the effort with the experiment is aimed at improving our understanding of the principal sources of systematic errors, reducing their magnitudes, measuring the residual amounts, and correcting the data for them to the extent required. This requires a very careful study of the polarized source and the reversal technique

### III. POLARIZATION AND POLARIZATION REVERSAL

The beam polarization is produced in the so-called spin filter region of the source, described by McKibben.[1] Leaving this region the beam is longitudinally polarized with a sense (helicity) specified by the sign of $B_S$, an axial field in the spin filter. Next, the helicity of the beam is reversed at 1 kHz by a precession field $B_T$ located between the Lamb-shift polarizer[2] of the source and the argon charge ionizer. $B_T$ is set to produce a precession angle of 180°. The usual time pattern of $B_T$, called alternating reversal, has $B_T$ successively +, 0, -, 0, and so on; resulting in the helicity pattern +, -, +, -.

After leaving the source a DC magnet and crossed electric field, below referred to as "precessor", is used optionally to insert a phase shift of 180° between the beam helicity and the reference oscillator. This is normally done every few hours during data-taking, as is the helicity phase governed by the spin filter field.

### IV. SYSTEMATIC ERRORS

Most of the unwanted modulation of the beam creating sources of systematic errors originate in processes occurring in the reversal region; they include:
- a. quenching of the 2s state of H, giving current modulation,
- b. ionization of a small part of the beam in the reversal region, giving position modulation and transverse polarization modulation,
- c. spatial misalignment of magnetic fields, giving transverse polarization.
- d. improper values of magnetic field gradients, giving residual transverse polarization modulation, and also giving rise to a spatial distribution of transverse polarization having a non-zero second moment at the detector.
- e. correlation between polarization and phase space of proton beam.

Each of these effects has been extensively studied and brought under control to a substantial degree. Table I illustrates the situation:

Another source of systematic errors is ground loops, i.e., electrical paths which transmit a spurious signal at the reference or spin modulation frequency. These have been reduced by using optical isolators. The program for reducing the effects of these and other errors is as follows: first, understand the physical processes going on in the source which give rise to the unwanted beam modulation; second, modify the source to reduce that modulation component; third, adjust the detector symmetry to reject that component; fourth and fifth, measure the amount of residual modulation, and store the result

TABLE I. TYPICAL SYSTEMATIC ERRORS

|  | Sensitivity | Amplitude | Correction to $A_L$ |
|---|---|---|---|
| Current Modulation | $4 \times 10^{-4}$ | $6 \times 10^{-7}$ | $2.4 \times 10^{-10}$ |
| Position Modulation (one coordinate) | $3.7 \times 10^{-4}$ mm | $9 \times 10^{-5}$ mm | $3.3 \times 10^{-8}$ |
| Transverse Polarization (one coordinate) | $2.7 \times 10^{-5}$ | $1 \times 10^{-3}$ | $2.7 \times 10^{-8}$ |

in the computer along with the PV signal at the standard 1-sec interval. From time to time measure the residual sensitivity of the source to that modulation component. Finally, make a numerical correction for that effect (off-line).

As an example, consider the error reduction program as it has been applied to one beam variable, namely, current modulation at the reference frequency $f_0$. The principal causes of this are: 1) undesired quenching of the 2s state, and 2) deflection of a prematurely ionized component, followed by interception of a portion of the beam at some aperture. The quenching is reduced by supplying a compensating electric field at right angles to the alternating magnetic field which precesses the spin. The magnitude of the E field is set so that for the atoms moving at the mean velocity of the beam, the sum of the Lorentz fields is zero. The reversal scheme is such that the fundamental component of current modulation is at $f_0 \div 2$, not $f_0$. The $f_0$ component is much reduced.

The unwanted ionization is reduced by cryopumping the critical regions of the source, and with appropriate sweeping fields. The Lorenz balancing of the fields also means that for the ions, synchronous forces are suppressed.

In the detector electronics system, the ratio of scattered to transmitted beam in first order is independent of current modulation. The remaining current modulation on the transmitted (B) signal is used as the error signal on a feedback servo to the source, where the amplified signal is applied to an auxiliary quench electrode. Finally, the residual current modulation on the B signal is monitored by the computer. The detector sensitivity to current modulation is measured from time, and an attempt is made to correct this residual, see (Table I). Similar techniques are used to suppress the other unwanted types of modulations on the beam.

At present, there are seven servo control channels applied to keeping the beam centered and unmodulated. Namely, when the servo feedback loops are closed, the tandem beam is very stable and quiet.

## V. DATA ACQUISITION

The experiment is operated under control of a small computer. At one-second intervals the data from the experiment are recorded. They include the following 1 kHz components, suitably digitized:

$Y = (S-B)/B$, $x = (B_L-B_R)/B_{center}$, $y = (B_u-B_D)$, $p_x = (S_u-S_D)/S_u+S_D)$,

$p_y = (S_L-S_R)/(S_L+S_R)$, $i = B_L+B_R+B_u+B_D+B_{center}$.

The data are recorded for 400 intervals, then the computer prints out the average values and standard deviations for $Y, x, y, p_x, p_y$ and other quantities of interest. The computer then selects a new combination of signs for the fields in the spin filter and Argon regions and starts another 400 point set. After sixteen 400 sec runs, the data taking is interrupted and the sensitivities measured. If any have become large, the detector is readjusted. Periodically during the run made in June '78 the signs of the precessor phase and of the spin filter field were reversed.

Table II gives the state of our knowledge at the end of 1977 of longitudinal asymmetry $A_L$ for 15 MeV protons on $^1H$ and $^2H$. Longitudinal asymmetry is defined as $[\sigma(+) - \sigma(-)] \div [\sigma(+) + \sigma(-)]$, where + and - denote the two helicity states of the beam. It is seen that $A_L$ was consistent with zero at the $10^{-7}$ level in H and D.

TABLE II. LONGITUDINAL ASYMMETRY DATA (1977)

|  | $D_2$ | $H_2$ |
|---|---|---|
| $A_L \times 10^7$ uncorrected | -1.2 ± .85 | +0.8 ± 1.4 |
| Sum of systematic corrections | -0.81 | 0.75 |
| $A_L \times 10^7$ corrected | -0.35 ± 0.85 | +0.05 ± 1.4 |

In January '78 we took data for about two weeks. Four sets of data were taken, corresponding to four different sets of "precessor" and spin filter combinations. They are listed in Table III as A, B, C, D. A minus one in the precessor column of Table III indicates that a phase shift of $\pi$ in the phase of the beam helicity has been inserted by the precessor. The helicity column entry is the product of the precessor and spin filter helicities; i.e., it is the beam helicity phase. Roughly equal time was spent taking data in each of the four configurations.

TABLE III.   JANUARY 1978 DATA

| Label | Prec. | Spin Filter | Helicity | Sec. | Data x $10^7$ |
|---|---|---|---|---|---|
| A | -1 | +1 | +1 | 68800 | -5.40 ± 1.32 |
| B | -1 | +1 | -1 | 68545 | -1.36 ± 1.35 |
| C | +1 | -1 | -1 | 60000 | -0.55 ± 1.24 |
| D | +1 | +1 | +1 | 58400 | -2.26 ± 1.25 |

We can form four linear combinations of A, B, C, and D, given in Table IV. The various types of systematic error contribute differently in these linear combinations. The current modulation contribution i, horizontal displacement x, vertical displacement y, and ground loop g are unaffected by the spin filter and precessor phases. Transverse polarization $p_x$ contribution correlates with helicity, as does the true parity violation effect PV. No known piece correlates with "spin filter", or with "precessor". This is because of special coils to compensate "spin filter" reversal on $p_x$ and $p_y$, and because of special retuning after "precessor" reversal to compensate $P_x$ and $P_y$. We compare these results with measured beam variables and measured sensitivities from data set "Y" (about 4% of the data), and show the estimated contribution as the last column.

TABLE IV.   COMBINED DATA

| Combination - Contributions | Data | Estimate |
|---|---|---|
| 1/4(A+B+C+D)  ⇒ g+i+x+y | -2.4 ± .64 | -2.8 ± .73 |
| 1/4(-A-B+C+D) ⇒ 0 | +.99 ± .64 | 0 |
| 1/4(-A+B-C+D) ⇒ 0 | +.58 ± .64 | 0 |
| 1/4(A-B-C+D)  ⇒ pv + px | -1.44 ± .64 | pv - .025 |

The contributions to the (A+B+C+D)/4 estimate are g = -3.0 ± .7, i = -0.4 ± .1, x = 0.6 ± 0.1  and y = 0.04 ± .05, for a net value of -2.8 ± 0.73.    The $p_x$ contribution to the helicity (bottom row Table IV) is -0.025, leading to a PV effect of (-1.4 ± 0.6) x $10^{-7}$. Dividing by the beam polarization, 85%, the longitudinal asymmetry is $A_L$ = (-1.7 ± 75) x $10^{-7}$.

This result is not in disagreement with the estimates of Brown et al.[2] and of Henley et al.[3]    Further improvement of the sensitivity by a factor of perhaps two seems practical.

## REFERENCES

*Work supported by U.S. Department of Energy and the National Science Foundation.
1. J. M. McKibben. Proc. 1976 ANL Conf. on Polarized Beams and Targets, AIP Conference Proceedings No. 35, p. 375.
2. V. Brown, E. M. Henley, and F. R. Krejs, Phys. Rev. C $\underline{9}$, 935 (1975).
3. E. M. Henley and L. Wolfenstein, Nucl. Phys. A300 (1978), p. 265.
   E. M. Henley, Nucl. Phys. A300 (1978), p. 273.

SEARCH FOR PARITY VIOLATION IN
POLARIZED PROTON SCATTERING AT 6 GEV/C*

D. E. Nagle, J. D. Bowman, C. M. Hoffman,
R. E. Mischke, J. M. Potter
Los Alamos Scientific Laboratory, Los Alamos, N. Mex. 87106

N. Lockyer, T. A. Romanowski
Ohio State University, Columbus, Ohio 43210

D. R. Moffett
Argonne National Laboratory, Argonne, Ill. 60439

B. Nelson
University of Chicago, Chicago, Illinois 60637

W. H. Sawyer
Elmhurst College, Elmhurst, Ill. 60126

E. C. Swallow
Elmhurst College, Elmhurst, Ill. 60126 and
University of Chicago, Chicago, Illinois 60637

R. L. Talaga
University of California, Berkeley, CA 94720

D. M. Alde
University of Illinois, Urbana, Ill. 61801

ABSTRACT

We have been searching for a small deviation from the parity symmetry in proton-nucleon scattering. For 6 GeV polarized protons on $H_2O$ we find for the longitudinal asymmetry a preliminary value of $A_y = \Delta\sigma/P2\sigma = (3.1 \pm 3.8) \times 10^{-6}$ (without systematic errors) using our ion chamber detectors, and $A_y = -(1.9 \pm 2.6) \times 10^{-6}$ (without systematics) using our scintillator detectors.

---

A series of experiments have been carried out in the 6 GeV polarized proton beam at ANL. The first, carried out in October '74 measured the transmission in a beryllium target,[1] with the result $\Delta\sigma/2\sigma = (5\pm9) \times 10^{-6}$. The second carried out in April '75,[2] showed a large parity violating effect, namely $(-15 \pm 2.4) \times 10^{-6}$ for $\Delta\sigma/2\sigma$. A plausible explanation for this effect is that the apparatus detected a small fraction of parity violating weak decays of polarized hyperons; e.g., $\Lambda \to p^+\pi^-$. To eliminate such effects, the apparatus was augmented by a magnetic filter or spectrometer, which rejected any hyperon decay products.
    The present arrangement is shown in Figs. 1 and 2.
    The proton beam enters from the left. The first bending magnet converts the beam to longitudinal polarization. The beam polarization

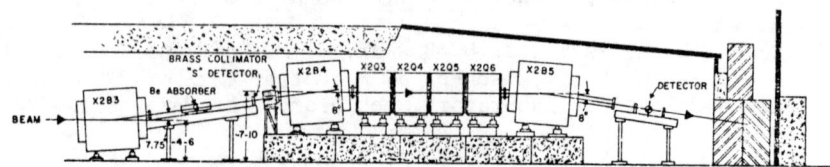

Fig. 1. Side View of PV apparatus.

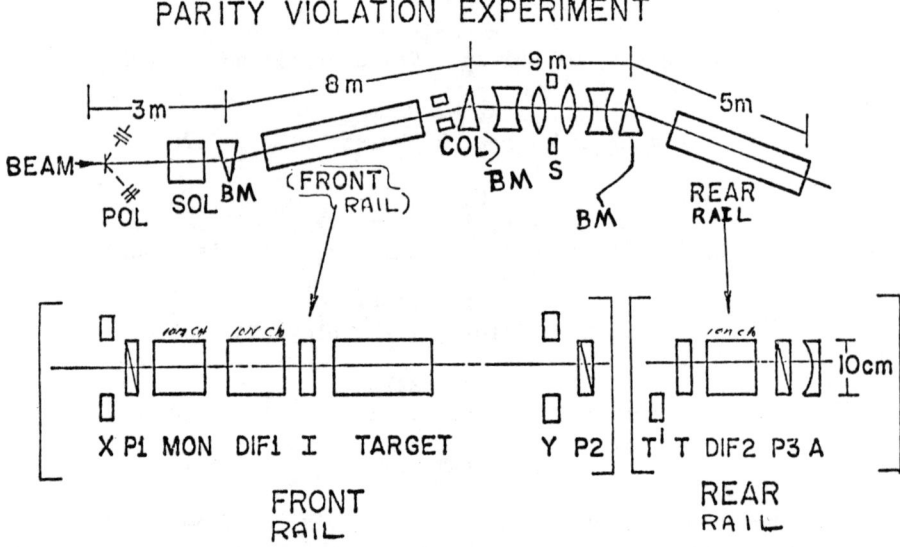

Fig. 2. Arrangement of detectors. X and Y are polarimeters; P1, P2 and P3 are position detectors, I and T are the scintillation detectors for measuring the transmission; Mon, DIF 1 and DIF 2 are the ion chambers for the transmission measurement; SOL is one of the spin precessing solenoids; COL is a collimator; A is a beam size determining scintillator, S is a detector for scraped beam, and $T^1$ a monitor of low momentum beam.

is about 70%. The incident beam detectors, the target, a polarization detector, and a transmission detector are installed on the first rail. Next is the spectrometer consisting of a bend, four focusing quads, and a final bend. The beam is again longitudinally polarized. The transmitted beam is detected in the array of counters and ion chambers mounted on the right hand rail. Finally the beam is dumped in a cavity in the shield.

The polarization is reversed between ZGS pulses. A fraction of the circulating beam is extracted, using targetted (Piccioni) extraction on a "front porch", and directed to our experiment. The remaining beam is accelerated to 12 GeV for other users. This mode of operation has proven to be essential for our obtaining quality data, because it makes us independent of the requirements of other users, and because targetted extraction produces an inherently quiet beam.

We make three measurements of the transmission, one using ion chambers, one using the Tand I detectors and a third using a scintillator pair on the front (left rail). Other detectors measure position of the beam centroid, residual transverse polarization, beam angle, and beam diameter. All quantities are integrated over the beam pulse, converted to digital form, stored in an on-line computer, and taped for further processing.

Extracting a very small helicity correlated PV signal on the transmission Z requires careful attention to noise supression and to elimination of spurious beam helicity correlated effects; i.e., systematics. The noise reduction program consists of designing detectors inherently insensitive to fluctuations in beam position, intensity, size, etc. These quantities are monitored by dedicated detectors (Fig. 2). Their signals are treated as variables in a linear regression analysis of the transmission, which reduces the noise by a factor of 2.5.

The helicity correlated, "systematic", effects arise from several sources: First is some small residual scraping and forward scattering of the beam before it enters our cave. Second, a small deviation of the beam polarization from longitudinal. These effects couple with beam displacement to give a spurious PV signal. These effects are measured in auxiliary experiments. Their corrections amount to $(0.50 \pm .70) \times 10^{-6}$ for scraping and to $(0.8 \pm 0.5) \times 10^{-6}$ for vertical polarization and to $(0.4 \pm 1) \times 10^{-6}$ for horizontal polarization, (scintillators).

The spectrometer has been designed to eliminate contributions from the hyperon decays to a level below $10^{-7}$. No evidence for hyperon decay was seen in the last run in the rear transmission detectors; however the detectors before the spectrometer reproduced the 1975 result, namely they gave $\overline{\Delta\sigma/2\sigma} = (-11 \pm 2) \times 10^{-6}$.

The relaxation analysis and systematic error corrections are being continued. The $\chi^2$ values per degree of freedom are good: 1.1 for scintillators and 0.8 for the ion chambers. The noise is 2.5 x proton statistics. Much of this noise presumably arises from fluctuations within the pulse about the mean values of position, intensity, etc. The net results are $A = -(1.9 \pm 2.6) \times 10^{-6}$ for the scintillators and $A = (3.1 \pm 3.8)$ for the ion chambers, (no systematics).

These values may be compared to an estimate by Henley and Krejs[3] of $-1 \times 10^{-7}$.

Future experiments: The sensitivity of the experiment could be improved to the order of $5 \times 10^{-7}$ in a month's running at the ZGS. This would be of considerable interest, since improved theoretical calculations could now be undertaken, based on better knowledge of the strong interaction amplitudes.[4]

REFERENCES

*Work performed under the auspices of the U.S. Department of Energy under Contract No. EY-76-S-02-2146.

1. J. D. Bowman, et al., Phys. Rev. Lett., 34, 1184, (1975).
2. H. L. Anderson et al., Proc. AIP Conf. No. 35 (1976) p. 273.
3. E. M. Henley and F. R. Krejs, Phys. Rev. D 11, 605 (1975).
4. E. M. Henley, private communication.

ESTIMATES OF W PRODUCTION WITH POLARIZED PROTONS

Frank E. Paige, T. L. Trueman and Thomas N. Tudron
Brookhaven National Laboratory, Upton, New York 11973

ABSTRACT

We discuss the measurement of parity violating asymmetries in polarized proton-proton scattering at high energy as a method of extracting the hadronic decays of the intermediate vector boson from the high-transverse-momentum hadronic background. In particular we present predictions of these asymmetries for jet production through the charged and neutral vector bosons. The asymmetries are very large and the method looks promising.

Although the W is expected to be produced copiously at very high energies, as will be provided by ISABELLE,[1] and to decay most of the time into hadrons, the problem of extracting its hadronic decays from the high $p_\perp$ hadronic background is very acute.[2,3] Some methods have analyzed,[4] relying on the fact that the hadrons from W decay should be richer in strange or charmed particles than the background, but the outlook for these is uncertain.

The possibility of storing polarized protons in ISABELLE[5] with high luminosity has led us to consider extracting the W signal from the background by measuring high $p_\perp$ jets produced by polarized protons with definite helicity. Because of the V-A coupling, only left-handed quarks produce charged W's. Suppose there is a correlation between the helicity of the proton and that of it quarks. Then the difference between the cross sections for high $p_\perp$ jets produced with left-handed and right-handed protons should show a W signal with no background from strong interactions. In fact such a correlation exists in many models[6] and has been confirmed experimentally.[7] Haber and Kane[8] have shown that one such model leads to a $W^+$ cross section which is about three times larger for left-handed protons than for right-handed ones. In this paper we calculate the expected asymmetry for the hadronic jet distributions.

Our calculations are based on the simple Drell-Yan model of W production and decay given in Ref. 3, supplemented by the SU(6) model for the quark-spin wave function. (See Ref. 3 for a complete set of references.) We realize that there may be many corrections to this model, but we feel that it is adequate for our purposes. We assume that the quark distributions can be broken up into valence plus sea quarks, the sea quarks being unpolarized and SU(3) symmetric. Thus we have

$$u(x) = u_v(x) + s(x), \quad d(x) = d_v(x) + s(x),$$

with

$$u_v(x) = u_+(x) + u_-(x), \quad d_v(x) = d_+(x) + d_-(x),$$

where + and − denote respectively quark helicities equal to and opposite to that of the proton. The SU(6) model[6] gives

$$u_+(x) = \frac{5}{6} u_v(x), \quad d_+(x) = \frac{1}{3} d_v(x),$$

$$u_-(x) = \frac{1}{6} u_v(x), \quad d_-(x) = \frac{2}{3} d_v(x).$$

At the values of x most important for us, all models and the experimental data roughly agree with the SU(6) limits.[6,7] As in Ref. 3, we set the Cabibbo angle to zero.

We estimate hadronic decays of the W by calculating its decay into quarks and assuming that these give rise to hadronic jets. In a sufficiently large detector the momentum of these jets should be measurable with only a small error from the background of soft particles. In particular, it would then be possible to measure the jet-jet mass. Because of the soft particles, however, it will not be possible to determine the charge or other quantum numbers of the jet, so we sum over these. In our calculations we assume that there are four flavors. The number of flavors enters only through the number of decay channels of the W, so that inclusion of a third lepton and a third quark doublet would increase the width of the W peak and lower its height by a factor of 1.5.

In Ref. 3, backgrounds for high $p_\perp$ jets from hard quark-quark scattering were estimated. In subsequent papers,[11] additional contributions from gluon-quark and gluon-gluon scattering were found to be quite large, mainly because there are a lot of gluons at small x. In addition, scaling violation is important in obtaining an overall fit to data. However, the scaling violation is not very large for $x \sim 0.1$, the area of most interest to us. Thus, in the following figures we show two background curves: the lower one is that estimated in Ref. 3; the higher one includes gluon scattering with the distribution

$$g(x) = 3(1-x)^5/x$$

(cf. Cutler and Sivers, Ref. 11.) From the results of Ref. 11 it is clear that the effects of scaling violation will be to tilt the curve so the background is smaller for high $p_\perp$ and larger for small $p_\perp$.

Consider first the hadronic jets from $W^+$ and $W^-$ production. It was shown in Ref. 3 that all quantities of interest can be expressed in terms of the invariant cross section $d\sigma/d\hat{s}d\hat{t}d\hat{y}$. Thus to calculate the parity violating asymmetries, we need

$$\frac{d\sigma^{--}}{d\hat{s}\,d\hat{t}\,d\hat{y}} - \frac{d\sigma^{++}}{d\hat{s}\,d\hat{t}\,d\hat{y}} = \frac{1}{s}\,\frac{8}{\pi\hat{s}^2}\,\frac{(G/\sqrt{2})^2 m_W^4}{(\hat{s} - m_W^2)^2 + 4m_W^2 \Gamma^2}\,(\hat{t}^2 + \hat{u}^2) \quad (1)$$

$$\times \left\{ \frac{1}{3} [u_v(x)s(x') + s(x)u_v(x')] - \frac{1}{6} [d_v(x)s(x') + s(x)d_v(x')] \right\}.$$

Here $\hat{s}$, $\hat{t}$ are the invariants for the elementary $q\bar{q} \to W \to q\bar{q}$ process, $\hat{y}$ is the rapidity of the W in the c.m. system and

$$x = e^{\hat{y}}\left(\frac{\hat{s}}{s}\right)^{1/2} , \qquad x' = e^{-\hat{y}}\left(\frac{\hat{s}}{s}\right)^{1/2} .$$

In Fig. 1, Eq. 1 is integrated to give the single jet cross section at 90° and $\sqrt{s}$ = 800 GeV, with $\sin^2\theta_W$ = 0.25 or $m_W$ = 76 GeV. Both the unpolarized cross section and the difference $\sigma^{--} - \sigma^{++}$ are shown. Since we are plotting $\sigma^{--} - \sigma^{++}$ and the unpolarized cross section $\sigma = \frac{1}{4}(\sigma^{--} + \sigma^{++} + \sigma^{-+} + \sigma^{+-})$, in principle the W signal in the former could be 4 times as big as the latter. In fact, because we assume the sea is unpolarized, the factor can be at most 2. Our calculated result shown in Fig. 1 is only a factor 2 below this limit. Thus, taking the asymmetry leaves behind most of the W signal. If the statistical error on the background can be made significantly smaller than the W signal, it should stand out clearly in the difference. Clearly a large, but we hope not impossibly large, number of events will be required.

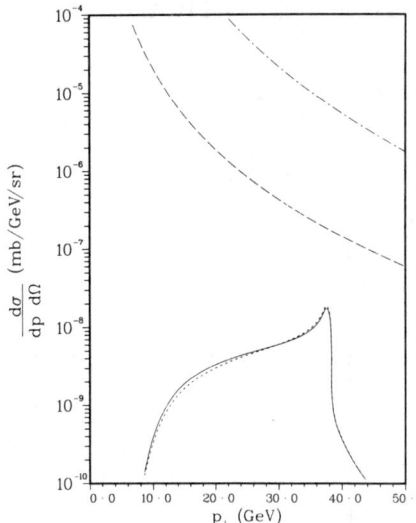

Fig. 1. Single jet cross sections for $W^{\pm}$ at $\sqrt{s}$ = 800 GeV, $\theta$ = 90°. Solid curve: Asymmetry for $W^{\pm}$. Dotted: Cross section for $W^{\pm}$. Dashed: Cross section for quark-quark scattering. Dot-dashed: Cross section for quark and gluon scattering.

The situation can be improved by observing both jets into which the W decays. In Fig. 2, the cross sections for back-to-back jets at 90° and $\sqrt{s}$ = 800 GeV are shown. The signal to background is improved and a much sharper signal shape is obtained. A further improvement is obtained by measuring two jets in the same hemisphere at a small angle to the beam. (In this configuration x is increased, emphasising the polarized valence quarks over the gluons.) This effect is shown in Fig. 3 for two jets each at 20° to the beam. Because of the size of the jets, it will probably be difficult to go to smaller angles than this.

One can apply the same considerations to $W^{\circ}$ production. The results are more model dependent, and the asymmetry is expected to be less because the $W^{\circ}$ couplings are not generally pure V-A. The di-jet asymmetry using the Weinberg-Salam model[12] is

$$\frac{d\sigma^{--}}{d\hat{s}\,d\hat{t}\,d\hat{y}} - \frac{d\sigma^{++}}{d\hat{s}\,d\hat{t}\,d\hat{y}} = \frac{1}{s}\,\frac{4}{\pi\hat{s}^2}\,\frac{(G/\sqrt{2})^2\,m_{W^\circ}^4}{(\hat{s}-m_{W^\circ}^2)^2 + 4m_{W^\circ}^2(\Gamma^\circ)^2}\,(\hat{t}^2 + \hat{u}^2)$$

$$\times \left\{ \frac{1}{3} a_u b_u [u_v(x)s(x') + s(x)u_v(x')] - \frac{1}{6} a_d b_d [d_v(x)s(x') + s(x)d_v(x')] \right\} \left( a_u^2 + b_u^2 + a_d^2 + b_d^2 \right), \qquad (2)$$

where $a_q$ and $b_q$ are the vector and axial vector couplings of the quark q:

$$a_u = \left(1 - \frac{8}{3} \sin^2\theta_W\right)/\sqrt{2}, \qquad b_u = 1/\sqrt{2}$$

$$a_d = -\left(1 - \frac{4}{3} \sin^2\theta_W\right)/\sqrt{2}, \qquad b_d = -1/\sqrt{2}$$

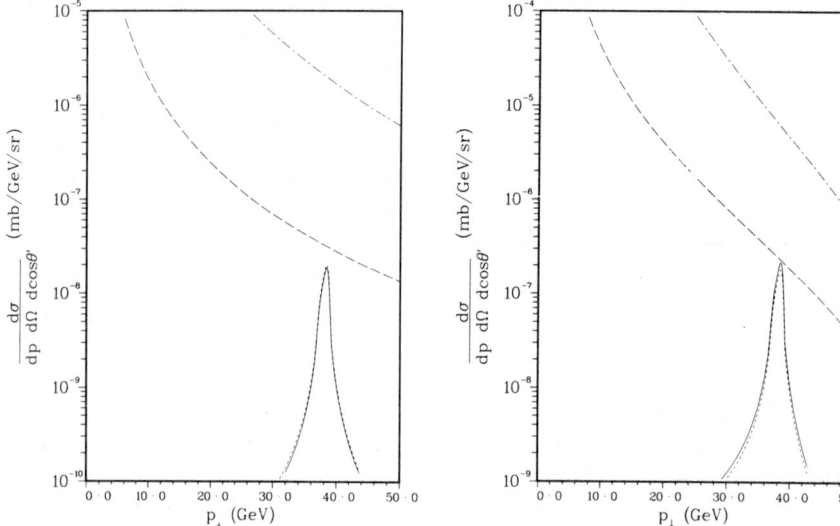

Fig. 2. Double jet cross sections for $W^\pm$ at $\sqrt{s}$ = 800 GeV, $\theta_1 = \theta_2 = 90°$. The curves have the same meanings as in Fig. 1.

Fig. 3. Same as Fig. 2, but at $\sqrt{s}$ = 800 GeV, $\theta_1 = \theta_2 = 20°$.

Fig. 4 shows the results for $\sin^2\theta_W$ = 0.2, 0.25 and 0.3, covering the range indicated by experimental data.[13] The results are seen to be very sensitive to $\sin^2\theta_W$. In fact, the asymmetry changes sign between 0.3 and 0.35. The signal to background is always worse than for the charged W's.

Measuring the asymmetry for leptonic decays of the W is also interesting. Here the background is not expected to be a problem. The asymmetry shows unambiguously that the peak in the spectrum is due to a parity violating process and not to some new hadron. Furthermore, it provides a test of the model which is used and a measure of the quark-spin wave function. The asymmetry in the single lepton spectrum coming from $W^+$ decay is shown in Fig. 5, along with the unpolarized Drell-Yan background.[3]

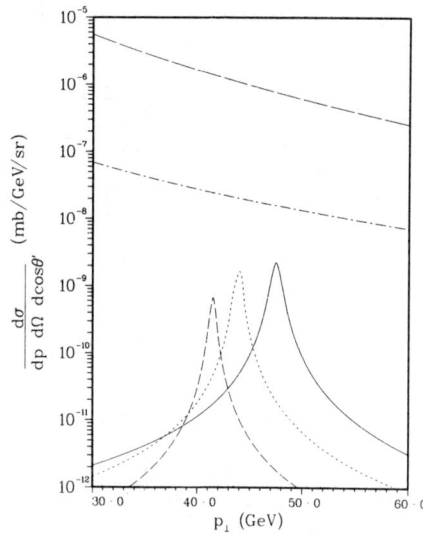

Fig. 4. Double jet cross sections for $W^\circ$ at $\sqrt{s}$ = 800 GeV, $\theta_1 = \theta_2 = 90^\circ$.
Solid curve: Asymmetry for $W^\circ$ with $\sin^2\theta_W = 0.20$.
Dotted: Asymmetry with $\sin^2\theta_W = 0.25$.
Dashed: Asymmetry with $\sin^2\theta_W = 0.30$.
Dot-dashed: Cross section for quark-quark scattering.
Long-dashed: Cross section for quark and gluon scattering.

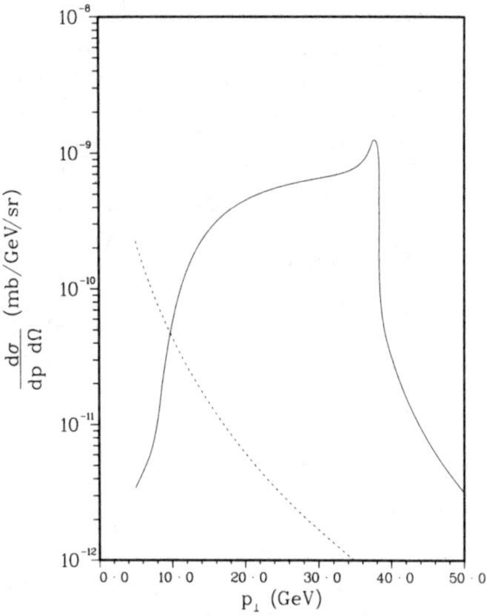

Fig. 5. Single $e^+$ cross section at $\sqrt{s}$ = 800 GeV, $\theta = 90^\circ$.
Solid curve: Asymmetry for $W^+ \to e^+\nu$.
Dotted: Cross section for Drell-Yan process.

In conclusion, it seems that polarized proton beams colliding at high energies may prove to be the most practical way to study the hadronic decays of the W.

We would like to thank Howard Weisberg for bringing the possibility of polarized proton beams in ISABELLE to our attention and Mark Sakitt for a helpful discussion. An account of this work has appeared previously as BNL-24919 and has been submitted to the Physical Review.

## REFERENCES

1. "ISABELLE, A Proton-Proton Colliding Beam Facility" Brookhaven National Laboratory BNL 50648, April 1977.
2. C. Quigg, Rev. Mod. Phys. **49**, 297 (1977).

3. R. F. Peierls, T. L. Trueman and L. L. Wang, Phys. Rev. D $\underline{16}$, 1397 (1977).
4. F. E. Paige and R. B. Palmer, ISABELLE Proceedings of 1977 Summer Workshop, BNL 50721 (1977); P. K. Williams et.al., <u>ibid</u>.
5. E. Courant in "Higher Energy Polarized Proton Beams," A. D. Krisch and A. J. Salthouse eds. AIP, N.Y. (1978).
6. L. M. Sehgal, Phys. Rev. $\underline{D10}$, 1663 (1974); G. W. Look and E. Fishbach, Phys. Rev. $\underline{D16}$, 211 (1977); F. E. Close, Nucl. Phys. $\underline{B80}$, 269 (1974); R. Carlitz and J. Kaur, Phys. Rev. Lett. $\underline{37}$, 673 (1977); J. Kaur, Nucl. Phys. $\underline{B128}$, 219 (1977).
7. M. J. Alguard et.al., Phys. Rev. Lett. $\underline{37}$, 1258 (1976); $\underline{37}$, 1261 (1976); $\underline{41}$, 70 (1978).
8. H. E. Haber and G. L. Kane, "Detection of Intermediate Vector Bosons and High Energy Weak Interactions from Decay of Hadron Resonances." U. M. HE 78-11 (1978).
9. F. Halzen and D. M. Scott, "Chromodynamics and the Experimental Signatures of Weak Bosons." COO-881-41 (1978).
10. I. Hinchcliffe and C. H. Llewllyn Smith, Phys. Lett. $\underline{66B}$, 281 (1977); J. Kogut and J. Shigemitsu, Nucl. Phys. $\underline{B129}$, 461 (1977).
11. R. Cutler and D. Sivers, Phys. Rev. $\underline{D17}$, 196 (1978); B. L. Combridge, J. Kripfganz and J. Ranft, Phys. Lett. $\underline{70B}$, 234 (1977); A. P. Contogouris, R. Gaskell and S. Papadopoulos, Phys. Rev. $\underline{D17}$, 2314 (1978); R. F. Feynman, R. D. Field and G. C. Fox, "A Quantum Chromodynamic Approach for the Large Transverse Momentum Production of Particles and Jets," CALT-68-651 (1978).
12. S. Weinberg, Phys. Rev. Lett. $\underline{19}$, 1264 (1967); A. Salam, in <u>Elementary Particle Theory: Relativistic Groups and Analyticity (Nobel Symposium No. 8)</u>, edited by N. Svartholm (Almquist and Wiksele, Stockholm, 1968) p. 367.
13. M. Holder, et.al., Phys. Lett. $\underline{72B}$, 254 (1977); A. M. Cnops et.al., Phys. Rev. Lett. $\underline{41}$, 357 (1978); C. Y. Prescott, et.al., Phys. Lett. $\underline{77B}$, 347 (1978).

# FINAL STATE POLARIZATIONS IN NEUTRINO INDUCED REACTIONS

Paul Langacker
University of Pennsylvania, Philadelphia, Pa. 19104

## ABSTRACT

The theoretical expectations for final baryon polarizations in neutrino baryon elastic and quasi-elastic scattering are discussed. In particular, a measurement of the transverse polarization of the final proton in $\nu(\bar{\nu})$p elastic scattering could effectively discriminate between two classes of competing gauge theory models and could also distinguish V,A neutral current interactions from S,P,T. A measurement of the polarization orthogonal to the scattering plane would test for time reversal violation and second class currents in the weak interactions.

## INTRODUCTION

Consider the elastic or quasi-elastic scattering of a $\nu$(or $\bar{\nu}$) from a nucleon, $1(k)+B(p) \to 1'(k')+B'(p')$, where 1 and 1' are the initial and final leptons, and B and B' are the initial and final baryons. Let E be the incident neutrino energy and $Q^2=-(k-k')^2>0$. The laboratory frame polarization of the final baryon will be denoted $P_i(Q^2)$, where i=L,T, and O refer to the longitudinal, transverse, and orthogonal directions, which are defined in Figure 1.

$P_T(Q^2)$ and $P_O(Q^2)$ can be measured in large detectors by observing the angular asymmetries produced when the final baryon rescatters in the target material. Such a determination would be of interest[1] because $P_T$ is a sensitive probe of the structure of the hadronic neutral current. A non-zero value of $P_O$ would imply time reversal violation in the weak interactions; in the neutral current case it would also imply either a second class current or an off-diagonal coupling at the neutrino vertex. $P_L$ is also a probe of the neutral current, but cannot be measured by a simple rescattering.

## THE NEUTRAL CURRENT

The transverse polarization $P_T(Q^2)$ for small $Q^2$ is a sensitive probe of the V,A structure of the interaction[1,2]. The matrix element for the reaction in Fig. 1 is

$$m = \frac{G_F}{\sqrt{2}} \bar{1}'\gamma^\mu(1+\gamma^5) 1 \langle B'|J_\mu^H|B\rangle \qquad (1)$$

If the hadronic matrix element is dominated by V-A for $Q^2 \approx 0$ then

---

*Work supported in part by the Department of Energy

the final baryon will have helicity -1 in the center of mass, which corresponds to $P_T(0)=+1$ in the lab. Similarly, if the hadronic matrix element is V+A, V, or A, then $P_T(0)=-1, 0$ or 0, respectively. These values hold for both $\nu$ and $\bar{\nu}$ induced reactions. For $Q^2>0$ one must carry out a full computation of the amplitude,[1] but one expects that the qualitative features of the $Q^2=0$ case will be preserved for moderate $Q^2$. That this is the case is illustrated in Fig. 2, which displays $P_T(Q^2)$ calculated[1] for quasi-elastic scattering at E=1.7 GeV (which is the average energy of the Brookhaven neutrino spectrum). One sees that for moderate momentum transfers (e.g. $Q^2 \approx 0.5 \text{GeV}^2$) the polarization is still large and positive (40-50%).

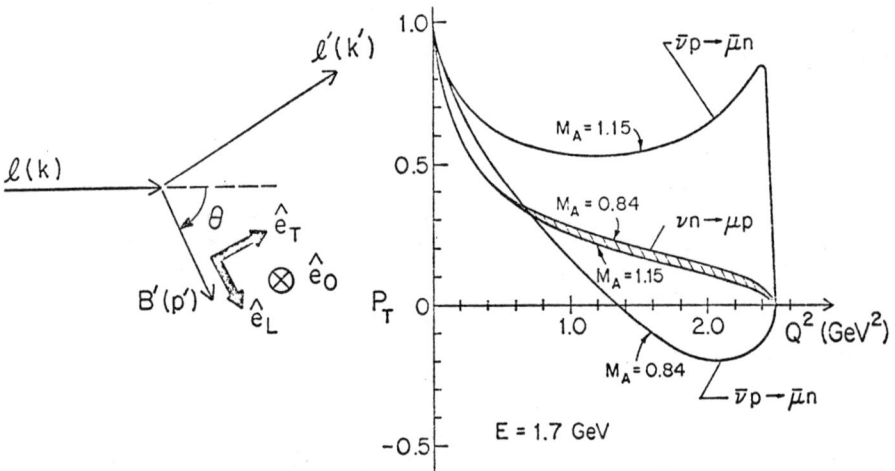

Fig. 1 - The Laboratory polarization directions $\hat{e}_L$, $\hat{e}_T$, $\hat{e}_0$

Fig. 2 - Transverse polarizations calculated[1] for quasi-elastic scattering. $M_A$ is the mass parameter in the axial vector dipole form factor.

The transverse polarization in the charged current case is of little interest except for calibration. In the neutral current case, however, a measurement of $P_T(Q^2)$ would be of considerable interest. Much effort has been expended recently[3-9] on determining the parameters of the hadronic neutral current. One can write,

$$J_\mu^H = \bar{u}\gamma_\mu[\epsilon_L(u)(1+\gamma^5) + \epsilon_R(u)(1-\gamma^5)]u$$
$$+ \bar{d}\gamma_\mu[\epsilon_L(d)(1+\gamma^5) + \epsilon_R(d)(1-\gamma^5)]d \qquad (2)$$

+ heavy quark terms,

where $J_\mu^H$ is the current in Equation 1 relevant to $\nu(\bar{\nu})$ hadron neutral current scattering. Sehgal showed that deep inelastic inclusive and semi-inclusive pion production data constrain the $\epsilon$'s to be in one of four regions of coupling parameter space, called A,B,C, and D by Hung and Sakurai.[3] These regions are displayed in Figure 3. It was subsequently argued that elastic $\nu(\bar{\nu})p\rightarrow\nu(\bar{\nu})p$ scattering rules out solutions[5,6] C and D and strongly favors[7] solution A over solution B. Furthermore, exclusive pion production weakly favors solution A over solution B[8,9].

It therefore appears that solution A, which corresponds to the Weinberg-Salam model, is the solution chosen by nature. However, in view of the extreme difficulty of the experiments and the great importance of the results, it is very important to perform other experiments to confirm this result. In particular, even a rough measurement of $P_T$ for small $Q^2$ could effectively distinguish[1] between solutions A and B, because the proton matrix elements of $J_\mu^H$ are dominated by V+A and V-A in the two cases, respectively, leading to large polarizations of opposite sign. The predicted values for $P_T(Q^2)$ at E=1.7GeV are shown in Figure 4. Specific predictions for the Weinberg-Salam model for various values of $x=\sin^2\theta_W$ are shown in Figure 5. The polarization is predicted to be negative for x>0.25, vanishes for x=0.25, and is slightly positive for x≳0.20. The model for which the $u_R$ is in a doublet (which corresponds to Solution B) predicts a large and positive value for $P_T(Q^2)$.

It should be emphasized that the neutral current effects observed in polarized electron baryon scattering[10] are directly related to the $\epsilon$ parameters measured in neutrino scattering only in theories involving a single Z boson. Hence, detailed measurements of the neutral current parameters in neutrino reactions are important for searching for the effects of additional Z bosons.

It has also been suggested that the final hadron density matrix elements in exclusive $\Delta$ production[11] and inclusive $\rho$ production[12] can effectively constrain the neutral current.

Finally, Fischbach, et al.[2] have argued that $P_T(Q^2)$ can distinguish between a pure V,A neutral current and a pure S,P,T interaction. The point is that the neutrino helicity flips in the S,P,T case, so for $Q^2=0$ angular momentum conservation requires that the final baryon must have helicity +1 or -1 in the center of mass for $\nu$ and $\bar{\nu}$ induced reactions, respectively. Therefore, $P_T(0)=-1$ for $\nu p\rightarrow\nu p$ and +1 for $\bar{\nu}p\rightarrow\bar{\nu}p$. This is to be contrasted to the V,A case, for which $P_T(0)$ is the same for $\nu$ and $\bar{\nu}$.

## TIME REVERSAL VIOLATION

One can also measure the orthogonal polarization $P_0(Q^2)$. Any non-zero value for $P_0$ (other than a tiny amount due to electromagnetic final state interactions in the quasi-elastic case) would indicate time reversal violation in the weak interactions (one is probing a kinematic region different from $\beta$ decay). This would mean that the hadronic current $J_\mu^H$ must be the sum of at least two pieces with different time reversal transformation properties. In the neutral current case there is the additional constraint[13] that a Hermitian first-class neutral current must have normal time reversal transformation properties. Hence, $P_O \neq 0$ for elastic scattering would also require either a second class current or an off-diagonal coupling to a new neutrino at the lepton vertex (which allows $J_\mu^H$ to be non-Hermitian).

The construction of time-reversal-violating second class currents out of quark fields requires[13,14] the introduction of unusual new quantum numbers. The detection of a non-zero $P_O(Q^2)$ is therefore very unlikely, but, if found, extremely important.

## CONCLUSION

It has been argued that measurements of $P_T(Q^2)$ and $P_O(Q^2)$ in elastic and quasi-elastic neutrino-baryon scattering could very effectively constrain the structure of the weak neutral current, as well as search for such exotic effects as large time reversal violations in the weak interactions.

## REFERENCES

1. J. E. Kim, P. Langacker, and S. Sarkar, Phys. Rev. D$\underline{18}$, 123 (1978).
2. E. Fischbach, J. T. Gruenwald, S. P. Rosen, H. Spivak, and B. Kayser, Phys. Rev. D$\underline{15}$, 97 (1977).
3. L. M. Sehgal, Phys. Lett. $\underline{71B}$, 99 (1977).
4. G. Ecker, Phys. Lett. $\underline{72B}$, 450 (1978).
5. P. Q. Hung and J. J. Sakurai, Phys. Lett. $\underline{72B}$, 208 (1977).
6. P. Langacker and D. P. Sidhu, Phys. Lett. $\underline{74B}$, 233 (1978).
7. D. P. Sidhu and P. Langacker, Phys. Rev. Lett. $\underline{41}$, 732 (1978).
8. L. F. Abbott and M. Barnett, Phys. Rev. Lett. $\underline{40}$, 1303 (1978).
9. E. H. Monsay, Phys. Rev. Lett. $\underline{41}$, 728 (1978).
10. C. Y. Prescott et al., Phys. Lett. $\underline{77B}$, 347 (1978).
11. A. Le Yaouanc et al., Nucl. Phys. B$\underline{125}$, 243 (1977).
12. J. F. Donoghue, Phys. Rev. D$\underline{17}$, 2922 (1978).
13. P. Langacker, Phys. Rev. D$\underline{14}$, 2340 (1976); $\underline{15}$, 2386 (1977), and references therein.
14. B. R. Holstein and S. B. Treiman, Phys. Rev. D$\underline{13}$, 3059 (1976).

245

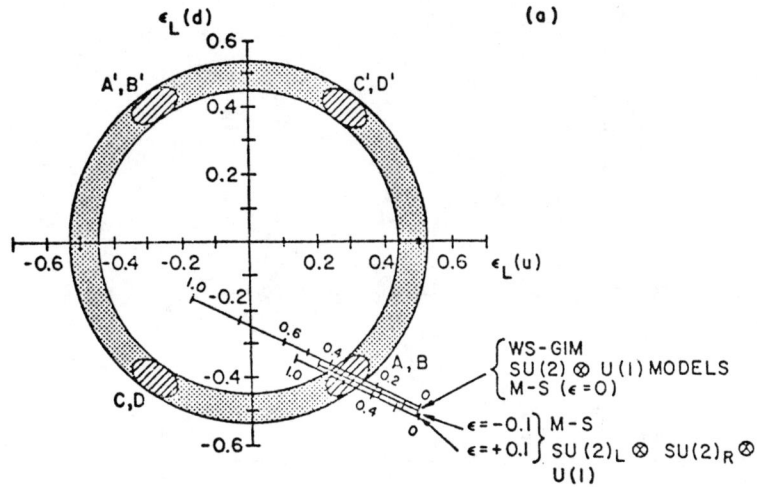

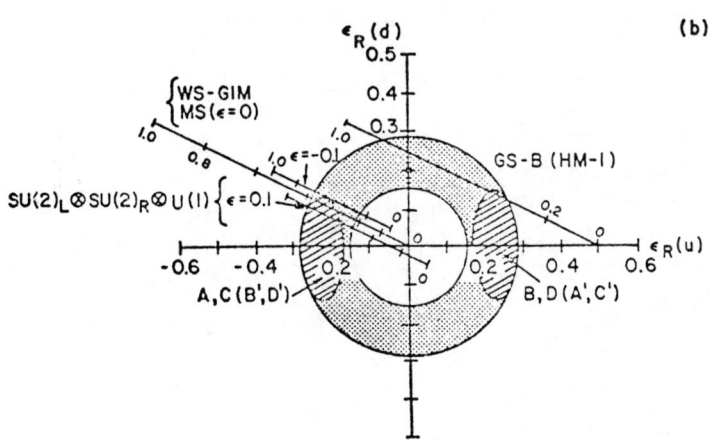

Fig. 3 - Solutions A,B,C, and D for the neutral current couplings.
Also shown are the predictions of the Weinberg-Salam model
(WS-GIM) and the model with a $u_R$ in a doublet (HM-I).

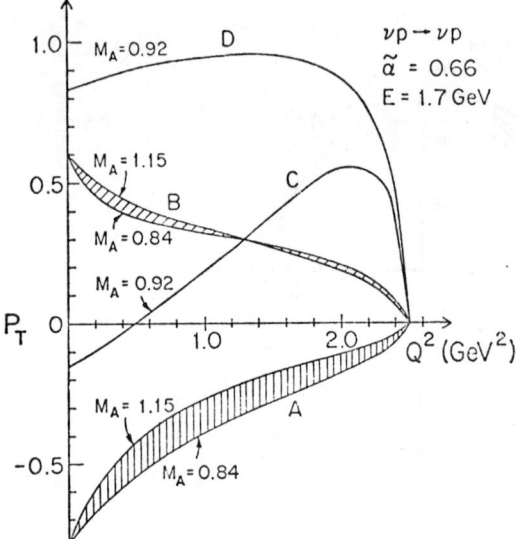

Fig. 4 - $P_T(Q^2)$ for the neutral current parameters in the centers of regions A,B,C, and D. The curves for $\bar{\nu}p$ are similar for $Q^2 < 0.4$ GeV$^2$.

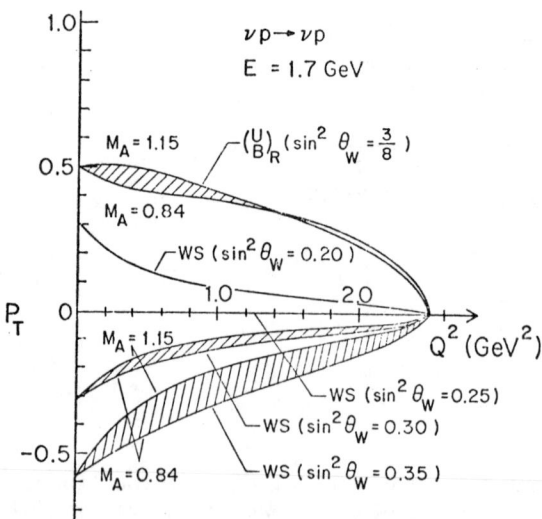

Fig. 5 - $P_T(Q^2)$ for the Weinberg-Salam model and HM-I. The curve for $\bar{\nu}p$ is similar for small $Q^2$. The curve for $\sin^2\theta_W = 0.20$ uses $M_A = 0.92$.

## COMMENT ON CONTRIBUTION OF P. LANGACKER

Louis Michel
Bures-Sur-Yvette

I would like to point out that, although quasi two-body reactions $\nu p^+ \to \mu^- \Delta^{++}$, $\bar{\nu} p^+ \to \mu^+ \Delta^\circ$, $\nu p^+ \to \nu \Delta^+$, etc. have a smaller cross section than the elastic reactions, the observable $\Delta$-polarization is much more easy to observe: it is known by the very observation of this resonance. From the angular distribution of the (strong, parity conserving) $\Delta$-decay one can measure only the quadrupole of the $\Delta$-polarization, i.e. five real components $q_i$ which must satisfy $\Sigma q_i^2 = 1/3$ (1 corresponds to complete polarization): for details, see M. Doncel, L. Michel, P. Minnaert: Analysis of $\Delta$-polarization, in "Physics from Friends" Papers dedicated to Ch. Peyrou on his 60th birthday; Multi Office S. A. Geneva 1978). Time reversal invariance of the weak currents can be tested since it implies that two of these five components vanish. For charge weak currents, in this quoted paper, we have compared the theoretical predictions with the only published data we know of (Schreiner, von Hippel, Phys. Rev. Lett. 30 (1973) 307. The agreement is not good; there must exist abundant (and more recent) data. As shown in the paper Le Yaouanc et al. Phys. Rev. D15 (1977) 2447, the polarization of $\Delta$'s produced by neutral weak currents give some information on their nature.

ISSN: 0094-243X/79/510247-01$1.50 Copyright 1979 American Institute of Physics

Chapter 4    Polarized Hadron Beams

## Acceleration of Polarized Ions in Synchrotrons

L.C. Teng

Fermi National Accelerator Laboratory, Batavia, IL 60510

This is a review of the problems involved in accelerating polarized ions (protons or deuterons) to high energies in synchrotrons and the resolutions of these problems. Nothing new has developed since the last workshop in Ann Arbor a year ago. This paper serves only to collect the available information in one convenient location.

1. Dynamic Equations of the Spin

In a fixed laboratory frame the spin $\vec{s}$ (defined only in the rest frame) obeys the equation

$$\frac{d\vec{s}}{dt} = \frac{g}{2}\frac{e}{m\gamma}\vec{s}\times(\vec{B}_{\parallel}+\gamma\vec{B}_{\perp})-\vec{s}\times\left[(\gamma-1)\frac{e\vec{B}_{\perp}}{m\gamma}\right]$$

$$= \frac{e}{m\gamma}\vec{s}\times\left[\vec{B}_{\parallel}+\vec{B}_{\perp}+G(\vec{B}_{\parallel}+\gamma\vec{B}_{\perp})\right] \qquad (1)$$

where

$$g = \text{gyromagnetic ratio} = \begin{cases} 5.586 & \text{for proton} \\ 1.714 & \text{for deuteron} \end{cases}$$

$$G = \frac{g}{2} - 1 = \begin{cases} 1.793 & \text{for proton} \\ -0.143 & \text{for deuteron} \end{cases}$$

$\vec{B}_\parallel$, $\vec{B}_\perp$ = Components of magnetic field parallel and perpendicular to velocity.

This is just the equation derived by Bargmann, Michel and Telegdi[1] and used by Froissart and Stora[2]. Several features are interesting to point out.

a. If the rest frame were inertial, we would have on the right-hand-side only the first term which is simply $\vec{s} \times \vec{B}$ in the rest frame.

b. The second term, the Thomas precession, arises because the rest frame is rotating and hence non-inertial. This makes the fixed inertial laboratory frame "look like" it is rotating with angular velocity $(\gamma-1) \times$(cyclotron frequency).

c. The Thomas precession is purely kinematic in origin. It is derived[3] simply by transforming at time t from the rest frame to the laboratory frame and back to the slightly rotated rest frame at t+Δt. It is also purely relativistic in origin, since it vanishes when $\gamma=1$.

d. The last form of the right-hand-side gives simply $\vec{s} \times \vec{B}$ in the laboratory frame when the anomalous moment G=0. The spin will then follow the velocity $\vec{V}$.

It is convenient to use a rotating laboratory frame for which one of the axes (say, 1-axis) always points along the velocity. Such a frame rotates with the cyclotron frequency and the spin equation becomes

$$\left(\frac{d\vec{s}}{dt}\right)_{rot} = \frac{d\vec{s}}{dt} - \vec{s} \times \left(\frac{e\vec{B}_\perp}{m\gamma}\right)$$

$$= \frac{g}{2} \frac{e}{m\gamma} \vec{s} \times (\vec{B}_\parallel + \gamma\vec{B}_\perp) - \gamma\vec{s} \times \left(\frac{e\vec{B}_\perp}{m\gamma}\right)$$

$$= \frac{e}{m\gamma} \vec{s} \times \left[(1+G)\vec{B}_\parallel + \gamma G\vec{B}_\perp\right] \qquad (2)$$

We will also transform the independent variable to the vertical rotation angle θ of the particle motion given by

$$d\theta = \frac{eB_v}{m\gamma} dt \qquad (\theta \text{ advances } 2\pi \text{ per turn})$$

where we have resolved $\vec{B}_\perp$ into the radial (r) and the vertical (v) components

$$\vec{B}_\perp = \vec{B}_r + \vec{B}_v.$$

The right-handed coordinate system (//, r, v) or (1, 2, 3) are shown in Figure 1. The spin equation is now

$$\frac{d\vec{s}}{d\theta} = \frac{m\gamma}{eB_v}\left(\frac{d\vec{s}}{dt}\right)_{rot}$$

$$= \vec{s} \times \left[\frac{(1+G)\vec{B}_{//} + \gamma G \vec{B}_r}{B_v} + \gamma G \hat{B}_v\right] \equiv \vec{s} \times \vec{F}. \qquad (3)$$

Generally $\vec{B}_v$ is large and slow-varying. Its main function is to deflect the particle beam to travel along a horizontal planar closed orbit. For the spin, it produces a precession around the vertical axis of angle $2\pi\gamma G$ per turn. The horizontal field components $\vec{B}_{//}$ and $\vec{B}_r$ are small and fast-oscillatory and arise either from a vertical closed-orbit distortion ("frequency" n=integer per turn) or from a vertical betatron oscillation ("frequency" N±ν, N = ring magnet lattice periodicity and ν = vertical betatron oscillation number per turn). Precession of the spin about the oscillatory horizontal field does not produce sizeable secular effect except when the frequency of the vertical precession coincides with one of the frequencies of the horizontal field. These are the depolarizing resonances.

Instead of the vector form (3) the spin equation can also be expressed in the spinor form. This is because of the isomorphism between the 3-dimensional rotation group R(3) and the 2-object unitary/unimodular group SU(2). Let us write

$$\vec{S} = \psi^+ \vec{\sigma} \psi \tag{4}$$

where $\psi$ is the 2-component spinor and $\vec{\sigma}$ are the 2x2 Pauli spin matrices. We shall use the representation in which $\sigma_3$ is diagonal, namely

$$\sigma_1 = \begin{pmatrix} 0 & 1 \\ 1 & 0 \end{pmatrix} \quad \sigma_2 = \begin{pmatrix} 0 & -i \\ i & 0 \end{pmatrix} \quad \sigma_3 = \begin{pmatrix} 1 & 0 \\ 0 & -1 \end{pmatrix}.$$

The dynamic equation for the spinor then becomes

$$\begin{aligned}\frac{d\psi}{d\theta} &= \frac{i}{2} \vec{\sigma} \cdot \vec{F} \psi \\ &= \frac{i}{2} \begin{pmatrix} \gamma G & \varepsilon e^{ik\theta} \\ \varepsilon^* e^{-ik\theta} & -\gamma G \end{pmatrix} \psi \end{aligned} \tag{5}$$

where

$$\varepsilon e^{ik\theta} \equiv (1+G) \frac{B_\parallel}{B_V} - i\gamma G \frac{B_r}{B_V}$$

gives the relative oscillatory horizontal field. If we write the two components of $\psi$ as

$$\psi = \begin{pmatrix} a e^{i\frac{\gamma G}{2}\theta} \\ b e^{-i\frac{\gamma G}{2}\theta} \end{pmatrix} \tag{6}$$

we get from Eq. (5)

$$\begin{cases} \frac{da}{d\theta} = \frac{i\varepsilon}{2} b e^{-i(\gamma G - k)\theta} \\ \frac{db}{d\theta} = \frac{i\varepsilon^*}{2} a e^{i(\gamma G - k)\theta} \end{cases} \tag{7}$$

The degree of polarization is then defined as

$$P \equiv \psi^+ \sigma_3 \psi = |a|^2 - |b|^2. \tag{8}$$

The spinor $\psi$ is nothing but the quantum mechanical wave function and Eq. (5) is just the wave equation in the Schrödinger formulation.

2. Resonances and Adiabatic Passage

Equation (3) shows that the states a and b are static when $\varepsilon = 0$ and are quasistatic even when $\varepsilon \neq 0$ as long as the coupling is rapidly oscillatory. They can undergo large secular variations only close to resonances when

$$\gamma G = k = \begin{cases} N \pm \nu & \text{intrinsic resonances} \\ n & \text{imperfection resonances} \end{cases}$$

The amplitude $|\varepsilon|$ gives the width or the strength of the resonance.

For the adiabatic passage of a resonance we assume (for intrinsic resonances)

$$\alpha \equiv \frac{d}{d\theta}(\gamma G - k) = \frac{1}{2\pi}(G\Delta\gamma \mp \Delta\nu)$$

$$= \text{approximate constant} \ll k \tag{9}$$

where $\Delta\gamma$ and $\Delta\nu$ are the changes per revolution. In this case Eq. (7) becomes a confluent hypergeometric equation having an asymptotic solution which is expressible in terms of elementary functions (see Ref. 2). From this we get

$$P_+ = P_-(2e^{-\frac{\pi}{2}\frac{|\varepsilon|^2}{\alpha}} - 1) \tag{10}$$

where $P_-$ and $P_+$ are the polarizations of the particle long before and long after passing the resonance.

Numerically integrated solutions are plotted in Fig. 2 for $|\varepsilon|/\sqrt{\alpha}$ = 0.2, 0.6, 1.2, and 1.8. The features of these solutions can be readily understood by a detailed tracing of the precession of the spin about the horizontal field.

To give an estimate of the width $|\varepsilon|$ of an imperfection resonance we start from the following rms relation between the vertical closed-orbit distortion z and the radial field $B_r$

$$z_{rms} = \frac{\pi}{\sqrt{2}|\sin \pi \nu|} \frac{1}{f} \frac{<\beta>}{\sqrt{M}} \left(\frac{B_r}{B_v}\right)_{rms} \simeq \pi \frac{<\beta>}{\sqrt{M}} \left(\frac{B_r}{B_v}\right)_{rms} \quad (11)$$

where

$$f \equiv \frac{\text{total length of dipoles}}{\text{total length of all magnets}} \simeq 1$$

M = total number of all magnets

$<\beta>$ = average amplitude function

and where in the last expression we have taken the usual quarter-integer value for $\nu$. In deriving this relation we have assumed that $\frac{B_r}{B_v}$ is random and uncorrelated between magnets. Assuming further that $B_\parallel$ is negligible we get for the resonance $\gamma G = n$ (n<<M)

$$|\varepsilon| = \gamma G \times (\text{amplitude of } n^{th} \text{ harmonic of } \frac{B_r}{B_v})$$

$$= \frac{\gamma G}{\sqrt{M}} \left(\frac{B_r}{B_v}\right)_{rms} \simeq \frac{n}{\pi} \frac{z_{rms}}{<\beta>} . \quad (12)$$

If the depolarization due to each resonance is individually small but all resonances from $n_1$ to $n_2$ are crossed, the total depolarization is given by

$$1-P_+/P_- = 1 - \prod_{n=n_1}^{n_2} \left[ 2e^{-\frac{n^2}{2\pi\alpha}(\frac{z_{rms}}{<\beta>})^2} - 1 \right]$$

$$= 1 - \prod \left[ 1 - \frac{n^2}{\pi\alpha}(\frac{z_{rms}}{<\beta>})^2 \right]$$

$$= \frac{1}{\pi\alpha}(\frac{z_{rms}}{<\beta>})^2 \left( \sum_{n=n_1}^{n_2} n^2 \right). \tag{13}$$

For a given maximum permissible depolarization this gives an upper limit for $z_{rms}$.

The widths of the intrinsic resonances are sensitively dependent on the magnet lattice structure. A computer program DEPOL was written by E.D. Courant[4] to compute $|\varepsilon|$ from given lattice parameters. The results for CPS, AGS, and the Fermilab main ring are given in Table 1. All these synchrotrons have a fairly strong resonance at N = 0. Other strong resonances occur for CPS at N = 50, for AGS at N = 60, and for Fermilab main ring at N = odd multiples of 96. These special values of N can be easily understood in relation to the periodicities of the ring magnet lattices.

The program DEPOL can also be used to compute $|\varepsilon|$ for imperfection resonances. The results are more precise than those given by Eq. (12). Furthermore, they exhibit the effects of the specific structure of the magnet lattice on the harmonic content of $B_r/B_v$.

3. Ameliorating the Effect of Resonance

    a. Reduce strength of resonance

        For imperfection resonances the only obvious way to reduce

depolarization is to reduce $|\varepsilon|$ by correcting for the vertical closed-orbit distortion $z_{rms}$. As an example, consider Fermilab main ring for which $<\beta> \cong 50$ m, $\alpha \cong 10^{-3}$ (3.3 MeV/turn) and $n_1 = 18$, $n_2 = 766$ (8 GeV to 400 GeV). For a depolarization of less than 10% ($P_+/P_- > 0.90$) Eq. (13) gives

$$z_{rms} < 0.07 \text{ mm}$$

which is extremely difficult, if not impossible, to achieve. This despite the fact that the widths of resonances for which n = N = (intrinsic lattice periodicity) tend to be much larger than those included here. If we take a feasible $z_{rms} = 0.5$ mm Eq. (13) then gives in reverse

$$\left( \sum_{n=n_1}^{n_2} n^2 \right) < 3.14 \times 10^6$$

or

$$n_2 = G\gamma_{max} < 210$$

which corresponds to a maximum energy of about 110 GeV.

If deuterons are accelerated in the Fermilab main ring over the same momentum range ($\gamma = 4.84$ to 214), with $|G| = 0.143$ we have only $n_1 = 1$ and $n_2 = 30$, and at the same energy gain per turn (3.3 MeV/turn) we have $\alpha \cong 4 \times 10^{-5}$. Equation (13) then gives, for a depolarization less than 10%

$$z_{rms} < 1.8 \text{ mm}$$

which is easy to achieve.

  b. Increase "speed" of passage

For intrinsic resonances we can increase $\alpha = \frac{1}{2\pi}(G\Delta\gamma \mp \Delta\nu)$ by changing the tune rapidly with pulsed quadrupoles inserted in the lattice. The tune change per turn $|\Delta\nu|$ can generally be made much larger than $G\Delta\gamma$. This method was employed successfully on the Argonne ZGS[5]. The "speed of passage" $\alpha$ is increased 30 to 100 fold by the $\nu$-jump and the depolarization $1-P_+/P_-$ of the strongest resonance is reduced from over 100% to less than 8%.

There is, however, a limit to the effectiveness of this method. The maximum safe range of $\nu$-jump is limited to around $\delta\nu = \pm 0.1$ by beam stability considerations. Even if this jump could be made instantaneous the width $|\epsilon|$ of the resonance should not be larger than $|\delta\nu|$. For $|\epsilon| = |\delta\nu|$ the maximum quasi-steady state polarization just before and after the jump is already down to $\frac{1}{\sqrt{2}} = 70\%$ (see Ref. 4). Hence the $\nu$-jump cure is effective only for $|\epsilon| \gtrsim 0.1$. Scanning over Table 1 we see that for AGS all resonances except $\gamma G = 60-\nu$ at $\gamma = 28.6$ are curable by $\nu$-jump. For Fermilab this method is effective only up to $\gamma \simeq 60$. Beyond this, some other method of cure must be employed.

c. Total flip

In crossing a strong intrinsic resonance the spin could be almost totally flipped to the opposite direction. For example, to get $P_+/P_- < -0.99$ we need $|\epsilon^2|/\alpha > 3.4$ which can be obtained by a sufficiently large vertical oscillation amplitude and/or a sufficiently low speed of crossing. To ensure that $|\epsilon|$ is large for all particles in the beam we should impart to the beam a large coherent vertical oscillation before the resonance by using a kicker magnet. This oscillation, if staying coherent, can be damped out by a feed-back damper loop after the resonance is crossed. In addition, one can reduce the speed of crossing, $\alpha$, by using the pulsed $\nu$-jump quadrupoles "in reverse". This last scheme could be applied to flip the polarization of the beam between alternate synchrotron pulses. To obtain the rather clean "flip" and "no flip" values of

$$\frac{P_+}{P_-} = \begin{cases} 0.99 \\ -0.99 \end{cases} \quad \text{we need} \quad \frac{|\epsilon|^2}{\alpha} = \begin{cases} 0.0032 \\ 3.4 \end{cases} \quad (14)$$

Hence a factor of ~1000 is required between the "no-flip" and "flip" values of $\alpha$. This can be obtained, for example, by the following values of $\nu$-jump:

$$\Delta\nu = \begin{cases} 50 \; G\Delta\gamma & \text{"no flip"} \\ -0.95 \; G\Delta\gamma & \text{"flip"} \end{cases} \quad (15)$$

For some values of $|\epsilon|$ and $\Delta\gamma$ these $\Delta\nu$ values are reasonable and practical.

Total flip, however, cannot be attained by reducing $\alpha$ indefinitely if $|\epsilon|$ is too small. Too small an $\alpha$ will make it impossible to totally cross the resonance (width $\pm |\epsilon|$) within the available stable range of $\delta\nu \cong \pm 0.1$. For example, to get 99% flip we need

$$\frac{|\epsilon|^2}{\alpha} = 3.4 \quad \text{or} \quad \alpha = \frac{|\epsilon|^2}{3.4} \; .$$

To get one width $|\epsilon|$ away from the exact resonance we have

$$\alpha\theta = |\epsilon| \quad \text{or} \quad \theta = \frac{|\epsilon|}{\alpha} = \frac{3.4}{|\epsilon|} \; .$$

Since $\alpha$ is so small we have essentially $\Delta\nu = G\Delta\gamma$ which gives at the above $\theta$ value a tune shift of

$$\delta\nu = \Delta\nu\frac{\theta}{2\pi} = \frac{3.4}{2\pi}\frac{G\Delta\gamma}{|\epsilon|} \cong \frac{\Delta\gamma}{|\epsilon|} \ . \tag{16}$$

This must be smaller than 0.1. With $\Delta\gamma = 10^{-3}$ for the Fermilab synchrotron, then, $|\epsilon|$ must be larger than 0.01.

The total flip scheme is planned[6] for the 3-GeV Saturne 2. The two intrinsic resonances encountered: $\gamma G = \nu$ (at $\gamma = 2.075$) and $\gamma G = 8-\nu$ (at $\gamma = 2.386$) will flip the spin twice and restore the polarization.

4. Eliminating Resonances by Modifying Precession (Siberian Snake)[7].

In the process so far discussed (described e.g. by Eq. (5)) the spin precesses only about the vertical axis (3-axis). The precession "frequency" is $\gamma G$ which increases monotonically and continuously with $\gamma$ and hence, crosses all frequency components of $\vec{B}_h$. By introducing precessions about horizontal axes it is possible to keep the "frequency" of the resultant precession from crossing resonant values. Horizontal precession can be introduced by adding a section of magnetic field in an otherwise field-free straight section of the synchrotron. The essential point is that since the additional field is not needed for the orbital motion of the particle the field strength can be adjusted independently of the particle momentum to produce any resultant precession desired. Of course, one has to make sure that for the resultant precession there still exist states of stationary polarization at some azimuthal location around the circumference of the synchrotron.

The simplest scheme requires the addition of only one $\pi$ horizontal precession. From a location diametrically opposite this horizontal precession, the resultant precession in one revolution is simply a rotation of angle $\pi$ about the horizontal precession axis. This can be seen from Figure 3. Starting at this location with an arbitrary radius vector at A, after a vertical rotation $\pi\gamma G$, a horizontal rotation $\pi$, and then the remaining vertical rotation $\pi\gamma G$, the vector arrives in succession at B, C and D. It is easy to see that, independent of $\gamma G$, the resultant motion from A to D is a simple horizontal rotation of angle $\pi$. The stationary spin states are along the horizontal rotation axis and the precession "frequency" is, now, a half-integer instead of $\gamma G$. Since the frequency components n and n±ν of the oscillatory field are never half-integers, they will never come in resonance with the precession.

Formally, the vertical precession of Eq. (5) (putting $\varepsilon = 0$) when integrated over one turn gives

$$\psi_1 = e^{i\pi\gamma G\, \sigma_3}\psi_0. \qquad (17)$$

The succession of 3 rotations described above gives (We arbitrarily take the horizontal axis to be the 1-axis. The result is the same with the 2-axis.)

$$\psi_1 = e^{i\pi\frac{\gamma G}{2}\sigma_3}\, e^{i\frac{\pi}{2}\sigma_1}\, e^{i\pi\frac{\gamma G}{2}\sigma_3}\psi_0$$

$$= (\cos\pi\tfrac{\gamma G}{2} + i\sigma_3 \sin\pi\tfrac{\gamma G}{2})(\cos\tfrac{\pi}{2} + i\sigma_1 \sin\tfrac{\pi}{2})$$

$$(\cos\pi\tfrac{\gamma G}{2} + i\sigma_3 \sin\pi\tfrac{\gamma G}{2})\psi_0$$

$$= \left[i\sigma_1 \cos^2\pi\tfrac{\gamma G}{2} - (\sigma_1\sigma_3+\sigma_3\sigma_1)\sin\pi\tfrac{\gamma G}{2}\cos\pi\tfrac{\gamma G}{2}\right.$$

$$\left. -i\sigma_3\sigma_1\sigma_3 \sin^2\pi\tfrac{\gamma G}{2}\right]\psi_0$$

$$= i\sigma_1\psi_0 = e^{i\frac{\pi}{2}\sigma_1}\psi_0 \qquad (18)$$

and is, therefore, equivalent to a simple horizontal rotation of angle $\pi$ as stated.

More generally, if the succession of all rotations is specified by $\psi_1 = S\psi_o$, to get the axis $\hat{n}$ and angle $\phi$ of the resultant rotation we write ($\wedge$ denotes unit vector)

$$S = e^{i\frac{\phi}{2}\hat{n}\cdot\vec{\sigma}} = \cos\frac{\phi}{2} + i\,\hat{n}\cdot\vec{\sigma}\sin\frac{\phi}{2} .$$

In terms of S we then have

$$\begin{cases} \cos\frac{\phi}{2} = \frac{1}{2}\,\mathrm{Tr}(S) \\ \hat{n} = \frac{1}{2\,i\,\sin\frac{\phi}{2}}\,\mathrm{Tr}(\vec{\sigma}S) . \end{cases} \qquad (19)$$

For the specific S given in Eq. (18), Eqs. (19) lead to the same conclusion stated above.

There are two choices for the horizontal axis.

a. The l-axis ( $/\!/$ -axis)

In this case, longitudinal polarization is stationary. The most straightforward way to introduce a longitudinal precession is by adding a longitudinal field. We see from Eq. (2) that the longitudinal precession angle is 1+G times the bending angle if the field were transverse. Thus, for a precession angle $\pi$ we need a longitudinal $B_{/\!/}\ell$ which if transverse, would bend the beam $\frac{\pi}{2.793} = 64°$ for protons and $\frac{\pi}{0.857} = 210°$ for deuterons. To supply such a large $B_{/\!/}\ell$ by a solenoid is impractical for energies above, at most, some 5% the peak energy of the synchrotron. Furthermore, the effects of this very large longitudinal field on the beam (linear tune shift and nonlinear effects) will be difficult to compensate.

It is possible to make up a longitudinal precession with a number of transverse precessions. The advantage is that the transverse precession angle is $\gamma G$ times the equivalent bending angle, namely

$$\phi_\perp = \gamma G \frac{eB_\perp}{m\gamma} \frac{\ell}{V} \cong G \frac{eB_\perp \ell}{mc} \qquad (20)$$

where V is the particle velocity and where in the last expression we have assumed $V \cong c$ at high energies. Thus, at high energies $\phi_\perp$ is approximately independent of the energy. To get $\phi_\perp = \frac{\pi}{2}$, for example,

$$B_\perp \ell = \begin{cases} 27.4 \text{ kGm} & \text{for proton} \\ 687 \text{ kGm} & \text{for deuteron.} \end{cases}$$

For proton this is a reasonable field to be supplied by a dipole. But a transverse field deflects the beam. One must, therefore, devise a combination of a series of dipoles such that the resultant precession is $\phi_\parallel = \pi$ and the sum total orbital effect is zero. Since the orbital deflection angle equals $\phi_\perp/\gamma G$ and is hence small at high energies, it can be approximately[*] represented by a transverse vector $\vec{\epsilon}_n$ (for the $n^{th}$ dipole). The conditions for zero orbital effect are, then

$$\sum_n \vec{\epsilon}_n = 0 \quad \text{and} \quad \sum_n L_n \vec{\epsilon}_n = 0 \qquad (21)$$

where $L_n$ is the longitudinal distance from the deflection center of the $n^{th}$ dipole to a point beyond the last dipole. One such combination of 8 dipoles is [8]

---

[*]Strictly, the deflection of the velocity is a rotation. The resultant of rotations about different axes does not scale as the individual rotations. Hence, exact cancellation of orbital effect is, in principle, impossible for all energies.

| n | $\hat{\epsilon}_n$ | $\gamma G \epsilon_n (= \phi_n)$ | $L_n$ (arbitrary unit) |
|---|---|---|---|
| 1 | $\hat{2}$ | $\pi/2$ | 9 |
| 2 | $\hat{3}$ | $\pi/2$ | 8 |
| 3 | $-\hat{3}$ | $\pi/2$ | 7 |
| 4 | $-\hat{2}$ | $\pi/2$ | 6 |
| 5 | $-\hat{3}$ | $\pi/2$ | 5 |
| 6 | $\hat{3}$ | $\pi/2$ | 4 |
| 7 | $-\hat{3}$ | $\pi/2$ | 3 |
| 8 | $\hat{3}$ | $\pi/2$ | 0 |

With conventional 18 kG magnets this series of dipoles will occupy, at the minimum, a length of about 15 m. At the low energy end, if the particle speed is sensibly different from the speed of light, the dipoles must be ramped to track the V-dependence given in Eq. (20). Furthermore, the apertures of the dipoles must be large enough to accommodate the transverse excursion of the orbit which is largest at low energy and varies as $1/\gamma$.

    b. The 2-axis (r-axis)

The following series of 5 dipoles will accomplish this:

| n | $\hat{\epsilon}_n$ | $\gamma G \epsilon_n (= \phi_n)$ | $L_n$ (arbitrary unit) |
|---|---|---|---|
| 1 | $(\hat{2}+\sqrt{3}\,\hat{3})/2$ | $\pi$ | 5 |
| 2 | $-\hat{2}$ | $\pi$ | 4 |
| 3 | $(\hat{2}-\sqrt{3}\,\hat{3})/2$ | $\pi$ | 3 |
| 4 | $-\hat{3}$ | $\pi$ | 2 |
| 5 | $\hat{3}$ | $\pi$ | $2-\sqrt{3}$ |

The stationary polarization is now radial. Compared to the longitudinal case (a. above) the total dipole length required here is 20% greater but the total length occupied by the series of dipoles is about the same.

    The need of the "Siberian Snake" for high energy synchrotrons such as the CERN-SPS or the Fermilab accelerator is obvious. Some experimental test of this clever scheme should be started as soon as possible.

References

1. V. Bargmann, L. Michel and V.L. Telegdi, Phys. Rev. Lett. $\underline{2}$, 435 (1959)
2. M. Froissart and R. Stora, Nucl. Inst. and Methods, $\underline{7}$, 297-305 (1960)
3. See e.g. "Classical Electrodynamics" by J.D. Jackson, John Wiley and Sons (1962)
4. E.D. Courant, AIP Conference Proceedings, No. 42, p. 94., (1977)
5. T. Khoe, R.L. Kustom, R.L. Martin, E.F. Parker, C.W. Potts, L.G. Ratner, R.E. Timm, A.D. Krisch, J.B. Roberts, J.R. O'Fallon, Particle Accelerators, $\underline{6}$, No. 4, pp. 213-236 (1975)
6. E. Grorud, J.L. Laclare, G. Leleux, "Resonances de Depolarisation dans Saturne 2", GOC-GERMA 75-48/TP-28 Saclay Report (July 1975)
7. Ya. S. Derbenev and A.M. Kondratenko, Proceedings of the 10th International Conference on High Energy Accelerators, Protvino, Vol. 2, p. 70 (1977)
8. E.D. Courant and L.G. Ratner, AIP Conference Proceedings, No. 42, p. 41 (1977)

Table 1. The strength $|\varepsilon|$ of intrinsic resonances in the CERN-CPS, the Brookhaven-AGS, and the Fermilab main ring. In all cases the normalized vertical oscillation emittance is taken to be $10\,\pi$ mm-mrad.

CPS ($\nu=6.32$)

| $N\pm\nu(=\gamma G)$ | $\gamma$ | $|\varepsilon|$ |
|---|---|---|
| 10−ν | 2.06 | .00005 |
| 0+ν | 3.53 | .00933 |
| 20−ν | 7.64 | .00045 |
| 10+ν | 9.12 | .00047 |
| 30−ν | 13.23 | .00087 |
| 20+ν | 14.70 | .00050 |
| 40−ν | 18.82 | .00077 |
| 30+ν | 20.29 | .00309 |
| 50−ν | 24.40 | .14192 |
| 40+ν | 25.88 | .00174 |
| 60−ν | 29.99 | .00195 |
| 50+ν | 31.64 | .16773 |

AGS ($\nu=8.75$)

| $N\pm\nu(=\gamma G)$ | $\gamma$ | $|\varepsilon|$ |
|---|---|---|
| 12−ν | 1.82 | .00030 |
| 0+ν | 4.89 | .01535 |
| 24−ν | 8.52 | .00059 |
| 12+ν | 11.59 | .00539 |
| 36−ν | 15.22 | .01373 |
| 24+ν | 18.29 | .00101 |
| 48−ν | 21.93 | .00148 |
| 36+ν | 25.00 | .02663 |
| 60−ν | 28.63 | .15666 |
| 48+ν | 31.70 | .00233 |

Fermilab main ring ($\nu=19.40$) (only dominant ones)

| $N\pm\nu(=\gamma G)$ | $\gamma$ | $|\varepsilon|$ |
|---|---|---|
| 0+ν | 10.84 | .0256 |
| 6+ν | 14.19 | .0060 |
| 12+ν | 17.54 | .0035 |
| 18+ν | 20.89 | .0016 |
| 24+ν | 24.25 | .0025 |
| 30+ν | 27.60 | .0049 |
| 36+ν | 30.95 | .0062 |
| 42+ν | 34.30 | .0057 |
| 48+ν | 37.65 | .0028 |
| 54+ν | 41.01 | .0021 |
| 60+ν | 44.36 | .0048 |
| 66+ν | 47.71 | .0098 |
| 72+ν | 51.06 | .0133 |
| 78+ν | 54.41 | .0198 |
| 84+ν | 57.77 | .0321 |
| 90+ν | 61.12 | .0518 |
| 96+ν | 64.47 | .1653 |
| 192+ν | 118.10 | .0560 |
| 288+ν | 171.73 | .2952 |
| 384+ν | 225.36 | .0921 |
| 480+ν | 278.99 | .2138 |
| 576+ν | 332.63 | .0998 |
| 672+ν | 386.26 | .2995 |
| 768+ν | 439.89 | .0244 |

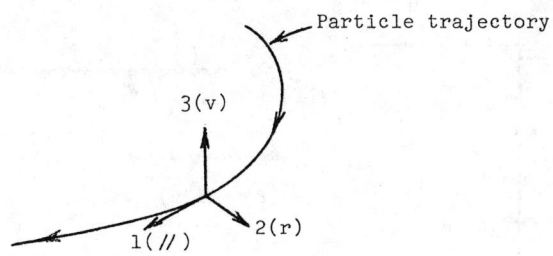

Figure 1. The coordinate system (1,2,3) or ($\parallel$,r,v)

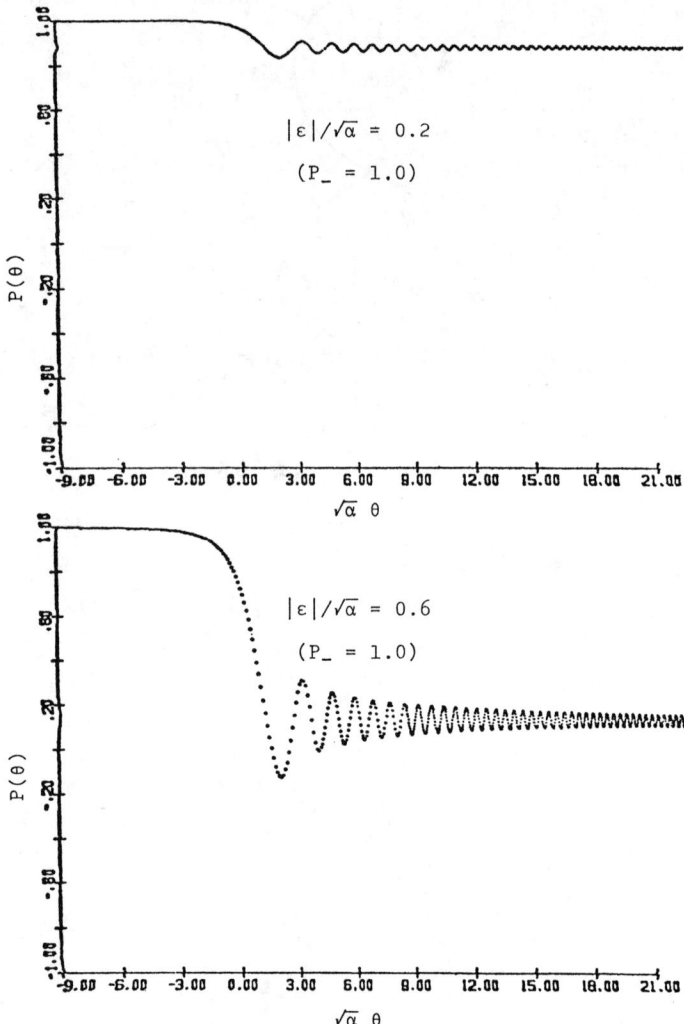

Figure 2. Polarization as function of "time" during adiabatic passage of resonances of various strengths.

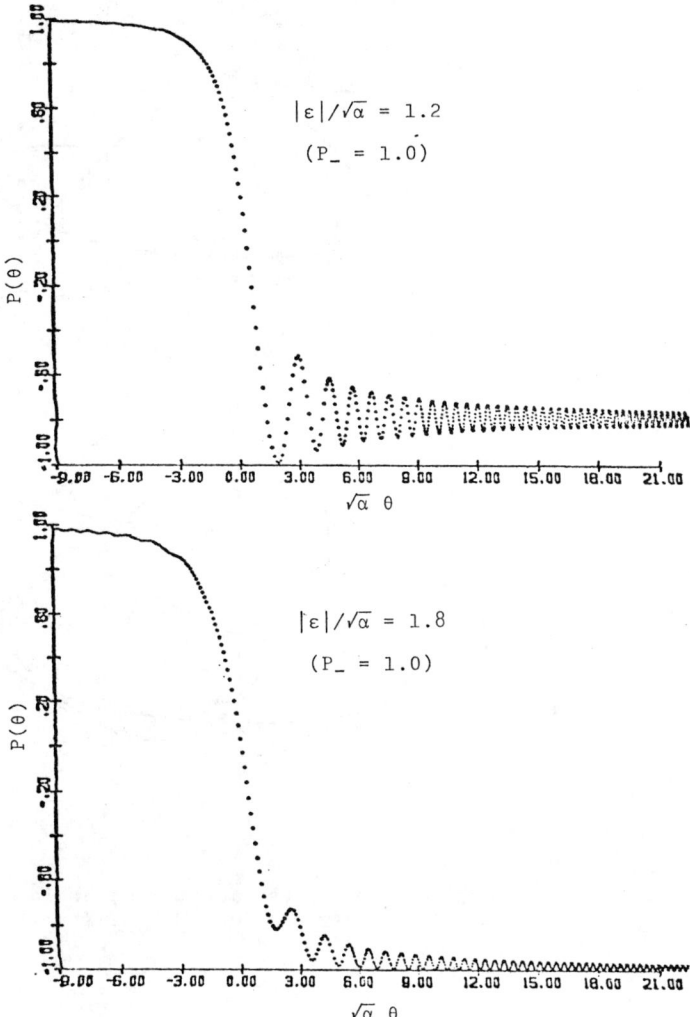

Figure 2. Polarization as function of "time" during adiabatic passage of resonances of various strengths.

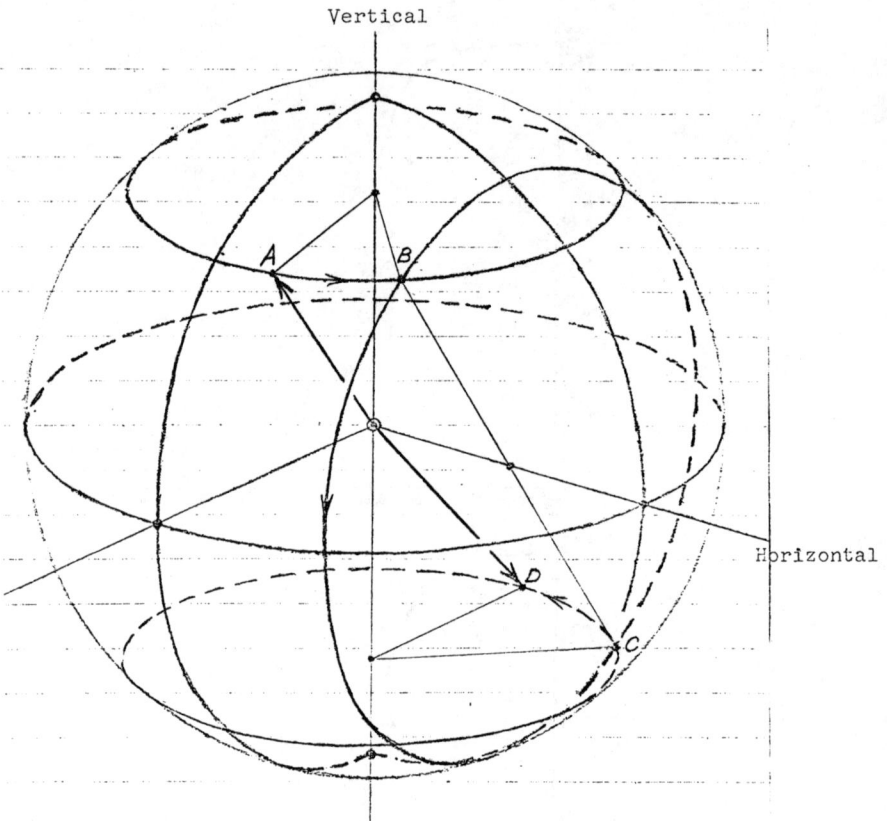

Figure 3. Spherical diagram illustrating the principle of eliminating resonances by adding a $\pi$ horizontal precession (Siberian Snake).

# POLARIZED-ION SOURCES

W. Haeberli
University of Wisconsin, Madison, Wisconsin 53706[†]

## ABSTRACT

The methods used to produce beams of polarized $H^+$ and $H^-$ ions are summarized and recent progress is reviewed. Sources based on the Lamb-shift method are useful primarily to produce $H^-$ (or $D^-$) beams. The best beam currents obtained are about 1 μA. Fast spin-switching schemes have recently been devised for Lamb-shift sources with spin filters. Atomic-beam sources have been improved to the point of yielding 50 μA $H^+$ and 100 μA in the pulsed mode. The best negative ion currents to date have been obtained with a new colliding-beam method in which thermal polarized $H^0$ atoms from an atomic beam source are bombarded with a beam of fast Cs atoms. A first test of this method yielded a DC current of 3 μA polarized $H^-$.

## INTRODUCTION

I have been asked by the organizing committee to present a summary of the methods used to produce polarized beams, and to report on recent progress and new ideas in the design of sources for polarized ions. I intend to take seriously the instructions I was given to address myself to the general audience of this symposium, and not to those of you who are working in this field. It will thus be necessary to first talk briefly about the methods employed to produce polarized beams, even though several reviews of this kind can be found in the literature.[1-6]

My talk will refer primarily to sources of polarized protons, but it should be understood that the same methods are used to provide polarized deuterons and, in principle, polarized tritons. Sources for polarized $^3$He ions or other even heavier ions (e.g. Li) have been constructed also but presumably are of less interest here.

In the next section I will summarize the principles on which the construction of polarized ion sources rests. The following two sections summarize the methods and the recent progress of the two types of polarized proton sources now in use, namely the Lamb-shift (or "metastable") sources in which H-atoms in the excited metastable 2S-state are first polarized and subsequently ionized, and the atomic-beam (or ground-state) sources which polarize H atoms by means of inhomogenous magnetic fields. For atomic-beam sources, an area of intense current interest is the study of new methods to ionize the beam by a colliding-beam method. This subject and a few comments about the production of polarized deuterons, will occupy the last two sections of the talk.

## GENERAL PRINCIPLES

In order to understand the functioning of polarized-ion sources

ISSN: 0094-243X/79/510269-21$1.50 Copyright 1979 American Institute of Physics

it may be useful to remind you that these devices always start out by polarizing the electron of the hydrogen (or deuterium) atom, and not the proton in the hydrogen atom. This is much easier because the electron magnetic moment is three orders of magnitude larger than the proton moment. The proton finds itself in the very strong magnetic field caused by the magnetic moment of the nearby electron. For the hydrogen ground state this field is 174 kG. Any externally applied field therefore has little direct effect on the proton. The proton is affected only indirectly through the coupling to the electron. On the other hand, the average field which the proton generates at the position of the electron is only a few hundred gauss. We speak of a weak or a strong external field depending on whether it is small or large compared to the "critical field" $B_c$. For hydrogen (ground state) $B_c$ = 507 G, while for the $2S_{1/2}$ excited state the critical field is reduced by a factor $2^3$ to $B_c$ = 63.4 G.

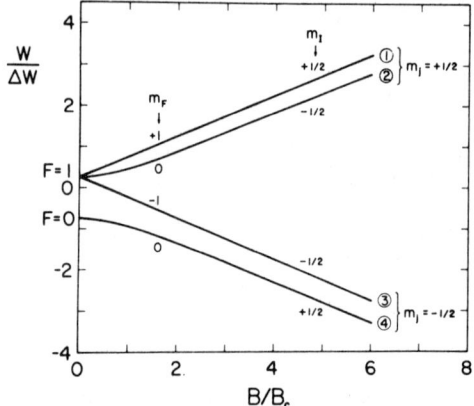

Fig. 1. Energy level diagram of the hydrogen atom in a magnetic field. The energy is measured in units of $\Delta w = 5.8 \times 10^{-8}$ eV. The critical field $B_c$ is 507 G for the ground state and 63.4 G for the $2S_{1/2}$ excited state (from ref. 1).

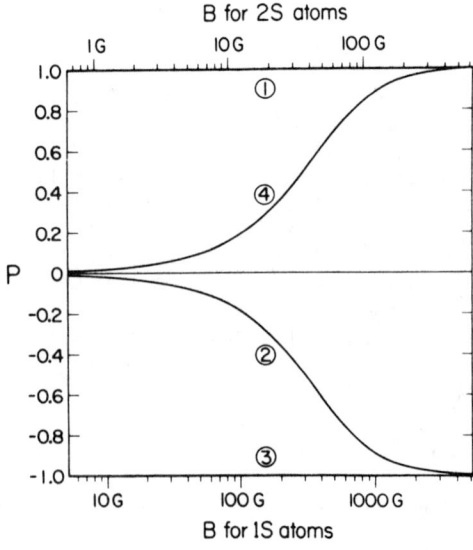

Fig. 2. Polarization P of protons in the hydrogen atom as a function of external magnetic field if any one of the four hyperfine components in Fig. 1 are occupied (from ref. 1).

The energy levels of hydrogen as a function of magnetic field strength are shown in Fig. 1. The labels $m_j$ and $m_I$ are the strong-field quantum numbers giving the spin projection of the electron and the nucleus, respectively. The nuclear polarization P as a function of external magnetic field is shown in Fig. 2. The variation of P with B arises from the hyperfine interaction. Polarized proton sources separate states 1 and 2 from states 3 and 4. Equal population of states 1 and 2 lead to P=0 in a strong field, while in a weak field P=0.5. All current sources make provision for some kind of transition between atomic levels to obtain, ideally, $|P| = 1$.

Ionization of the atoms is carried out in an external field which defines the polarization direction. In practically all sources the field is parallel to the momentum of the extracted ion beam, so that the primary beam has longitudinal polarization. Transverse polarization is obtained by 90° electrostatic deflection of the beam, or by a spin precessor (e.g. crossed-field analyzer, see ref. 6).

## LAMB-SHIFT SOURCES

The famous paper by Lamb and Retherford on the discovery of the Lamb-shift[7] pointed out an interesting new method to produce polarized hydrogen atoms. Polarizing a beam of hydrogen atoms is extremely easy if the atoms are in the n=2 excited state. The Lamb-shift removes the degeneracy between the $2S_{1/2}$ and the $2P_{1/2}$ states (Fig. 3a). The $2S_{1/2}$ state is metastable ($\tau$=0.1 sec) but of course the $2P_{1/2}$-state decays rapidly by dipole radiation. If an electric field is applied, decay of the $2S_{1/2}$-state is induced by Stark-mixing with the $2P_{1/2}$-state. This effect is referred to as "quenching" of the metastable atoms. An electric field of 10V/cm already shortens the lifetime to 3.7 μsec, which illustrates that the metastable atoms are fragile and in the ion source must not be exposed to large electric fields. For a given electric field strength,the 2S lifetime is the shorter the smaller the energy difference between the states being mixed. Thus in a magnetic field of 575 gauss, where the $m_j$ = -1/2 components of the 2S state cross the $m_j$ = +1/2 components of the 2P state (Fig. 3a) the lifetime is about a factor $10^3$ shorter for $m_j$ = -1/2 atoms (electron spin opposite to the magnetic field) than for $m_j$ = +1/2 atoms (electron spin parallel to the field, Fig. 3b). It is thus very easy to polarize a beam of metastable hydrogen atoms in electron spin by passing the beam through a region of static magnetic and electric fields. The tricky part of the problem is to produce the beam of metastable atoms in the first place, and in the end to ionize only the H(2S) atoms and not also the ground state atoms.

We will now discuss the various components of a source based on this principle (Fig. 4). First we will consider the production and selective ionization of the H(2S) atoms. In all current ion sources, the metastable atoms are produced by the method proposed by Donnally et al., i.e. electron pick-up of 500 eV $H^+$ in Cs vapor.[8] The proton beam is produced by a conventional ion source (duoplasmatron or rf ion source). Cesium is used because the ionization

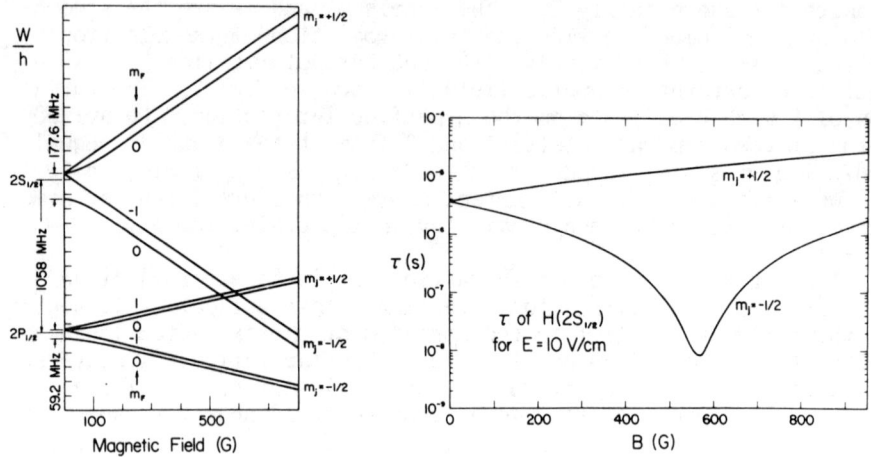

Fig. 3. Energy level diagram of the hydrogen atom in the n=2 excited state (left). Life time of $2S_{1/2}$ hydrogen atoms in the presence of a 10V/cm static electric field (right).

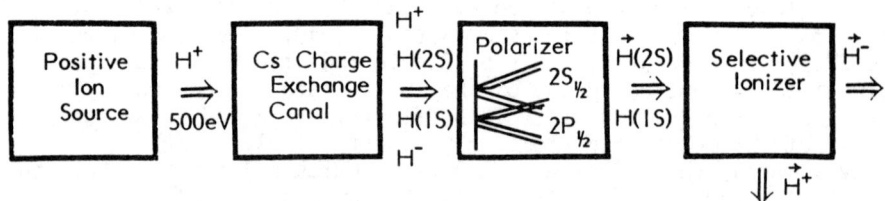

Fig. 4. Principle of the Lamb-shift polarized-ion source (from ref. 4).

energy of Cs (3.89 eV) is almost the same as the binding energy (3.4 eV) of H(2S). The very small energy defect for $H^+ + Cs \rightarrow H(2S) + Cs^+$ results in a large probability (∼32%, ref. 9) for pickup into the metastable state. The cross section has a maximum at 500 eV (Fig. 5a).

After passing through electromagnetic fields which polarize the H(2S) atoms, the metastable atoms are turned into ions by passing the beam through a gas or a vapor. In argon gas roughly 10% of the H(2S) atoms can be converted to negative ions, while iodine vapor is used to produce positive ions. In either case the charge exchange medium is chosen to ionize the 2S atoms but not the ground state atoms (selective ionization). It is this requirement that dictates the choice of 500 eV as the beam energy in the apparatus. For Ar the charge transfer for H(2S) has a resonance at 500 eV and the cross section compared to H(1S) is enhanced by roughly two orders of magnitude (Fig. 5b).

We are now returning to the step we skipped: the fields used to polarize the H(2S) atoms. The simplest way to polarize the atoms is to pass the beam through a coil a few cm long which provides a short region of axial magnetic field somewhere near 575 G. A transverse electric field of a few V/cm quenches the $m_j = -1/2$ atoms, i.e. states 3 and 4 in Fig. 1. The remaining H(2S) atoms equally populate states 1 and 2. As we saw in Fig. 2, ionization in a weak field would yield, ideally, P=0.5. All current sources use one of three methods to enhance the proton polarization. The simplest is by sudden reversal of the axial magnetic field (field reversal fast compared to Larmor period of atom in the field). This scheme, proposed by Sona,[11] is shown in Fig. 6. The beam of 2S atoms is assumed to

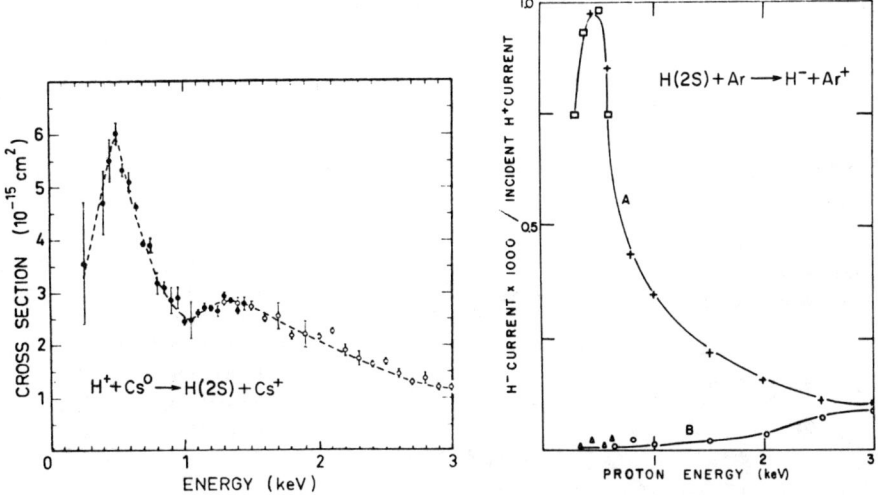

Fig. 5. Production of metastable H atoms by charge exchange of protons in Cs vapor (left, from ref. 9), and selective ionization of H(2S) atoms in Ar to form H⁻ ions (right, from ref. 10). Curve B shows the H⁻ current when the H(2S) atoms are quenched.

travel in the +z direction. After leaving the quenching region A, the axial magnetic field is decreased to zero and then reversed at point B. If the field reversal is sudden the angular momentum F will maintain the same spatial orientation, which means that with respect to the field direction we have reversed the projection $m_F \rightarrow -m_F$, in Fig. 1. After the field reversal the hydrogen states 2 and 3 are equally occupied. Ionization of the hydrogen atoms in a strong field yields a proton polarization P = -1.

Fig. 7 illustrates another method, called the "spin filter", developed at Los Alamos. The device permits quenching of all but one magnetic substate. The method requires application of three

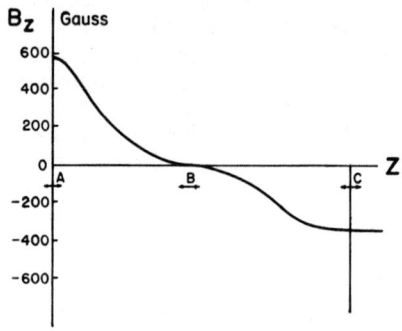
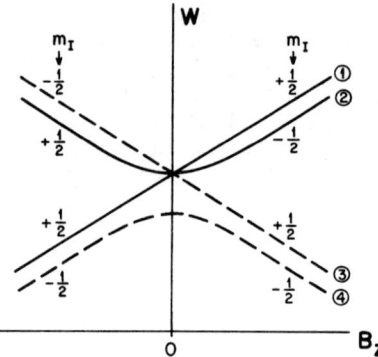

Fig. 6. Sudden field reversal to enhance the proton polarization in Lamb-shift sources.

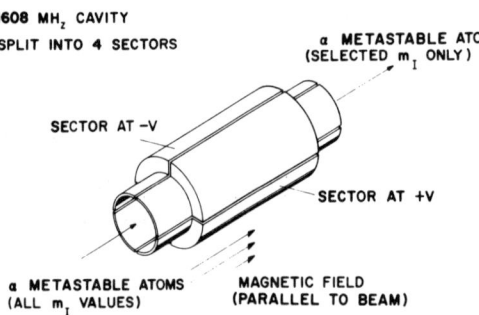

Fig. 7. Electric and magnetic fields used in the spin filter (ref. 12). All but one hyperfine state are quenched when 500 eV H(2S) atoms pass through the device.

fields: a very uniform magnetic field near 575 G, a static transverse electric field and a radio frequency field near 1600 MHz. The method involves the simultaneous interaction of three energy levels. It has been shown that the spin filter can be tuned to pass at will H(2S)-atoms in states 1 or 2. Besides these two schemes, it is also possible to produce RF transitions between 2S and 2P states in a weak magnetic field to quench all three F=1 hyperfine states of hydrogen, being left with only state 4 in the $2S_{1/2}$-state (see ref. 3).

The highest intensity H⁻ polarized beams that have been reported from Lamb-shift sources are about 1 μA. The source being built for the 12 GeV KEK proton synchrotron uses the dipole transitions mentioned above. The small length of the transition unit [cavity length = 5 cm] permits a relatively small distance between Cs and Ar canal which is beneficial to the intensity. Recently 1.1 μA max with P=0.7 was obtained in pulsed operation.[13] The Giessen source recently described in the literature[14] operates with Sona transitions and produces a DC H⁻ beam of 0.7 μA with P=0.7, but if it were designed only for H⁻ and not also for D⁻ a beam of $\sim$ 1 μA would be expected. Similarly, the source with Sona transition on the TRIUMF cyclotron is reported to produce 1 μA in regular operation.[15] From the intensity point of view the Sona scheme has an advantage because two, rather than only one, hyperfine components are retained. On the other hand the condition that the field reversal be sudden sets limits to the permissible beam diameter in the zero-crossing region. High beam polarizations (P$\simeq$0.85) have been reported with sources employing a spin filter but the best beam intensity obtained with this type of source is below 0.5 μA (ref. 5). An advantage of the spin filter is that ionization of state 1 can take place in either a weak or a strong magnetic field.

While the beam intensity obtained with Lamb-shift sources does not approach the H⁺ beam currents from atomic-beam devices, Lamb-shift sources continue to be of interest because they readily provide negative ions. Negative ions are of particular interest for cyclotrons, where the negative ion beam can be readily extracted by passing the beam through a thin foil which strips the electrons from H⁻ to form H⁺. Similarly, by stripping at injection into synchrotrons, it is possible in principle to use multiturn injection up to the space charge limit of the machine even with beams of low intensity. In practice, the method has limits, but it is estimated that as far as the beam intensity from the accelerator is concerned, 1 μA of H⁻ is equivalent to 10-30 μA H⁺.

The development of this type of source spans a twelve-year period, starting with the feasability-demonstrations at Yale and Milan in 1966 and the installation of the first source on an accelerator at Wisconsin in 1967. Despite much activity, only a modest increase in the beam intensity ($\sim$ factor 2) was achieved during the last five years. The intensity limitation probably arises at least in part from a fundamental problem: quenching of the metastable atoms by the charged component of the beam emerging from the Cs canal. Even though the charged component is usually deflected away by electric fields, this cannot be done quickly because the deflecting field too induces quenching. The other basic problem arises from the difficulty of producing 500 eV H⁺ beams of high brightness.

Recent improvements on Lamb-shift sources include the development of means to switch the sign of polarization rapidly (e.g. 1 kHz) and to eliminate the intensity modulation which usually is associated with switching.[16] Both of these improvements are important for parity experiments because one wishes to avoid modulation of the event rate with reversal of the helicity. Rapid switching is accomplished by following a spin filter which selects state 1 with a sudden field-reversal and spoiling the sudden reversal by the application of a small transverse magnetic field in the region where the longitudinal field changes sign.

For the production of positive ions Lamb-shift sources cannot compete with the beam intensity obtained from atomic-beam sources and thus are rarely used.

Lamb-shift sources with a guaranteed beam output of 0.3 μA

## ATOMIC-BEAM SOURCES

Atomic-beam sources use spatial separation of ground state hydrogen atoms. A schematic diagram is shown in Fig. 8. Hydrogen atoms from a gas discharge emerge through an aperture ($\sim$ 3 mm diameter) in the discharge vessel. The beam passes along the axis of a six-pole magnet. At the pole tips $B_m \simeq 10$ kG, so that $B \gg B_c$ for most of the magnet aperture. In a six-pole magnet, the potential energy of the atoms is proportioned to $r^2$ (harmonic oscillator potential), because $|B|$ is proportional to $r^2$ and the magnetic moment of the atoms is nearly constant since $B \gg B_c$ over most of the magnet aperture. The potential is attractive for $m_j = +1/2$ [states 1 and 2] and repulsive for $m_j = -1/2$. Since $\mu_B B_m \ll kT$ only atoms in a narrow cone ($\sim$ 2° half-angle) around the six-pole axis pass

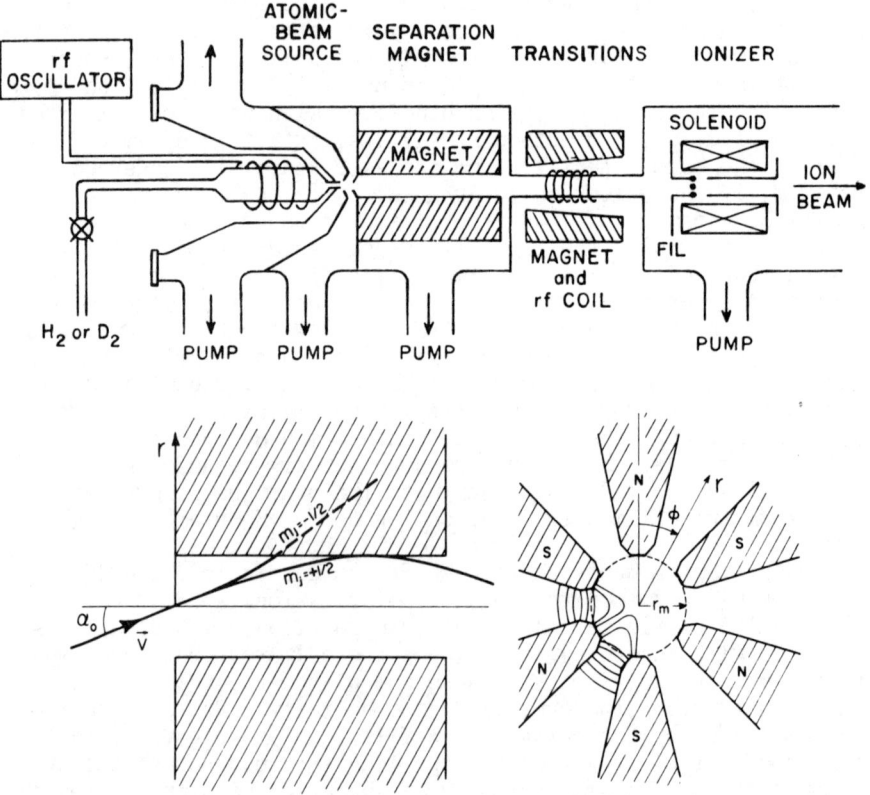

Fig. 8. Schematic diagram of an atomic-beam polarized-ion source (top). The drawing is not to scale. Trajectories of hydrogen atoms with $m_j = +1/2$ and $m_j = -1/2$ in the field of a six-pole magnet (bottom).

through the magnet. The intensity of $m_j = +1/2$ atoms at the magnet exit depends on the field at the pole tips, $B_m$, and the intensity and temperature of the atoms emerging from the discharge. The gas pressure in the discharge tube has an optimum value of about 3 mm Hg. For higher pressure the useful atomic beam intensity decreases because of scattering by the gas in the nozzle and in the region immediately beyond the opening. Large pumps are necessary to reduce scattering between the nozzle and the magnet. Usually this region is divided into two separately pumped compartments, in order to decrease the path length over which the atomic beam must travel in a relatively high pressure region. Typically one obtains $10^{16}$ polarized atoms per second at the exit of the magnet with a beam diameter of 1 cm at a distance of 50 cm from the end of the magnet.

At the exit of the separation magnet the atoms are polarized with respect to the six-pole field but they have no preferred orientation in space. To line up the spins in a given direction, a uniform field region is provided some distance from the magnet exit. Between the six-pole field and the uniform field region the atoms reorient themselves, following the field direction at each point along the path, (adiabatic field variation). For atoms of thermal velocity the condition of adiabatic field variation is easily met.

In all modern sources, the atomic beam is exposed to an rf field prior to ionization in order to enhance the polarization. The method used is the adiabatic passage method (see ref. 1) which should be viewed as a means to turn the atoms in a particular substate slowly upside down in space as they pass through the transition region. By using two rf transition units, one which operates in a weak field and a second one operating in a strong field, one can produce protons with polarization of either $P = +1$ or $P = -1$.

A typical ionizer consists of a 20-cm long solenoid coaxial with the atomic beam; electrons of $\sim$ 200 eV are injected along the magnet axis and are confined by the strong field. The electron current density is limited by space charge. The positive ion beam is accelerated to a few keV energy immediately after it emerges from the solenoid.

Prior to 1975 the maximum DC beam current was roughly 10 μA, obtained with a commercial source built by ANAC.[18] This beam current corresponds to an ionization efficiency of about $3 \times 10^{-3}$. At this symposium two years ago Parker reported modifications to the commercial ion source that resulted in peak currents of 70 μA. The largest single improvement came from pulsing the dissociator rf and pulsing the gas flow to the dissociator. There is a continuing program at ZGS to improve beam intensity by accumulation of improvements, each of which relatively minor, but the total of which is leading to impressive gains. By now peak pulsed beams above 100 μA have been obtained. In a program at CERN in conjunction with ANAC, improvements to various source components have been studied. The CERN source produces a DC polarized $H^+$ beam of 60 μA and over 100 μA when the dissociator is pulsed.[19] This source employs a longer ionizer and a short so-called "compressor" six-pole at the exit of the separation magnet. The compressor[20] acts as an achromatic lens which increases the atomic beam density in the ionizer.

At the University of Bonn, Kruger et al.[21] have shown that a high ionization efficiency can be obtained by a Penning discharge in the strong field of a superconducting solenoid (75 kG). Their arrangement is shown in Fig. 9. The solenoid is about 30 cm long. Electrons, produced in the region of electrode E8 by residual gas ionization are accelerated into the solenoid. The increasing magnetic field acts as a mirror. It is estimated that the oscillating electron current is about 5A. The estimate is based on measurements of the potential depression, obtained by measuring the energy of the ion beam. Polarized deuteron beams of 60 μA were extracted

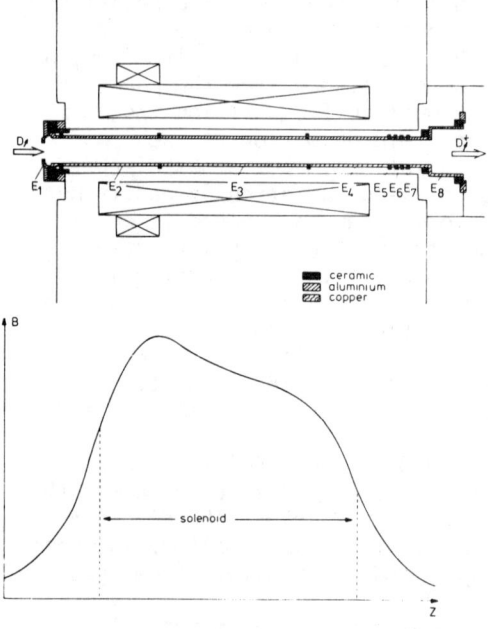

Fig. 9. Schematic view of the Penning ionizer developed at Bonn (ref. 21).

without special refinements of the atomic beam source. However because the extracted ions have to pass through the strong fringe field at the solenoid exit, the beam emittance (estimated to be $2\pi$ cm rad $eV^{1/2}$) may be a problem in some applications.

Measurements of the velocity distribution of atoms from the CERN source have shown that the beam is slightly supersonic. Attempts have been made for many years to produce supersonic atomic hydrogen jets since in principle a large increase in intensity should be possible. Improvements of polarized beam intensity are also expected if the gas in the dissociator is cooled, since a lower average velocity of the atomic beam results in a larger acceptance angle of the six-pole magnet. Some increase in ion beam has been obtained in various laboratories from cooling the exit nozzle of the dissociator with liquid nitrogen. Microwave dissociators of small volume, cooled with liquid nitrogen, are being studied at CERN and ZGS.

The proton polarization from atomic-beam sources is in the vicinity of P=0.7. The sign of polarization is readily switched by switching the RF transitions on the neutral beam prior to ionization. In contrast to Lamb-shift sources, where one is dealing with fragile H(2S) atoms, in atomic-beam sources the number of atoms in the beam is not affected by the RF fields. There is, nevertheless, a measurable intensity modulation of the ion beam associated with switching the sign of the proton polarization. This variation arises because the probability of ionization depends on the time the atom spends in the ionizer. The velocity of the atoms, in turn, is changed slightly when the magnetic moment of the atoms entering the ionizer field is changed. The intensity modulation has been measured at the SIN injector in connection with the 50 MeV p-p parity experiment[22] and is found to be $\sim 2 \times 10^{-5}$, in rough agreement with calculations.

Atomic-beam sources have been used on tandem accelerators for many years. The negative ions which these machines require are produced by charge exchange of $H^+$ in Na vapor at a $H^+$ energy of roughly 10 keV. The best $H^-$ beams so obtained are about 0.3 μA, but an order of magnitude improvement should be expected if recent technology in the production of polarized $H^+$ were exploited.

## COLLIDING BEAMS

As mentioned already, because of the gain achieved in multi-turn injection, 1 μA of polarized $H^-$ is equivalent to 10-30 μA $H^+$. Correspondingly, 200 μA $H^-$ would correspond to at least 2 mA $H^+$ and this clearly would go a long way to promote the general use of polarized beams in high-energy physics.

While nobody has approached this kind of beam intensity, I should like to explain the reason for current excitement and optimism.

First a few words about background. Polarized negative ions were produced for the first time at the University of Wisconsin[23] when we showed in 1964 that the nuclear polarization survived charge exchange in a carbon foil from $H^+$ to $H^-$. We also showed that negative ions could be accelerated and stripped without depolarization provided the stripping of both electrons took place within a short enough time.[24] The atomic-beam source we had was a rudimentary device, home-built on a small budget. One of the weakest parts was the charge exchange from $H^+$ to $H^-$ and it occurred to me that more intensity might be obtained by avoiding the positive ion as an intermediate step. One version of direct charge transfer from $H^o$ to $H^-$ which I proposed[25] in 1968 makes use of fast neutral Cs atoms as the donor of electrons:

$$\vec{H}^o + Cs^o \rightarrow \vec{H}^- + Cs^+ \qquad (1)$$

The cross section for this process[26] is shown in Fig. 10, while Fig. 11 shows a schematic diagram of a possible experimental arrangement.

The expected $H^-$ current is given by

$$I = jn\sigma V \qquad (2)$$

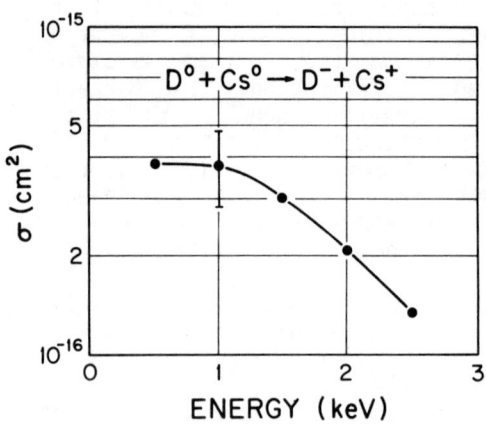

Fig. 10. Cross section for reaction (1). The energy scale applies to $D^0$ (ref. 26).

where V is the volume common to both beams, n is the density of $H^0$ target atoms, j is the current density of the donor atoms $D^-$ or $Cs^0$ [particle-amperes in the latter case] and $\sigma$ is the charge exchange cross section. For the sake of discussion, in the following we will assume $n = 10^{11}$ cm$^{-3}$ [e.g. $2 \times 10^{16}$ atoms/sec with $v = 3 \times 10^5$ m/sec and atomic beam diameter 1 cm], $V = 25$ cm$^3$ [i.e. length of ionization region $\simeq 33$ cm].

Eq. (3) gives

$$I = 1 \; \mu A \; H^- \; per \; mA/cm^2 \; Cs^0 \tag{3}$$

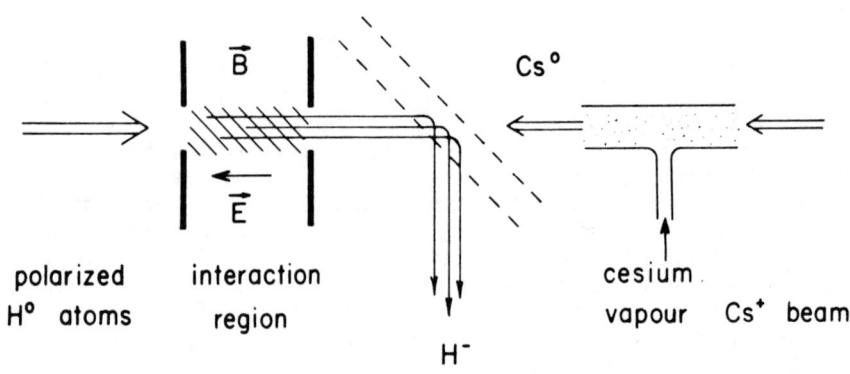

Fig. 11. Schematic diagram of arrangement for direct conversion of a polarized atomic beam to polarized negative ions.

This looked promising since intense Cs beams are readily produced by surface ionization on a hot ($\sim 1200°$) tungsten surface. The other reaction I proposed[25] looked even more interesting, namely the use of the hydrogen (or deuterium) negative ion as the donor:

$$\vec{H}^0 + D^- \rightarrow \vec{H}^- + D^0 \tag{4}$$

The cross section[27] for this process is shown in Fig. 12. At 1 keV H⁻ energy or 2 keV D⁻ the cross section is $2 \times 10^{-15}$ cm² and one should expect

$$I = 5 \text{ μA H}^- \text{ per mA/cm}^2 \text{ D}^- \tag{5}$$

Since DC arc sources producing several mA D⁻ were available at that time, it seemed possible to produce several μA polarized H⁻ ions with this reaction.

Of these two reactions, only reaction (1) has been tried so far and this only very recently. At our laboratory, work on a cesium test source was underway twelve years ago but was dropped because the then new Lamb-shift source provided a simpler solution to improve the source on our tandem. We took up the problem again a few years ago with the intention of trying reaction (4) first. We soon became aware of the basic problem that the electric field of the beam of donor negative ions repels the newly-formed low-energy polarized negative ions unless they are confined by a very strong magnetic field. While such fields could be produced by a superconducting solenoid, the beam emittance was expected to be too poor for our purposes. We then worked with Cs⁰ as the donor and at the end of last year[28] obtained a polarized negative beam of 3 μA. For pulsed operation of the dissociator the current would be at least 6 μA, which already competes with the best positive sources as far as acceleration in synchrotrons is concerned.

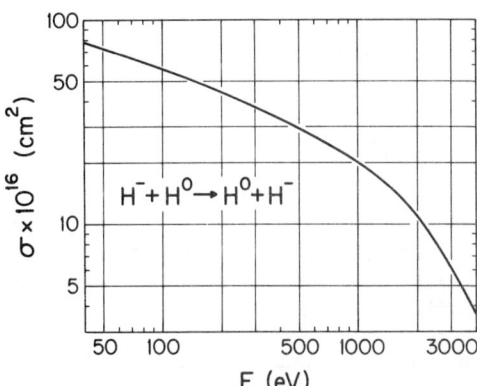

Fig. 12. Cross section for reaction (4). The curve is a smooth line drawn through the data points of ref. 27.

The arrangement which we used in our test is shown schematically in Fig. 13. The collision region was 33 cm long. The Cs⁰ beam had an energy of 40 keV and a current density (measured calorimetrically) of 2 particle-mA/cm². According to Eq. (3) this should have yielded 2 μA which is compatible, within the uncertainties, with the measured beam intensity of 3.0 μA. The emittance of the beam and the polarization of the beam both were better than those obtained with the conventional atomic-beam source for negative ions. A simplified scale-drawing of the new ionizer is shown in Fig. 14. A better engineered version of this source is presently being constructed for our accelerator. A source of this type is being offered for sale by ANAC.[18]

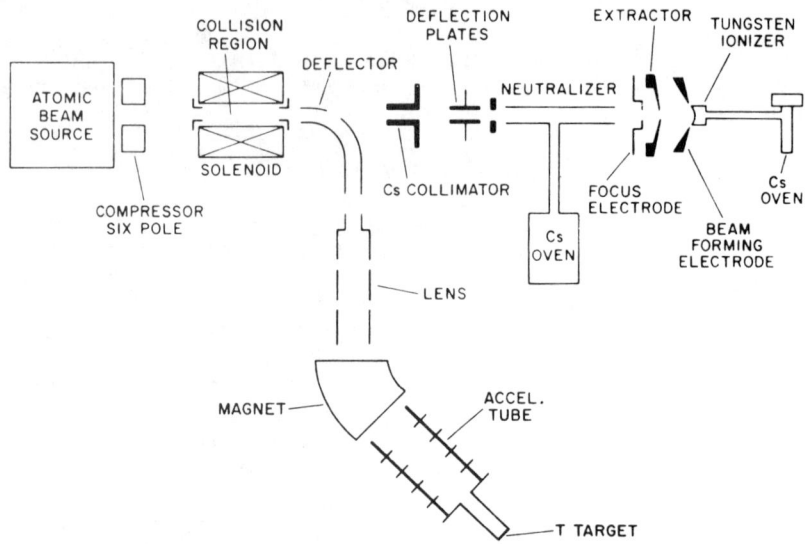

Fig. 13. Schematic diagram of the arrangement used to produce polarized H⁻ and D⁻ by direct charge exchange between polarized thermal H⁰ and a fast Cs⁰ beam (ref. 28).

Already this source is producing considerably more negative polarized beam than other types of ion sources, even though the new source has not been developed at all in a technical sense. The fact that both colliding beams are neutral is a pleasant feature of the source, since the two beams are entirely decoupled from each other. We know that the output beam increases in proportion to the Cs⁰ beam. The place to look for large gains in polarized-beam output is in the development of the Cs⁰ source. Another benefit of neutral collision partners is the fact that the magnetic and electric fields in the ionization region can be adjusted for best negative ion extraction and best polarization, without affecting either of the beams. This has some potentially interesting advantages. Since one needs no strong magnetic field to confine the electrons, it is possible to ionize in a weak magnetic field. To get full polarization one would need two six-pole magnets and an RF transition unit between them. One advantage of weak-field ionization is the lower beam emittance. Another advantage, of interest for parity experiments, is that the beam polarization should be more uniform in direction across the entire beam, because one avoids the transverse polarization components which are introduced by precession of the spin when the ions exit through the fringe field of a strong-field ionizer. Another advantage of a weak magnetic field to parity experiments is that the intensity modulation with spin reversal will be reduced [2 ↔ 4 transition between the two six-poles, weak field transitions after the second six-pole, see ref. 30].

Even though fast Cs⁰ atoms have advantages over H⁻ or D⁻ as an

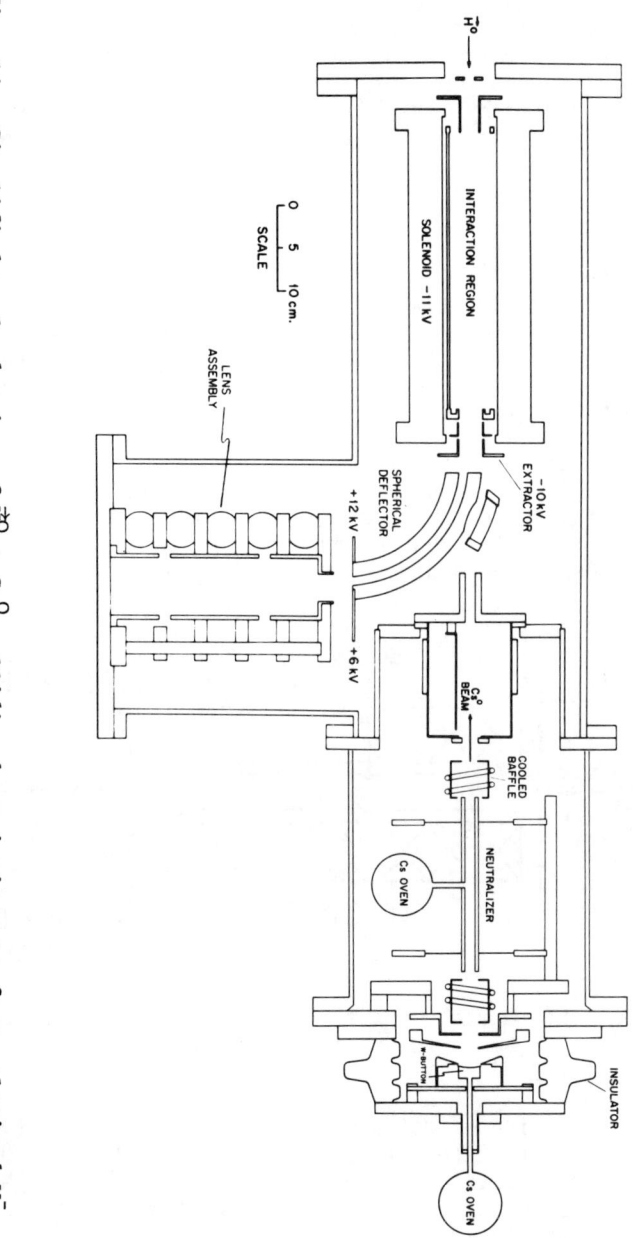

Fig. 14. Simplified scale drawing of $\vec{H}^0 + Cs^0$ colliding beam ionizer to form polarized $H^-$.

electron-donor in colliding beam experiments, it is tempting to speculate what could be achieved with reaction (4). The difficulties with reaction (4) that I mentioned can be overcome to some extent by applying a very strong magnetic field in the interaction region to confine the low-energy $\vec{H}^-$ ions. The fact that this works has been illustrated by work of the Bonn group.[31] They studied the production of positive ions by a crossed beam method, as proposed by Beurtey:[32]

$$H^+ + \vec{H}^0 \rightarrow H^0 + \vec{H}^+ \qquad (6)$$

Even though the polarized $H^+$ intensity was not high ($\sim 1$ μA) they showed that in a 75 kG field the polarized $H^+$ ions could be confined and extracted in the presence of a $\sim 1$ mA $H^+$ beam.

Another possibility is to make explicit provisions to compensate the $D^-$ space charge in reaction (4). This might be done by injecting positive ions ($D^+$, $Cs^+$) into the interaction region (Fig. 15). At Wisconsin we can readily produce 5 mA $Cs^+$ at 20 keV over 10 mm diameter about 30 cm from the source. If this beam is slowed for instance to 2 keV upon entering the collision region, it suffices to neutralize the space charge of 40 mA $D^-$ at 2 keV. A number of $H^-$ or $D^-$ sources that produce pulsed currents in the 100 mA range have recently been described (see e.g. ref. 33). If

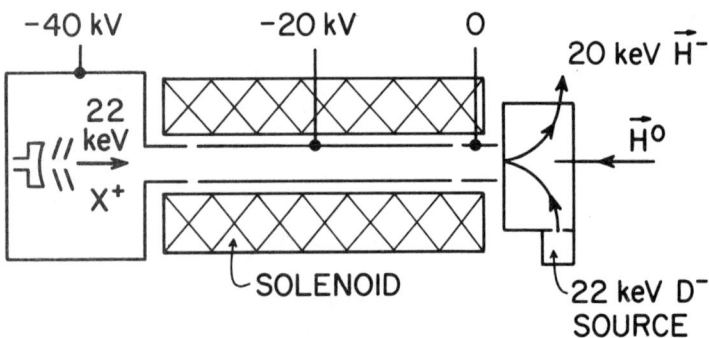

Fig. 15. Possible arrangement to provide for space-charge compensation of the $H^-$ beam in the charge exchange reaction $H^0 + D^- \rightarrow H^- + D^0$.

this kind of beam is deflected magnetically and decelerated from $\sim 20$ keV to 2 keV to interact with the polarized $H^0$ beam over some 20 cm length, a polarized $H^-$ beam in the 100 μA range might be possible. The presence of a deflection magnet between atomic-beam source and interaction region poses no problems. This in fact is the arrangement we used in our first tests of the $H^0 + Cs^0$ source.

Other schemes to solve the space charge associated with large $D^-$ beams were proposed in the proceedings of the Ann Arbor Workshop on High Energy Polarized Proton Beams.[30] It was pointed out that there is no need to inject a beam of positive ions as in Fig. 15 if instead one traps whatever positive ions are present in the interaction region by placing electrodes which are positive with respect to the center electrode at both ends of the solenoid. Another interesting idea to be tried at ZGS, is to let the $D^-$ beam cross the $H^0$ beam, as shown in Fig. 16. The interaction volume is now smaller but the distances over which the low-energy beams have to travel are small also. The smaller interaction volume can be offset by using a pulsed beam of $H^-$ ions of 1A or more. Sources that meet this requirement are available. A preliminary test of this scheme is soon to be carried out at ZGS.[34]

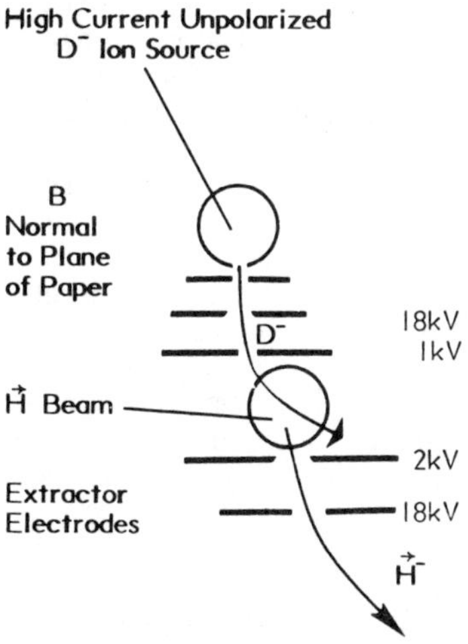

Fig. 16. Proposed scheme[30] to produce intense beams of polarized $H^-$.

## POLARIZED DEUTERONS AND OTHER IONS

Acceleration of polarized beams to high energies in circular accelerators and storage in storage rings poses particular problems because of depolarizing resonances. One solution under consideration is the use of polarized deuterons instead of protons.[35] Thus I should like to make a few comments about sources for polarized deuterons.

All the ion sources discussed above have been used to provide polarized deuterium ions as well. For Lamb-shift sources, the energy of the positive ions is increased from 500 eV for H to 1 keV for D to maintain the same velocity of the hydrogen atoms in the two charge exchange collisions. The increased beam energy has the effect that in most sources a somewhat higher beam output ($\sim$ factor 1.2 to 1.5) is obtained for $D^-$ than for $H^-$. The improvement is more pronounced for sources which use sudden field reversal than for sources with spin filter. The reason lies in the fact that in the spin filter only 1/3 of the $m_j$ = +1/2 atoms pass through the filter for D compared to 1/2 for H.

The most important difference between operation of polarized ion sources with D and H arises from the fact that the deuteron has spin 1 and a smaller magnetic moment than the proton. Thus there are three closely spaced $m_j$ = +1/2 hyperfine components for D vs. the two more widely spaced components for H. Still, in Lamb-shift sources, the spin filter can be tuned to pass any one of the three components, leading, after ionization in a strong field, to polarized deuterons in any one of the three substates with nuclear spin projection $m_I$ = +1, 0, or -1. Because of the lower magnetic moment, the critical field for deuterium in the 2S state is 14.6 Gauss vs. 64 Gauss for H. The sudden field-reversal method also is applied to deuterium, leading to a maximum vector polarization of 2/3, but the weak-field dipole transitions cannot be used for deuterium since the energy separation is no longer large compared to the line width. In atomic-beam sources, RF transition units have been devised that permit deuteron vector polarization of either +2/3 or -2/3. There is no reason why the vector polarization for atomic-beam sources can not be increased from 2/3 to 1 by applying RF transitions also between two six-pole separation magnets.[29] The fact that for deuterium in the ground state the critical field is only $B_c$ = 120 Gauss vs. 507 Gauss for H is of no consequence for electron bombardment ionizers since a field of $\sim$ 2k Gauss is required to confine the electrons. However, for the $\vec{H}^o$ + $Cs^o$ source the field in the interaction region can be reduced for deuterons with a corresponding improvement in the emittance of the extracted beam.

Sources for polarized ions other than H and D are presumably only of academic interest to high-energy physics. It is practical to produce polarized tritium ions in Lamb-shift sources but not so far in atomic-beam sources which require a much larger gas throughput. The Lamb-shift principle has been applied in one case[36] to produce polarized $^3He^+$. Another method to obtain a polarized $^3He^+$ beam is by extraction of $^3He$ ions from an optically pumped $^3He$ gas cell.[37] The atomic-beam method has been applied[38] to produce polarized heavy ions, such as $^6Li$, $^7Li$ and $^{23}Na$.

## CONCLUSIONS

We have available technically well developed sources which produce highly polarized hydrogen or deuterium ions suitable for injection into linear accelerators, cyclotrons and synchrotrons. Pulsed proton beams of about 100 μA with a polarization of about 75% are available from well engineered atomic-beam sources. The work at ZGS has shown that with proper care excellent reliability of these sources is obtained. Further refinements may yield another factor of two or so in beam current. Besides the ZGS, atomic beam sources are installed at SIN and on several cyclotrons used for nuclear physics research.

Negative polarized ions are of interest because of the ease with which they can be inserted into circular machines by stripping, yielding an advantage of maybe a factor 20 over positive ions. The production of negative polarized ions by charge exchange of positive ions in a Na vapor canal is standard practice in nuclear physics laboratories which use tandem accelerators equipped with atomic-beam polarized sources. These sources produce at best about 0.3 μA DC, but a factor ten improvement should readily be achieved if a 100 μA pulsed $H^+$ undergoes charge exchange in Na vapor. Nevertheless this approach is unexciting for synchrotrons because one is no better off than accelerating the positive ions in the first place.

Lamb-shift sources can produce polarized positive or negative ions, but for positive ions can not compete with atomic-beam sources. Sources of this type for H⁻ are installed at TRIUMF, LAMPF and is also planned for the 12 GeV KEK synchrotron. At best these sources produce about 1 µA of polarized H⁻ with roughly 80% nuclear polarization. The work on Lamb-shift sources in a number of laboratories over several years suggests that large increases in beam intensity are unlikely.

The application of the colliding-beam principle to charge exchange between polarized $H^0$ from an atomic beam source and a fast $Cs^0$ beam is in its infancy. No engineered source of this type exists, but the first tests of the principle yielded a H⁻ beam of 3 µA DC of high polarization (90% of maximum possible value). Production of at least 6 µA pulsed beam of good emittance should be straightforward. The source requires engineering to improve the reliability of the $Cs^0$ source and to increase the $Cs^0$ beam intensity. Since the source has not been developed systematically, large gains in intensity can be expected. The colliding-beam principle using D⁻ as an electron-donor to convert polarized $H^0$ to polarized H⁻ holds promise of high beam current. This scheme has not been tried, but may well produce pulsed H⁻ beams of 100 µA if a serious development program is undertaken. On the other hand, it is quite safe to say that the exuberant expectation at the 1977 Ann Arbor Workshop on Higher Energy Polarized Proton Beams of a several mA polarized H⁻ source within six months was not founded on sound technical judgment!

Ultimately, the goal of polarized-ion source development for synchrotrons is to fill the machine to the space-charge limit. The application of the colliding-beam method to produce H⁻ is but one promising avenue. Another is the storage of polarized particles prior to injection in a separate circular or linear device. The work at Saclay on the storage of ions in a superconducting solenoid, which was reported at this conference by Beurtey,[39] is an interesting example of progress in this direction. Only additional development work will show the best way to reach the ultimate goal of overcoming the intensity limitations in high-energy experiments with polarized beams.

## REFERENCES

† Work supported in part by the U.S. Department of Energy.
1. W. Haeberli, Ann. Rev. Nucl. Sci. 17, 373 (1967).
2. H.F. Glavish, Proc. Third Int. Symp. on Polarization Phenomena in Nuclear Reactions (Univ. Wisconsin Press, 1971), p. 267.
3. B.L. Donnally, Proc. Third Int. Symp. on Polarization Phenomena in Nuclear Reactions (Univ. Wisconsin Press, 1971), p. 295.
4. H.F. Glavish, Proc. Symp. on Ion Sources and Formation of Ion Beams, Brookhaven 1971, p. 207
5. T.B. Clegg, Proc. Fourth Int. Symp. on Polarization Phenomena in Nuclear Reactions (Birkhäuser, Basel, 1970), p. 111.
6. W. Haeberli, in Nuclear Spectroscopy and Reactions (Academic Press, 1974) Part A, p. 151.
7. W.E. Lamb, Jr. and R.C. Retherford, Phys. Rev. 79, 549 (1950).

8. B.L. Donnally, T. Clapp, W. Sawyer and M. Schultz, Phys. Rev. Lett. 12, 502 (1964).
9. A. Pradel, F. Roussel, A.S. Schlachter, G. Spiess and A. Valance, Phys. Rev. A10, 797 (1974).
10. B.L. Donnally and W. Sawyer, Phys. Rev. Lett. 15, 439 (1965).
11. P.G. Sona, Energ. Nucl. 14, 295 (1967).
12. J.L. McKibben, G.P. Lawrence and G.G. Ohlsen, Phys. Rev. Lett. 20, 1180 (1968).
13. Y. Mori, A. Takagi, M. Kobayashi, Y. Wakuta, S. Fukumoto, K. Kikuchi and A. Katase, Proc. Symp. on Ion Sources and Application Technology (Tokyo 1977) p. 87.
14. W. Arnold, H. Berg, H.H. Krause, J. Ulbricht and G. Clausnitzer, Nucl. Instr. Meth. 143, 441 (1977).
15. P. Schmor, private communication.
16. J.L. McKibben, High Energy Physics with Polarized Beams, AIP Conf. Proc. No. 35, p. 375.
17. General Ionex Corp., Newburyport, Ma. 01950
18. ANAC, 3067 Olcopt St., Santa Clara, Ca. 95050
19. W. Kubischka, B.A. MacKinnon and H.F. Glavish, to be published.
20. H.F. Glavish, Proc. Fourth Int. Symp. on Polarization Phenomena in Nuclear Reactions (Birkhäuser, Basel, 1970) p. 844.
21. A. Kruger, H.-G. Mathews, S. Penselin and A. Weinig, Nucl. Instr. Meth. 138, 201 (1976).
22. M. Simonius, private communication.
23. W. Grüebler, W. Haeberli and P. Schwandt, Phys. Rev. Lett. 12, 595 (1964).
24. W. Haeberli, W. Grüebler, P. Extermann and P. Schwandt, Phys. Rev. Lett. 15, 536 (1965).
25. W. Haeberli, Nucl. Instr. Meth. 62, 355 (1968).
26. C. Cisneros, I. Alvarez, C.F. Barnett and J.A. Ray, Phys. Rev. A14, 76 (1976); A.S. Schlachter, P.J. Bjorkholm, D.H. Loyd, L.W. Anderson and W. Haeberli, Phys. Rev. 177, 184 (1969).
27. D.G. Hummer, R.F. Stebbins, W.L. Fite and L.M. Branscomb, Phys. Rev. 119, 668 (1960).
28. D. Hennies, R.S. Raymond, L.W. Anderson, W. Haeberli and H.F. Glavish, Phys. Rev. Lett. 40, 1234 (1978).
29. D. von Ehrenstein, D.C. Hess and G. Clausnitzer, Phys. Lett. 19, 114 (1965).
30. H.F. Glavish, Higher Energy Polarized Proton Beams (Ann Arbor 1977) AIP Conf. Proc. No. 42, p. 47.
31. W. Hammon and A. Weinig, Nucl. Instr. Meth. 130, 23 (1975).
32. R. Beurtey and M. Borghini, J. Phys. C2, 56 (1969).
33. G.I. Dimov, G. Ye. Dereviankin and V.G. Dudnikov, IEEE Trans. Nucl. Sci. 24, 1545 (1977); K. Prelec, Th. Sluyters and M. Grossman, IEEE Trans. Nucl. Sci. 24, 1521 (1977); P.W. Allison, IEEE Trans. Nucl. Sci. 24, 1594 (1977).
34. E.F. Parker, proceedings of this conference.
35. L.C. Teng, proceedings of this conference.
36. O. Karban, S. Oh and W.B. Powell, Phys. Rev. Lett. 33, 1438 (1974).
37. D.O. Findley, S.D. Baker, E.B. Carter and N.D. Stockwell, Nucl. Instr. Meth. 71, 125 (1969).

38. E. Steffens, W. Dreves, H. Ebinghaus, M. Köhne, F. Fiedler, P. Egelhof, G. Engelhardt, D. Kassen, R. Schafer, W. Weiss and D. Fick, Nucl. Instr. Meth. 143, 409 (1977).
39. R.M. Beurtey, proceedings of this conference.

POLARIZED ION SOURCE DEVELOPMENT AT ARGONNE NATIONAL LABORATORY*

Everette F. Parker
Argonne National Laboratory, Argonne, Ill. 60439

In 1968, Haeberli[1] suggested the direct conversion of polarized atomic hydrogen beams to negative ion beams by processes of the type

$$H^0\uparrow + X^n \rightarrow H^-\uparrow + X^{n+1} \quad . \quad (1)$$

This idea was not vigorously pursued at that time, however, due to the development of Lamb shift sources with useful $H^-$ currents. It now appears that the Lamb shift source is limited to currents of a few µA at best, so attention is now shifting to the direct charge exchange technique as a possible way to produce more intense $H^-$ ion beams. Recently, Haeberli et al.[2] obtained a $D^-$ current of 3 µA by colliding a 40 KeV $C_s^0$ beam with a $D^0\uparrow$ beam. Three µA of negative beam on an initial attempt using less than optimum equipment is a very impressive result, considering that it represents the highest $\bar{D}\uparrow$ current ever produced.

The success of the Wisconsin group, plus the value of a negative ion beam for charge exchange injection into high energy proton synchrotrons led us to conclude that we should attempt to develop a high current pulsed polarized negative ion source for high energy physics applications. The Argonne National Laboratory (ANL) approach will be based on the reaction

$$D^0\uparrow (H^0\uparrow) + H^- (D^-) \rightarrow D^-\uparrow (H^-\uparrow) + H^0 (D^0) \quad . \quad (2)$$

This reaction was selected over the $C_s$ reaction used by Haeberli because (1) the charge transfer cross section for hydrogen is much larger ($\sim 3 \times 10^{-15}$ cm$^2$ vs $4 \times 10^{-16}$ cm$^2$), (2) much higher $H^-$ ($D^-$) beam currents densities (>100 mA/cm$^2$ vs 20 mA/cm$^2$), and (3) with hydrogen, there are none of the surface erosion and handling difficulties associated with $C_s$ beams and charge exchange cells. These positive aspects of hydrogen over $C_s$ will be offset to some extent by the very severe space charge problems associated with very low energy, high current density $H^-$ ($D^-$) beams. Indeed, the primary technical problem for this program is to determine the extent to which we can circumvent space charge problems.

The ANL polarized negative ion source test stand will consist of an ANAC, Inc., Model 2101 Atomic Beam System, a magnetron-type $H^-$ ion source [3] on loan from Fermi National Accelerator Laboratory, a charge exchange chamber, and a beam splitter magnet which will deflect the $H^-$ beam into the charge exchange chamber and deflect

---

*Work supported by the U. S. Department of Energy.

the D⁻↑ beam extracted from the charge exchange chamber into the diagnostic region. The design of the charge exchange chamber is, of course, the goal of the project.

The magnetron source is installed and operating on the test stand. The atomic beam stage and splitter magnet are due to arrive at ANL in early February 1979. The initial charge exchange chamber, which consists simply of a two-gap deceleration stage (2 aperture and a cylinder lens with 1.5 cm diameters), is also installed on the test stand. If the atomic beam system and splitter magnet arrive in February, the first colliding beam experiments should begin by early spring.

## REFERENCES

1. W. Haeberli, Nuclear Instruments and Methods $\underline{62}$, pp. 355-357, (1968).
2. D. Hennies et al., Phys. Rev. Letters, Vol. 40, No. 19, pp. 1234-1236 (8 May 1978).
3. C. Schmidt and C. Curtis, Proc. 1976 Proton Linear Accelerator Conf., AECL-5677, Chalk River Nuclear Lab., Chalk River, Ontario, Canada, Sept. 14-17, 1976, pp. 402-404 (Nov. 1976).

## ON THE POSSIBILITIES TO OBTAIN HIGH ENERGY POLARIZED PARTICLES IN ACCELERATORS AND STORAGE RINGS

Ya.S.Derbenev, A.M.Kondratenko

Institute of Nuclear Physics, Novosibirsk 90

### Abstract

In the storage rings and traditional cyclic accelerators the obtaining of the beam polarization is hampered at high energies by depolarization effects connected mainly with the energy dependence of a spin precession frequency. In this work the ways of controlling this dependence are considered, using additional fields introduced into straight sections. In particular, use of spin flips around the directions in the orbit plane allows elimination of the energy dependence of the effective precession frequency. Possible schemes providing the conservation of polarization during acceleration are presented. Under stationary conditions for electrons and positrons it is possible to suppress substantially the depolarizing influence of quantum energy fluctuations during radiation by using the method of spin flips in straight sections.


Last years are marked by the growing interest to experiments with polarized high energy particles [1]. According to modern ideas such experiments can provide the important information on the nature of interactions and the structure of elementary particles.

The only way to obtain high energy polarized photons (antiprotons) which seems available today, is to obtain such particles in a polarized state at comparatively low energies with the subsequent acceleration. The sources of polarized photons, which exist or under design can provide practically interesting intensities[1-3]. One can speak of the obtaining polarized proton beams with intensities close to those of unpolarized ones. Such a possibility appears when one uses charge exchange injection of polarized protons into a cyclic accelerator and their accumulation with the help of electron cooling [3]. One can hope to obtain polarized antiprotons (protons) with the aid of nuclear scattering by a polarized target applying the electron cooling to compensate for multiple Coulomb scattering [4].

For electrons and positrons the natural mechanism of the radiative polarization exists. In some cases the more efficient way of polarization can be that using colliding polarized photons, for example, those of laser radiation [5,6]. One can't also exclude the possibility to accelerate these particles in the polarized state from a source*.

The common problem for accelerators and storage rings of both light and heavy particles is the elimination of depolarizing effects connected with distortions of a magnetic system and the field inhomogeneity. Recent theoretical and experimental studies of this problem provide the foundation for optimism. This report is devoted to a short review of known and new possibilities of incresing the polarization stability.

## Acceleration of polarized particles

Depolarization during the acceleration occurs due to passing of a large number of resonances between the spin precession frequency in the main field (proportional to the energy) and frequencies of the perturbing fields on a particle trajectory.
In the usual situation with the field vertical over the whole orbit the following resonances are passed during the acceleration (at the moment we neglect small synch-

---

\* Recently a method of obtaining polarized positrons and electrons by pair production in collisions of polarized photons has been proposed [7].

rotron oscillations):

$$\nu_o \equiv \gamma G \approx \nu_k \equiv K_z \nu_z + K_x \nu_x + K_{//} \qquad (1)$$

where $2\tilde{\pi}\gamma G$ is an angle of spin rotation around the vertical direction with respect to a velocity per turn ( $\gamma$ is a relativistic factor, $G=(g-2)/2$ is ratio of an anomalous gyromagnetic ratio to a normal one), $\nu_z$ and $\nu_x$ are the numbers of vertical and radial betatron oscillations per turn, $K_x, K_z, K_{//}$ are integers.

After passing each resonance the coupling of average vertical projections is determined by the following formula [9,10] (the condition of intermixing of the spin precession around the field over the phases usually holds*):

$$\langle S_z^+ \rangle = \langle (2 e^{-2 J_k} - 1) S_z^- \rangle ,$$

where the parameter $J_k = \tilde{\pi} w_k^2 |\dot{(\nu_o - \dot{\nu}_k)}|^{-1}$ is determined by the rate at which the resonance, $\dot{\nu}_o - \dot{\nu}_k$ is passed and its strength $w_k$ ( $\tilde{\pi} w_k^{-1}$ is the time of vertical polarization flip in an exact resonance).

The variation of the polarization degree can be small at fast $(J_k \ll 1)$ or at slow passing ( $J_k \gg 1$ ). The condition of the polarization conservation after acceleration is:

$$4 \sum_{J_k \ll 1} J_k + 2 \sum_{J_k \gg 1} e^{-2 J_k} \ll 1 . \qquad (2)$$

It is necessary to exclude intermediate resonances because they lead to depolarization of the beam.

---

\* It may happen that during the time of acceleration at electron accelerators for small energies the intermixing of horizontal spin components doesn't occur. In this case after passing coherent resonances $\nu_o = K_{//}$ the polarization direction changes rather than its degree decreases [8].

At low energies, for example, one can use sufficiently fast passings by compensating for harmonics of the integer resonances $\nu_0 = K_{//}$ and applying a system which provides jumps of betatron frequencies at the moments when the resonances with betatron harmonics are passed[11]. Using such methods the argonne group succeeded in acceleration of polarized protons up to maximum accelerator energy of 12 GeV [12].

One can also avoid depolarization using slow passings $(\mathcal{J}_K \gg 1)$. In this case synchrotron oscillations of the precession frequency are important. Taking these oscillations into account results in the splitting of each resonance (1) in a series of modulation resonances. Their strengths decrease with the increase of their number "m". If a resonance is passed quickly without the precession frequency modulation, then the modulation resonances are passed quickly as well. Therefore at $\mathcal{J}_K \ll 1$ the account of the synchrotron modulation does not lead to notable effects. At slow passing of the resonances (1) the situation is more complicated. At $\mathcal{J}_K \gg 1$ the modulation resonances with low values of "m" are passed slowly, while these with sufficiently high "m" quickly. In this case the simple decrease of the rate at which a resonance is passed is not useful as due to non-idealities of the magnetic system a large probability of intermediate passings exists all the resonances including modulation ones.

Another method of proceeding to adiabatic passing consists in increasing resonance strengths by the special introduction of perturbing fields, so that the resonance is in fact tuned off by a distance equal to the strength of harmonics of the introduced field [13]. For example, it is sufficient to increase additionally the resonance strength $W_K$ so that to provide the low rate of passing resonances (1), and taking into account the synchrotron modulation $\nu_0$ one has

$$w_\kappa^2 \gg \sigma_\nu \nu_f \qquad (3)$$

where $\sigma_\nu$ is an amplitude of the frequency synchrotron modulation $\nu_o$, $\nu_f$ is a frequency of synchrotron oscillations. If the condition (3) holds, the strengths of modulation resonances are exponentially small. It is clear that in this case one can choose parameters so that during the passing with a low rate averaged over synchrotron oscillations modulation resonances are passed sufficiently fast.

The simplest case when a necessary increase of resonance strengths can be achieved is that of integer resonances $\nu_o = \kappa_{\shortparallel}$. If, for example, a constant magnetic field along the particle velocity is introduced into the straight section of the accelerator then the strengths of all integer resonances due to this field are equal:

$$w_\kappa = (1+G) H_o \ell/L.$$

Here $H_v$ is measured in units of the average guiding magnetic field of the accelerator, $\ell/L$ is a fraction of the orbit which is occupied by the introduced field. This method has been used at the Novosibirsk electron-positron storage ring VEP-2M to avoid the beam depolarization at the resonance $\nu_o = 1$ during the energy variation in the experiment [14].

At higher energies ($\nu_o \gg 1$) a transverse with respect to the orbit field can be useful, rather than a longitudinal one. For simultaneous compensation of the orbit distortion it is possible to introduce into the straight section the helical field with an integer number of periods transverse with respect to the velocity. This field results in the recovery of the particle velocity direction. The spatial shift can easily be compensated for at the next section by a unidirectional field with a zero average value, not distoring the spin motion.

It is possible that using the whole set of these methods (jumps of betatron frequencies at the corresponding resonances, additional perturbations for elimination of the depolarizing influence at integer resonances, compensation of the dangerous harmonics) one will be able to increase the energy of polarized protons and go higher than 12 GeV achieved now.

Such methods will be also useful for the acceleration of deuterons up to high energies. Due to the small value of its anomalous magnetic moment (approximately by a factor of 25 lower than that of a proton) a number of resonances appears considerably lower for deuterons than for protons during the acceleration up to the same energy. It is easier to eliminate the depolarizing influence of the resonances with betatron frequencies by jumps of betatron frequencies. The requirements to the throughness of the compensation for harmonics of integer resonances are less severe. For elimination of the depolarizing influence of integer resonances by introducing additional fields the required value of longitudinal fields is slightly less than that for proton of the same energy (by a factor of 3), while transverse fields must be considerably higher (for example, for the helical perturbating field the value must be larger by a factor of about 10).

2. The difficulties of the acceleration of polarized particles by the methods described above will grow fast because of the increase of the number of resonances, their strengths and scatter of precession frequencies.

The methods based on the substantial rebuilding of the spin motion appear more universal and perspective in the region of high energies. Introduction of special sufficiently high magnetic fields in the sections of an accelerator or a storage ring allows to be tuned off not only from the integer resonances, but from those with betatron frequencies as well [15].

As follows from the general theory of the spin motion in cyclic accelerators and storage rings, at any stationary configuration of the magnetic field in each place of the orbit the equilibrium polarization direction always exists $\vec{n}(\theta) = \vec{n}(\theta + 2\pi)$.

This direction is generally varying along the orbit and is not less stable than that along the field during the motion in a homogeneous field. The generalized precession frequency $\nu$ (in units of the revolution frequency) is an angle of spin rotation around $\vec{n}$ during one turn, devided by $2\pi$ [16].

Consider the following example (Fig.1). Introduce into a straight section I a longitudinal magnetic field rotating spin by an angle  around the velocity. To this end, a field $H_\nu = 21 E/g\ell$ is needed at the length $\ell$, where $H_\nu$ is the field in teslas, $E$ is the energy in GeV, $\ell$ is the length in m, $g$ is a particle $g$-factor. Let us change the longitudinal field $H_\nu$ proportionally to $E$ during the energy variation. The longitudinal field doesn't distort an equilibrium orbit, the coupling of radial and vertical oscillations due to it can be compensated for (if necessary) by four thin lenses (two at each side of the section with the field $H_\nu$).

One can easily see that the longitudinal direction is the stable polarization direction in an opposite section II. In fact, a spin directed along the velocity in the section II appears again longitudinal after one turn of the particle. Outside the section I the equilibrium polarization $\vec{n}$ is the plane of the particle motion and appears energy dependent everywhere but the section II. In the example considered the vertical polarization direction (outside the section I) appears unstable and gradually disappears due to the scatter of particle orbits. The po arization is stable if during the beam injection it is oriented along $\vec{n}$.

To determine the spin precession frequency $\nu$ it

is sufficient to trace the motion of the spin oriented transversely to $\vec{n}$. The spin oriented vertically in the section II (transversely to the orbit) after one revolution is flipped, i.e. turns by angle $\tilde{n}$ around $\vec{n}$. Thus, the fractional part of the precession frequency is always equal to $1/2$ at any energy of the particle: $\cos\tilde{n}\nu = 0$. This circumstance distinguishes sharply the considered example from a usual case of the vertical magnetic field for which the precession frequency $\nu_o = \gamma G$ continuously grows with energy and has a scatter $\delta\nu = G\delta\gamma$.

Let us underline that small deviations of the magnetic system from the described ideal one can result only in the insignificant perturbation of the direction of equilibrium polarization and a deviation of the precession frequency.

Additional fields can be switched on either before the injection (in the example given the particles are injected with longitudinal polarization in the section II) or during the acceleration.

Thus, in this example there are no spin resonances and the polarization degree is conserved adiabatically ( $\dot{\nu}_o \ll 1$ ) during acceleration.

Note that these spin resonances can also be avoided at smaller values of the angle $\varphi$ of spin rotation in the section I. Resonances $\nu = K_z \nu_z + K_x \nu_x + K_{//}$ appear impossible if the following condition is held [15]:
$$|\cos(\varphi/2)| < |\cos\tilde{n}(K_z \nu_z + K_x \nu_x)|.$$

3. At high energies ( $\gamma G \gg 1$ ) for spin rotation in the section I around the horizontal direction a transverse with respect to the velocity field must be used, rather than a longitudinal one. To rotate spin by an angle $\tilde{n}$ by a homogeneous field $H$ at the length $\ell$, 54 kG x m are needed for protons and 45 kG x m for electrons independently of the energy. However, the introduction of a radial field distors essentially the orbit at main sections of an accelerator. One can meet the requirements of

spin rotation by a given angle and the orbit recovery at the end of the section if the fields varying along the orbit are used. Some examples of spin rotation by an angle $\tilde{\pi}$ by fields consisting of homogeneous field sections are shown in Figs. 2,3. One can also use to this end the helical transverse magnetic field with an integer number of periods $N$. To rotate spin by an angle $\tilde{\pi}$ in the section I, $H\ell = 54 \cdot \sqrt{1 + 4N}$ kGs·m are needed for protons and $H\ell = 45 \cdot \sqrt{1 + 4N}$ kGs·m for electrons. In particular, at $N = 1$ it makes 120 kGs·m for protons and 100 kGs·m for electrons. A stable periodical direction in the section II lies in the orbit plane at an angle $arctg$ ($\sqrt{1 + 4N} / 2N$) with respect to the velocity. The precession frequency $\nu$ is again equal to $1/2$ at any N.

In the schemes presented the particle velocity is recovered to $(\gamma G)^{-3}$ (exact recovery is possible by small field corrections if necessary). The arising spatial shift of the orbit ($\sim \ell/\gamma G$) in Figs. 2,3 and $\sim \ell(\sqrt{N}\gamma G)^{-1}$ for a helical field) can easily be compensated for at the next section by a unidirectional field with a zero mean value.

Switching on and off the rotating fields can be done adiabatically during the acceleration with conservation of the stability of spin and orbital motion.

Note also that if it more convenient to conserve the vertical direction of the equilibrium polarization in main sections one can use two flips around orthogonal horizontal axes in two symmetrical sections (e.g., applying the schemes of Figs. 2,3). Here $\vec{n}(\theta)$ changes its sign but $cos\tilde{\pi}\nu = 0$.

The methods described can be applied to the acceleration of polarized particles up to maximem energies of the accelerators and storage rings (existing or under design) providing that spin resonances do not overlap, i.e. until the spin rotation during one revolution remains small (one means spin rotation by perturbating fields due to the non-ideal features of the magnetic system and by free vertical oscillations of the particles).

One can estimate the accuracy with which an angle of magnets is known with respect to the vertical in the plane perpendicular to the orbit. For proton accelerators of 1000 GeV, with a radius of 1 km, a frequency of betatron oscillations equal to 20, a number of independent elements of the magnetic system equal to 1000 this value is approximately $10^{-4}$. Possible vertical shifts of focusing lenses mustn't exceed 0.1 mm, the vertical beam size mustn't exceed 1 mm at the maximum energy. For each specific facility more accurate criteria can be obtained. Note that these requirements vary proportionally to the energy at a fixed value of the maximum accelerator field.

The question about the possibility to accelerate polarized particles for the energies where resonances overlap remains unanswered. One can't exclude today that further studies will lead to positive results in this energy region as well.

## Suppression of quantum depolarization of electrons and positrons

Last years are characterized by the growing interest to obtaining radiative polarization of the beams in high energy storage rings of electrons and positrons (energies of the order of dozens GeV and more). In traditional storage rings at the energy about hundred GeV to eliminate the depolarizing influence of the quantum fluctuations of synchrotron radiation rather serious requirements to the accuracy of magnetic systems are needed [17]. Due to energy dependence of the precession frequency radial magnetic fields appear the most dangerous arising, for example, at random vertical deviations of focusing lenses or at random inclinations of the vertical field magnets. As follows from ref.[17], the depolarization rate $\lambda_d$ in storage rings with everywhere vertical field is equal to the following value if non-correlated perturbing radial magnetic fields are present:

$$\lambda_d = \lambda_0 \frac{11\tilde{\pi}^4}{54} \frac{\nu_0^4(1+2\cos^2\tilde{\pi}\nu_0)}{\sin^4\tilde{\pi}\nu_0} \sum_{n=1}^{Q} \eta_n^2 \overline{H_n^2} |F_n^{[\nu_0]}|^2, \quad (4)$$

where $\lambda_0 = \frac{9}{11} \frac{d}{dt} \overline{\left(\frac{\Delta E}{\mathcal{E}}\right)^2}$ is a radiative polarization decrement, $Q$ is a number of sections with radial fields $H_n$ (in units of the average magnetic field of the storage ring), $\eta_n$ is a fraction of the orbit occupied by the n-th section, $F_n^{[\nu_0]}$ is a characteristic function of the storage ring determined by the energy and vertical focusing. In storage rings with a smooth vertical function the quantity $F_n$ is approximately equal to $\nu_0^2([\nu_0]^2 - \nu_z^2)$ ($[\nu_0]$ is an integer part of $\nu_0$, $\nu_z$ is a reduced frequency of vertical betatron oscillations). The formula above is valid as far as the energy spread of the precession frequency $\sigma_\nu = \Delta\nu_0$ is small if compared to the distance between resonances ( $\sigma_\nu \ll 1$ ). In the region $\sigma_\nu \gtrsim 1$ the depolarization rate sharply increases by a factor equal at least $3(2\tilde{\pi}^2 \nu_f^3)^{-1}$ if, as usual, the reduced frequency $\nu_f$ of synchrotron oscillations is much less than unity.

One of the possible ways to enhance the role of polarizing processes is application of magnetic snakes ("wigglers") with compensation of the energy dependence of the direction of equilibrium polarization in the wiggler region [18]. A more universal way of depolarization suppression can be suggested which uses flips of the vertical polarization in an even number of sections along the orbit. Using of multiple flips for increasing spin stability can be compared to using of strong focusing instead of the week one for betatron oscillations of the particles. A stable direction $\vec{n}$ of the equilibrium polarization remains vertical in main sections providing that a number of flips is even.

However, each spin flip changes its sign. Here depolarizing effects are again due to the non-ideal character of the magnetic system*.

Sections of spin flips must be optimally placed in such a way, that an angle of spin rotation around $\vec{n}$ be equal to zero in sections with the main field. It means that in case of two sections they must be opposite lying, in case of 4 must be placed in a quarter of the orbit length, etc. Thus, the effective precession frequency $\nu$ is determined only by sections with additional fields and is energy independent. For a precession frequency not to be in a resonance with a revolution frequency, the resulting axes of spin rotation lying in the orbit plane must be equal at all sections.

If a storage ring is symmetric then there is no polarizing influence of the main field due to spin reorientation. However, polarization can be provided by a laser or introducing magnetic wigglers.

For a storage ring with 2 M spin flip sections ( $M \ll \nu_x, \nu_z$ ) in an optimal case at $\cos\tilde{n}\nu = 0$ the formula (4) is modified):

$$[\lambda_d]_{2M} = \lambda \frac{11\tilde{n}^4}{54} \frac{\nu_o^4}{M^2} \sum_{n=1}^{Q} |n^2 H_n^2 / F_n^{[\nu_o]}|^2. \qquad (5)$$

Here $\lambda = \frac{9}{11} \frac{d}{dt} \overline{(\frac{\Delta \mathcal{E}}{\mathcal{E}})^2}$ is a polarization decrement taking into account radiation at the bending sections. By contrast with (4) this formula remains valid at $6\nu \gg 1$. Its applicability condition is practically always fulfilled:

$$\nu_o^2 \frac{\lambda}{\omega_o} \ll M.$$

---

\* At add number of flips the equilibrium polarization in the main sections appears lying in the orbit plane. Thus, $\vec{n}$ is strongly energy dependent and fast depolarization occurs due to energy quantum fluctuations.

Thus the influence of depolarizing factors is decreased by at least a factor of $M^2$ and hence the requirements to the magnetic system are M times weaker. After two flips the resonant energy dependence of the depolarition rate is removed as well as the resonant spin diffusion at large energy spread (providing that $\sigma_\nu \gtrsim 1$).

It may prove convenient to use for polarization the fields already introduced in straight sections for spin flips. For example, the magnetic field of Fig.2 can be used. The sections III where the field is parallel to $\vec{n}$ are made short enough so that the polarizing influence of radiation at these sections is dominant. The degree of radiative polarization is then close to

$$\zeta = \frac{8}{5\sqrt{3}} \langle (\vec{H}\vec{n})^3 \rangle / \langle |H|^3 \rangle$$

($\vec{H}$ is a magnetic field transverse with respect to the velocity) its maximum value $8/(5\sqrt{3})$, whereas the polarization time decreases by a factor of $\lambda/\lambda_0 = \langle |H|^3 \rangle / \langle |H_0^*|^3 \rangle$ ($H_0$ is a main field).

As the spin flip due to fields of Fig.2 occurs with respect to the velocity, then tune off a resonance one must perform a flip around another axis at one or several sections. For example, $\cos\hat{n}\nu = 0$, if one uses in one of the sections the scheme of Fig.3 rotating a spin around the vertical direction.

---

\* The energy spread increases by a factor of about $\sqrt{\lambda/\lambda_0}$.

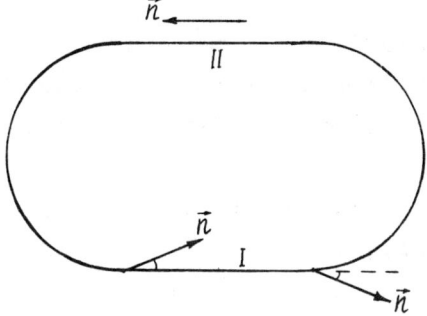

Fig.1   Scheme of spin motion with the rotation by 180° around the velocity in section I.

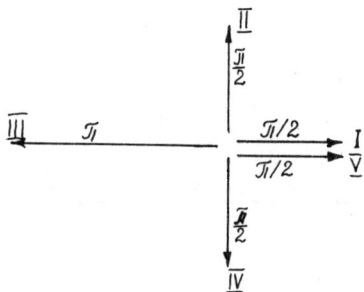

Fig.2   Scheme of spin rotation around the velocity in a section with the recovery of the particle velocity. The figure plane is perpendicular to the velocity. At sections I, II, IV, V spin is rotated at 90° around field directions, at the section III with a radial field – at 180°. At this section the equilibrium polarization is directed along the field.

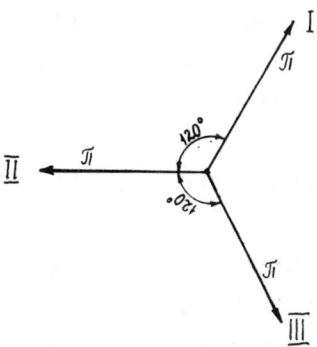

Fig.3   Scheme of spin rotation by 180° around the radial direction. The field in the section II is radial.

## References

1. Higher Energy Polarized Proton Beams, AIP Conference Proceedings (Ann Arbor, 1977) N°42, New York, 1978.
2. Ya.A.Pliss and L.M.Soroko. Usp. Fiz. Nauk 107, 28 (1972).
3. Yu.I.Belchenko et al., Proc. of X Intern. Conf. on High Energy Accel., V.I, p.287, Protvino, 1977.
4. Ya.S.Derbenev et al., ibid. v.II, p.55, Protvino, 1977.
5. Ya.S.Derbenev, A.M.Kondratenko, E1L.Saldin, Preprint INP 78-64, Novosibirsk (1978).
6. Ya.S.Derbenev, A.M.Kondratenko, E.L.Saldin, Preprint INP 78-68, Novosibirsk (1978).
7. E.L.Saldin, Preprint INP 78-69, Novosibirsk (1978).
8. H.A.Simonyan, Proc. of IV Intern. Conf. on Accelerators, p.915, Dubna, 1963.
9. M.Froissart, R.Stora. Nucl. Instrum. Methods 7, 297 (1960).
10. Ya.S.Derbenev, A.M.Kondratenko, A.N.Skrinsky, Zh. Eksp.Teor.Fiz. 60, 1216 (1971).
11. D.Cohen. Rev. Sci. Instrum. 33, 161 (1962).
12. R.L.Martin, Proc. of X Intern. Conf. on High Energy Accel., V II, p.64, Protvino, 1977.
13. Ya.S.Derbenev, A.M.Kondratenko. Dokl. Akad. Nauk SSSR 223, 830 (1975).
14. Ya.S.Derbenev et al., Proc. of X Intern. Conf. on High Energy Accel., v.II, p.76, Protvino, 1977.
15. Ya.S.Derbenev, A.M.Kondratenko, ibid, v.II, p.70, Protvino, 1977.
16. Ya.S.Derbenev, A.M.Kondratenko, A.N.Skrinsky, Dokl. Akad. Nauk SSR 192, 1255 (1970).
17. Ya.S.Derbenev, A.M.Kondratenko, A.N.Skrinsky, Preprint INP 77-60, Novosibirsk (1977).
18. Ya.S.Derbenev et al. Particle Accelerators, 8, 115 (1978).

# THE POSSIBILITY OF POLARIZED BEAMS AT THE AGS*

E.D. Courant and R.D. Ruth
Brookhaven National Laboratory
Associated Universities, Inc.
Upton, New York 11973
and
State University of New York at Stony Brook
Stony Brook, New York 11794

## INTRODUCTION AND SUMMARY

The physics done with high energy polarized proton beams has generated considerable interest during the last few years. Recent data[1,2] suggest that spin effects are growing with increasing $p_\perp^2$ and energy, and to understand the significance of these effects, it is necessary to extend the available polarized proton beam energy.

The success of the Argonne ZGS in accelerating polarized protons[3] to 12 GeV through many intrinsic and imperfection depolarizing resonances has developed confidence that a similar approach will work at the Brookhaven AGS. The 1977 Ann Arbor Workshop[4] concluded that acceleration of polarized protons to about 25 GeV does look possible, so a two week study was held at Brookhaven this summer to investigate polarized proton acceleration at the AGS in more detail and to produce a preliminary design and cost estimate.[5] The principal participants in the study are listed under Reference 5; however, a number of other staff members from Brookhaven were involved contributing ideas, information, advice, and design work.

The Brookhaven study discovered no new problems which cannot be solved. We prefer a polarized proton ion source of the H⁻ type, which could yield pulses of 75% polarized H⁻ ions with an intensity of 10-100 μamp and a length of 1 to 3 msec. Upon injection this would result in an AGS intensity of $3 \times 10^{10}$ to $10^{12}$ polarized protons per pulse which, together with the 2 sec repetition rate and the high extraction efficiency of the AGS, would yield an extracted beam intensity 5 to 150 times larger than that of the ZGS. Twelve new pulsed tune-shift quadrupoles will be necessary to jump the intrinsic resonances while the existing 96 correction dipoles can be used to tune out the imperfection harmonics. Most of the polarization monitors necessary are simply extensions of existing polarimeters; however, a fast internal polarimeter with an associated thin internal target would be useful for rapid tuning during the acceleration cycle. With these modifications it should be possible to accelerate polarized protons through the 8 intrinsic and 47

---

* Work performed under the auspices of the U.S. Department of Energy.

ISSN: 0094-243X/79/510307-11$1.50 Copyright 1979 American Institute of Physics

imperfection resonances in the AGS up to 23 GeV/c by late 1980. Although no decision has yet been reached with regard to the implementation of such a program, it is presently being considered together with other options for future AGS operation.

The purpose of this paper is to highlight some of the findings of the workshop while leaving the details of the cost estimates to the preliminary design study.[5] However, the results of the preliminary cost estimates are included in the following table:

Table I Cost Estimate AGS Acceleration of Polarized Protons

| | $ Thousands | |
|---|---|---|
| Injection System: | | |
| Polarized Ion Source | 405 | |
| Pre-accelerator Modification | 180 | |
| 750 KeV Beam Line | 85 | $ 670K |
| Pulsed Quadrupole System for Intrinsic Resonances: | | |
| Magnets | 300 | |
| Power Supply and Switching System | 220 | |
| Other Components, Labor | 100 | |
| | | $ 620K |
| Pulsed Dipole System for Imperfection Resonances: | | |
| Power Supplies and Control System | 150 | |
| Other Components, Labor | 80 | |
| | | $ 230K |
| Polarimeters: | | |
| 200 MeV Polarimeter | 25 | |
| Internal Polarimeter | 25 | |
| Internal Target (Gas Jet $75K or Rotating wheel $15K) | 45 | $ 95K |
| Absolute HE Polarimeter | | |
| Magnets and Power Supplies | 660 | |
| LH$_2$ Target System | 55 | |
| Counters, Electronics, etc. | 45 | |
| | | $ 760K |
| TOTAL | | $2375K |
| Contingency (20%) | | $ 475K |
| | | $2850K |
| Transfer of ZGS Equipment | | $ 750K |
| Total Cost to DOE | | $2100K |

## POLARIZED ION SOURCE, $H^+$ v.s. $H^-$

Both positive ($H^+$) and negative ($H^-$) polarized ion sources (PIS) exist, with the highest intensities (pulsed operations) presently available being ~100 μA ($H^+$) and ~5 μA ($H^-$). These sources are ground state atomic beam devices commercially available from ANAC, Inc.[6] The $H^+$ source uses an electron beam to ionize the atomic hydrogen, while the $H^-$ source[7] uses a 40 keV Cesium ($Cs^0$) charge exchange cell to ionize the atomic hydrogen, i.e.:

$$H^0\uparrow + Cs^0 \rightarrow H^-\uparrow + Cs^+ \qquad (1)$$

Source development activities presently underway at several laboratories should result in significant current increases, particularly for the $H^-$ sources, within the next year or so. ANAC is presently redesigning their entire ion source. The atomic beam stage will contain a more powerful dissociator, and the single large sextupole will be replaced by three smaller independently adjustable sextupoles. This should increase the amount of atomic hydrogen available for ionization, either $H^+$ or $H^-$. The length of their electronic ionizer will be increased, and its optics improved to give it a higher ionization efficiency. These improvements should lead to an $H^+$ PIS with a 100-200 μA output current. The $H^-$ sources will also benefit from the atomic beam stage improvements and the $Cs^0$ charge exchange $H^-$ source current could reach 10 μA. Significantly higher $H^-$ source currents could result from an $H^-$ source development program presently underway at ANL.[8] In this program a deuterium ($D^-$) charge exchange cell ($H^0\uparrow + D^- \rightarrow H^-\uparrow + D^0$) is being tried as an ionizer. Since its cross section is a factor of 10 larger than the $Cs^0$ cross section and high current Dimov-type $D^-$ sources are now available, this technique is expected to yield $H^-$ currents of several hundred microamperes within the next 12 to 18 months, and currents approaching 1 mA may be possible.

Although the cost and complexity of both types of PIS's ($H^+$ or $H^-$) are about the same, their relative value per microampere of beam current is quite different. Present thinking at BNL is that 2 mA of $H^-$ current from the linac will, with charge exchange injection, produce the same circulating intensity in the AGS as 65 mA of $H^+$ (~$10^{13}$ p/p). Thus, the 5 μA of $H^-$ presently available with a $Cs^0H^-$ PIS will produce more beam than the 100 μA of $H^+$ presently available. With $H^+$ injection the best one might hope for is circulating beam of $2-3 \times 10^{10}$ p/p, while with $H^-$ the expected intensity range is $3 \times 10^{10}$ to $10^{12}$ polarized protons per pulse or even more if the $H^-$ source can produce milliampere currents.

It is not unreasonable to have both $H^+$ and $H^-$ injection on the AGS. The ZGS ran for several years this way. Today, however, with both polarized and unpolarized $H^-$ ion sources available which can produce as much or more circulating beam intensity as $H^+$ ion sources, this is not required or desired. Thus, if the AGS is given a polarized beam capability, it might be converted to $H^-$ injection for both its unpolarized and polarized operation. At ANL, when the ZGS ran for over two years with H⁻ injection, the operating efficiency, stability, and intensity were much better than that ever achieved with H⁺ injection. Preliminary results from FNAL, where H⁻ injection is now standard on the 8 GeV booster, likewise indicate H⁻ injection is a better mode of injection.

If the AGS is converted to H⁻ injection, ion sources capable of producing 25 to 50 mA of linac current exist, so the 250 μsec of available linac beam pulse width will still be adequate to operate the AGS at full intensity. To take full advantage of polarized H⁻ injection, however, the linac rf system must be modified to allow beam pulse widths of 1 to 3 milliseconds. The additional energy required to support the copper losses associated with increasing the rf pulse length by a factor of ~10 is offset to some extent by the absence of any significant beam power requirements, but some additional energy storage may be required. Of course, to make use of the H⁻ beam, a charge exchanging stripper and orbit bump system must be designed and installed.

To summarize, significant but straightforward modifications are required on the injector if the AGS is to be given a polarized beam capability. The utility and flexibility of this facility will be much greater if the AGS is converted to H⁻ injection. If H⁻ injection is used and the polarized beam development activity begins in early 1979, the injector could be ready to provide 10 to 100 μA (H⁻↑) for injection into the AGS by late 1980. The injected beam polarization will be about 75% and rapid spin reversal on each AGS cycle will be possible. Assuming a 50% beam transmission efficiency in the LINAC and no beam loss in the AGS during acceleration, the expected AGS beam intensities for various ion source and LINAC conditions are as follows:

| H⁻ ↑ Source Intensity | LINAC Pulse Length | AGS Intensity |
|---|---|---|
| 10 μamps | 1 millisec | $3 \times 10^{10}$ |
| 10 μamps | 3 millisec | $10^{11}$ |
| 100 μamps | 1 millisec | $3 \times 10^{11}$ |
| 100 μamps | 3 millisec | $10^{12}$ |

## DEPOLARIZATION IN THE AGS

### The Causes

Particles undergoing vertical betatron oscillations experience horizontal depolarizing magnetic fields from the quadrupole fields in an alternating gradient synchrotron. The horizontal field frequencies seen by the particle are $kP \pm \nu$ where k is an integer, P is the machine periodicity and $\nu$ is the vertical betatron tune. Depolarization can occur during acceleration when the spin precession frequency, $\gamma(g/2 - 1) \equiv \gamma G$, becomes equal to one of these frequencies. Thus, the resonances are given by

$$\gamma G = kP \pm \nu \quad \text{"Intrinsic Resonances"} \quad (2)$$
{frequency in terms of the turning angle, $\theta$}

On the other hand all accelerators have horizontal imperfection field components of frequency k; thus, resonance will also occur when

$$\gamma G = k \quad \text{"Imperfection Resonances"} \quad (3)$$

Either of these types of resonance may be characterized by an effective strength, $\epsilon$, calculable from the machine lattice for a given beam emittance.[9]

Consider, for example, the case in which the perturbing fields and the particles precession frequency differ by a constant amount $\delta$ (a beam on a flat top near a resonance). Then the vertical component of the polarization is given by

$$\frac{P}{P_o} = \frac{\delta}{\sqrt{\delta^2 + \epsilon^2}} \quad (4)$$

If the beam has vertical polarization $P_o$ very far from the resonance, and we accelerate "slowly" to within $\delta$ of the resonance, then we will measure the polarization given above. However, this is not yet a true depolarization since if we reverse the above process after a flat top, the spin reorients itself along the vertical direction.

On the other hand the effect of traversing a resonance at a uniform rate, $\alpha$, from $\delta = -\infty$ to $\delta = +\infty$ was calculated by Froissart and Stora[10] to be

$$P/P_o = (2e^{-\pi\epsilon^2/2\alpha} - 1) \quad (5)$$

where

$$\alpha \equiv \begin{cases} Gd\gamma/d\theta & \text{for imperfection resonances} \quad (6) \\ Gd\gamma/d\theta \pm d\nu/d\theta & \text{for intrinsic resonances} \end{cases}$$

This relation clearly indicates the relative importance of $\epsilon$ and $\alpha$, however, in practice we would like to approach a resonance slowly, jump it quickly, and leave it again slowly. In this case, if we let the crossing be instantaneous, the depolarization is [11]

$$\frac{P}{P_o} = \frac{\delta^2 - \epsilon^2}{\delta^2 + \epsilon^2} \quad \begin{array}{l} \alpha \to \infty \text{ jump from} \\ -\delta \text{ to } \delta \end{array} \quad (7)$$

This relation provides us an upper limit to the polarization when a finite fast jump is performed. In order to estimate the effect of $\alpha$ being finite but large we simply construct the product of (5) and (7) and obtain

$$\frac{P}{P_o} = \frac{\delta^2 - \epsilon^2}{\delta^2 + \epsilon^2} (2e^{-\pi\epsilon^2/2\alpha} - 1) \qquad \begin{array}{l} \text{finite fast jump} \\ \text{"slow" approach and} \\ \text{departure} \end{array} \qquad (8)$$

The above result is useful for intrinisic resonances in which one can change the tune abruptly to increase $\alpha$ and thus decrease the depolarization. However, the effect of imperfection resonances is calculated with Eq. (5).

Figure 1 shows values calculated for the AGS intrinisic and imperfection resonance strengths, $\epsilon$,[9] and the resulting depolarization for a complete traversal of each resonance at the normal AGS acceleration rate $(d\gamma/dt = 60/\text{sec})$ calculated with Eq. (5). Clearly fast resonance jumping is essential to minimize polarization losses from the AGS intrinsic resonances.

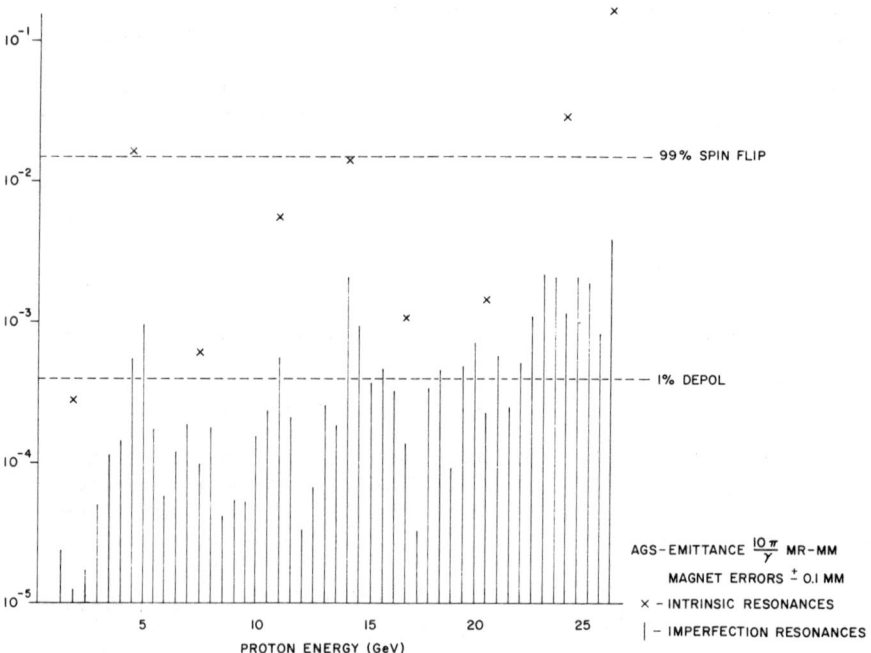

Fig. 1. AGS Resonance Strengths, $\epsilon$.

## The Cures

The standard method[3] to accomplish a resonance jump is the pulsing of quadrupoles to change the tune abruptly which by Eq. (6) increases $\alpha$ at the resonance. It is also necessary to decrease the depolarization due to Eq. (7) so we choose a total tune shift of .25 at 14 GeV/c. This change can be accomplished with twelve 1/2 meter long "unit" quadrupoles with equal gradients of 11.7 KG/m placed in a region where the vertical betatron function is a maximum ($\beta \simeq 22$m). If we select a rise time of 2 $\mu$sec for these magnets, we obtain the estimates for depolarization given in Table II.

Table II

DEPOLARIZATION DUE TO AGS INTRINSIC RESONANCES
Magnet full field risetime = 2 $\mu$sec ($4\pi/3$ radians)
The numbers in parentheses refer to the fixed "unit"
Quadrupoles. (tune shift of .25 at 14 GeV/c).

| Resonance $G\gamma_{res}$ = $kP \pm \nu$ | $\gamma_{res}$ | $\epsilon$ Resonance Strength Parameter | $\Delta\nu = 2\delta$ Total Tune Shift Assumed | Relative Pulsed Quad Strength | $\alpha = d\nu/d\theta$ Resonance Crossing rate | Depolarization [Eq. (8)] |
|---|---|---|---|---|---|---|
| 12-$\nu$ | 1.81 | .0054 | .25 | 0.10 | .0597 | .994 |
| 0+$\nu$ | 4.88 | .0154 | .25 | 0.32 | .0597 | .958 |
| 24-$\nu$ | 8.51 | .0006 | .25 | 0.56 | .0597 | 1.000 |
| 12+$\nu$ | 11.57 | .0054 | .25 | 0.76 | .0597 | .994 |
| 36-$\nu$ | 15.20 | .0137 | .25 | 1.0 | .0597 | .966 |
| 24+$\nu$ | 18.26 | .0010 | .25(.208) | 1.20 (1.0) | .0597(.0498) | 1.000(1.000) |
| 48-$\nu$ | 21.89 | .0015 | .25(.173) | 1.44 (1.0) | .0597(.0413) | 1.000(1.000) |
| 36+$\nu$ | 24.96 | .0266 | .25(.152) | 1.65 (1.0) | .0597(.0363) | .880( .734) |
| 60-$\nu$ | 28.86 | .1576 | .25(.132) | 1.90 (1.0) | .0597(.0315) | ----* |
| 48+$\nu$ | 31.65 | .0023 | .25(.120) | 2.08 (1.0) | .0597(.0287) | .999( .996) |
| Resultant Depolarization after acceleration up to: | | | | | | |
| 48-$\nu$ | 21.89 | | | | | .914( .914) |
| 36+$\nu$ | 24.96 | | | | | .804( .671) |

*An effective fast passage through this resonance is impossible; however, slow spin flip may be possible.

In the above calculations we have optimistically assumed that the spread in $\delta$ is much less than the total tune shift. The range of $\delta$ from the $\gamma$ spread is

$$\Delta\delta = G\Delta\gamma \simeq G\gamma \frac{\Delta P}{P} \qquad (9)$$

Since the full beam has $\Delta P/P \sim .15\%$ at $\gamma = 25$, this yields

$$\Delta\delta_{full} = .07 \qquad (10)$$

This is certainly smaller than .25 but not much smaller than .152 the tune shift for the chosen "unit" strength pulsed quadrupole at $\gamma=25$.

On the other hand, these calculations have also ignored the spread in $\epsilon$, since they apply to a representative beam in which all the particles have the same vertical excursion (that of the outside of the beam envelope). The beam really contains a distribution of $\epsilon$'s proportional to the distribution of the beam in the vertical direction which means that the effective ε of the beam is somewhat less than that indicated. So we feel that the optimism in some parts of the calculation is balanced by a corresponding pessimism in other parts.

We have one other problem to address in that the values for the polarization obtained in Table 2 assume that the effects of the imperfection resonances have been eliminated. These imperfection resonances occur when

$$G\gamma_k = k \qquad (11)$$

Since the value of $\gamma_k$ is independent of the betatron oscillation frequency, $\nu$, jumping the resonances with a rapid betatron tune shift does not work here as it did for the intrinsic ($\nu$ dependent) resonances. The rate of traversal through an imperfection resonance is determined only by $d\gamma/dt$, while the strength of a particular imperfection resonance, k, depends on the strength of the synchrotron's imperfection field component of harmonic k and on its proximity to $\nu$.

The properties of the various AGS imperfection resonances assuming uncorrected, random magnet misalignments of ± 0.1 mm were shown in Fig. 1. Most imperfection resonances cause depolarization of less than 1% and can almost be ignored; however, several cause depolarization of 10% or more and must be corrected.

To eliminate depolarization at these resonances one could measure the vertical orbit distortions accurately enough to determine the field imperfections at the 0.1 mm magnet displacement level and correct these imperfections directly. However, this precision would be difficult with the present AGS beam position detection systems. The approach used at the ZGS is to apply a horizontal field correction pulse which covers the resonance crossing period with the correct field strength to minimize the polarization loss. A similar technique can be used at the AGS. The horizontal field correction required for the $k^{th}$ resonance can be written in the form:

$$B_k(\theta) = \alpha_k \sin k\theta + \beta_k \cos k\theta \qquad (12)$$

The two independent parameters, $\alpha_k$ and $\beta_k$, can then be experimentally determined to minimize the polarization loss. Fortunately there are 96 correction dipoles currently installed in the AGS (an adequate number to generate the required harmonics for all the 47 resonances up to 26 GeV); however, new power supplies, a control system, and considerable additional software will be necessary to generate the finesse required to tune out these imperfection harmonics.

## AN INTERNAL POLARIMETER

As we have indicated, most of the polarization monitors necessary are extensions of those already in use at the ZGS.[3] However, in order to facilitate tuning through depolarizing resonances during the acceleration cycle, it is useful to have a polarimeter which is capable of a rapid relative measurement of the beam polarization before and after each depolarizing resonance. To insure that each resonance has been optimally jumped, an <u>absolute</u> knowledge of $P_B$ at each energy is also necessary; therefore, such a polarimeter must be calibrated against an absolute polarimeter. At the ZGS the CERN polarimeter, consisting of two identical scintillator range telescopes, is used for such tuning; however, it is situated in the extraction line, so the beam must be extracted to measure the polarization. The measurement would be much more efficient if the polarimeter were situated in the ring so that it could be electronically sampled to obtain a value of the beam polarization at a number of points during the acceleration cycle. The polarimeter target could be a wheel of $CH_2$ or metal fibers or possibly a hydrogen gas jet. The internal polarimeter itself could be similar to the CERN polarimeter and consist of two identical left and right scintillation counter telescopes which each detect the recoil proton in proton-nucleon elastic scattering at small $P_\perp^2$ ($P_\perp^2 \approx .15$ $(GeV/c)^2$). At this value of $P_\perp^2$ the scattering angle ($\sim 77°$) and momentum ($\sim 400$ MeV/c) of the recoil particle are almost independent of the beam momentum so the polarimeter arms can be fixed. The low recoil momentum allows the elastic signal to be separated by time of flight, dE/dx pulse height discrimination, and ranging. In this region the cross section is large ($d\sigma/dt$ is about 20 mb/$(GeV/c)^2$) and independent of energy from 4 GeV/c to 26 GeV/c; thus, the event rate is quite high. However, over this momentum range the analyzing power falls with momentum P according to the empirical formula $A_{pp} = .75/p$ (see, for example, ref. 5). Therefore, the time necessary to obtain a given precision on the beam polarization varies by a factor of 20 over the range 4 to 26 GeV/c.

To estimate the time necessary to obtain a given precision on the beam polarization measurement at various representative momenta, we choose an internal polarimeter looking at a rotating $CH_2$ fiber wheel target with the following parameters:

| | |
|---|---|
| Acceptance in Each Arm $\Delta t [\Delta\phi/2\pi]$ | $10^{-3} (GeV/c)^2$ |
| Internal Beam Intensity $[I_o]$ | $10^{11}$ protons/pulse |
| AGS Turn Time | 2.7 μsec |
| Polarimeter Sample Time | 5 millisec [1850 passes] |
| "Average" Thickness of $CH_2$ Target ($\ell$) | $6 \cdot 10^{-5}$ cm [$10^{-6} L_{coll}$] |
| Target Time in Beam | 250 millisec [$10^5$ passes] |
| $d\sigma/dt$ (at $P_\perp^2 = .15$) | $2 \cdot 10^{-26}$ cm$^2$/$(GeV/c)^2$ |

Such a wheel might contain $CH_2$ fibers of 0.05 mm diameter (or metal fibers of 0.02 mm diameter) with a mean spacing of 5 mm. During the 250 millisec that this target is in the beam, it would absorb about 10% of the beam. The polarimeter event rate during the suggested 5 millisec electronic sampling time is:

Events = $2 I_o$ [5 millisec/2.7 μsec] $(N_o \rho \ell)$ [$d\sigma/dt$] $\Delta t \Delta \phi / 2\pi$

= $2 \cdot 10^{11}$ [1850] $(3.4 \cdot 10^{19})$ [$2 \cdot 10^{-26}$] [$10^{-3}$]  (24)

≈ $10^6$ events/2 sec pulse

The corresponding analyzing power, data time and precision in $P_B$ at various momenta are:

| $P_{lab}$ | 6 GeV/c | 14 GeV/c | 24 GeV/c |
|---|---|---|---|
| $A_{pp}[P_\perp^2 = 0.15]$ | 12.5% | 5.5% | 3% |
| Time | 10 sec | 60 sec | 120 sec |
| Events | $5 \cdot 10^6$ | $3 \cdot 10^7$ | $6 \cdot 10^7$ |
| Error in $P_B$ | ± .6% | ± .6% | ± .8% |

The beam polarization was calculated using:

$$P_B = A_M / A_{effective} \quad (13)$$

while the error in $P_B$ was obtained by assuming somewhat pessimistically that

$$\Delta P_B = \frac{1}{(.5 A_{pp}) \sqrt{Events}} \quad (14)$$

Scattering from a metal or $CH_2$ fiber target is dominated by heavy nuclei which reduces the effective analyzing power. This can be eliminated by using a hydrogen gas jet target, however, this gives a factor of about 100 lower luminosity and may require perhaps 25 times more running time to acquire similar precision in the polarization. Since the gas jet is also technically more complex and more expensive, it was not studied in detail during the workshop.

## REFERENCES

1. J.R. O'Fallon et al., Phys. Rev. Lett. **39**, 733 (1977); D.G. Crabb et al., Phys. Rev. Lett. **41** (1978).
2. G. Bunce et al., Phys. Rev. Lett. **36**, 1113 (1976).
3. T. Khoe et al., Particle Accelerators **6**, 213 (1975); and in II International Symposium on HEP with Polarized Beams and Targets (Argonne 1976) M.L. Marshak, Editor AIP Conf. Proc. No. 35.
4. Higher Energy Polarized Proton Beams (Ann Arbor 1977); A.D. Krish and A.J. Salthouse, Editors, AIP Conf. Proceeding No. 42.

5. B. Cork, E.D. Courant, D.G. Crabb, A. Feltman, A.D. Krisch, E.F. Parker, L.G. Ratner, R.D. Ruth, K.M. Terwilliger, Preliminary Design Study for Acceration of Polarized Protons in the Brookhaven AGS (Contact A.D. Krisch, University of Michigan for copies).
6. H.F. Glavish, pp 47-66 in Ref. 4; and ANAC Technical Reports ANAC Inc., P.O. Box 7453, Menlo Park, CA 94025.
7. D. Hennies et al., Phys. Rev. Lett. $\underline{40}$, 1234 (1978).
8. E.F. Parker, Source Development at Argonne Proc. of III International Symposium on HEP with Polarized Beams and Targets, G.H. Thomas, Ed. (to be published AIP Conf. Proc.).
9. E.D. Courant, pp 94-100 in Ref. 4; and private communication.
10. M. Froissart and R. Stora, Nucl. Instr. and Meth. $\underline{7}$, 297 (1960).
11. R.D. Ruth, subsequent elaboration of this result is forthcoming.

# HYPERON BEAMS AS A SOURCE OF POLARIZED PROTONS*

David G. Underwood
Argonne National Laboratory, Argonne, IL. 60439

## ABSTRACT

A high energy polarized proton beam which would utilize lambda decays as a source of polarized protons has been proposed. We discuss the operation of such a beam and related physics experiments.

## INTRODUCTION

Before I go into detail about obtaining a polarized beam from lambda decay, I would like to briefly mention some physics motivation for such a beam.

Many spin effects at high energy and high $p_\perp$ have been discussed at this symposium. There are exciting new results, both experimental and theoretical, pertaining to the study of hadron structure and interactions through spin effects.

Deep inelastic e-p asymmetry measurements done at SLAC with longitudinally polarized electrons and protons indicate that the leading quark of a proton may carry most of the polarization information.[1]

There are predictions for the asymmetry $A_{LL}$ in inclusive production of pions or jets by polarized proton beams on polarized proton targets. The calculations have been done both for the effective-gluon model[2] and for QCD[3,4] with various assumptions about the spin structure of the proton.

We are proposing to measure this asymmetry to check the extent to which such models give a quantitative picture of hadron structure. Checking for fermion quarks and vector gluons could best be done with longitudinal polarization, of both beam and target fermions.

It has been proposed that the sea quarks and anti-quarks in a polarized proton are also polarized.[5] It would be nice to check this by measuring $A_{LL}$ (proton-proton) for di-lepton production. Assuming a Drell-Yan mechanism, annihilation of quark and anti-quark to give a virtual photon would have a much higher cross section for spins aligned than for spins opposed.

Another problem that could be studied with a polarized beam is the rise of the total cross section at high energies. One approach to the problem is to consider it in terms of s-channel helicity amplitudes. $\Delta\sigma_L$ is a linearly independent combination of the amplitudes involved in the rising cross section.

I will now go through some of the basic ideas of obtaining a polarized beam from lambda decay and give examples from a proposed design.

---

* Work supported by the U.S. Department of Energy.

More complete descriptions of how to obtain polarized beams and a proposed design are given in the proceedings of a June 1977 symposium[6] and a preprint[7] is available from Argonne.

## A Summary of Principles of the Polarized Beam

1. The polarization comes from the parity violating decays of lambdas.
2. Spin direction is almost unchanged in transforming from the lambda center-of-mass frame to the laboratory.
3. There are two simple ways to obtain polarized proton beams from lambda decays by using properties of beam transport systems. We call these the longitudinal and transverse modes.
   a. In the longitudinal mode, high momentum protons are chosen which necessarily come from forward decays of lambdas near the end point of the lambda production spectrum.
      Polarization reversal can be accomplished with minimal disturbance of the beam by using a snake of 8 magnets, 4 of which are reversable.
   b. In the transverse mode, a correlation exists between transverse polarization and position in phase space for a beam of protons or anti protons from lambda decay for momenta below the end point. This correlation is enhanced by a target system which eliminates decay near the lambda production point. Collimating off part of the proton beam at a focus can predominantly remove one polarization component while leaving the other.
      The sign of polarization can be selected on a spill-to-spill basis. The direction of polarization can be transformed with a 4 magnet snake of non-reversing magnets.
4. A high intensity proton beam from lambda decay will have large divergence, large effective source size, and large momentum spread. A beam designed for reasonable intensity will have spin rotations of up to 90° in the quadrupoles and up to 360° in the dipoles. It is possible to transport such a beam with very little depolarization or chromatic abberation.
   The beam transport involves -
   a. Global cancellation of spin precession by quadrupoles (Fig. 1) and
   b. Local cancellation of spin precession by dipoles (Fig. 2).

The basic point in cancelling spin precession is that the field integral must be zero. This means that a ray at the final focus must have the same direction it had before entering the transport system. The overall transver matrix is then $\begin{pmatrix}1 & 0\\ 0 & 1\end{pmatrix}$ in each dimension. The beam we are proposing[8,9] is mirror symmetric about an intermediate focus with a transfer matrix of $\begin{pmatrix}-A & D\\ 0 & -\frac{1}{A}\end{pmatrix}$ at that focus.

In order for the global cancellation of precession in the quadrupoles to occur, the spin directions entering one set of quadrupoles must be the same as the directions exiting the preceeding set. Precessions by bends must cancel between quadrupole sets.

Net bends between quadrupoles can lead to depolarization because rotations about orthogonal axes do not commute.

To summarize, the operation of the beam in transverse mode is as follows:

1. Downstream of a production target, all charged particles including the primary beam and lambda decay products are swept away for a distance of some fraction of a lambda decay length (about 6 meters).

2. Most of the remaining lambdas decay before the first set of quadrupoles and the forward collimated neutral beam continues through the quadrupoles to the neutral beam dump. The protons are carried around the dump by a series of dipoles but with no net bend and no momentum dispersion.

3. A vernier bend at a telescopic focus can displace the beam spot further downstream at the intermediate focus for polarization selection.

4. At the intermediate focus the momentum dispersed beam is momentum selected in one transverse dimension and polarization selected in the other. A hole collimator is used so that there is no beam motion downstream.

5. The beam is momentum recombined with all bend field integrals cancelled before the final two quadrupole triplets.

6. In order to transform the polarization direction to the direction required in a particular experiment, precession magnets may be used downstream of the final quadrupoles where the polarization directions for various phase space elements are again coherent.

Parameters of a proposed design are shown in Fig. 3.

## REFERENCES

1. M. J. Alguard et al., SLAC-PUB-2110, April 1978.
2. Hai-yang Cheng and E. Fischbach, "Polarization Asymmetry in High -$p_\perp$ Inclusive Production, Purdue University preprint, West Lafayette, Indiana.
3. J. Babcock, E. Monsay, D. Sivers, Argonne National Laboratory preprint ANL-HEP-PR-78-39, September 1978.
4. K. Hidaka, E. Monsay, D. Sivers, Argonne National Laboratory preprint ANL-HEP-PR-78-47, October, 1978.
5. F. E. Close and D. Sivers, Phys. Rev. Lett. $\underline{39}$, 1116 (1977).
6. A. Yokosawa, Argonne National Laboratory conference proceedings, ANL-HEP-CP-77-45, (1977).
7. D. Underwood et al., Argonne National Laboratory preprint ANL-HEP-PR-78-05, December 1977.
   Earlier work on this subject includes:
   a. O. E. Overseth, "NAL 1969 Summer Study Report", SS-118, Vol. I.
   b. O. E. Overseth and J. Sandweiss, "NAL 1969 Summer Study Report, SS-120, Vol. I.
   c. P. Dalpiaz, J. A. Jansen, and G. Coignet, CERN/ECFA/72/4. Vol. I, p. 284.

d. CERN proposal SPSC/p. 87, July 1977.
8. I. P. Auer et al., "Construction of Polarized Beams and an Enriched Antiproton Beam Facility in the Meson Laboratory and Experiments Using Such a Facility," Fermilab Proposal No. 581.
9. I. P. Auer et al., "Measurement of the Asymmetry in High $p_\perp$ Events Using a Polarized Proton Beam and Target, Fermilab Proposal No. 582.

322

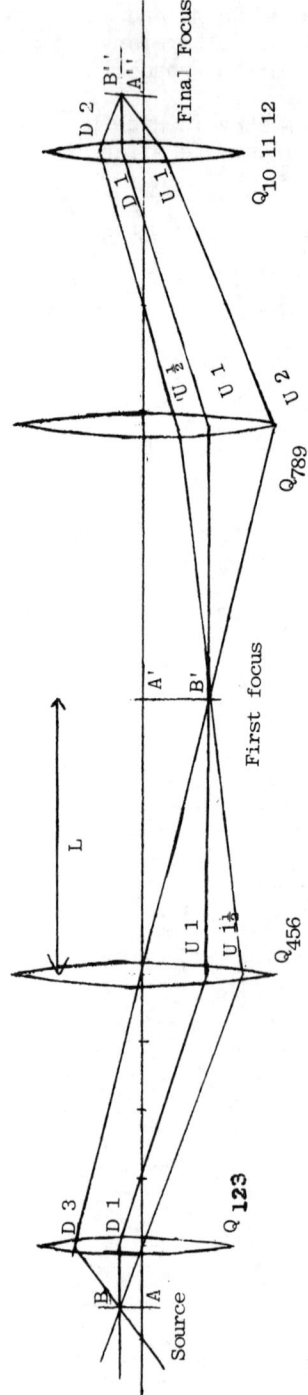

Fig.1

Diagram showing that there is no net spin precession for any off-axis or large divergence ray. This diagram is for a beam with $\begin{pmatrix} 1 & 0 \\ 0 & 1 \end{pmatrix}$ transfer matrix, symmetrically placed elements with spacing L, and idealized thin quadrupoles. The strength of $Q_{123}$ is three times that of $Q_{456}$. The labels "D 1", "U 1½", etc. refer to one unit of downward-bending field integral, one and one-half units of upward-bending field integral, etc.

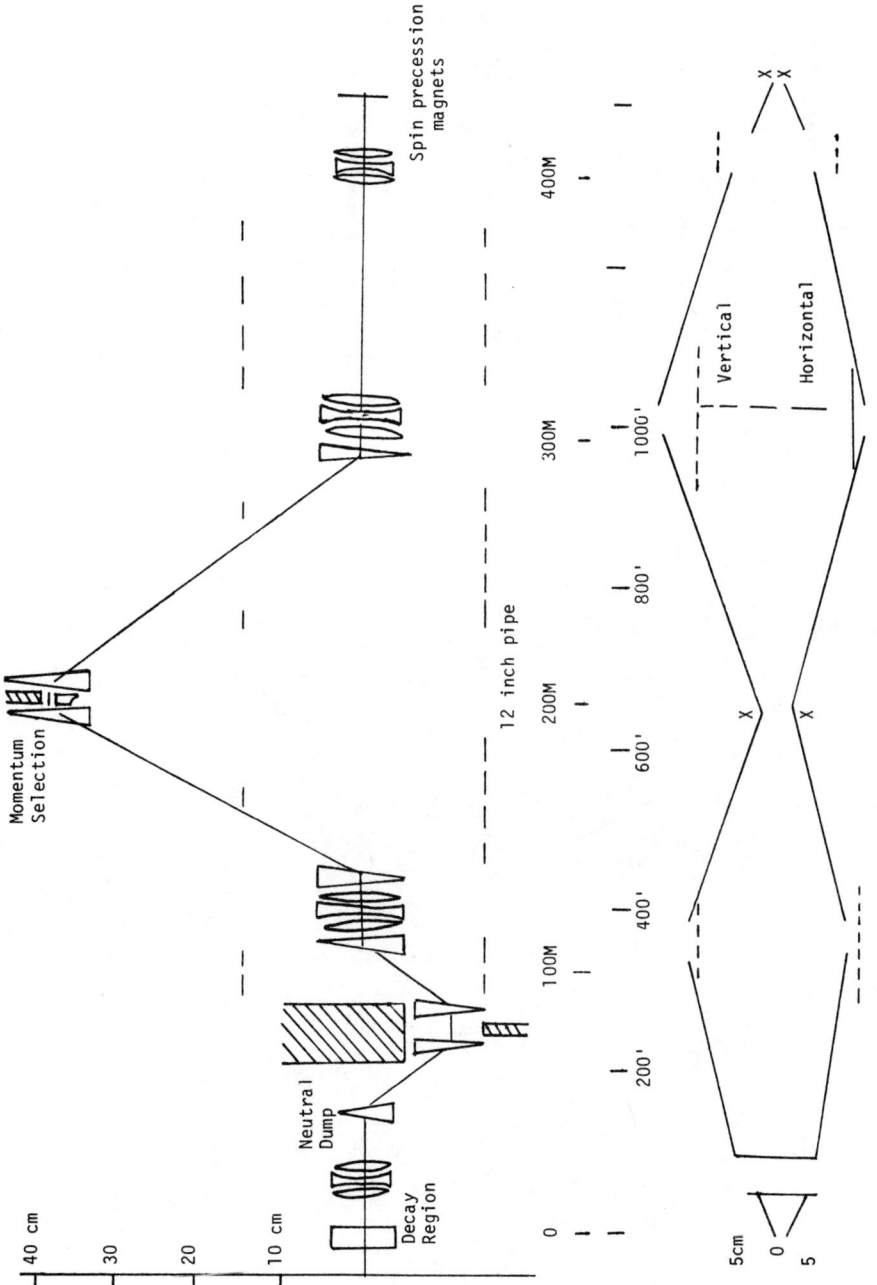

Fig. 2  Beam Envelopes.  The crosses (x) show the effect of a ± 5% momentum bite.

324

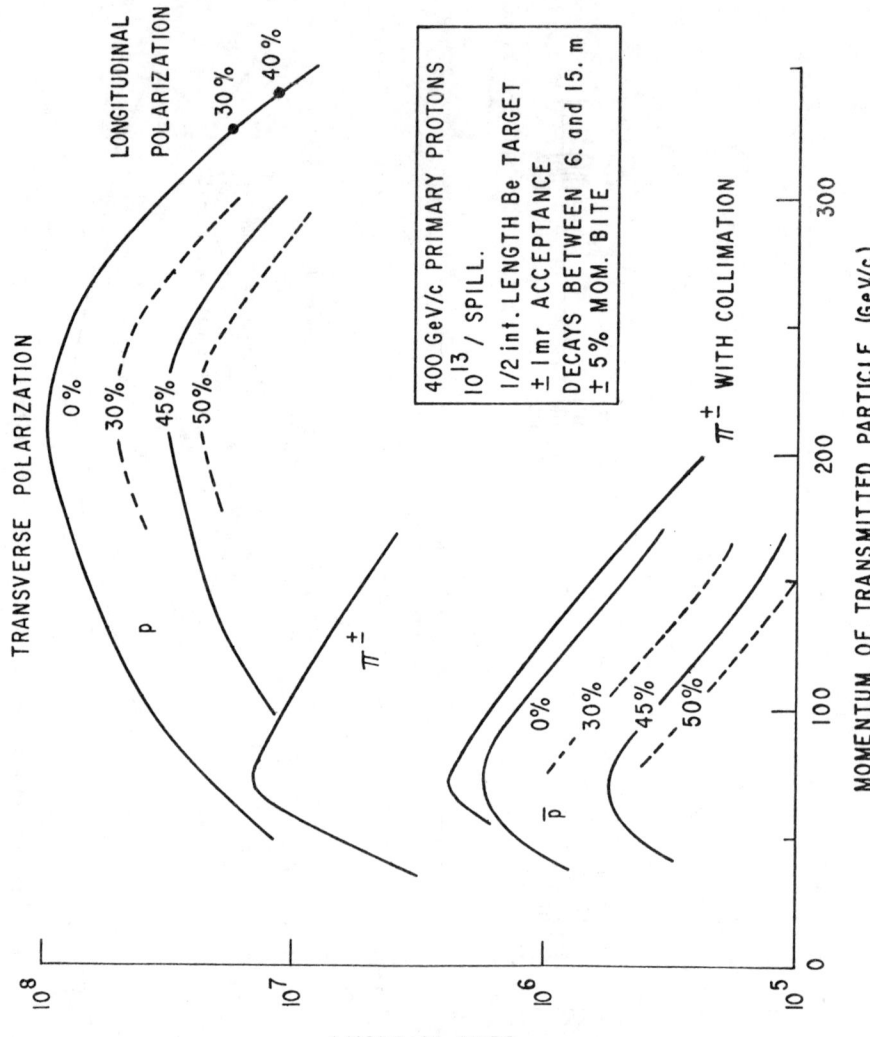

FIG. 3  Estimated polarized proton and antiproton beam-intensities.

# STUDIES ON POSSIBILITY OF ACCELERATING POLARIZED PROTONS AT KEK

S. Suwa
National Laboratory for High Energy Physics (KEK), Tsukuba, Japan

## INTRODUCTION

This report presents the work done to investigate the possibility of accelerating polarized protons at KEK. The KEK 12 GeV proton synchrotron consists of four stages: a 750-KeV Cockcroft-Walton preinjector, 20-MeV linac, a 500-MeV, 20-$H_z$ booster synchrotron, and a main ring. Nine pulses from the booster are injected in each 2-sec cycle of the main ring. Acceleration of the polarized protons is a subject which has been considered from early stages of the accelerator construction, although we have not definitely decided to incorporate the polarized-proton beam facility into the synchrotron.

Thus, a polarized-proton ion source has been developed to explore future possibility. Also problems of depolarizing resonances in the synchrotron, which was previously investigated,[1] are being reexamined particularly for the booster.

## POLARIZED ION SOURCE DEVELOPMENT

In order to accelerate polarized protons in the booster we need an intensity of $\gtrsim 5 \times 10^9$ ppp for beam control during acceleration. This means for one turn injection 1 mA or higher beam intensity will be needed. For negative $H^-$ beams, since some 100-turn charge exchange injections seems feasible,[2] a 10 μA or higher intensity beam will be useable. To explore such a possibility, a pulse-operated polarized $H^-$ source of Lamb-shift type was developed by Y. Mori and other pre-injector group members.[3] Figure 1 shows the set-up of the source. A characteristic feature is the use of a spin selector of r.f. electric field[4] crossed with a weak magnetic field at a proper angle. Also the system was made as short as possible. They obtained 1.1 μA (50 μsec pulsed) $H^-$ beam with 70% polarization.

Although this seems the highest intensity ever achieved by existing Lamb-shift sources, they have found that increase of intensity of order of magnitude or even of a factor seems very difficult because of the quenching of the metastable $2S_{1/2}$ states due to the space charge effect of the $H^+$ beam. Therefore, an atomic beam type source with a charge transfer scheme, such as $D^- + H^0 \rightarrow D^0 + H^-$,[5] has been suggested as a possibility for future development to obtain high intensity $H^-$ ion beams.

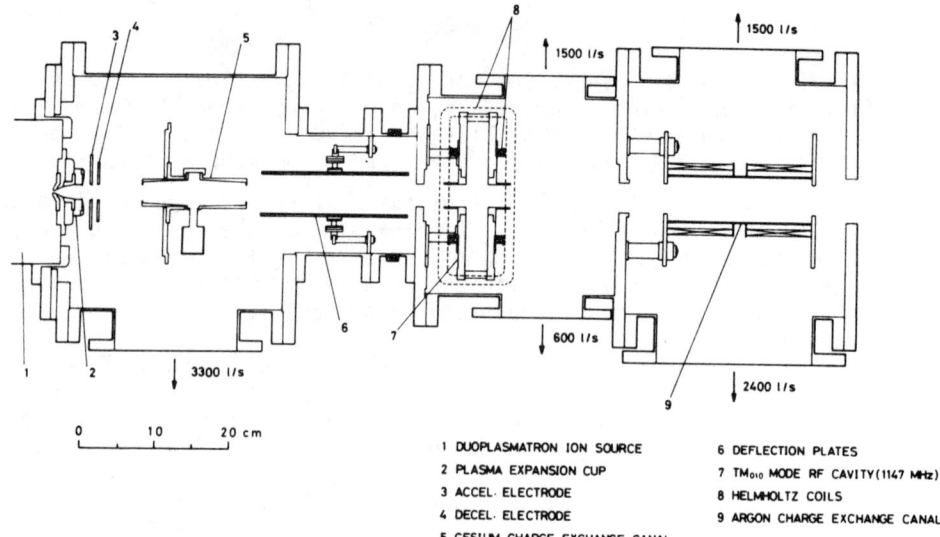

Fig. 1. Polarized H⁻ Source

## STUDY ON SPIN-FLIP IN THE BOOSTER

As is well-known, depolarization resonances appear at $\gamma G = nN \pm \nu_z$ (intrinsic resonances) and $\gamma G = n$ (imperfection resonances), where $G = g/2 - 1 = 1.7926$, N is the number of periodicity of the accelerator structure and n is any integer. Since the parameters of our booster are: $\gamma_{max} = 1.533$ (T = 500 MeV), $\gamma_{min} = 1.021$ (T = 20 MeV), $\nu_z = 2.25$ and N = 8, we have only one intrinsic resonance $\gamma G = \nu_z$ at $\gamma = 1.255$ and one imperfection resonance $\gamma G = 2$ at $\gamma = 1.115$.

Here are briefly given the results so far obtained by S. Hiramatsu and Y. Mori on computer calculation of the adiabatic passage through the intrinsic resonance with synchrotron oscillation being taken into account. The procedures are similar to the one of Reference 6. Starting with the standard spin equation of motion, we transform them to the ones in the coordinate system (x', y', z) moving with the particle and rotating around the vertical (z) axis with the spin precession frequency,

$$\Omega = \frac{e}{m_0 \gamma} (1 + \gamma G) B_z . \qquad (1)$$

The spin equation of motion is given by

$$\frac{d\vec{s}}{dt} = \vec{s} \times \vec{\Omega} , \qquad (2)$$

where $\Omega_z = 0$ and $\Omega_h = \Omega_{x'} + i\Omega_{y'}$ is expressed by

$$\Omega_h = \frac{e}{m_0\gamma} \sum_{n=-\infty}^{\infty} [F_n \exp i \{\int \epsilon_n dt + \Psi_z\} + G_n \exp i \{\int \delta_n dt + \Psi_z\}] \quad (3)$$

and $\epsilon_n = (\gamma G + \nu_z + nN)\omega_0$, $\delta_n = (\gamma G - \nu_z + nN)\omega_0$. $(\omega_0 = \frac{d\theta}{dt} = \frac{1}{R}\frac{ds}{dt})$ (4)

The factors $F_n$ and $G_n$ are Fourier expansion coefficients of the structure function of the lattice with periodicity of N and are evaluated numerically. They are proportional to the oscillating horizontal fields and then these are proportional to the vertical betatron oscillation amplitude.

In solving the equation (2) with (3) and (4), only the one resonant component in which $\epsilon_n$ or $\delta_n$ is close to zero were taken into account. The other non-resonant rapidly oscillating terms do not cause any secular effect and can be neglected. Thus taking a resonant term, say with $G_n$, one obtains

$$\frac{dS_h}{dt} = i \frac{e}{m_0\gamma} |G_n| S_z \exp i \{\int \delta_n dt + \Psi_z + \alpha_n\},$$

where $S_h = S_{x'} + iS_{y'}$.

When $\gamma$ and hence $\delta_n$ (also Gn) varies with t, the equation can not be in general solved analytically. It was solved by computer calculation for the booster.

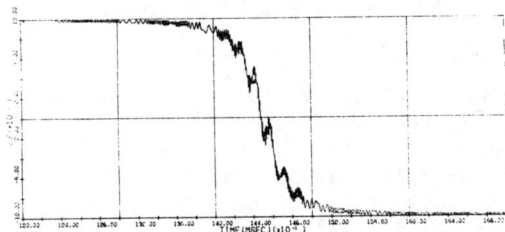

Fig. 2 Polarization Behavior in Adiabatic Passage throng $\gamma G = \nu_z$ Resonance in KEK Booster Calculated with Synchrotron Oscillation.

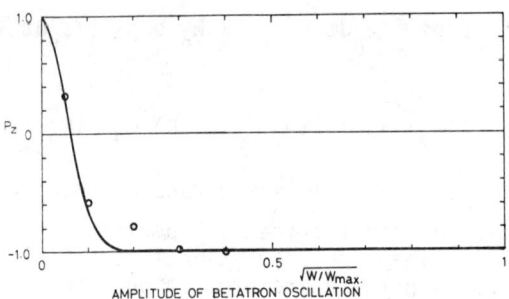

Fig. 3  Spin-flip vs Betatron Phase Space at $\gamma G = \nu_z$ Resonance
Solid Curve:  Without Synchrotron Oscillation
Circles:      With Synchrotron Oscillation

Without synchrotron oscillation, $\delta_n$ is given by $\delta_n = \dot{\gamma} G \omega_0 t = \Delta\gamma G \omega_0^2 t/2$, where $\Delta\gamma$ is the energy gain per turn. In this case the spin-flip of protons with initial $S_z = 1$ is analytically given by,[7]

$$S_z = 2 \exp\{-\frac{\pi^2}{\Delta\gamma G} \frac{|G_n|^2}{|<B_z>|^2}\} - 1 ,$$

where $<B_z> = e/m_0\gamma\omega_0 = R/B_z\rho$.

The solid curve in Fig. 3 is given by this formula for the $\gamma G = \nu_z$ resonance of the booster. Computer calculations were found to agree well with this curve.

For the $\gamma G = \nu_z$ resonance in the booster, design parameters of the booster are used; $\Delta\gamma = 7.1 \times 10^{-6}$ (6.68 kv/turn), synchrotron oscillation frequency $f_s \sim 7$ kHz, $\Delta p/p = \pm 0.3\%$ and $\omega_0 = 2\pi f = 30.2 \times 10^6$ sec$^{-1}$. Wmax corresponds to the maximum beam emittance of 12.6 $\pi$ mm mrad. Figure 2 shows as an example the complete spin-flip with synchrotron oscillation for W = W max. Circles plotted in Fig. 3 are values of spin-flip for various betatron amplitudes with the synchrotron oscillation. From Fig. 3, one can conclude that more than 90% of the initial polarization is preserved, unless an unreasonable phase space density distribution is assumed. This demonstrates that if a resonance is strong enough almost complete spin flip can occur even with synchrotron oscillations in a strong-focussing synchrotron.

As for imperfection resonance although it has not yet been investigated, we expect it should be easy to suppress by correcting the closed orbit distortion, if necessary, because we have only one resonance, n = 2. Thus it appears that acceleration of polarized protons in the booster is feasible.

We have not fully investigated the depolarization in the main ring yet. A. Ando and E. Takasaki of KEK made rather tedious calculations with a simulation program. The spin motion was traced step by step along the beam circulating around the machine simulating actual particles without synchrotron oscillation. According to their results, we have many strong $(4n + \nu_z)$ and medium strong intrinsic resonances. In general, $4n + \nu_z$ resonances are stronger than $4n - \nu_z$ (N = 4 for the main ring). These will be investigated more in detail also by the method as described above in the case of the booster in the near future.

## REFERENCES

1. T. K. Khoe, KEK-Report KEK-73-8, 1973; also H. Sasaki, Proc. U.S. - Japan Seminar, KEK, p. 247, (1973).

2. K. Kobayashi, KEK-Report, KEK-76-2, 1976.

3. Y. Mori et al., Second Symposium on Ion Source and Technology, Tokyo, (1977); Y. Mori et al., Nuclear Instr. and Methods 141, 383 (1977).

4. B. L. Donnally, Bull. Am. Phys. Soc. 12, 509 (1962).

5. H. F. Glavish, AIP Conference Proceedings No. 42, p. 42; D. Hennies et al., Phys. Rev. Lett. 40, 1234, (1978).

6. E. Grovud et al., COC-GERMA 75-48/TP-28.

7. E. C. Courant, AIP Conference Proceedings No. 42, p. 94. Also see Ref. 1.

# POLARIZED PARTICLES AT SATURNE

(Progress report on Source, beams and experimental program)

R. M. BEURTEY

Laboratoire National Saturne CEN-Saclay - France

## ABSTRACT

This paper is a status report on the production, acceleration and future programs with polarized protons and deuterons at Saturne. The cryebis source is described, as well as the way to accelerate polarized protons without depolarization. Some indications are given on the future experimental program, mainly nucleon-nucleon and experiments proposed to test reaction mechanisms.

## INTRODUCTION

By building a new accelerator in the range (0.3-3 GeV), the french physicist community working with intermediate energy hadrons, had for first goal to obtain classical projectiles (p,d,α) of moderate intensity ($\sim 10^{12}$) but of very good quality, reliability and flexibility, in order to perform fine nuclear spectroscopy. At the very starting point of the reconstruction of Saturne, it was wanted a possible production and acceleration of heavy ions (A $\lesssim$ 40) and polarized particles (p,d). A new set : (Source + Preinjector), named CRYEBIS[1] was designed in 1976 for this purpose, and is now under test. Meanwhile, the applications for new experiments were sent by physicists to the new "Laboratoire National Saturne"*. About 60 % of the 36 proposals or letters of intention are devoted to experiments using heavy ions and polarized particles. The whole system Cryebis + transport system will be installed summer 79 and will be in operation at the end 79.

The new ring, Saturne captured the first protons the 26th of July this year. The next steps are :
- 1st extraction, and beginning of the physics at the end of November 78
- 2nd extraction, and beginning of beam sharing in may-june 79 (sharing on succesive pulses, or <u>inside</u> the same pulse)
- Installation of CRYEBIS and of its transport system during summer 79. First tests, and acceleration during autumn 79
- Full operation, all particles and beam sharing, automatic control in spring 80.

(1) - CRYEBIS is built at ORSAY by a collaboration of IPN-ORSAY and CEA - Dr CABRESPINE is in charge of this facility

*The National Laboratory "SATURNE" is financed half and half by the universities (IN2P3) and the atomic energy (CEA)

# POLARIZED SOURCE AND BEAMS

## PRODUCTION

The production of polarized projectiles (and heavy ions) will be very specific to Saturne, using the "CRYEBIS" source. I remind you the principle of CRYEBIS on fig.1. In a high solenoïdal magnetic field (35 kilogauss cryogenic) an intense beam of electrons is focussed into a very small diameter cylinder near the axis. An adjustable electrostatic potential is distributed by elementary isolated cylinders all along the axis. A well is created which stores the positive ions created inside. For heavy ions a very high current density is necessary to produce a number of high charge state ions in a limided time ($J > 1000$ A/cm2). For polarized protons and deuterons, only the intensity of the electron beam and the overlap with the atomic jet are essential.

This atomic beam is produced by a "classical" jet source (Glavish or Saclay type), focussed by a sextupole followed by ad-hoc electromagnetic transitions, and goes through a hole made in the heart of the cathode along the axis of Cryebis. The focussing of the electrons will be adjusted to optimize the overlapping along the ionization length (1.5 meter). The electron beam diameter will be 3 to 4 mm. The maximum number of produced ($\vec{p}$, $\vec{d}$) is estimated to be between 1 and $3.10^{11}$ per pulse.

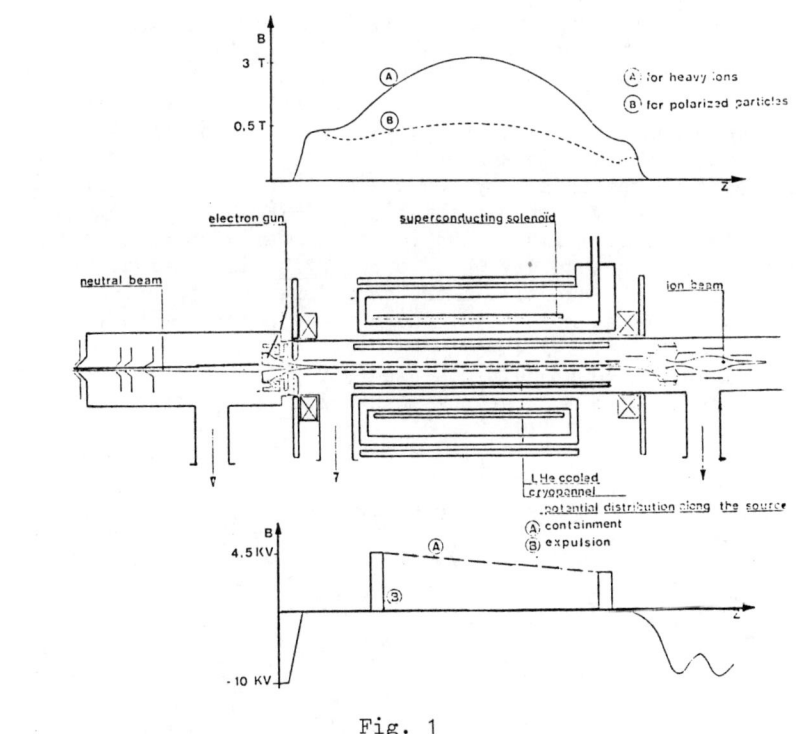

Fig. 1

I show you on fig.2 the average cycle of Saturne.

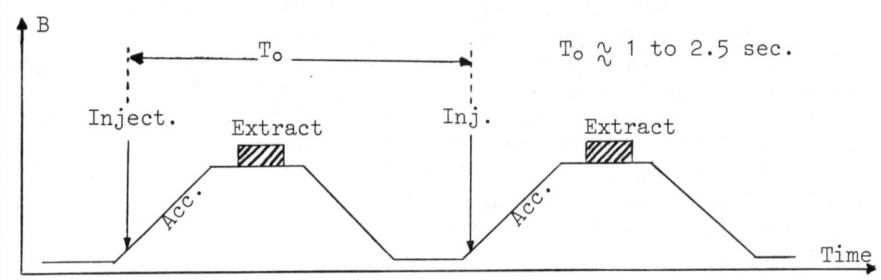

Fig. 2 <u>Cycle of SATURNE</u>

One sees on this (fig. 2) that almost one second could be used to prepare the source, ionize and store, then extract from Cryebis and send the beam to Saturne. (an adjustable slope B, see fig. 1, allows to match the extraction time from Cryebis with the injection time of the machine).

STATE OF CRYEBIS (october 78)

With a low intensity (0.5 A) low voltage (2 kV), Cryebis produced high charge states of heavy ions in 15 msec. ($Ne^{10+}$, $K_r$, Xe). The first tests with polarized ions will run on beginning 79. Using diaphragms, the atomic jet enveloppe at the center of Cryebis was simulated. The densities along the solenoïd are such that $10^{11}$ ions would be easily obtained in a storage time $\tau_s$ < 0,1 second.

The <u>residual gas</u> under cryogenic conditions is of very low pressure. This is a specific characteristic of CRYEBIS : a preliminary electron pulse, by its high ionisation rate, can clean-up the interior of the source such a way that <u>no pollution</u> is observable on the following pulse. Consequently, there should be no depolarization by scattering on residual molecules, and a final polarization very near the theoretical value.

Table I shows the transitions used for polarizing the protons and deuterons (we also look at the possibility of polarizing $^6Li$ and $^7Li$).

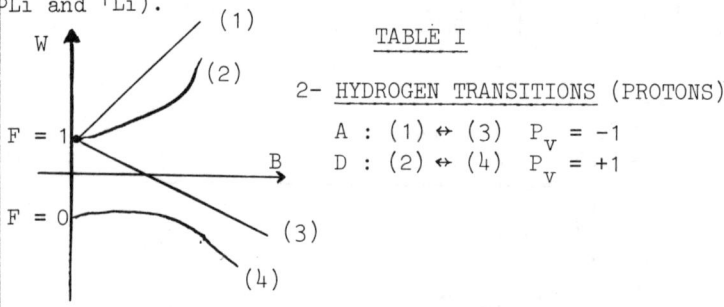

TABLE I

2- <u>HYDROGEN TRANSITIONS</u> (PROTONS)

A : (1) ↔ (3)  $P_v$ = -1
D : (2) ↔ (4)  $P_v$ = +1

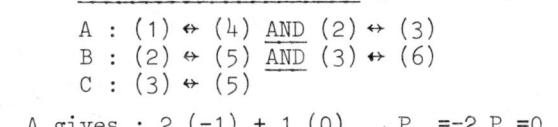

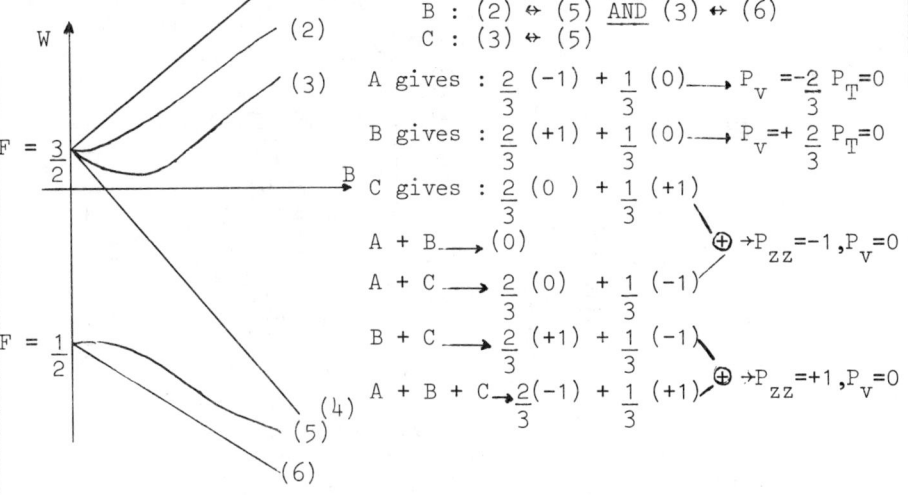

It is possible to plan, cycle after cycle, sets of <u>vector</u> polarizations (± 1 for $\vec{p}$ ; ± 2/3 for $\vec{d}$) or <u>tensor</u> (four different states in succession) or eventually <u>8</u> states (vector, tensor, and two "zero" reference states).

ONE MORE REMARK : The same Cryebis system, with a slightly different potential shape, would be able to trap simultaneously protons <u>and</u> electrons ionized from an atomic jet. If electrons are polarized (and <u>they are</u> easily using a sextupole), one should obtain <u>polarized electrons</u> of a good average intensity. (this would be especially fitted with pulsed machines with a repetition period from 1 millisecond to 1 second).

FROM THE CYEBIS Source to the physicists

Fig. 3 shows a complete map of Saturne (Source, injector, accelerator, experimental areas). Beam lines (1) (4) (5) will be ready for operation january 79.

Particles produced inside Cryebis will be injected into the 20 MeV - Linac by a specialized transport line. First, at 10 KeV on the high voltage platform, an electrostatic deviator followed by a solenoïd gives the spin a vertical direction. Second, deuterons and protons are accelerated to 400 KeV (200 KeV resp.) and transferred to the linac entrance. Deuterons are injected into the synchrotron at 10 MeV, protons at 5 MeV. Unfortunately, at these energies the whole transparency : Linac + injection + capture + acceleration + extraction is <u>small</u> (2 to 4 %), and no more than 2 to 4 $10^9$ particles at the targets will be obtained.

To compensate for this loss, and use storage of few CRYEBIS pulses a project exists for another preaccelerator, called MIMAS -

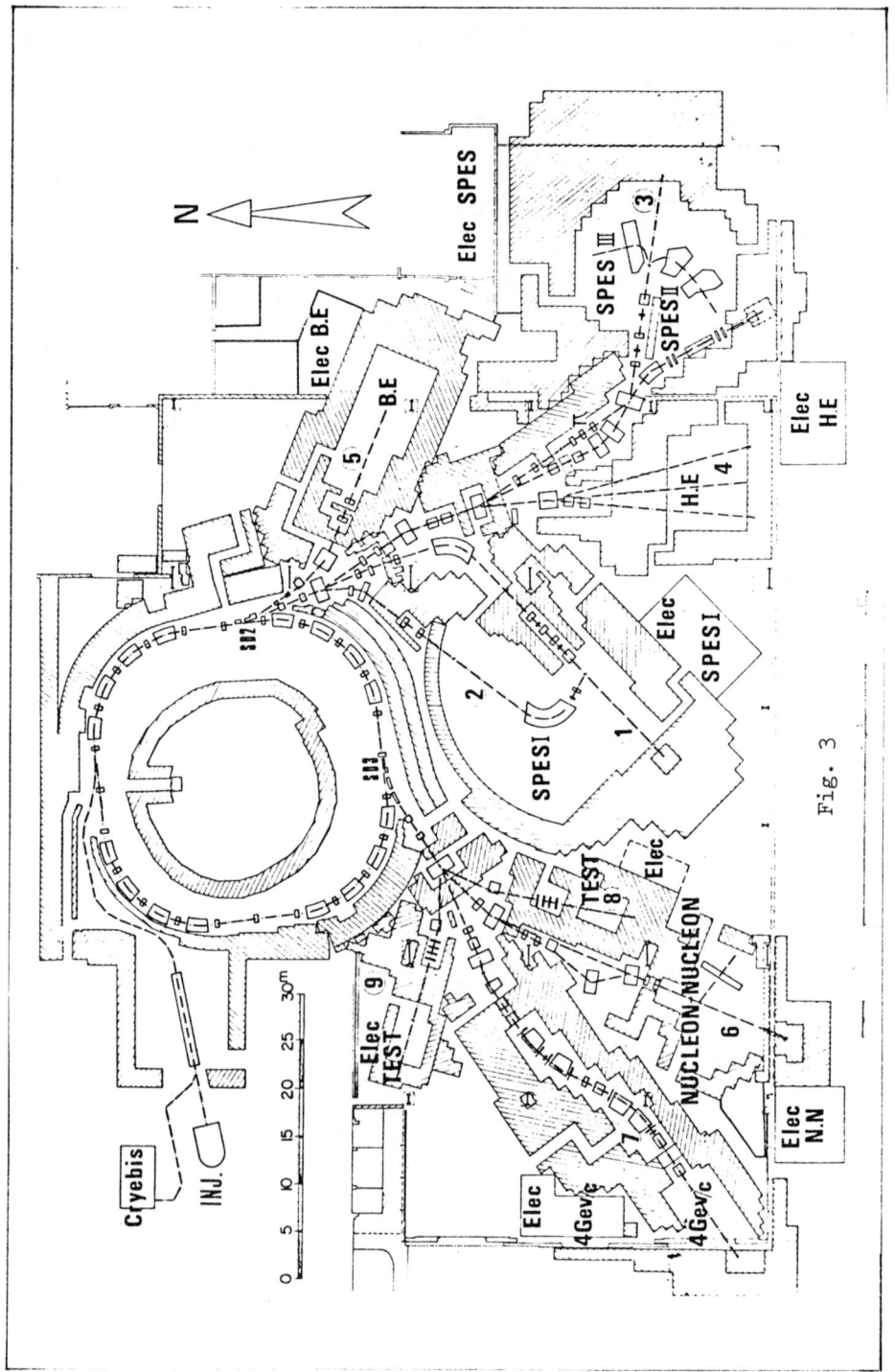

Fig. 3

Cf. at the end of the paper.

A specific low intensity detection system inside the machine will be installed, the sensitivity of which will be a few $10^7$ per pulse. The polarization of extracted beams will be mesured and controlled on a polarimeter installed at the exit of the machine, allowing to follow the variation of the polarization as a function of the energy.

Polarized neutrons will be obtained by break-up of the deuterons ($P_N \sim 60$ %).

NO DEPOLARIZATION INSIDE SATURNE II

It is well known that polarized deuterons are difficult to depolarize due to their small anomalous magnetic moment. This is true also at the new Saturne.

Concerning protons the strong focussing produced by the quadrupoles induces large magnetic components of first order ($\gamma G = \nu_z$ and $8-\nu_z$). Do these magnetic components produce depolarization ? (This is also the case for integer or "closed orbit" resonances $\gamma G$ = Integer).

Following the A. Abragam's text book notations (1) I give on fig. 4 two extreme situations described in the reference system of the particle rotating at a resonance frequency ($\nu$).

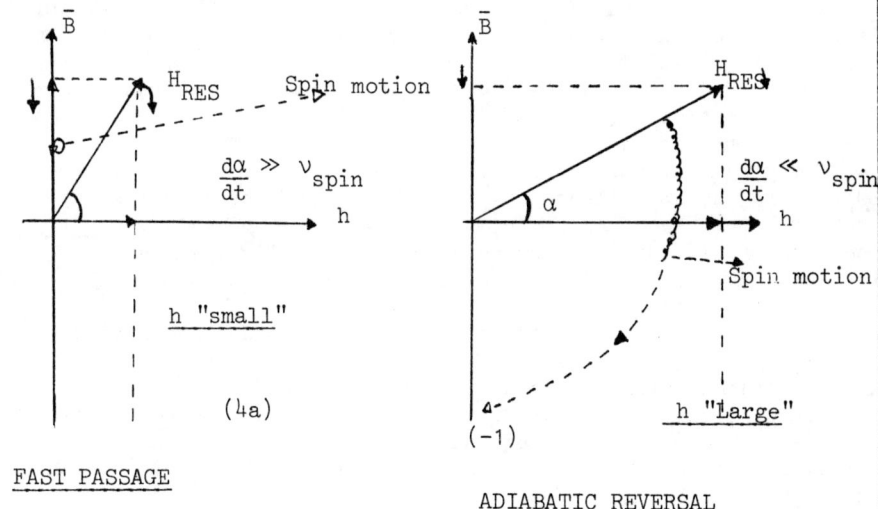

FAST PASSAGE

ADIABATIC REVERSAL

Fig. 4 - <u>Two extreme situations in going through resonances</u>

The resultant magnetic field $H_{RES}$ is given by two components :
- The Fourier component h, which is <u>fixed</u> in this rotating frame.
- The main magnetic field H seen by the particle, which goes through zero at resonance and reverse.

If ($H_{RES}$) reverses with an instantaneous rotation speed always very large compared to the spin rotation frequency around $H_{RES}$, we are in the case of a " FAST PASSAGE ". The spin makes a fraction of turn during the full reversal of $H_{RES}$ (fig. 4a).

"SPINS REMAIN VERY NEAR THE INITIAL (Vertical) DIRECTION"

On the contrary (fig. 4b) if the Fourier component is so large that the spin frequency is at every time very large compared to the instantaneous rotation speed of $H_{RES}$, then the spin will follow $H_{RES}$. This is the extreme case of <u>adiabatic passage</u>.

"SPINS FOLLOW ADIABATICALLY THE REVERSAL of the Field"

A simple formula (2) gives the final value of the polarization :
$$P_{fin} = P_{in} (2e^{-A} - 1)$$

Fast passage : $A \ll 1$  $P_{fin} = P_{in} (1-\varepsilon)$

Adiabatic reversal : $A \gg 1$  $P_{fin} = -P_{in} (1-\varepsilon')$

By construction (decided at the very beginning) Saturne is faced only with those two <u>extreme</u> situations.
1 - <u>First order</u> resonances ($\gamma G = \nu_z$ and $\gamma = 8 - \nu_z$) and also integer (closed orbit) resonances are "adiabatic" - $A \gg 1$. The full reversal is obtained except for a very small part of the "heart" of the emittance ($\lesssim 3.10^{-3}$).
2 - <u>Second order</u> resonances ($\gamma G = \pm \nu_z \pm \nu_x \pm 4n$) are related to non linear transverse components, i.e. :
 - hexapolar components inside dipoles
 - fringing fields of quadrupoles
BUT due to the small length of the Q-poles, of the specific nature of the enveloppe, and finally because sextupole components are <u>very weak</u> one has for all those resonances $10^{-4} < \bar{A} < 1,3.10^{-3}$.

As a result, the whole acceleration cycle will produce : $\frac{\Delta P}{P} < 2\%$.

Fig. 5 gives an exhaustive list of first order and closed orbit resonances, with the "width" of the "polarization stop-bands" (energy band inside which the polarization goes from + 90 % to - 90 %). It should be easy to measure this reversal extracting and sending the beam on a polarimeter just before and after the resonance field value. It will be difficult to use beam at energies <u>inside</u> the stop-bands.

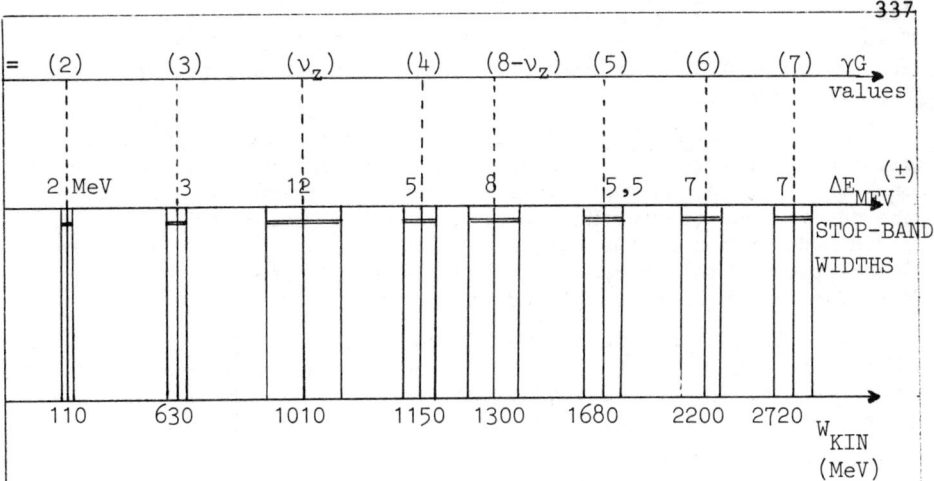

Fig. 5

Adiabatic Resonances and "Polarization stop-bands"
($\gamma G$ = Integer produced by C.O. deformation - $\delta$ - function type, 4mm)

EXPERIMENTAL PROGRAM AND FUTURE DEVELOPMENTS

Proposals for experiments at Saturne are mostly submitted for running 79 - middle 80 - Few of them are approved by the Scientific Comittee. I shall then give only general goals and methods concerning experiments willing use polarized beams - New proposals are waited at the beginning of 79 - (polarized beams are not provided till autumn 79).

NUCLEON-NUCLEON

An ambitious program has been proposed in the energy range 300 MeV-3 GeV, working on (pp) and (np) scattering. It is intended to measure :
- Polarizations (essentially n-p, partly pp)
- Wolfenstein parameters (D,R,A) with rescattering
- Spin correlations

All those experiments require no more than two scatterings, with no additional magnetic field between $1^{st}$ and $2^{nd}$ scattering.

a - THE BEAM will be prepared such a way that the incident polarizations could be directed along three possible orthogonal directions. A set of movable magnets, with a solenoïd which rotates the spin into the horizontal plane, will give all possible orientations over the full range of energies of Saturne. Neutrons obtained by break-up of polarized deuterons will also be oriented in all directions (The polarization will be approximately 60 %, the intensity at the target $\sim 10^5$ per cm2) Fig. (6a, 6b) gives the set-up of the beam for each case.

a2 - THE POLARIZED TARGET is of the frozen spin type, using a dilution refrigerator[3]. After the maximum polarization is reached, one can decrease the magnetic field and rotate the spin direction to the desired one (deuterons and protons can be polarized). The maximum dimension of the target will be (5 x 5 x 5) cm3. A complete description of this target has been presented elsewhere[4]. This facility will be completed at the end of 78.

a3 - THE PROGRAM is not yet fixed in all details. It is not sure on the other hand, that the complete detection system would be ready at the end of 79. In that case, the first experiments will be simpler (single scattering, polarizations and correlations).

OTHER EXPERIMENTS with ($\vec{p},\vec{d}$)

b1 - A large part of the first proposals are related to experiments using polarized projectiles to test reaction mechanisms. For instance :

- ELASTIC SCATTERING OF POLARIZED PROTONS around 1 GeV will be a complement to previous scattering experiments with unpolarized beams. The accuracy on neutron matter distribution is very dependent on the accuracy of the nucleon-nucleon amplitudes, extracted from angular distributions through (KMT) or (Glauber) sophisticated models. Some ambiguities could be removed by measuring the polarization up to reasonable momentum transfers.

- BACKWARD $\vec{D}$ SCATTERING ON P (and $\vec{P}$ on D)
It is a challenge to understand the mechanism : large momentum single scattering ? transfer of a nucleon, isobar ? multiple scattering effects ? Vector and tensor polarization could give a few more observables to understand the dominant process.

- ARE THERE RESONANCES ($\Delta^{++}$) IN $^3$HE ?
An experiment made at the end of Saturne I p + He$^3 \rightarrow \ ^3$H + "X" on SPES I, has shown a "bump" in the missing mass ressembling a $\Delta$ with a smaller width. Polarized protons could help to understand the mechanism.

b2 - A special experiment will be used at the end 79 (approved). It consists in crossing an intense molecular beam ($^2$H, $^2$D, He, $CO_2$..) with the beam of projectiles circulating inside the machine. Recoils can be measured by silicon junction detectors in coïncidence. Polarization effects will be followed (and cross-sections also), at small transfers, as a function of the energy (almost continuous study of N-N, ND,.... scattering as functions of the energy).

Other experiments need the large spectrometer (SPES III). They will probably necessitate polarized projectiles (production of dibaryons ? D-state of the deuteron?...).

FUTURE

A new projet is under development, in order to overcome the very low efficiency of the actual injection scheme for polarized

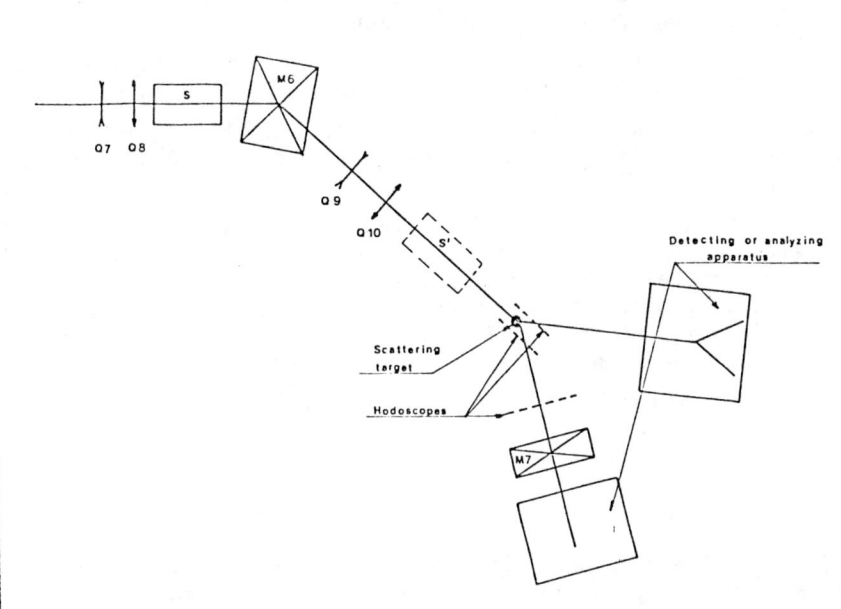

Fig 6a

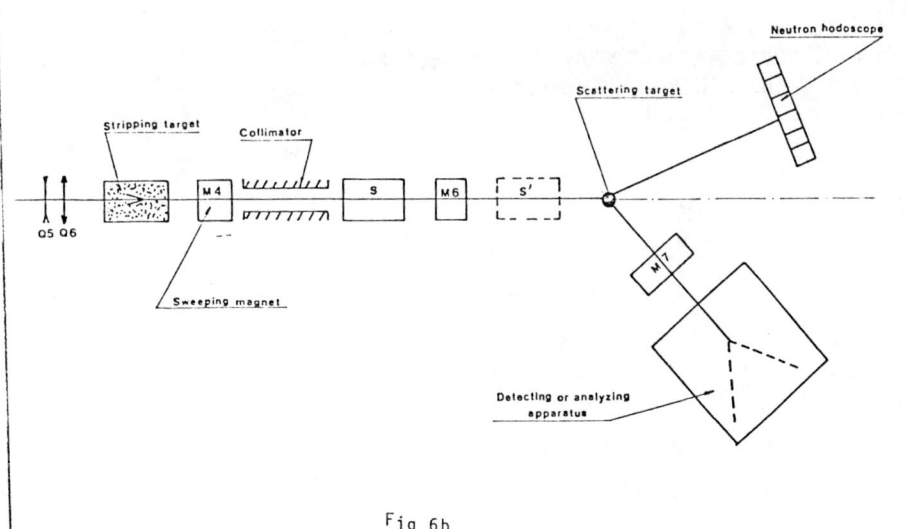

Fig 6b

particles and heavy ions. This project, called "MIMAS"* would be a small synchrotron with two essential properties :

c - Matching at best CRYEBIS → MIMAS, i.e. capturing a maximum number of particles from the source, accumulating 5 to 10 pulses.

c2 - Matching at best MIMAS → SATURNE, i.e. having a final six-dimensional phase space inside MIMAS exactly fitted to the six-dimensional space of Saturne.

In such a scheme, it should be possible to fill in at the limit of the space charge one of the three buckets circulating in Saturne. And get intense beams of heavy ions, and polarized particles. This is a long term project, which is not yet approved.

The situation at the end 79 will be anyhow very exciting at Saturne concerning polarization studies between 300 MeV and 3 GeV.

Saturne I is dead - Welcome to Saturne II !.

*MIMAS is the nearest satellite of the Saturne planet.
(Maximum Intensity Machine by Acceleration and Storage)

REFERENCES

1 - A. ABRAGAM - Principles of Nuclear Magnetism
2 - M. FROISSART, RSTORA NIM 7 (1960) 297
3 - T.O. NIINIKOSKI - Communication to this conference
4 - S. BREHIN et al - Communication to the GRATZ conference

# THE 200 nA VARIABLE ENERGY POLARIZED PROTON BEAM AT TRIUMF

J.L. Beveridge, G. Dutto, and P.W. Schmor
TRIUMF, Vancouver, B.C., Canada V6T 1W5

G. Roy
Dept. of Physics, University of Alberta, Edmonton, Alta., Canada T6G 2E1

## ABSTRACT

A Lamb-shift type polarized H⁻ source has been recently upgraded to produce a 1 µA beam current within a normalized emittance of about $0.3\pi$ mm-mrad. The polarization, enhanced through a diabatic field reversal, is approximately 80%. The 500 eV ions are accelerated and focused via a 5 kV gap lens to match the acceptance of a 300 kV acceleration tube. At 300 keV, at the entrance of a 45 m long electrostatic transport line, the spin is rotated via a Wien filter to compensate for the precession caused by the cyclotron's stray magnetic field along the line. Spin aberrations along the injection path, including the electrostatic spiral inflector, have been calculated to be less than a few degrees. A slight loss in polarization has been measured in the cyclotron between 200 MeV and 520 MeV; however, the polarization of the 200 nA extracted beam is greater than 70% at all energies. The extracted beam is delivered primarily to two beam lines in one experimental area. One is equipped with a ~500 keV magnetic spectrometer. The other, with a liquid deuterium target, can produce a 50-65% polarized neutron beam with a flux of the order of $10^6$ n/sec through a 5 cm diam collimator. An additional beam line is being installed to utilize the capability of extracting two beams simultaneously.

## INTRODUCTION

The layout of the TRIUMF facility is given in Fig. 1. The polarized source was installed adjacent to the high intensity H⁻ source

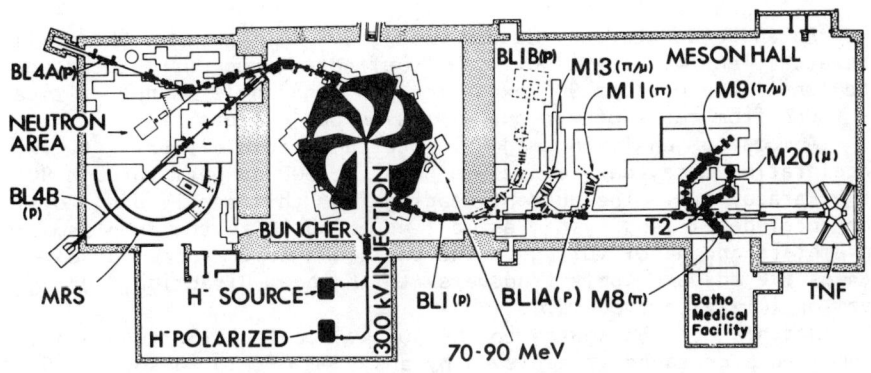

Fig. 1. Layout of the TRIUMF facility.

during the fall of 1975. By March 1976 a polarized beam had been
transported at ~300 keV along the 45 m long injection line, accelerated to 520 MeV in the cyclotron, and extracted. Experiments requiring polarized protons were located, in the proton hall, along both
beam line 4A (BL4A) and beam line 4B (BL4B) and, in the meson hall,
at the first target position in beam line 1A (BL1A). During 1976 the
polarized current extracted from the cyclotron varied from 10 to
30 nA. Higher intensities were required by experiments utilizing a
polarized neutron beam produced from a liquid deuterium target in
beam line 4A. A program to upgrade the source was begun in 1977 with
the result that by April 1978 a maximum current of 1 µA of polarized
H⁻ was produced by the source, with the required emittance, and
200 nA were extracted at 500 MeV.

## SOURCE

The polarized source, shown schematically in Fig. 2, is of the
Lamb-shift type[1] and uses the diabatic field reversal technique of
Sona[2] to enhance the polarization. Through the use of a close-
coupled geometry the overall length from the duoplasmatron to the
entrance of the 300 kV acceleration tube has been kept to 1.5 m.
Helmholtz coils, not shown, are used to reduce the cyclotron's fringe
field in the zero-cross region between solenoids 1 and 2. The spin
direction can be altered, in about 2 sec, by reversing the field in
all three solenoids simultaneously, either manually or automatically
through the cyclotron computer.

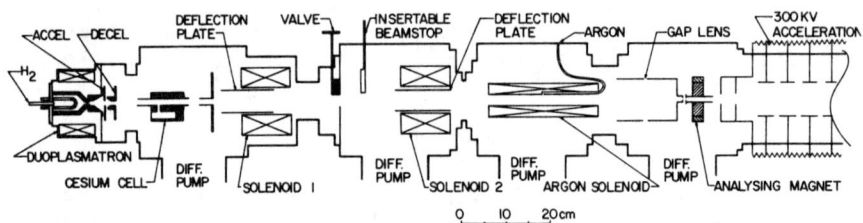

Fig. 2. Schematic diagram of the Lamb-shift source.

An anode aperture of approximately 0.4 mm minimizes the spill on
the accel lens without significantly affecting the current through
the lens. The present 90° expansion cup cone angle is not critical
as found from tests of various expansion cup designs.

A momentum analyzing magnet, located immediately prior to the
acceleration tube, can be used to bend the 500 eV beam through 90° onto a Faraday cup. The current reading from this cup is used to align
the accel-decel lens system as well as to optimize the duoplasmatron
parameters and accel voltage. The positions of the accel-decel
lenses are adjusted only transverse to the beam direction, during the
beam optimization process.

Matching of the source to the 300 kV acceleration tube and the
beam line acceptance is achieved by a system of two lenses. A 5 kV
gap lens focuses the H⁻ beam through a 3 mm aperture. This is followed by a second gap lens, at approximately 0.1 kV, which is used to
make small adjustments to the virtual position of the focus.

Alignment of the ground cylinder of the 5 kV lens is crucial. Mechanical alignment has not been adequate, and the final position has been determined using the beam. Horizontal and vertical steering prior to the acceleration tube is achieved by the analyzing magnet and by voltages applied to insulated plates on the pole faces.

The beam emittance has been measured, at the Wien filter location, 40 cm downstream of the acceleration tube. The vertical emittance, shown in Fig. 3, is $10\pi$ mm-mrad at 300 keV ($\sim 0.3\pi$ mm-mrad normalized). The theoretically expected emittance ellipse is also shown and indicates that the emittance orientation is much as expected. Since the system has axial symmetry, the horizontal emittance has not been measured and it has been assumed to be as calculated. The beam centroid changes slightly with spin direction; however, the beam line can be tuned to accept this larger effective emittance.

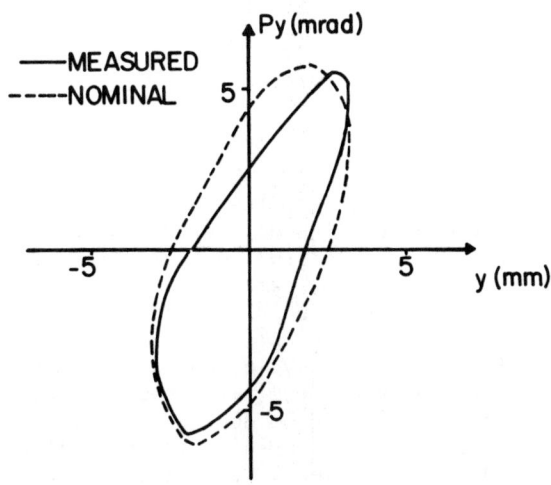

Fig. 3. The vertical emittance, as measured (solid curve), and as theoretically expected (dashed curve).

## INJECTION LINE

A 45 m long injection line, shown in Fig. 4, transports the 300 keV H⁻ beam from the source into the cyclotron. The beam enters the cyclotron axially and is bent into the median plane by a spiral inflector. The line contains 90 electrostatic quadrupoles, 66 electrostatic steering plates, a Wien filter, and three RF devices, namely, a beam buncher, a 1:5 selector and a beam chopper.

The Wien filter, which has a 1.1 cm aperture between the electrostatic plates and an overall length of 64 cm, is used to precess the spin of the H⁻ beam from the axial orientation at the source to an orientation which will be vertical at the end of the injection line. The three RF devices are phase locked to the 23 MHz RF accelerating system in the cyclotron. The beam buncher is capable of

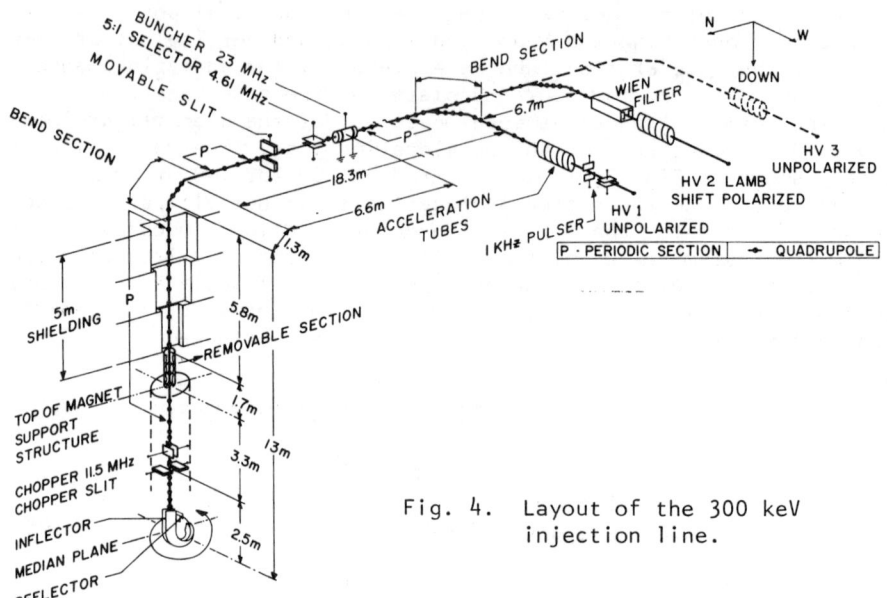

Fig. 4. Layout of the 300 keV injection line.

injecting 40% of the dc beam into the 45° wide cyclotron phase acceptance. The 1:5 selector eliminates four out of the five beam bunches being accelerated in the cyclotron, so that at extraction the time between beam bursts is 215 nsec rather than 43 nsec. The chopper can be used to reduce the width of each beam burst from approximately 4 to 1 nsec.

The fringe field from the cyclotron extends over the full length of the injection line varying from a few gauss to over 100 G in the horizontal section and from 100 to 3000 G along the vertical section. This field is essentially transverse to the beam line along the horizontal section and parallel along the vertical line. To minimize the effects of this field, mild steel cylindrical shields, 25 cm in diameter and 4 mm thick, were placed around the horizontal line wherever possible. Small ferrite dipoles were placed in these shields to compensate for the field in the unshielded regions. The total spin precession from the uncompensated stray magnetic field was calculated to be 63°, which compares well with the 73° inferred from the optimum settings for the Wien filter. Calculations indicate that the spin aberrations due to the uncompensated stray magnetic field is negligible along the horizontal line and that along the vertical line the depolarization due to aberrations should be less than 2%. It is estimated that the inflector could introduce spin aberrations of the order of 1°, which would produce negligible depolarization.

## CYCLOTRON

The properties of the TRIUMF H⁻ accelerator, a six-sector, strong focusing cyclotron, have been described in a number of recent papers.[3,4] Two or more proton beams may be simultaneously extracted, over an energy range variable from 183 to 520 MeV, by inserting

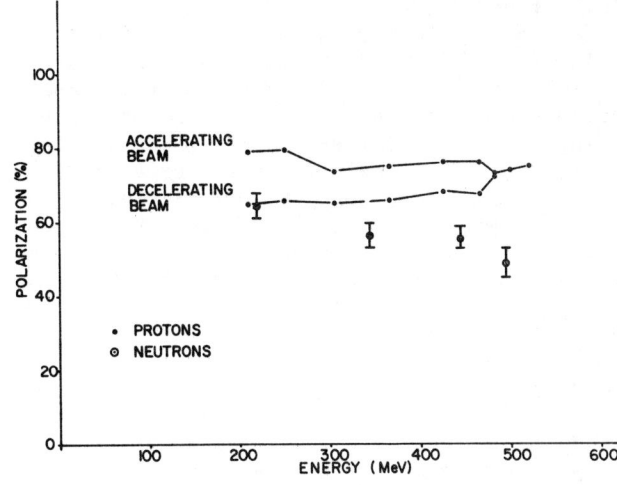

Fig. 5. The polarization of the proton beam (closed circles) and the beam line 4A neutron beam (open circles) as a function of energy.

0.025 mm thick carbon foils into the median plane of the cyclotron to strip the two electrons from the circulating H⁻ ions. The energy resolution of the extracted beam is normally about 1 MeV FWHM. With the use of internal slits this can be reduced to 0.5 MeV and is expected to reach 0.1 MeV with the advent of third harmonic RF flat-topping which will permit separated turn operation up to full energy. Approximately 25% of the current injected into the cyclotron can be accelerated to 520 MeV. Transmission through the injection line is typically about 80% (1/2 of the loss is due to gas stripping). Consequently, with 1 µA at the source it was possible to extract 200 nA at 500 MeV.

The polarization of the extracted beam has been measured as a function of energy. The H⁻ beam was accelerated up to 520 MeV and decelerated back towards the machine centre. A narrow foil was used to partially strip both the accelerating and the decelerating beams. The polarization of the two beams was measured and the results are shown by the closed circles in Fig. 5. There is a slight loss in polarization (~3%) in the region 460 to 480 MeV and (~6%) in the region 250 to 300 MeV. The reason for this loss in polarization is not understood and a program to investigate possible causes is planned.

## EXPERIMENTAL AREAS

Along beam line 4A (Fig. 1), a 10 cm long deuterium target can produce a monoenergetic polarized neutron beam with a neutron flux of approximately $2 \times 10^6$ neutron/sec through a 5 cm aperture for the 200 nA proton beam. A superconducting solenoid, capable of precessing the spin of 500 MeV protons through 270°, can be located in front of this target. The measured neutron polarization, from 50 to 65%, is shown at four neutron energies as open circles in Fig. 5. This polarized neutron beam has been used to do $\vec{n}p$ scattering in order to measure the Wolfenstein polarization transfer parameters.

Beam line 4B is a low intensity line designed for currents less than 100 nA. This line has two experimental stations. The first is a general purpose station with four movable arms for detection apparatus. The second station has a medium resolution (~0.5 MeV) magnetic spectrometer. Experiments using the polarized beam have examined elastic scattering on helium and deuterium, quasi-free scattering on calcium and oxygen, and inclusive scattering on helium. In all experiments the beam polarization is continuously monitored by polarimeters measuring the $\vec{p}p$ scattering asymmetry. These polarimeters have been calibrated to an accuracy of about 1% by two double scattering experiments, one of which involved polarization pumping from the $\vec{p}^4 He$ reaction.

An experiment, studying $\vec{p}\pi$ reactions on various targets using a 65 cm Browne-Buechner magnetic spectrograph to measure the angular dependence of spin-dependent effects, has until recently been located in beam line 1A. The increase in unpolarized proton beam intensity, with the associated increase in residual activity, has eliminated experimental stations using polarized beam along BL1A. This experiment, with an upgraded spectrograph, is moving to a low intensity ($\leqslant$10 nA) beam line 1B, which is being constructed so that two polarized proton experiments can, once again, run simultaneously.

## FUTURE IMPROVEMENTS

An improved beam bunching system is presently being considered which should be able to improve the overall beam transmission by about 20%. Considerable work has gone into matching the beam from the unpolarized source to the cyclotron with the result that the beam transmission from this source to extraction has exceeded 30%. With a comparable effort, a 50% increase in the overall transmission and extracted intensity can be expected for the polarized beam. New extraction ports have been installed in the cyclotron which can extend the variable energy range down to 65 MeV. Proton beams from these ports could be transported to the beam line 1B target position. In addition, preliminary studies have been made of a post-accelerator complex which could raise the energy of a beam from the TRIUMF cyclotron to a few GeV.

## REFERENCES

1. W. Haeberli, these proceedings.
2. P.G. Sona, Energ. Nucl. 14, 295 (1967).
3. J.R. Richardson, in Proc. 7th Int. Conference on Cyclotrons and Their Applications (Birkhäuser, Basel, 1975), p.41.
4. M.K. Craddock, E.W. Blackmore, G. Dutto, C.J. Kost, G.H. Mackenzie, and P.W. Schmor, IEEE Trans. NS-29(3), 1615 (1977).

# PROPOSAL FOR A SIBERIAN SNAKE CONFIGURATION

A. Turrin
INFN, Laboratori Nazionali, 00044 Frascati (Italy)

Abstract: One apparently efficient configuration for a Siberian Snake is found. It allows one to inject low-energy polarized beams and provides fully polarized beams even at energies that are far from its reference energy.

As it is well known, a technique for avoiding depolarization effects on particles that are being accelerated through cyclic accelerators has recently been suggested by Derbenev and Kondratenko[1]. This method to accelerate without depolarization provides continuous longitudinal polarization in the straight section ("privileged" straight section) diametrically opposite to a straight section where a suitable arrangement of magnetic fields is added to the usual ring lattice. Since application of this scheme involves severe changes in the geometry of the ring as well as new beam-dynamics problems connected with the resulting (unit) superperiodicity, the arrangement of magnetic fields doing the above-mentioned job was called[2] the "Siberian Snake".

Using 2-component spinor algebra, Montague[3] was the first to consider in detail the spin motion when a Siberian Snake is inserted in a ring. He discusses a "snake" configuration (represented in Fig. 1 (see Fig. 4(a) of Ref.[3])) composed of three transverse-field magnets. (We will adopt throughout the same notation as that used by Montague).

The results of Montague's calculation for this configuration can be summarized as follows:

i) The extrema for the half trace $\cos(\pi\nu)$ of the transfer matrix around one revolution are given as a function of the beam energy, E, by the equation

$$\left|\cos(\pi\nu)\right|_{\text{extr}} = \cos(\phi/2)\left[1 + \sin^2(\phi/2)\right]^{1/2}, \qquad (1)$$

where $\nu$ is the effective precession wave number; $\phi = \pi E/E_0$ is the precession angle around the y-direction (parallel to the momentum direction) suffered by the particle after the "snake" is passed through, and $E_0$ is the reference energy, i.e., the energy corresponding to a spin rotation $\phi$ equal to $\pi$, exactly.

ii) The minima for the longitudinal polarization in the privileged straight section are given as a function of E by the equation

$$\left|P_y\right|_{\min} = \sin^2(\phi/2). \qquad (2)$$

Then, by inspection of Eq. (1) he concludes that, with this "snake" configuration, serious depolarization due to the integer resonances (corresponding to $\cos(\pi\nu) = \pm 1$) cannot occur providing the injection energy be greater than $\approx 35\%$ of the reference energy, $E_0$.

From here downwards, we will show briefly that such a constraint on the magnitude of the injection energy becomes considerably weaker if use of the "snake" configuration represented in Fig. 2 is made. Moreover, we will show that the performance of such a "snake" configuration maintains its maximum value even for $\sim 0.6 < E/E_0 < \sim 1.4$.

For this "snake" configuration the half trace of the matrix around one revolution is given by

$$\cos(\pi\nu) = \cos^2(\phi/2)\cos(\chi/2), \quad (3)$$

where $\chi = 2\pi\gamma a$ is the free precession phase per revolution ($\gamma$ is the energy of the particle in units of its rest energy and a is the particle's magnetic moment anomaly).

The extrema of $|\cos(\pi\nu)|$, are

$$|\cos(\pi\nu)|_{extr} = \cos^2(\phi/2). \quad (3a)$$

The expression for the longitudinal polarization, $|P_y|$, in the privileged straight section can be constructed by using the (unit) normed eigenvectors. We obtain

$$|P_y| = \sin(\phi/2)\left[1+\cos^2(\phi/2)\right]^{1/2}\left[1-\cos^2(\chi/2)\cos^4(\phi/2)\right]^{-1/2}. \quad (4)$$

The minima of $|P_y|$ are

$$|P_y|_{min} = \sin(\phi/2)\left[1+\cos^2(\phi/2)\right]^{1/2}. \quad (4a)$$

Our results readily reveal that no complications due to the $\cos(\pi\nu) = \pm 1$ resonances are predicted, even for polarized beams that are injected at energies $\approx 15\%$ of the reference energy, $E_o$. Finally, it is worth drawing attention to the flat maximum of $|P_y|_{min}$ in the vicinity of $E = E_o$.

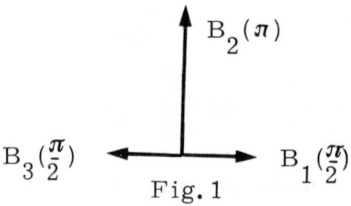

Fig. 1

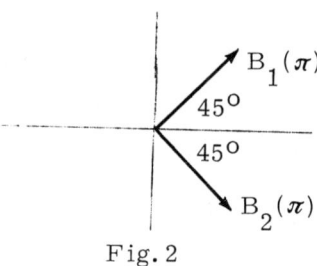

Fig. 2

### REFERENCES

1. Ya. S. Derbenev and A. M. Kondratenko, Proceedings 10-th International Conference on High-Energy Accelerators, Protvino, Moscow, July 1977, Vol. 2, p. 70.
2. E. D. Courant and L. G. Ratner, Workshop on Possibilities for Higher Energy Polarized Proton Beams, Ann Arbor, 1977, Edited by A. D. Krisch and A. J. Salthouse (AIP Conference Proceedings No. 42, 1978), p. 41.
3. B. W. Montague, Polarized $e^+$-$e^-$ in LEP with the Siberian Snake, LEP-70/76 (1978).

# THE CRYOGENIC SOURCE "CRYPOL 2" OF POLARIZED HYDROGEN AND DEUTERIUM ATOMS

A. A. Belushkina, V. P. Ershov, V. V. Fimushkin, G. I. Gaj,
L. S. Kotova, Yu. K. Pilipenko, V. V. Smelyansky, A. I. Valevich
High Energy Laboratory, Joint Institute for Nuclear Research
Dubna, USSR

## I. INTRODUCTION

The development of pulsed sources of polarized atoms and ions is being performed at the Joint Institute for Nuclear Research, Dubna. A source of atoms can be used as a jet polarized target to operate in a main accelerator ring*[1,2], and a source of ions can be used to obtain an accelerated polarized beam of high energy deutrons.

At present ground state atomic-beam sources have the greatest intensity[3]. The beam intensity of such sources is largely dependent on the magnetic field of a sextupole separation magnet and the pumping speed of a vacuum system. At the synchrophasotron a source of polarized ions should be installed on a special column with a voltage of 750 kV. This gives rise to strict limits on the space dimensions of the source and its power used. In this case it is convenient to use a cryogenic version of a source with superconducting magnets and cryopumps. Such a source is compact and requires low power. The cryopumps and superconducting magnets should be cooled by liquid helium.

The experimental sources of hydrogen and deuterium atoms with polarization by electron spin ("Crypol"[4] and "Crypol 2") have been developed to study their possibilities.

## II. DESIGN AND THE PRINCIPLE OF OPERATION OF "CRYPOL 2"

### 1. Cryostat and Vacuum Pumps

The general view of the source is shown in Fig.1. A source chamber (4), a skimmer chamber (5) and a sextupole superconducting magnet (10) are placed inside a liquid helium cryostat (3). Part of the walls of the chambers and the sextupole magnet channel is washed by liquid helium and serves as cryopanels to pump out the gas. Their areas are equal to 1020 cm$^2$, 110 cm$^2$ and 120 cm$^2$, respectively. It is known that using cryopumps one can easily get large pumping speeds in a broad range of pressures. For example, the specific speed of pumping out the hydrogen can achieve 20-30 $\ell/s \cdot cm^2$ at a surface temperature of 4.2°K. In this case the pressure of saturated hydrogen vapours is about $10^{-6}$ torr and for deuterium $10^{-10}$ torr (Fig.7).[5]

In order to reduce the evaporation of liquid helium, the cryostat has a shield cooled by liquid nitrogen. During cooling and filling the cryostat with liquid helium, the source was pumped out by means of a 500 l/s diffusion pump.

---
\* The idea of development of such a target was proposed by Yu. K. Pilipenko when carrying out the joint p-p experiment at the FNAL accelerator in 1972.

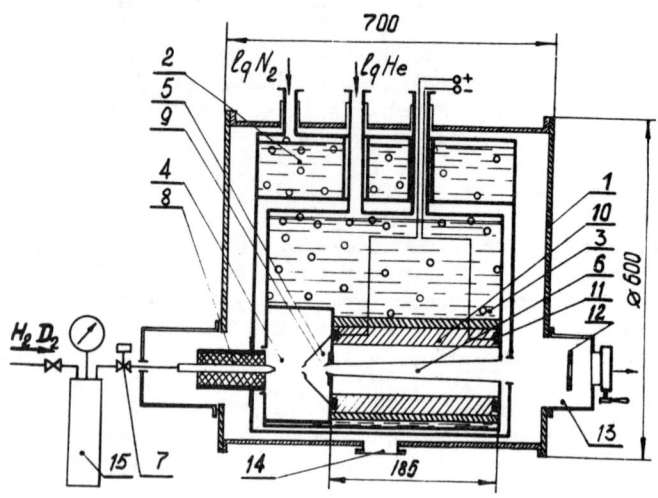

Fig.1. General view of "Crypol 2".

1. Vacuum jacket.
2. Nitrogen shield.
3. Helium cryostat.
4. Source cryochamber.
5. Skimmer cryochamber
6. Channel of the sextupole magnet.
7. Gas supply valve.
8. Dissociator.
9. Collimator ⌀ 10 mm.
10. Pole.
11. Coil.
12. Iris.
13. R.f. transition section
14. Diffusion pump connections.
15. Gas container.

## 2. Dissociator and System of Atomic Beam Formation

Two types of dissociator have been tested in "Cryopol 2": a U-shaped dissociator 130 mm long and 10 mm in diameter and a cylindrical one ⌀ 15 mm, l = 105 mm (Fig.2). The dissociators were made of pyrex glass. The glass pipes were rigidly mounted in the teflon bodies. The dissociators had a thermocontact with a nitrogen shield. This permits the dissociators to be kept cold. For better cooling, the glass end of the nozzle was wrapped in indium foil and was wound with copper wire. The foil was connected to the nitrogen shield.

Power to the U-shaped dissociator was supplied from the secondary circuit of the r.f. generator. In the case of the cylindrical dissociator, r.f. power was supplied by means of a 145 MHz resonator (6). In the operating conditions 200-250 W of the r.f. power were released at a pulse duration of 75 ms and a frequency of 0.15 Hz. According to the measured ion current of the polarized beam, the efficiency of the U-shaped dissociator was larger by 30-50%.

Deuterium or hydrogen passes pulsewise from a vessle (15)(Fig.1) through an electromagnetic valve (7) into the dissociator. The gas supply line was 3.5 mm in diameter and ~50 cm in length. The gas pressure in the vessel was measured by a precise differential gauge

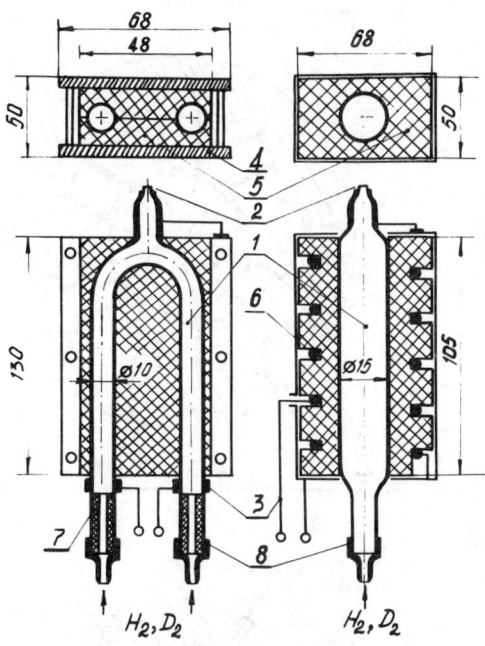

Fig.2. U-shaped and cylindrical dissociators.

1. Glass pipe (pyrex).
2. Nozzle Ø 3 mm.
3. R.f. power supply.
4. Supporting plate.
5. Teflon body.
6. Resonator.
7. Glass pipe.
8. Teflon sleeve.

and was usually equal to 5 torr.

The atomic beam formation was performed by a nozzle 3 mm in diameter, a skimmer Ø 4 mm and a collimator at the input of a magnet Ø 10 mm. The distance between the nozzle and the skimmer was 8 mm, between the skimmer and the collimator 50 mm.

3. Superconducting Sextupole Magnet

The following requirements are imposed upon a separation magnet of the source: a high gradient of the magnetic field and large channels between the poles for better gas pumping.

Two versions of sextupole magnet construction have been developed (Fig.3a,b). A magnet of version a) is used in "Crypol 2". A stainless steel pipe Ø 50 mm of the cryostat (5) is placed between the poles (2) and the pole tips (3). Thus, the magnet coils are in liquid helium and the pole tips in vacuum[6]. The part of the internal surface of the separation pipe (5), free of the pole tips, serves as a cryopump. The yoke and the poles of the magnet are made

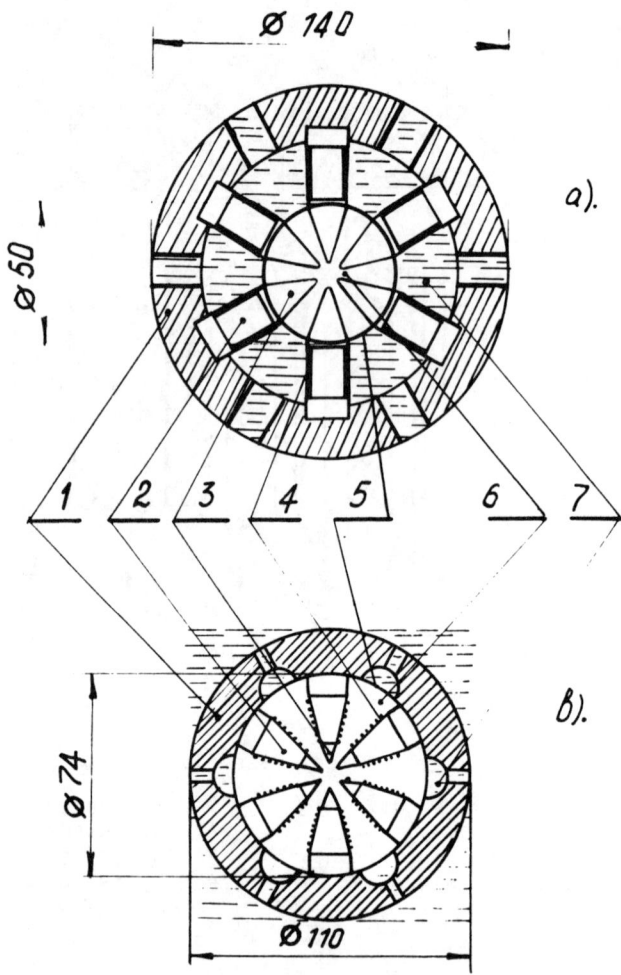

Fig. 3. Superconducting sextupole magnets:

a) magnet coils in liquid helium,
b) magnet coils in vacuum.

1. Yoke.
2. Pole.
3. Pole tip.
4. Coil.
5. Cryostat separation pipe.
6. Magnet vacuum channel.
7. Liquid helium.

of steel and the pole tips of 49K2V permendur. The coils are made
of NbTi superconducting cable 0.5 mm in diameter. Each coil has 55
turns. The magnet aperture is a taper: ∅ 10 mm at the beam entrance
and ∅ 14 mm at its exit. The magnetic field on the pole tips versus
current is shown in Fig.4. The magnet operates in a persistent current state.

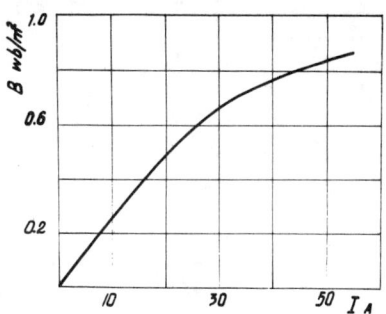

Fig.4. Magnetic field vs. the current of the sextupole magnet of "Crypol 2".

Version b) is a further step in the development of sextupole
magnets. In contrast to the previous case, the magnet coils in
version b) are placed in vacuum and cooled due to mechanical
contact of the poles with the cryostat and their thermoconductivity.
The yoke is made of Armco steel, and the pole tips of permendur.
The one-layer coils of the magnet are made from NbTi superconducting
fine filament conductor 0.5 mm in diameter. Each coil has 28 turns.
The diameter of the magnet aperture is 7 mm at the entrance and 10 mm
at the exit. The length of the poles is 135 mm. 35 mm of the length
is a taper; the rest is cylindrical. The magnet operated at a current of 30 A. The magnet quench occured at a current of 40-45 A due
to insufficient cooling of the coils. To test the magnet, a special
experimental setup was used. A 9 ÷ 10 mm image of the atomic deuterium beam was observed on a $MoO_3$ plate installed at a distance of
80 mm from the exit. Estimates show that the field with a high
gradient can be obtained using such a magnet construction. A disadvantage of this type of magnet is that it is difficult to have a
good cooling of the coils.

### III. IONIZER

To detect the atomic beam in "Crypol 2", a strong field ionizer,
based on the concept of Glavish[7], has been used. The internal diameter of the electron optics was 14 mm and of the ion one 19 mm.
A hollow filament was a tantalum ribbon 30 mm long, 1.5 mm wide by
0.1 mm thickness. Figure 5 shows the optical structure of the ionizer and the potentials of the electrodes. The magnetic field at
the solenoid center is 1.6 kG at a current of 250 A. The solenoid
length is 160 mm. The distance from the exit of the sextupole magnet
to the filament is 405 mm.

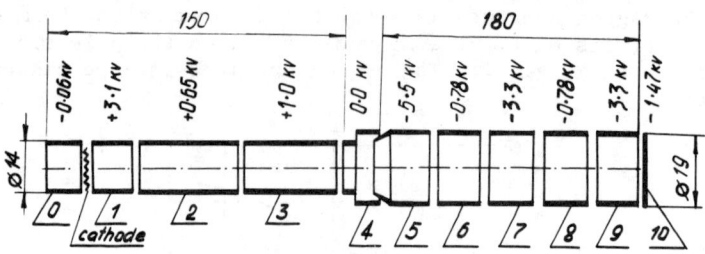

Fig.5. Optic structure of the ionizer.

The ionizer was synchronized with the source and operated in a pulsed mode with a pulse duration of 100–300 ms and a frequency of 0.15 Hz. Each pulse the ionizer was turned on by grounding the filament and by supplying a positive potential to anode 1. The filament emmision was 10–12 mA.

Vacuum in the ionizer was provided by means of the "Crypol 2" cryopumps. The ionizer was pumped out on both sides: through a vacuum valve ∅ 40 mm (filament side) and through a pipeline ∅ 50 mm (ion optics side). For preliminary pumping the ionizer and during its start-up, when the filament and all the optics were intensively degassed, a 100 1/s diffusion pump was used.

## 5. Experimental Results

The source has been tested using a beam of hydrogen and deuterium atoms. When "Crypol 2" operated with a deuterium beam, the vacuum in the source was $1 \cdot 10^{-7}$ torr and in the ionizer: $1.5 \cdot 10^{-7}$ torr in the vicinity of the filament and $2.5 \cdot 10^{-7}$ torr around the ion optics.

The pressure of deuterium before an electromagnetic valve was usually 5 torr; the duration of a gas supply pulse to the dissociator was 45 ms. The gas discharge per cycle was 50 torr. After each gas pulse from the nozzle, the pressure inside the ionizer increased by $8 \cdot 10^{-8}$ torr and then reduced to the initial value.

Figure 6 presents a picture of the pulses detected by an ion current collector (10). Curve 1 is the background ion current which is due to ionization of the residual gas.

The ion current of the beam, polarized by electron spin, was determined by the difference of the amplitudes on curves 3 (the dissociator and sextupole magnet are on) and 2(r.f. power to the dissociator is off). It was equal to 15 – 17 µA at a maximum. A rise in the deuterium pressure before the valve up to 6 torr and above resulted in increasing absolute values of the signals 3, 2. But at first the difference between them remained constant and then began to reduce. Probably, this showed evidence for decrease in dissociation. Estimates show that the dissociation was about 50%. The intensity of the atomic and molecular beams sharply dropped with decreasing the gas pressure below 3 torr.

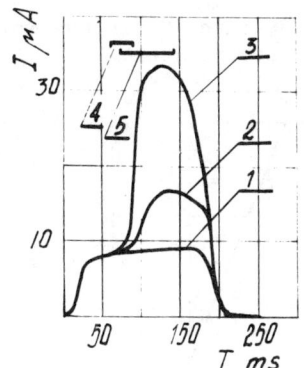

Fig.6. Ion current signals detected on the ionizer collector (10).
1. Background ion current due to the residual gas ionization.
2. Ionization current of the molecular beam (the r.f. generator of the dissociator is off.)
3. Ion current when the dissociator and the sextupole magnet are on.
4,5. Valve and r.f. generator pulses.

The ion current of the molecular beam (2) practically corresponds to the process of increasing the gas pressure before the nozzle.

The delay of the beam signal relative to the gas pulse is due to the process of gas propagation through the supply line when filling the dissociator.

As would be expected (Fig.7), when "Crypol 2" operated with the hydrogen beam, vacuum worsened to $3 \cdot 10^{-6}$ torr in the source and to $8 \cdot 10^{-7}$ torr in the ionizer while 100 l/s auxiliary diffusion pump was running. In this case the background ion current was increased to 30-40 μA making the determination of the polarized beam ion current difficult.

For normal running it is necessary to improve the vacuum in the ionizer: additional vacuum pumps should be installed and a parasitic gas flow from the source should be reduced.

The atomic beam intensity is to a great extent dependent on the vacuum inside the source. Therefore it is important to know that the speed of the cryopumps is sufficient to provide good conditions for atomic beam formation.

It is difficult to measure the pressure in the cryochambers of the nozzle, of the skimmer and in the sextupole magnet channel during the gas pulse. It is evident that at this time the pressure in the chambers increases and depends on the amount and temperature of input gas, the chamber size and the speed of gas condensation on the cryopanel.

The specific speed of gas pumping by the cryopanel is equal to

$$S_g = 3.67c\ (T_g/M)^{1/2}\ 1/s \cdot cm^2,$$

where $T_g$ is the gas temperature, M is the molecule weight, and c is the molecule capture coefficient.

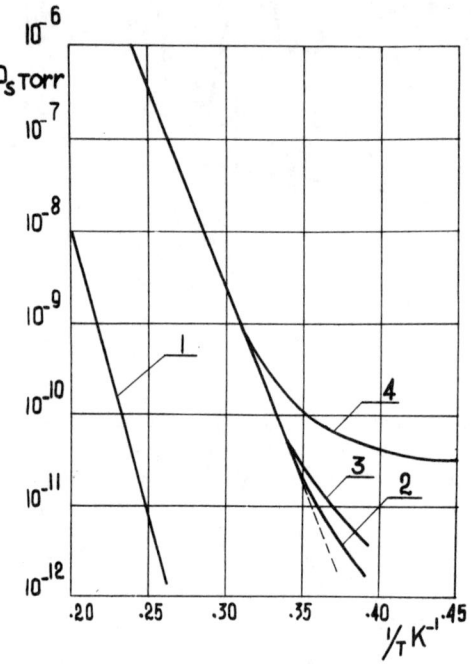

Fig.7. Vacuum pressure over the layer of solid gas[5].
1. $1.08 \cdot 10^{24}$ mol. $D_2/m^2$, the temperature of condensed gas $T_g$ is 290°K;
2,3. $7 \cdot 10^{20}$ mol. $H_2/m^2$ on copper and argon, $T_g = 77°K$;
4. $19 \cdot 10^{20}$ mol. $H_2/m^2$ on copper, $T_g = 400°K$.

If the pressure of condensed gas $P_g$ near the cryopanel is much larger than the equilibrium vapour pressure $P_s$ and the gas temperature $t_g \lesssim 300°K$, the gas evaporation from the cryopanel is slow. In this case the capture coefficient for hydrogen and deuterium $c \geq 0.5$.[8]

Despite the fact that $P_s$ for deuterium equals $\sim 10^{-10}$ torr, the pressure measured between the vacuum jacket and the nitrogen shield was $3 - 5 \cdot 10^{-8}$ torr. This was observed when the cryostat was completely filled with liquid helium, and the gas cycle was off. The cryostat design was not intended for super-high vacuum. Its seals were made using rubber gaskets. The nitrogen shield was insulated with 10 layers of aluminized mylar which is a source of gassings. The cryostat was not baked. Very often vacuum non-uniformities can be observed in cryogenic setups. One can assume that when the gas cycle is off, the vacuum inside the closed cryochambers is better than the measured one.

In order to reduce the vapour pressure $P_s$, liquid helium in the cryostat was pumped out to 25 torr which corresponded to the temperature of the cryopanels ~2.0 K and the pressure $P_s \lesssim 5 \cdot 10^{-11}$ torr for $H_2$ and $10^{-12}$ torr for $D_2$. Thus, the vapour pressure was reduced by several orders of magnitude.

The cryostat temperature decreased while the source and ionizer were running. The ion current of the polarized beam D was observed at the same time. The ion current was not changed, and the vacuum in the source was improved from $1 \cdot 10^{-7}$ torr to $6 \cdot 10^{-8}$ torr, i.e. it achieved the smallest pressure observed in the absence of pumping out and the gas cycle. As before, the vacuum drop during a pulse was about $8 \cdot 10^{-8}$ torr.

After the source had been switched from deuterium to hydrogen, the vacuum was not changed. Ion current signals looked approximately like those for deuterium. The ion current was reduced by 30%.

We have every reason to assume that in the case of deuterium vacuum conditions for the atomic beam are good and the speed of gas condensation is large enough. This is the reason why the decrease in the cryopanel temperature did not affect the atomic beam intensity.

For hydrogen at a cryostat temperature of $4.2°K$ this question is not so obvious and requires an additional study.

In order to measure the beam size, an iris was installed at a distance of 195 mm from the exit of the sextupole magnet. Figure 8 shows the values of ion current for the atomic and molecular beams of deuterium at different openings of the iris. It is seen that the ion current of the atomic beam fastly increases up to an opening 15 mm in diameter and then weakly changes repeating the shape of the curve for the molecular beam. The diameter of the polarized beam is likely not to be larger than 15 mm. A slow rise of the signal with subsequent increasing the iris opening is due to gas compression in the vicinity of the filament.

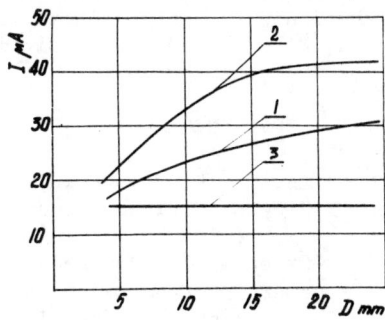

Fig.8. Ion current vs. the iris opening.
1. Molecular beam (the r.f. generator of the dissociator is off).
2. Atomic beam (the dissociator and sextupole magnet are on).
3. Background.

To increase the atomic beam intensity, comparative experiments have been carried out at various configurations of its formation system.

The molecule background was significantly reduced by installing a skimmer with a corresponding cryopump chamber. This permitted one to give up the mass-spectrometer with a bending magnet used previously and to detect the polarized beam ion current on an ionizer collector.

The increase of the collimator diameter from 8 to 10 mm at the input of the sextupole magnet and the increase of the opening diameter to 14 mm, when the collimator was removed, had no influence on the intensity of the atomic and molecular beams.

The change of the nozzle diameter from 3 mm to 4.5 mm and the separator one from 3.5 mm to 6 mm lead to the ion current decrease.

## 6. Conclusion and Further Development

"Crypol 2" has been tested for a long period of time. As usual, the duration of runs was 2-3 weeks without deuterium sublimation. The operation with the beam took about 8 hours a day. The cryostat was filled with liquid helium 3 times a week. For this purpose 100-110 $\ell$ of helium were required.

The source operation was stable. When the level of liquid helium was substantially reduced, the vacuum drop could be sometimes observed during several seconds. This is likely to be due to the sublimation of solid deuterium from the cryopanel.

At present one more setup, consisting of a cryogenic source and a cryogenic ionizer, is being constructed. The source differs from "Crypol 2" in sextupole magnet dimensions ($\ell$ = 300 mm). The ionizer outwardly is similar to the source. A cold bore of the ionizer superconducting solenoid, including a nitrogen shield, has a diameter of 50 mm and a length of 400 mm. The solenoid has been designed for a magnetic field of 6 T.

The authors are grateful to V. L. Stepanyuk and V. P. Vadeev for the development of the r.f. generator and modulator, V. I. Ilyuschenko for the development of the ionizer at the initial stage of operation, A. G. Zel'dovich and Yu. A. Plis for their useful advices and assistance.

## REFERENCES

1. L. S. Zolin, V. A. Nikitin, Yu. K. Pilipenko. "Forming and Trapping of a Gaseous Hydrogen Jet in a Vacuum," Cryogenics, June, 143, 1968.
2. V. D. Bartenev et al., "Cryopumped, condensed hydrogen jet target for the National Accelerator Laboratory main accelerator", Advances in Cryogenic Engineering, 18, 460, 1973.
3. E. F. Parker, "A review of the ZGS polarized ion source development activity", High Energy Physics with Polarized Beams and Targets, 382, AIP, NY, 1976.
4. A. A. Belushkina, V. P. Vadeev, A. I. Valevich et al., "Cryogenic source of polarized hydrogen and deuterium atoms", Pribory i tekhnika experimenta, 6, 31, 1976.

5. T. J. Lee, "The condensation of $H_2$ and $D_2$: Astrophysics and Vacuum Technology", J. Vac. Sci. Tech., 9, 1, 257 (1972).

6. C. Ekström, M. Olsmats and B. Wannberg, "Construction of a focussing atomic beam magnet resonance apparatus", Nucl. Instr. Meth., 103, 13 (1972).

7. H. F. Glavish, "A strong field ionizer for an atomic beam polarized ion source", Nucl. Instr. Meth., 65, 1 (1968).

## NN SCATTERING BELOW 1 GEV

D. V. Bugg
Queen Mary College, Mile End Rd, London E1 4NS

### ABSTRACT

The NN experimental programmes in progress at TRIUMF and SIN are reviewed, and related to measurements at the Argonne. From np measurements of P, $D_t$, $R_t$, and $A_t$ at TRIUMF, a unique and accurate set of I=0 phase shifts is now available up to 500 MeV. These show large systematic deviations from theoretical predictions in central, spin-orbit and tensor combinations, and a major revision of coupling constants appears to be required. Real parts of I=1 phase shifts are defined accurately up to 515 MeV by elastic pp data. It is desirable to take inelasticities from models of the inelastic channels based on NN-N$\Delta$ via $\pi$ and $\rho$ exchange. However, at 425 and 515 MeV this leads to a serious conflict with Argonne measurements of $\Delta\sigma_L$. Either these data or the model of the inelastic channels must be wrong.

### THE BASQUE PROGRAMME AT TRIUMF

The BASQUE group is a collaboration from Bedford College London, the University of Surrey, Queen Mary College London, UCLA, and the Universities of British Columbia and Victoria. The main objective of this group has been a determination of I=0 phase shifts up to 500 MeV by measuring Wolfenstein parameters in free np elastic scattering. In these experiments, the polarisation transfer from a polarised neutron beam to final state protons has been analysed using a carbon polarimeter. The steps in this programme will be reviewed taking a logical progression rather than the strictly historical order.

The first step was to calibrate the polarimeter using protons of known polarisation. Polarised protons were obtained by scattering an unpolarised beam elastically from liquid hydrogen at 24° lab. The polarisation of these protons was determined with an absolute accuracy of ±1.5% in a double scattering experiment illustrated schematically on Fig. 1[1]. Scattered protons were focussed at a $CH_2$ target viewed by telescopes selecting elastic scatters from hydrogen at 24°. The small background from carbon was measured with an equivalent C target. A solenoid precessed the spin of the secondary proton beam to the horizontal, so that the second scatter was in the vertical plane. Instrumental errors in the polarimeter were eliminated by reversing the solenoid periodically. Results of this experiment and a similar one by a QMC-Alberta-UBC collaboration[2] are shown in Fig. 2. The importance of these measurements is that they provide the absolute normalisation of all polarisation phenomena over the energy range 200 to 500 MeV. A measurement at 430 MeV at LAMPF[3] agrees to the third decimal place with the fitted curve. The small inflection at 450 MeV is physically reasonable, since this is the energy at which a diffraction peak develops, due to the onset of inelastic scattering.

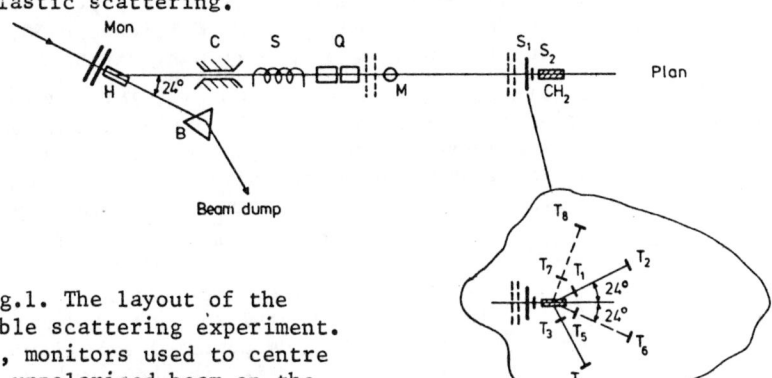

Fig.1. The layout of the double scattering experiment. Mon, monitors used to centre the unpolarised beam on the liquid hydrogen target H; C, collimator; S, superconducting solenoid; Q, quadrupoles; M, vertical steering magnet. Proportional chambers are shown dashed. $CH_2$ is the polythene target used for the second scatter. $S_1 S_2$ are the scintillators defining the secondary beam striking $CH_2$. $T_{1-8}$ are scintillation counters defining the second scatter.

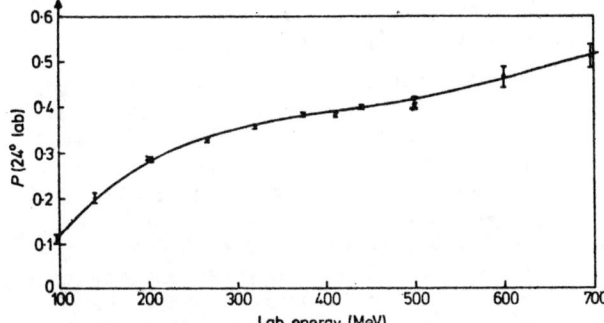

Fig. 2. Values of P(24° lab) for pp elastic scattering, and an empirical fit.

Fig. 3. The experimental layout for pp work. The polarimeter T monitors the primary beam polarisation. SS, superconducting solenoid; $LH_2$, liquid hydrogen target. $S_0$-$S_4$ are scintillation counters, and V a veto counter eliminating particles not scattered in carbon C.
Multiwire proportional chambers are shown as broken lines.

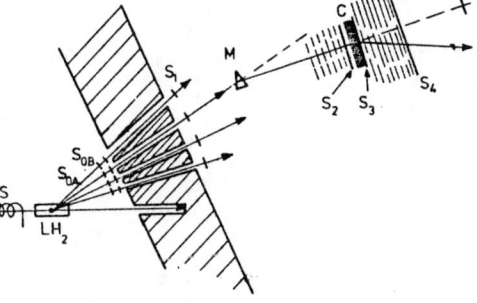

The polarimeter[4] is illustrated schematically at the right of Fig. 3. It consists of a block of carbon, 6 or 3 cm thick and 50 cm square, sandwiched between six multi-wire proportional chambers 50 x 50 cm$^2$ and another six at the back 100 x 100 cm$^2$. This array is mounted with trigger scintillators on a moveable carriage. Scatters in the range 4 to 30° are accepted without any attempt to select elastic events. The analysing power has been calibrated in the set-up of Fig. 1 at 14 energies, and is shown on Fig. 4. The accuracy of the calibration is ±2%, except near the extreme energies. When 6 cm carbon is used, 1 proton in 14 suffers a useful scatter, and the mean analysing power, averaged over scattering angle, is about 35%.

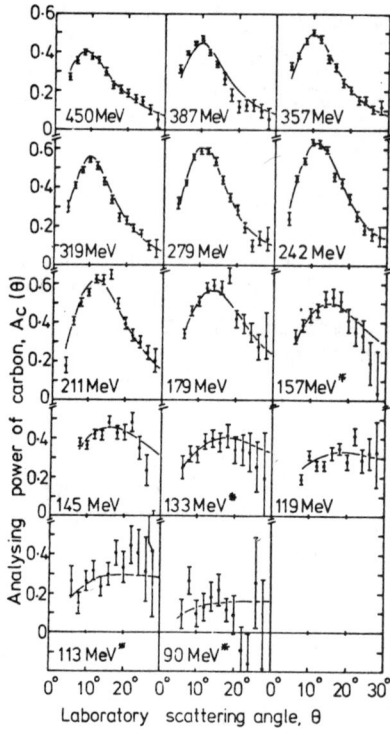

Fig. 4. The analysing power of the carbon polarimeter for protons. Energies are given at the centre of the carbon.

The first physics measurement[5] was of P, D, R and R' for pp elastic scattering from 210 to 515 MeV in the lab angular range 6 to 24°, where previous data were sparse. The set-up is shown on Fig. 3. This was a very quick measurement completed in less than 5 days of running. The objective was to improve I=1 phase shifts by determining the $\pi^0$pp coupling constant $g_0^2$ accurately, and fixing high partial waves. Previously, uncertainties in high partial waves spilled over into increased errors in low partial waves. some results are shown in Fig. 5 for the parameters R ≡ (S,0;S,0) and D ≡ (N,0;N,0), in the notation (beam, target; scattered, recoil) for measured spins. The curves are phase shift fits. Let me at this stage draw attention to the slightly untidy variation with energy of the fits near 90°, indicating that further data in the range 60 to 120° are still desirable.

The D parameter in the range 5 to 20° is very sensitive to $g_0^2$, and yields the value $g_0^2$ = 14.10 + 0.65; the error is intended to cover both statistical and systematic errors. This result agrees well with the value 14.27 ± 0.18 deduced in $\pi$N scattering for the exchange of charged pions.

These data and SIN data described below have been included in a phase shift analysis, and results for high partial waves are shown on Fig. 6. They are compared with the predictions of the Paris[7] group. Closely similar predictions have been made by several other

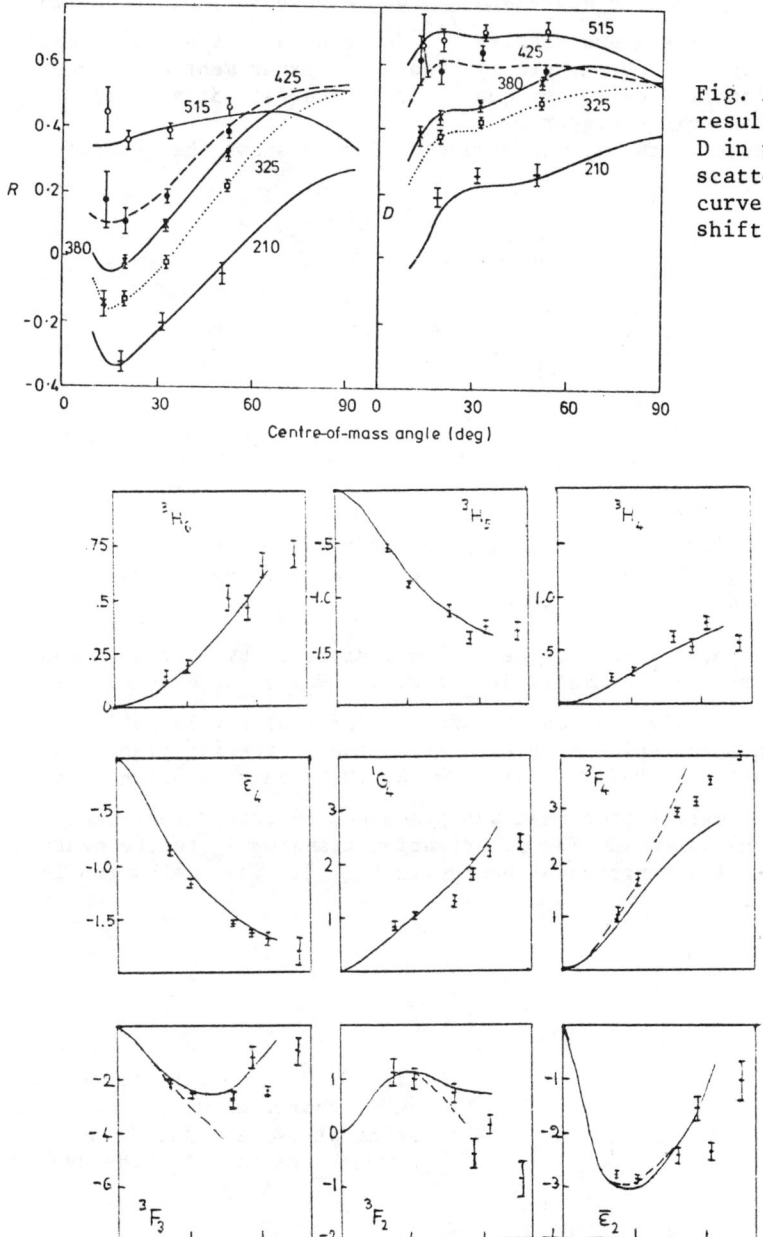

Fig. 5. BASQUE results for R and D in pp elastic scattering. The curves show phase shift fits.

Fig. 6. I=1 phase shifts for high partial waves. The full curves are predictions of the Paris group from the long-range forces ( 0.8 fm); dashed curves include the effects of phenomenological short-range potentials.

authors, notably Chemtob and Riska[8], Brown and Durso[9], Bohannon and Signell[10], and Epstein and McKellar[11]. One can see a small systematic disagreement in 1G4, but generally theory and experiment agree very well, giving confidence in our quantitative understanding of pp phase shifts in this energy range.

Next I turn to the np measurements. Fig. 7 shows the general

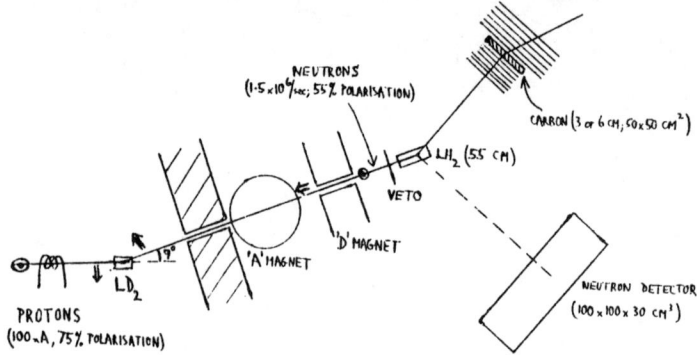

Fig. 7. The experimental layout for measurement of Wolfenstein parameters in np elastic scattering. Arrows indicate spin orientations.

layout. The polarised neutron beam[12] was produced by charge exchange of polarised protons in liquid deuterium at 9° lab. As we shall see in a moment, the polarisation transfer is particularly large[13] at this angle when the spin is orientated in the scattering plane and transverse to the beam (i.e. the $R_t \equiv (S,0;0,S)$ configuration). The spin of the primary proton beam was precessed to this direction by a superconducting solenoid. The polarisation transfer $R_t$ to the neutron beam is shown as a function of energy in Fig. 8. The mechanism is

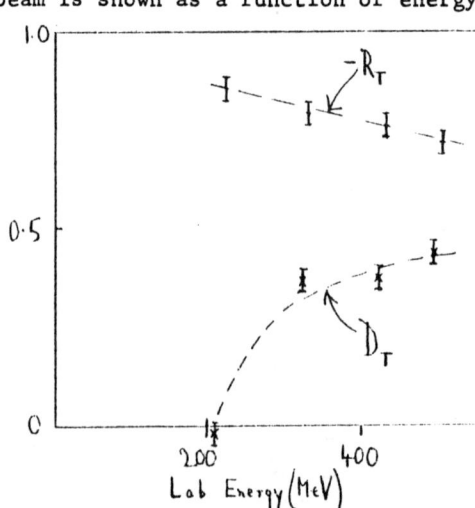

Fig. 8. Polarisation transfers in np charge exchange scattering at 9° lab. The dashed curves are to guide the eye.

essentially single pion exchange, which induces a spin-flip between
the initial proton and final neutron. Two magnets were used to precess the neutron spin to any desired direction and sweep charged
particles from the beam. The first precessed the spin in the
horizontal plane to either the transverse (S) or longitudinal (L)
direction; in the latter case, the second magnet could be used to
precess the spin to the vertical (N). Under normal running conditions
the proton beam had an intensity of 100 nA, an energy spread $\leqslant 1$ MeV,
and a polarisation of 75-80%. This polarisation was continuously
monitored by a polarimeter, as in Fig. 3. The energy spectrum of the
neutron beam had a sharp spike about 10 MeV wide near full energy and
a low energy tail starting 150 MeV lower; this tail was easily
removed by time of flight with respect to the cyclotron RF. The
neutron beam was collimated to a diameter of 9 cm at 12 m from the
production target, and then had an intensity of about $1.5 \times 10^6$/sec
and a polarisation of about 58%.

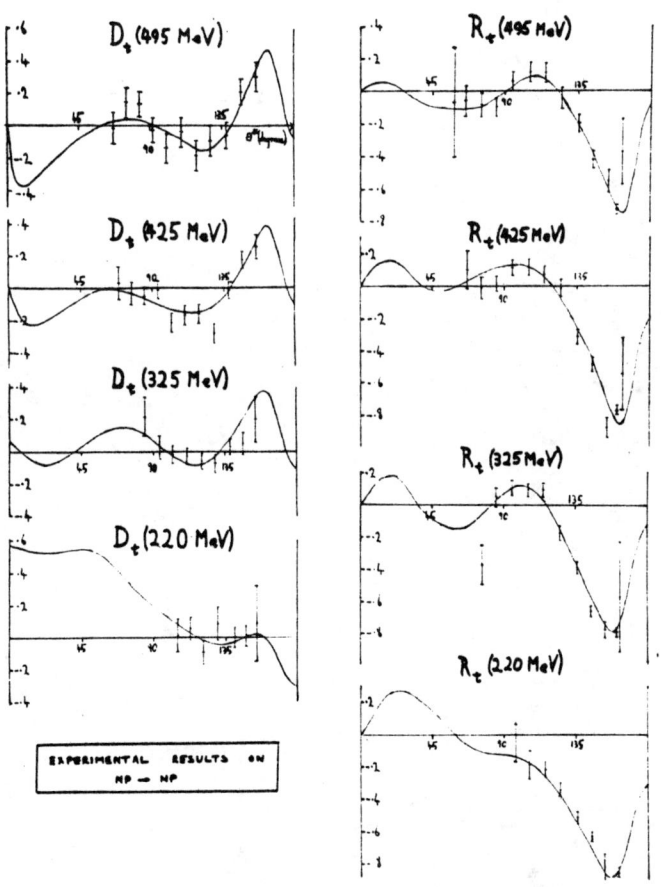

Fig. 9. BASQUE results for $D_t$ and $R_t$ in np elastic scattering.

Neutrons were scattered from a 55 cm long liquid hydrogen target and coincidences were recorded between the polarimeter and a large neutron detector consisting of 14 scintillators each 15 x 15 cm$^2$ and 100 cm wide. Time of flight was recorded to photomultipliers at each end of the scintillators, hence the location of the neutron interaction could be determined ±5 cm horizontally x ±7.5 cm vertically. This geometry cleanly separated elastic events in the hydrogen target from inelastic events. The background from the empty target was typically 2%.

The left- right asymmetry in scattering from liquid hydrogen determined the polarisation parameter $P \equiv (N,0;0,0)$ with very high statistics; the accuracy of the result was in fact limited to ±1.5% by monitor stability. The polarisation of the recoil proton determined $D_t \equiv (N,0;0,N)$, $R_t \equiv (S,0;0,S)$ and $A_t \equiv (L,0;0,S)$. Preliminary results [14] are shown in Figs. 9 and 10, together with phase shift

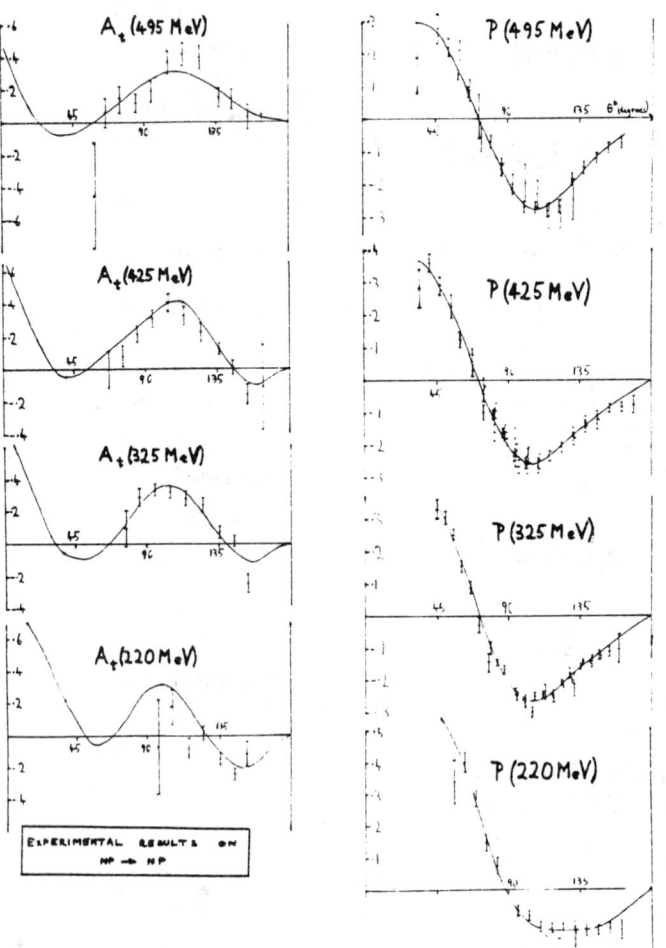

Fig. 10. BASQUE results for P and $A_t$ in np elastic scattering.

fits; some further statistics will be added in the final results. It is perhaps worth drawing attention to the rise in $D_t(9° \text{ lab})$ with energy (Fig. 8). It is possible that this may compete with $R_t$ at higher energies as a mechanism for making polarised neutron beams.

Before going on to the physics, I wish to clarify how the polarisation $\langle\sigma_n\rangle$ of the neutron beam was determined. There are three handles on this, and all agree closely. Firstly, Cheng et al [5] measured $P_{pp}$ and $P_{pn}$ simultaneously in quasi-free elastic scattering from liquid deuterium. BASQUE results agree within their ±3% errors with their normalisation of $P_{pp}$. Hence we can confidently use their values of $P_{pn}$ to normalise $\langle\sigma_n\rangle$ from the rate asymmetry, which measures $\langle\sigma_n\rangle P_{np}$. Secondly, $\langle\sigma_n\rangle = \langle\sigma_p\rangle R_t(pd\rightarrow n, 9° \text{ lab})$, where $\langle\sigma_p\rangle$ is the polarisation of the initial proton beam. In the $R_t$ configuration, the polarisation of the recoil proton is then given by $\langle\sigma_p\rangle R_t(pd\rightarrow n, 9° \text{ lab})R_t(np)$. If we assume that $R_t$ from deuterium is the same as from free protons (it should be independent of final state interactions at this angle), the phase shift fit at 9° lab then determines $R_t$ absolutely, and hence $\langle\sigma_n\rangle$. Results agree within ±3% with the first method. However, since the second method measures $R_t^2$, it is statistically the better. Thirdly, the polarisation of the recoil proton for unpolarised incident neutrons measured P=(0,0;0,N) absolutely, and hence the normalisation of the P curves. This third method is statistically less precise than the first two, but agrees.

All world pp and np data have been processed through phase shift analyses centred at 210, 325, 425 and 515 MeV. This analysis assumes zero inelasticity in I=0 states, and also allows for the Coulomb barrier in pp scattering. The phase shifts are unambiguously determined up to 515 MeV. The $D_t$ data are particularly helpful in fixing $\bar{\epsilon}$ parameters and $R_t$ data vital in fixing singlet phases, particularly the elusive 1P1 phase shift. Results for I=0 phases are shown on Fig. 11, and compared with predictions of the Paris group. Tensor, spin-orbit and central combinations of D waves are very revealing. Tensor and spin-orbit combinations are now quite smoothly varying with energy, and show quite large systematic discrepancies with the predictions. Similar discrepancies are evident, with bigger errors, in the G waves. These discrepancies have not yet been fully analysed in terms of coupling constants. However, one can see qualitatively the changes required if one remembers that in the Born approximation s-channel phase shifts are related to the isospin $I_t$ of the exchanged meson by:

$$\delta(I_s=0) = \tfrac{3}{2}\delta(I_t=1) - \tfrac{1}{2}\delta(I_t=0) \qquad (1a)$$

$$\delta(I_s=1) = \tfrac{1}{2}\delta(I_t=1) + \tfrac{1}{2}\delta(I_t=0) \qquad (1b)$$

Thus the discrepancy in the spin-orbit combination indicates that ρ exchange must be stronger; the discrepancy in the tensor combination may be partly due to this, but probably also indicates a significant

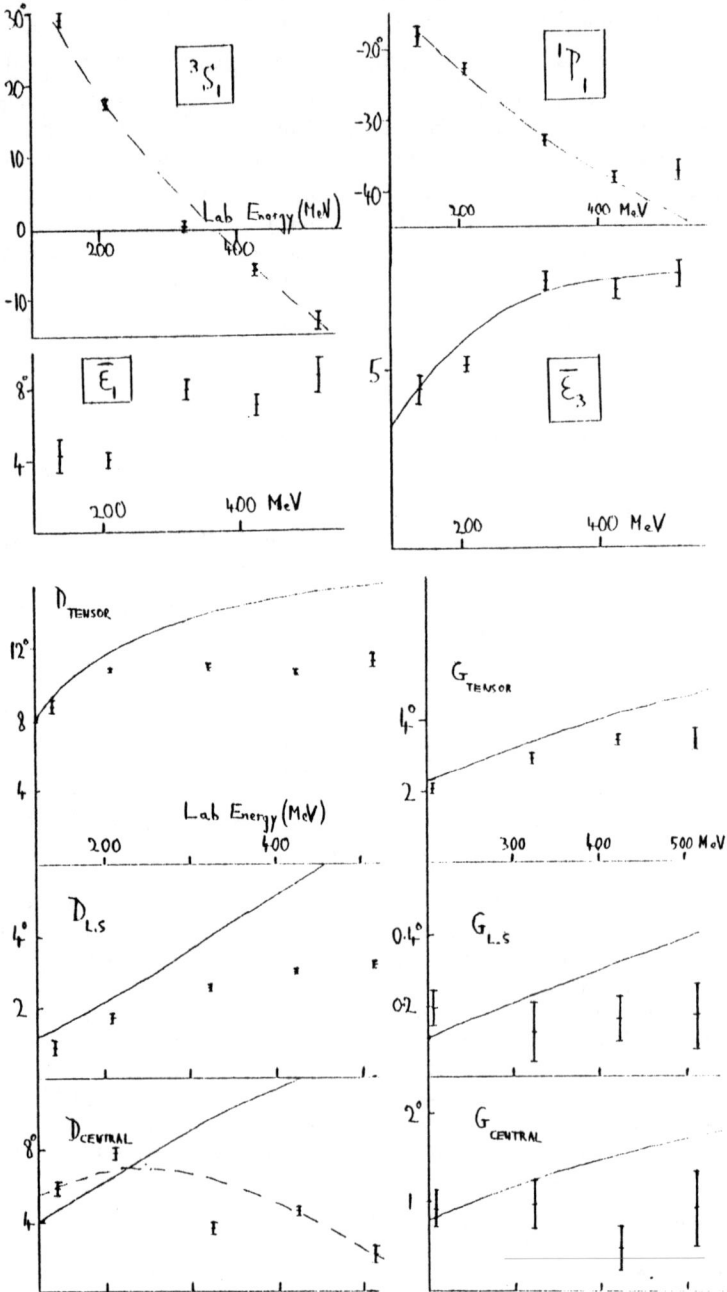

Fig. 11. I=0 phase shifts. Full curves are predictions of the Paris group. Dashed curves are to guide the eye.

η coupling. It is also important to remember that current models fit pp scattering very well, and it is clear now that this must be due to some extent to an arbitrary choice of coupling constants. Thus an increase in the strength of ρ exchange to fit I=0 phase shifts implies a corresponding reduction in the strength of ω exchange. It is past experience[16] that coupling constants depend in a strongly correlated way on the phase shifts, and one can anticipate a major reshuffle of coupling constants will be required.

The central combination of phase shifts is not smooth with energy. My guess as to the best curve is the dashed curve. This central combination is determined largely by $d\sigma/d\Omega$, and one can trace fluctuations in the phase shifts directly to fluctuations in current data, shown on Fig. 12. Measurements in the backward hemisphere at Saclay[17] and LAMPF[18] have a smooth energy dependence. However, LAMPF data are unnormalised and there is considerable flexibility in the way the backward hemisphere is linked to measurements[19] in the cm range 10-55°. In particular, good data in the range 50-120° are absent except at 210[20] and 400 MeV[21]. There is, for example, a ±20% uncertainty in $\sigma(70°)/\sigma(120°)$ at 325 MeV. Further measurements over the <u>whole</u> angular range with good absolute normalisation are clearly required. The BASQUE group is now making such measurements, and results are anticipated on a time-scale of about a year.

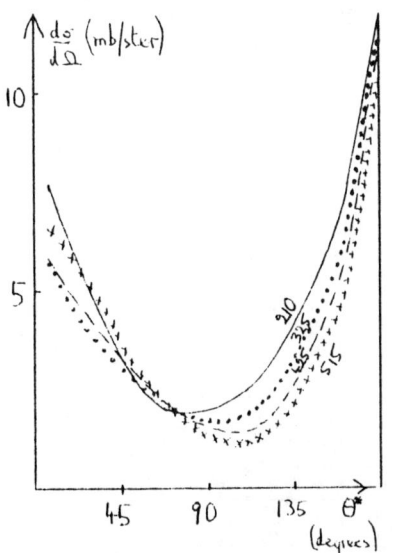

Fig. 12. Present $d\sigma/d\Omega$ data for np elastic scattering at 210, 325, 425, and 515 MeV.

## PP MEASUREMENTS AT SIN

The Geneva group working at SIN has a polarimeter very similar to the BASQUE one. Their attention has been directed to the pp system particularly in the range 400 to 600 MeV. They have measured D, R and A at very small angles (4 to 30°)[22], but with lower statistical accuracy than BASQUE. Their measurements of A are helpful in fixing one combination of phase shifts. These data and the phase shift fit are shown in Fig. 13. They have recently measured $C_{NN}$ using a polarised target of 6 cc volume. Results are shown in Fig. 14. These data are a considerable improvement on earlier ones, and at 515 MeV have the gratifying result of pulling 3H4 from an earlier anomalous value[5] into good agreement with theory. The Geneva group is now going on with measurements of pp→dπ⁺ in $C_{NN}$, $C_{LL}$ and $C_{LS}$ configurations.

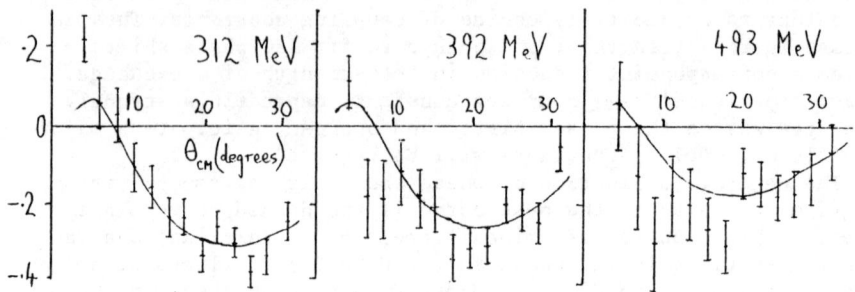

Fig. 13. Geneva measurements of A(pp→pp), and the phase shift fits.

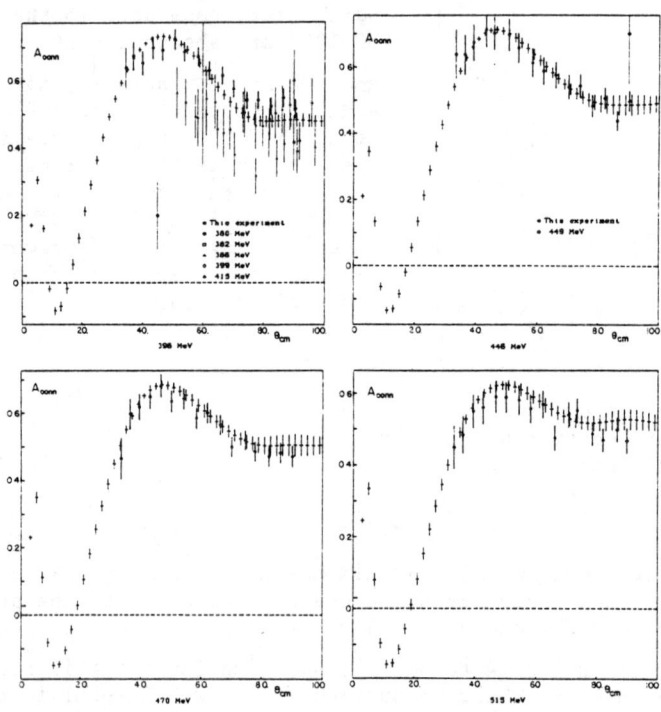

Fig. 14. Geneva results for $C_{NN}$ in pp elastic scattering. Phase shift predictions without these data are shown by the lighter crosses.

## MEASUREMENTS AT LAMPF

Up to now, the LAMPF programme has been largely directed to work with unpolarised neutron beams. Bonner et al[18] have measured $d\sigma/d\Omega$ from 120 to 180°, and the New Mexico group has reported measurements at 800 MeV in the forward hemisphere[23]. Further measurements are planned of D, R, A, R' and A' in pp elastic scattering and pn quasi-elastic scattering from deuterium[24], and also of $D_t$[25] in np scattering at 180°.

## ARE FURTHER DATA NEEDED UP TO 500 MEV?

Present phase shift solutions are unique and reasonably precise, so further data would be unlikely to yield qualitatively new physics. However, np data are now actually superior to pp, and some of the impact of the np data is lost in fixing I=1 phase shifts as well as I=0. This could be remedied quickly and easily at TRIUMF or SIN by measuring D, R and A from 60 to 120° with a precision of ±0.02 in pp elastic scattering. This would reduce errors on several I=1 phase shifts, notably 3F2 and 3F3 by a factor 3, and would improve the resolution of the F wave splitting between central, tensor and spin-orbit combinations. This would then allow np data to be utilised fully in fixing I=0 phase shifts. An alternative would be to measure $C_{LL}$, $C_{SS}$ and $C_{LS}$ from 30 to 90° with the same precision. However these parameters are slightly less sensitive, and in my opinion more difficult because of problems in tracking particles through the field of the polarised target.

In the np system, present measurements of Wolfenstein parameters fix qualitatively similar combinations of amplitudes, and off-diagonal elements of the error matrix are sizeable. There is the danger that systematic error in any one piece of data could spread its effect to many partial waves. $C_{NN}$ measures a quite different combination of amplitudes, and data from 30 to 150° with a precision of ±0.03 would dramatically reduce such correlations. One can view such a measurement as a cross-brace, stiffening the phase shift solution against systematic errors.

## A COMPARISON WITH ARGONNE DATA

DeBauer et al[26] have measured $\Delta\sigma_T$ from 1.2 to 6 GeV/c. Auer et al have measured $C_{LL}$[27] in the range 1.0 to 3.0 GeV/c and $\Delta\sigma_L$[28,29] from 1 to 6 GeV/c. Fig. 15 shows $C_{LL}$ at 1.0 and 1.1 GeV/c and the predictions of phase shift solutions around 425 and 515 MeV obtained without these data. The small discrepancy is easily taken up by very small adjustments in the phase shifts, resulting in the fit indicated by the dash-dot curve.

The $\Delta\sigma_T$ data (Fig. 16) show a sharp peak at 1.5 GeV/c, which Yokosawa[30] has interpreted as evidence for a 3F3 resonance. However,

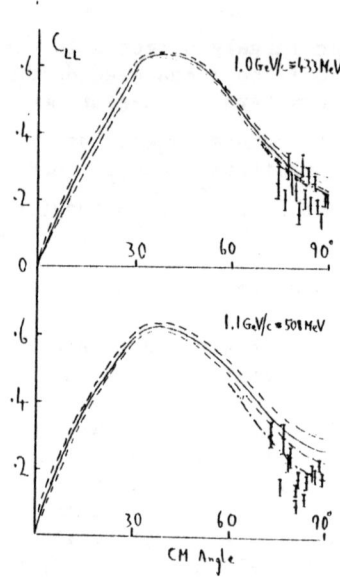
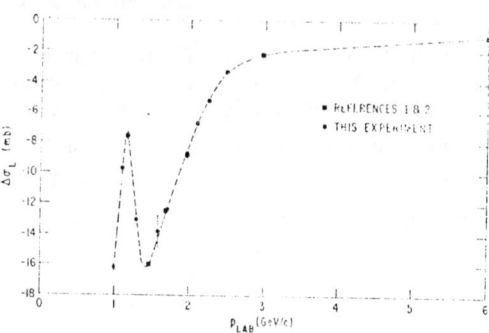

Fig. 16. Argonne measurements of $\Delta\sigma_L$. The dashed curve is to guide the eye.

Fig. 15. Argonne values of $C_{LL}$ folded about 90°. The curves are phase shift predictions excluding these data. Dash-dot curves include these data in the fit.

there appears to be a serious conflict at 1.0 and 1.1 GeV/c between these measurements of $\Delta\sigma_L$, the elastic data, and conventional wisdom about the inelastic scattering.

To understand this conflict, let us start with the inelastic channels. Near threshold $pp \to d\pi^+$ dominates. The cross section for this process peaks at 600 MeV ≡ 1200 MeV/c, and it has long been agreed that the main process is $pp \to N\Delta_1$ in a final s state, i.e. from an initial 1D2 state. Mandelstam[31] fitted production data in terms of π exchange, and found significant contributions from other partial waves. Above 600 MeV, $pp \to pn\pi^+$ dominates, and Ferrari and Selleri[32] showed that one can account for the experimental data at 800 and 970 MeV by π exchange. However, the data are even more peripheral than π exchange predicts, and they found it necessary to include a pion form factor. More recently, Green[33] has emphasised that ρ exchange will cut off π exchange below 1 fermi in much the same way as a form factor. Riska et al[34] and Niskanen[35] have fitted $pp \to d\pi^+$ using π + ρ exchange. Green, Niskanen and Sainio[36] have calculated inelasticities for every partial wave from this model. When the Argonne data on $\Delta\sigma_T$ and $\Delta\sigma_L$ appeared, Arik and Williams[37] made similar calculations, as did Kloet et al[39]. The essential outcome is that all authors agree within 20% or so on the inelasticities and hence on inelastic contributions to $\Delta\sigma_T$ and $\Delta\sigma_L$. The reason for the agreement is that π exchange can be calculated without arbitrary parameters, except the form factor, and ρ exchange can be estimated from photoproduction within at worst a factor 2; the total inelastic cross section is a useful over-riding factor on these last two degrees of freedom, and Green and Sainio[40] have recently updated the calculation of inelasticities with this constraint. The concensus has been that the inelastic channels are well understood, and that the structure in $\Delta\sigma_L$ must lie to a considerable extent in the elastic channel. The difficulty now apparent is that the elastic data around 425 and 515 MeV rule this out. The problem is not with the peak at 1.5 GeV/c, but the dip at 1.17 GeV/c.

Now let us turn to the elastic channel. The reason for the lengthy discussion of the inelastic channels above is that the elastic data will not fix very small inelasticities. For all waves except 1D2, the elasticity η is ≳0.96 up to 515 MeV. Because of this insensitivity, Arndt et al[40] and Bugg et al[5] initially adopted the oversimplified approach of lumping all the inelasticity into 1D2. Then η(1D2) was fixed essentially by the total inelastic cross section. However, although this introduces little error into real parts of the phase shifts, it gives seriously misleading results for $\Delta\sigma_L$ and $\Delta\sigma_T$. If one instead uses η values predicted by Green and Sainio, one gets the phase shift solutions shown in columns 1 and 3 of Table I.

Table I: Phase shift solutions (degrees) at 425 and 515 MeV, (a) with theoretical η, (b) η adjusted to fit $\Delta\sigma_L$, $\Delta\sigma_T$, Re $F_2$, and Re $F_3$.

| Energy (MeV) | 425(a) | 425(b) | 515(a) | 515(b) |
|---|---|---|---|---|
| 3P0 | −20.54±0.61 | −19.84±0.56 | −26.23±0.83 | −24.15±0.78 |
| 1S0 | −19.32±0.45 | −19.68±0.44 | −24.25±0.83 | −24.08±0.80 |
| 3P1 | −35.20±0.33 | −34.98±0.28 | −43.93±0.70 | −41.07±0.52 |
| 3P2 | 18.37±0.19 | 18.70±0.19 | 18.71±0.46 | 19.44±0.42 |
| ε2 | −2.57±0.17 | −1.81±0.13 | −1.30±0.34 | −0.65±0.31 |
| 3F2 | 0.35±0.21 | 0.29±0.20 | −0.71±0.32 | −0.22±0.28 |
| 1D2 | 11.93±0.17 | 12.16±0.14 | 12.67±0.18 | 13.35±0.15 |
| 3F3 | −2.50±0.15 | −2.86±0.13 | −1.51±0.56 | −1.11±0.48 |
| 3F4 | 3.57±0.12 | 3.65±0.11 | 4.03±0.09 | 4.05±0.09 |
| ε4 | −1.69±0.05 | −1.62±0.05 | −1.74±0.12 | −1.60±0.09 |
| 3H4 | 0.82±0.08 | 0.68±0.08 | 0.60±0.09 | 0.59±0.09 |
| 1G4 | 2.11±0.12 | 2.25±0.10 | 2.60±0.14 | 2.85±0.11 |
| 3H5 | −1.23±0.08 | −1.38±0.07 | −1.33±0.09 | −1.35±0.10 |
| 3H6 | 0.71±0.08 | 0.59±0.07 | 0.74±0.07 | 0.71±0.07 |
| η(3P0) | 0.9877 | 1 | 0.9700 | 1 |
| η(1S0) | 0.9857 | 1 | 0.9733 | 1 |
| η(3P1) | 0.9835 | 0.9979±0.0080 | 0.9603 | 0.9996 |
| η(3P2) | 0.9822 | 1 | 0.9608 | 0.9873 |
| η(3F2) | 0.9960 | 0.9960 | 0.9876 | 0.9876 |
| η(1D2) | 0.9673±0.0064 | 0.9495±0.0091 | 0.8965±0.0093 | 0.8279±0.0056 |
| η(3F3) | 0.9942 | 0.9942 | 0.9794 | 0.9794 |
| η(3F4) | 0.9980 | 0.9980 | 0.9944 | 0.9944 |
| η(3H4) | 1 | 1 | 0.9990 | 0.9990 |
| η(1G4) | 0.9982 | 0.9982 | 0.9937 | 0.9937 |
| η(3H5) | 1 | 1 | 0.9984 | 0.9984 |
| η(3H6) | 1 | 1 | 0.9996 | 0.9996 |
| $\chi^2$ | 286.9 | 299.9 | 479.8 | 507.3 |

It will be convenient to display results in the form of the amplitudes $F_2$ and $F_3$ related to $\Delta\sigma_T$ and $\Delta\sigma_L$ by:

$$\Delta\sigma_T = -(4\pi/p_L) \operatorname{Im} F_2 \qquad (2)$$

$$\Delta\sigma_L = (4\pi/p_L) \operatorname{Im} F_3 . \qquad (3)$$

Partial wave decompositions at t=0 are:

$$q^2 F_2/p_L = \sum_J \left[ -(2J+1)R_J + JR_{J-1,J} + (J+1)R_{J+1,J} + 2\sqrt{J(J+1)} R^J \right] \qquad (4)$$

$$q^2 F_3/p_L = \sum_J \left[ (2J+1)R_J - R_{J-1,J} - (2J+1)R_{JJ} + R_{J+1,J} + 6\sqrt{J(J+1)} R^J \right] \qquad (5)$$

where $R_i = (\eta_i e^{2i\delta_i} - 1)/2i$ \qquad (6)

and $R^J = \tfrac{1}{2}\sqrt{\eta_{J+1,J}\, \eta_{J-1,J}}\, \sin 2\bar{\epsilon}_J\, e^{i(\delta_{J+1,J} + \delta_{J-1,J})}$ . \qquad (7)

Values of $F_2$ and $F_3$ from the phase shift solutions are shown as crosses on Fig. 17. There is no problem with $\operatorname{Im} F_2$ ($\Delta\sigma_T$). However, the phase shift solutions suggest that $\operatorname{Im} F_3$ has a broad peak extending from 500 MeV/c to 2 GeV/c (dotted curve) rather than the narrow dip indicated by Argonne $\Delta\sigma_L$ data.

Next one asks whether the Argonne points could be fitted with different inelasticities. This is indeed so, but the values required do violence to the models described above for the inelastic processes. One now has $\Delta\sigma_T$, $\Delta\sigma_L$ and $\sigma_{tot}$ to fit with any three inelasticities. There is no unique solution, but all solutions have two properties in common:
 (a) 1D2 accounts for $\geq 70\%$ of the inelasticity, instead of the 35% predicted by theory,
 (b) inelasticity in 3F3 must be very small up to 1.17 GeV/c in order to fit $\Delta\sigma_L$, as one sees from equation 5.

Columns 2 and 4 of Table I show solutions which seem to me remotely plausible, achieved by retaining nearly theoretical inelasticities in high partial waves and setting them to low values in 1S0, 3P0, and 3P1, where the NN repulsive core (if it is not inelastic in origin) might inhibit the interaction. These and all other solutions predict inelastic contributions to $\Delta\sigma_L$ and $\Delta\sigma_T$ shown by the squares in Fig. 18, a factor 3 greater than theoretical predictions. It is not yet clear whether such solutions could be compatible with inelastic data, since in this energy range only pp→dπ⁺ has been studied carefully. In that reaction, Niskanen finds that the theoretical predictions fit $d\sigma/d\Omega$ and P well, and in particular that significant values of 3F3 are required to fit P. A valuable review of pp→dπ⁺ has been given by Jones[41]. He shows that the sin θ term ($\lambda_0$) in the polarisation is strong near threshold, and this almost certainly implies a strong 3P1 amplitude interfering with 1D2. One would expect pp→pnπ⁺ to resemble pp→dπ⁺ closely, and so one is reluctant to believe that current theories of the inelastic processes are wildly wrong.

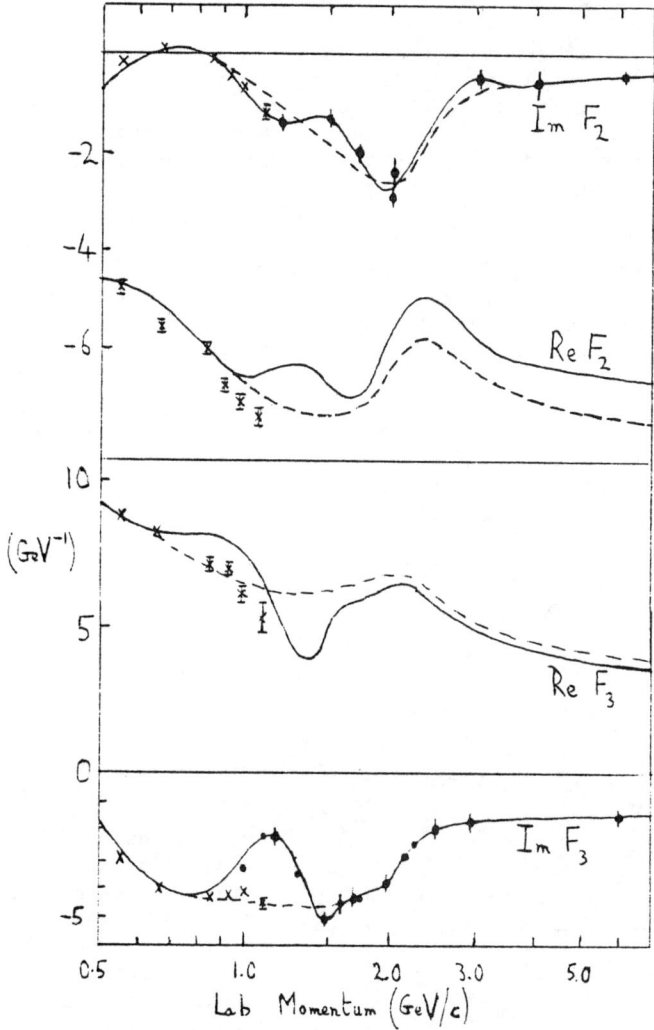

Fig. 17. Real and imaginary parts of $F_2$ and $F_3$. Circles are experimental points. Crosses are predictions of the phase shift analysis with inelasticities calculated by Green and Sainio.

One further check is possible. Grein and Kroll[42] have calculated Re $F_2$ and Re $F_3$ from forward dispersion relations using the full curves on Fig. 17, fitting the Argonne data. The phase shift solutions with the inelasticities of Green and Sainio give the values shown by the crosses. Re $F_2$ would agree with theory if the latter were renormalised downwards a small amount over the whole energy range 200 to 500 MeV by an adjustment to the coupling constants in the unphysical region. But Re $F_3$ shows no sign of the dispersion shape required by the dip in $\Delta\sigma_L$ at 1.17 GeV/c. Instead values agree rather better with the dashed curves. The phase shift solutions do agree well at all energies with the real part of the spin-averaged forward amplitude $F_1$.

In summary, either current prejudices about the mechanism of the inelastic channels or Argonne values of $\Delta\sigma_L$ must be wrong. To resolve this issue, it is desirable to have (i) values of $\Delta\sigma_L$ remeasured below 1.4 GeV/c, (ii) further data on the spin dependence of $pp \to d\pi^+$ and $pp \to pn\pi^+$, and (iii) a proper coupled channel analysis of elastic and inelastic data together. Yokosawa[30] has suggested a 3F3 resonance with a mass of 2260 MeV, a width of 200 MeV, and an elasticity of 0.2 to 0.3. This would result in a 6.5° contribution to the real part of 3F3 at 425 MeV; the results on Fig. 6 rule this out, and indeed show no sign of even 1° above the Paris prediction. If there is a resonance, its width must be much smaller, probably <60 MeV. This is required so that $\eta$(1D2) pushes $\Delta\sigma_L$ up sharply from 900 to 1200 MeV/c where it saturates, and $\eta$(3F3) then pulls it down again very rapidly by 1500 MeV/c. It is difficult to understand such a narrow resonance as being driven by attraction in the N$\Delta$ channel: firstly one would expect the width of the $\Delta$ should make the resonance at least 120 MeV wide, secondly one would expect inelasticity in the 3F3 channel to develop immediately from threshold, since there is no long-range barrier.

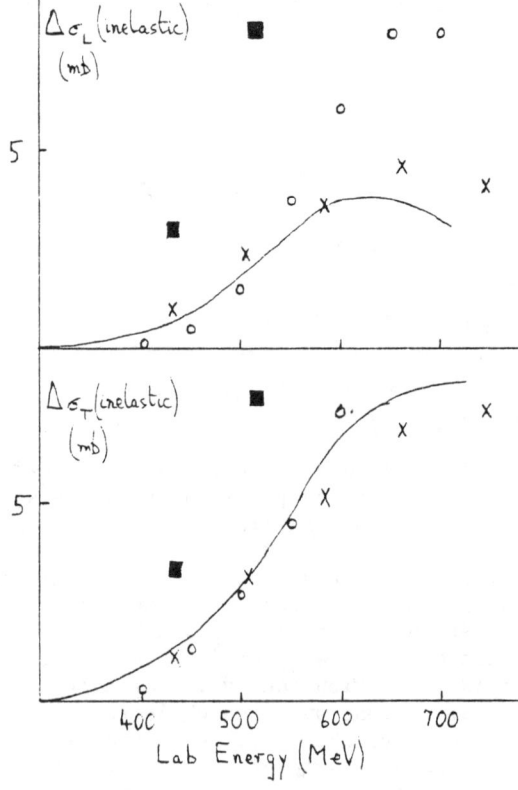

Fig. 18. Inelastic contributions to $\Delta\sigma_L$ and $\Delta\sigma_T$. Full curves are the calculations of Green and Sainio, crosses those of Arik and Williams, and circles those of Kloet et al. The squares are values required by a fit to Argonne values of $\Delta\sigma_L$ and $\Delta\sigma_T$.

REFERENCES

1. C. Amsler et al., J. Phys. G 4, 1047 (1978).
2. R. Dubois, M. Sc. Thesis, University of British Columbia (1978).
3. M. MacNaughton, private communication.
4. G. Waters et al., Nucl Instr. 153, 401 (1978).
5. D. V. Bugg et al., J. Phys. G 4, 1025 (1978).
6. D. V. Bugg, A. A. Carter and J. R. Carter, Phys. Lett. 44B, 278 (1973).
7. R. Vinh Mau, Mesons in Nuclei, eds. M. Rho and D. Wilkinson, to be published (1978).
8. M. Chemtob and D.O. Riska, Phys. Lett. 35B, 115 (1971).
9. G. E. Brown and J. W. Durso, Phys. Lett. 35B, 120 (1971).
10. G. E. Bohannon and P. Signell, Phys. Rev. D10, 815 (1974).
11. G. E. Epstein and B. H. J. McKellar, Phys. Rev. D10, 1005 (1974).
12. C. Amsler et al., Nucl. Instr. 144, 401 (1977).
13. F. Folkmann and D. F. Measday, CERN MSC report C-17/675 (1968).
14. D. V. Bugg et al., Proceedings of the Graz conference on Few Body Systems and Nuclear Forces, p. 134, (Springer-Verlag, 1978).
15. D. Cheng et al., Phys. Rev. 163, 1470 (1963).
16. D. V. Bugg, Nucl. Phys. B5, 29 (1968).
17. G. Bizard et al., Nucl. Phys. B85, 14 (1975).
18. B. E. Bonner et al, Proceedings of the Graz Conference on Few Body Systems and Nuclear Forces, p. 3, (Springer-Verlag, 1978).
19. A. J. Bersbach, R. E. Mischke and T. J. Devlin, Phys. Rev. D13, 535 (1976).
20. Yu. M. Kazarinov and Yu. N. Simonov, Soviet Phys. JETP 16, 24 (1963).
21. A. J. Hartzler, R. T. Siegel and W. Opitz, Phys. Rev. 95, 591 (1954).
22. D. Besset et al., Proceedings of the Graz Conference on Few Body Systems and Nuclear Forces, p. 130, (Springer-Verlag, 1978).
23. B. Dieterle, Proceedings of the Second International Conference on the Nucleon-Nucleon Interaction, Vancouver (1977).
24. H. A. Thiessen et al., LAMPF experimental proposal 392.
25. B. E. Bonner et al., LAMPF experimental proposal 403.
26. W. deBoer et al., Phys. Rev. Lett. 34, 558 (1975).
27. I. P. Auer et al, Argonne preprint ANL-HEP-PR-78-33.
28. I. P. Auer et al., Phys. Lett. 70B, 475 (1977).
29. I. P. Auer et al., Phys. Rev. Lett. 41, 354 (1978).
30. A. Yokosawa, Proceedings of the Second International Conference on the Nucleon-Nucleon Interaction, Vancouver (1977).
31. S. Mandelstam, Proc. Roy. Soc. A244, 491 (1958).
32. E. Ferrari and F. Selleri, Nu. Cim. 27, 1450 (1963).
33. A. M. Green, Reports on Progress in Physics 39, 1109 (1976).
34. D. O. Riska, M. Brack and W. Weise, Phys. Lett. 61B, 41 (1978).
35. J. A. Niskanen, Nucl. Phys. A298, 417 (1978).
36. A. M. Green, J. A. Niskanen and M. E. Sainio, J. Phys. G 4, 1055 (1978).
37. M. Arik and P. G. Williams, Nucl. Phys. B136, 425 (1978).
38. W. M. Kloet et al., Phys. Rev. Lett. 39, 1643 (1977).

39. A. M. Green and M. E. Sainio, private communication.
40. R.A. Arndt, R. H. Hackman and L. D. Roper, Phys. Rev. $\underline{C15}$, 1002 (1978).
41. G. Jones, Proceedings of the Second International conference on the Nucleon-Nucleon Interaction, Vancouver (1977), and TRIUMF report TRI-PP-77-13 (1977).
42. W. Grein and P. Kroll, Nucl. Phys. $\underline{B137}$, 173 (1978).

## Comments on Dr. D. V. Bugg's Talk

N. Hoshizaki
31 October 1978

1. I disagree with blindly using results of the OBE models of the inelastic channels for the partial wave states in which dibaryon resonances might exist. There is no reason to believe that these models can work well in the resonance region. None of them have so far succeeded in reproducing the structures of the $\Delta\sigma_L$ data correctly.

2. For $p_{lab} \gtrsim 1.1$ GeV/c, it is not proper to assume that the $\pi$-production occurs only in the pp $^1D_2$ state. In order to discuss resonances, it is necessary even at $\sim 1.1$ GeV/c to consider absorptions in the initial pp states at least up to F waves as free parameters. In order to relate information obtained from the pp $\to \pi^+$d process to the pp elastic phase shifts, effects arising from pp $\to \pi^+$np process must be adequately considered, because the latter process cannot be neglected even at $p_{lab} \sim 1.1$ GeV/c as we know since Mandelstam's 1958 paper.

3. Before one criticizes the existing experimental data, those mentioned above should be clarified.

NH/rl

## PURE SPIN TOTAL CROSS SECTIONS*

H. Spinka
Argonne National Laboratory, Argonne, Illinois 60439

### ABSTRACT

Measurements of the pure spin total cross sections in the NN system are described. Evidence for a dibaryon resonance interpretation of the structure in the I=1 cross sections at $P_{LAB} \lesssim 3$ GeV/c is discussed. New data on the total cross section difference for longitudinal beam and target in the I=0 system are presented.

### INTRODUCTION

Two major topics will be discussed in this talk. These topics deal with structure observed in pure spin total cross sections and various pp elastic scattering parameters between $P_{LAB}$ = 1 and 11.75 GeV/c. The first topic deals with the I=1 or pp case, to which a total of roughly 6 months running time was devoted by several experimental groups at the Argonne ZGS during the past few years. A considerable amount of theoretical work has been done to try to understand the observed structures near 1.2, 1.5 and 2.0 GeV/c. In particular, evidence for a dibaryon interpretation of this structure will be discussed. The second topic deals with the I=0 case, where preliminary data from this past August and September will be presented.

Before describing the data and its interpretation, consider briefly some motivation for these experiments. In the past, some of the first quantities measured at a new high energy proton accelerator were the spin-averaged total cross sections. Families of resonances have been found in this way in the $\pi$p and the $K^-$p systems. The understanding of these resonances, that is, their quantum numbers and their classification into families or multiplets, would have been considerably hampered if there were no easy way to experimentally distinguish particles according to charge ($\pi^+$, $\pi^\circ$, $\pi^-$) or according to particle type (p, $\pi^+$, $K^+$). The spin quantum number is unimportant for $\pi$N and $K^-$N total cross sections because the pions and kaons are spin -0 particles and because the strong interactions conserve parity to high accuracy. On the other hand, spin is important in the NN case, but until recently only spin-averaged total cross sections were measured. Thus, there is a good possibility that the physics may be easier to understand if the total cross sections for pure spin states were measured.

For the NN system, there are three independent total cross sections which can be measured. In terms of pure spin total cross sections, the measured quantities are

---

*Work supported by the U.S. Department of Energy.

$$\sigma^{Tot} = 1/2 \, [\sigma^{Tot}(\vec{\rightleftarrows}) + \sigma^{Tot}(\vec{\rightrightarrows})]$$

$$= 1/2 \, [\sigma^{Tot}(\uparrow\downarrow) + \sigma^{Tot}(\uparrow\uparrow)]$$

$$\Delta\sigma_T = \sigma^{Tot}(\uparrow\downarrow) - \sigma^{Tot}(\uparrow\uparrow)$$

$$\Delta\sigma_L = \sigma^{Tot}(\vec{\rightleftarrows}) - \sigma^{Tot}(\vec{\rightrightarrows})$$

where $\sigma^{Tot}$ is the spin-averaged total cross section, previously measured at many accelerators. The quantity $\Delta\sigma_T$ is the difference in total cross sections for beam and target transversely polarized. It has been measured by the Michigan[1,2] and the Rice[3,4] groups at the Argonne ZGS. The quantity $\Delta\sigma_L$ is the difference in total cross sections for beam and target longitudinally polarized (the arrows denote the spin directions in the lab frame). It has been measured in a series of measurements by the Argonne in-house polarization group at the ZGS. The physicists who have been involved in the $\Delta\sigma_L$ experiments are: I. P. Auer, A. Beretvas, E. Colton, W. R. Ditzler, H. Halpern, D. Hill, K. Nield, B. Sandler, H. Spinka, N. Tamura, G. Theodosiou, D. Underwood, R. Wagner, Y. Watanabe and A. Yokosawa[5,6,8,9] with assistance in the Legendre analysis and interpretation by K. Hidaka[7] and many others.

1. Structure in the I=1 NN System

In the past, before the measurement of pure spin total cross sections, there was no convincing evidence for a dibaryon in the pp system. Below the $\pi$-production threshold at $P_{LAB} = 0.77$ GeV/c, there is no indication of a dibaryon resonance (except the deuteron) from phase shift analyses in either the pp or pn systems. There are no peaks in $\sigma^{Tot}$(pp) comparable to those produced by the low lying resonances in $\sigma^{Tot}(\pi p)$ or $\sigma^{Tot}(K^- p)$. The rapid rise in $\sigma^{Tot}$(pp) near 1.2 GeV/c has been interpreted as a threshold effect from the N$\Delta$ channel (see Fig. 1). (The threshold for pp $\rightarrow$ NN$\pi\pi$ is 1.22 GeV/c, for pp $\rightarrow$ N$\Delta$ (1236) is 1.28 GeV/c, for pp $\rightarrow$ NN* (1400) is 1.74 GeV/c, and for pp $\rightarrow$ $\Delta\Delta$ is 2.12 GeV/c.) For these reasons and others, the pp system was considered "exotic" and without resonances.

The three I=1 total cross sections are shown in Fig. 1. The measurements employed the standard "good geometry" transmission technique. The experimental details for $\Delta\sigma_T$ and $\Delta\sigma_L$ are described in Refs. 1-6, 8, 9. Spin effects are seen to be a sizeable fraction of $\sigma^{Tot}$ at low energies and to exhibit striking structure as a function of $P_{LAB}$. The $\Delta\sigma_L$ results between 2.75 and 5.0 GeV/c and at 11.75 GeV/c are preliminary. Additional measurements of $\Delta\sigma_T$ between 1.1 and 2.75 GeV/c, reported at this conference by T. Mulera,[4] have recently been performed by the Rice University group. Their preliminary data confirms the structure previously seen in $\Delta\sigma_T$.

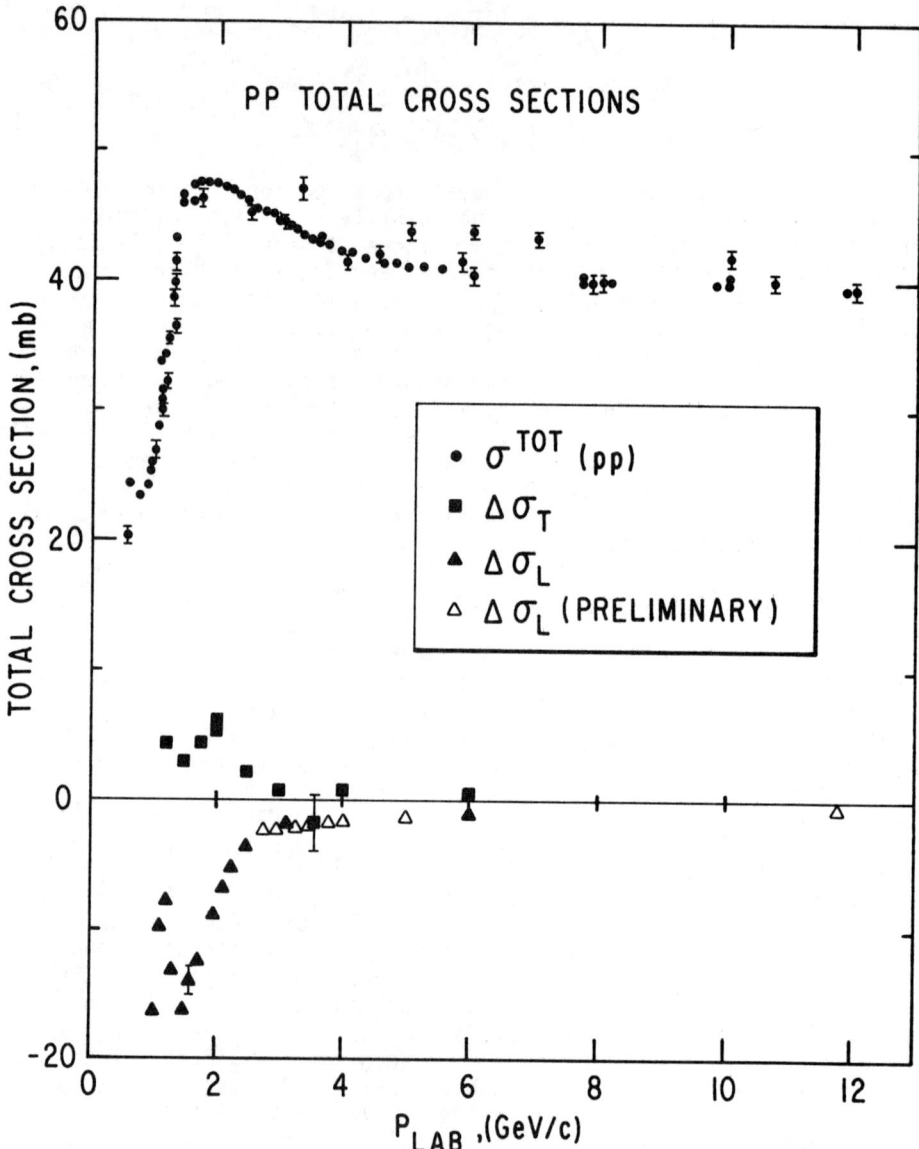

Fig. 1. PP Total Cross Section. The $\Delta\sigma_T$ results are from Refs. 1-4 and the $\Delta\sigma_L$ data (solid triangles) are from Refs. 5, 6, 8. Preliminary $\Delta\sigma_L$ results are shown as open triangles. (See also Fig. 6 and Ref. 4.)

The pp total cross sections can be expressed in terms of partial waves as follows:

$$\Delta\sigma_T \sim \text{Singlet} - \text{Triplet} \; (J=L\pm1 \text{ even})$$

$$\sigma^{Tot}(\rightleftarrows) = \sigma^{Tot} + 1/2 \, \Delta\sigma_L$$

$$\sim \text{Singlet} + \text{Triplet} \; (J=L\pm1, \text{ even})$$

$$\sigma^{Tot}(\rightrightarrows) = \sigma^{Tot} - 1/2 \, \Delta\sigma_L$$

$$\sim \text{Triplet} \; (J=L, \text{ odd}) + \text{Triplet}' \; (J=L\pm1, \text{ even}),$$

where Singlet refers to a sum of positive contributions from spin-singlet (S=0) partial waves and Triplet refers to a sum of positive contributions from S=1 partial waves. Note that a resonance in a spin-singlet partial wave would be characterized by a peak in $\Delta\sigma_T$ or $\sigma^{Tot}(\rightleftarrows)$ or $\Delta\sigma_L$. Likewise, a resonance in a spin-triplet J=L partial wave would be characterized by a peak in $\sigma^{Tot}(\rightrightarrows)$ or a dip in $\Delta\sigma_L$. Structure caused by either spin-singlet or J=L spin-triplet partial waves may be easier to see in $\Delta\sigma_L$ than in $\sigma^{Tot}(\rightleftarrows)$ or $\sigma^{Tot}(\rightrightarrows)$ because the coupled triplet (J=L±1) partial wave contributions somewhat cancel in $\Delta\sigma_L$.

## A. Structure Near 1.5 GeV/c

One of the most striking features of the data is the deep dip in $\Delta\sigma_L$ near 1.5 GeV/c. If the pure spin total cross sections are computed from $\sigma^{Tot}$ and $\Delta\sigma_L$, it appears that the structure occurs in $\sigma^{Tot}(\rightrightarrows)$ but not in $\sigma^{Tot}(\rightleftarrows)$ or $\Delta\sigma_T$ (see Fig. 2). As described above, this behavior is characteristic of partial waves with S=1, J=L=odd.

There are also indications of structure near 1.5 GeV/c in the elastic channel. The total elastic cross section seems to show a bump near that momentum. The polarization parameter averaged over the range $0.1 < |t| < 0.2 \; (\text{GeV/c})^2$ is shown in Fig. 3. All data shown were collected by one group[10] in order to minimize any systematic differences between the points. The differential cross section in this t-range changes slowly with lab momentum. Since the polarization times the differential cross section ($Pd\sigma/d\Omega$) contains contributions from spin-triplet terms only, then the rapid change in polarization indicates structure in the S=1 partial waves near 1.5 GeV/c.

The numerous differential cross section and polarization data in this momentum range were fit with Legendre polynomials as a function of angle:[7]

$$d\sigma/d\Omega = (1/k_{cm})^2 \; \Sigma \; A_N P_N (\cos\theta)$$

$$Pd\sigma/d\Omega = (1/k_{cm})^2 \; \Sigma \; B_N P_N^1 (\cos\theta)$$

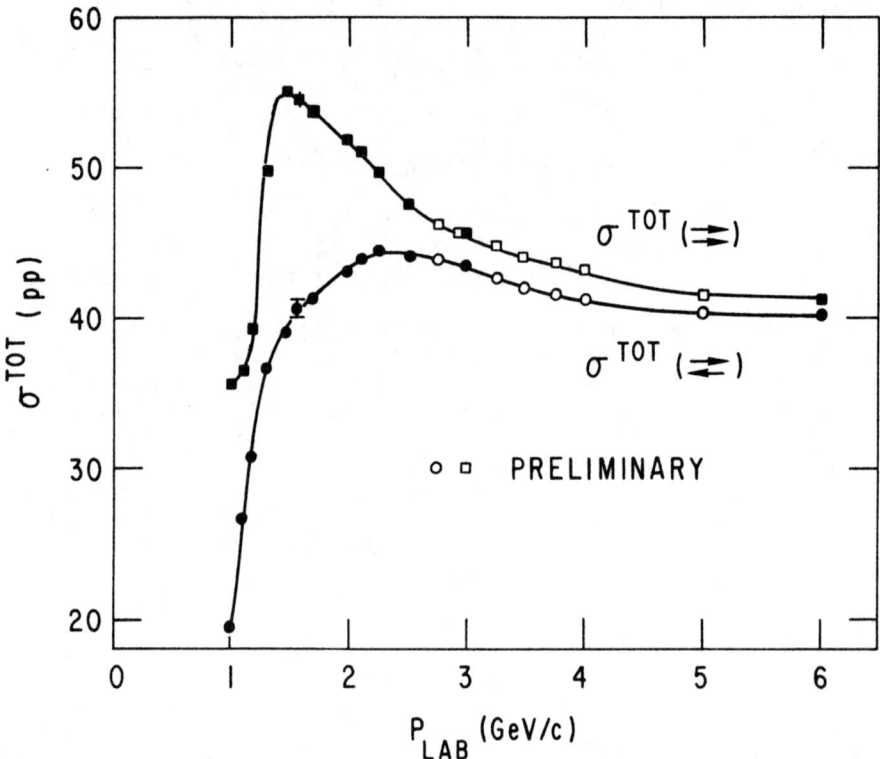

Fig. 2. Pure Spin pp Cross Sections. The curves are to guide the eye.

The coefficients $A_N$ and $B_N$ can be related to individual partial waves. The variation of the coefficients with momentum indicates that the partial wave $^3F_3$ is probably responsible for the structure seen in the polarization (Fig. 3). This partial wave (S=1,J=L=3=odd) is consistent with the structure observed in $\Delta\sigma_L$ and $\sigma^{tot}(\rightleftarrows)$ as described before.

At the same time the $\Delta\sigma_L$ data were taken, the elastic scattering parameter $C_{LL}$ (see Fig. 4)

$$C_{LL} = [d\sigma/dt\ (\rightrightarrows) - d\sigma/dt\ (\rightleftarrows)]/[d\sigma/dt\ (\rightrightarrows) + d\sigma/dt\ (\rightleftarrows)]$$

was measured near $\theta_{c.m.} = 90°$ for momenta between 1 and 3 GeV/c.[11] (Additional data up to 5 GeV/c were recently measured, but the $C_{LL}$ analysis is not yet complete.) It was shown[11,12] that the rapid change in $C_{LL}$ (90°) with $P_{LAB}$ is consistent with a Breit-Wigner resonance in the $^3F_3$ partial wave and a slow variation of the other partial waves near 1.5 GeV/c.

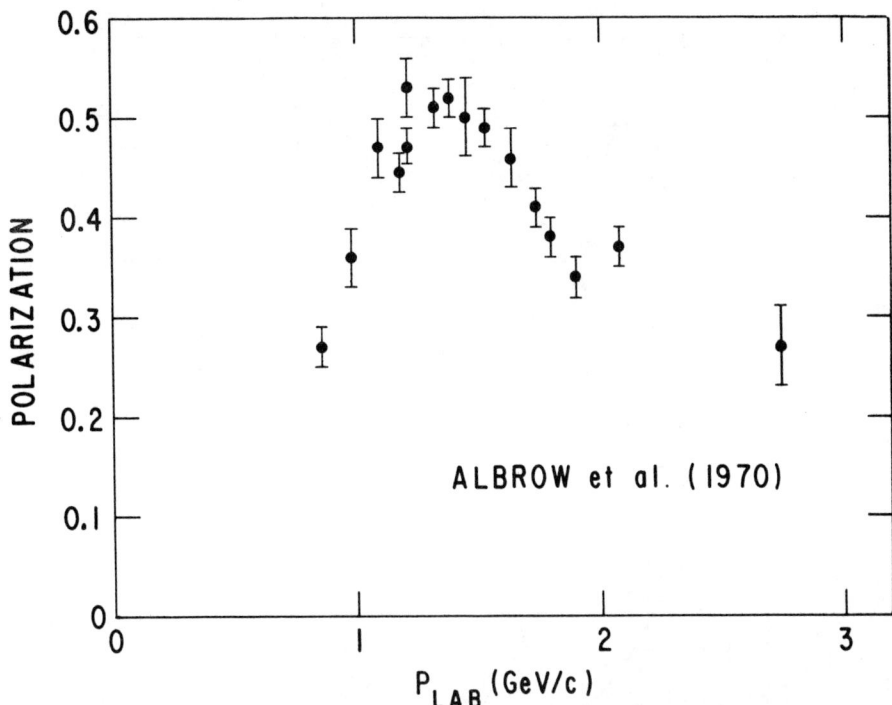

Fig. 3. Polarization in pp Elastic Scattering (from Ref. 10).

From the experimental results described, <u>assuming</u> the structure in $\Delta\sigma_L$ near 1.5 GeV/c is caused by a dibaryon resonance, then its properties are

| | |
|---|---|
| Mass | $\sim$ 2260 MeV |
| Width | $\sim$ 200 MeV |
| Charge | 2 |
| $J^P$ | $3^-$ |
| Elasticity | $\sim .2$ |

where the mass, width and elasticity are approximate values only. These parameters were obtained by a fit to the shape of $\Delta\sigma_L$[7] and were used to check the behavior of $C_{LL}(90°)$[11,12]

Additional evidence on the structure in $\Delta\sigma_L$ near 1.5 GeV/c is available from fixed -t dispersion relations by Grein and Kroll.[13,14] They define amplitudes $F_2$ and $F_3$ whose imaginary parts are

$$\mathrm{Im}F_2 = - (P_{LAB}/4\pi) \, \Delta\sigma_T$$

$$\mathrm{Im}F_3 = (P_{LAB}/4\pi) \, \Delta\sigma_L \; .$$

Grein and Kroll use phase shift predictions at low energies, the measured $\Delta\sigma_T$ and $\Delta\sigma_L$ values between 1 and 6 GeV/c and some assumptions

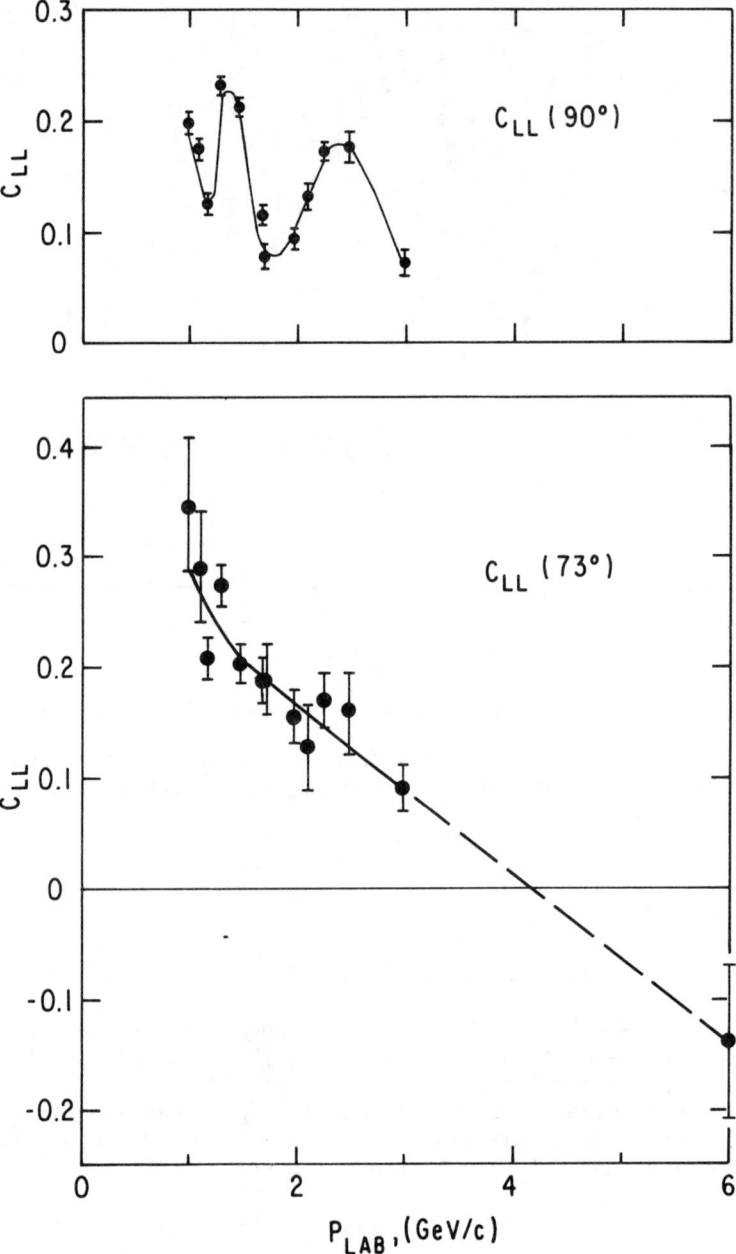

Fig. 4. The pp Elastic Scattering Parameter $C_{LL}$ vs. $P_{LAB}$. The results shown are from fits to the data presented in Ref. 11. The curves are to guide the eye.

about the high energy behavior to obtain $ImF_2$ and $ImF_3$ as a function of momentum. The real parts of $F_2$ and $F_3$ are then computed from dispersion relations. The amplitude $[-P_{cm}^2 (F_2 + F_3)/P_{LAB}]$, which contains no spin-singlet contributions, is displayed in an Argand diagram[14] in Fig. 5a. The minus sign is included since the structure seems to occur in $\sigma^{Tot}(\rightleftarrows)$ and since $ImF_3 \sim \sigma^{Tot}(\rightleftarrows) - \sigma^{Tot}(\rightrightarrows)$. A clear resonance-like behavior is seen in Fig. 5a. With the assumption that this behavior is caused by a dibaryon resonance, Grein and Kroll obtain a slightly larger mass and width and a slightly smaller elasticity (assuming J=3 for the resonance) than discussed above.

Finally Hoshizaki has performed a phase shift analysis for $P_{LAB}$ = 1.1 to 3 GeV/c. The input data include many elastic scattering results, the total cross sections $\sigma^{Tot}$, $\Delta\sigma_T$, $\Delta\sigma_L$, the results of Grein and Kroll's dispersion analysis, and some theoretical inputs for the behavior of the high spin partial waves. He concludes that the results indicate the existence of a diproton resonance in the $^3F_3$ partial wave (and possibly in the $^1D_2$ partial wave at a lower momentum). He obtains a slightly smaller mass and width and the same elasticity as quoted above for the $^3F_3$ resonance. Further details will be given by the next speaker (N. Hoshizaki).

B. Structure Near 1.2 GeV/c

The sharp peak in $\Delta\sigma_L$ near 1.2 GeV/c is possibly associated with a $^1D_2$ dibaryon resonance according to Hoshizaki's phase shift analysis. Its properties would be

| | |
|---|---|
| Mass | $\sim$ 2170 MeV |
| Width | $\sim$ 50 - 100 MeV |
| $J^P$ | $2^+$ |
| Elasticity | $\sim$ 0.1 |

A spin-singlet resonance would be expected to appear as a peak in $\Delta\sigma_T$ and $\sigma^{Tot}(\rightleftarrows)$ as well as in $\Delta\sigma_L$. There is an indication of a rise in $\Delta\sigma_T$ (see Fig. 1 and Ref. 4) as the momentum decreases to about 1.2 GeV/c, as would be expected. However, in this momentum range $\sigma^{Tot}(\rightleftarrows)$ is changing rapidly (see Fig. 2) and no peak is apparent. It is possible that the rapid rise in the total cross section, which is usually associated with the opening of the $N\Delta$ channel (also fed from the $^1D_2$ partial wave), might conceal any resonant structure in $\sigma^{Tot}(\rightleftarrows)$.

Additional arguments, similar to those used for the structure near 1.5 GeV/c, cannot be made in this case. The Legendre coefficient analysis of $Pd\sigma/d\Omega$ gives no information on the $^1D_2$ partial wave since $Pd\sigma/d\Omega$ contains contributions from spin-triplet partial waves only. In addition, Grein and Kroll[13] find no evidence for a resonance near 1.2 GeV/c from their dispersion relation analysis of the forward amplitudes. There appears to be structure in $C_{LL}$ (90°) and possibly in $C_{LL}$ (73°) near 1.2 GeV/c (see Fig. 4), but additional

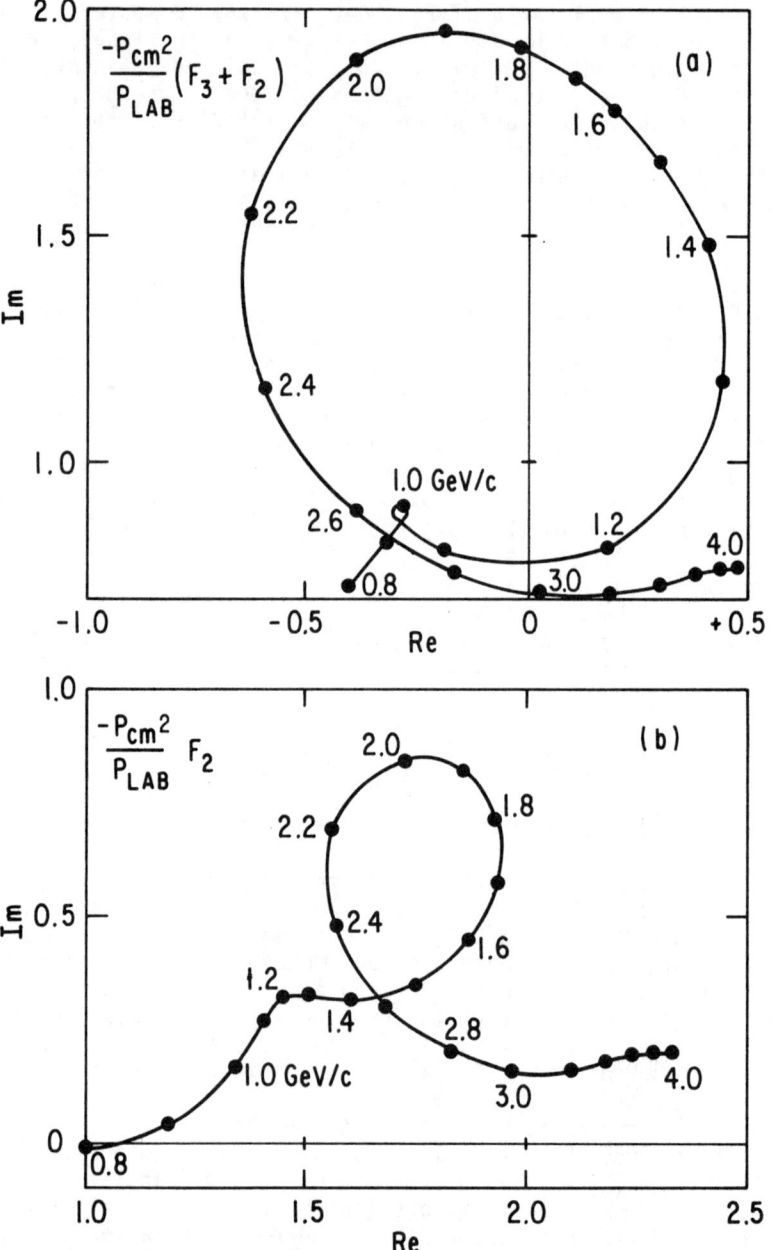

Fig. 5. Argand Diagram for the Amplitudes $[-P_{cm}^2(F_2 + F_3)/P_{LAB}]$ and $[-P_{cm}^2 F_2/P_{LAB}]$ from Ref. 14. Spin-singlet terms are absent in a) and spin-triplet J=L terms are absent in b).

measurements of both elastic and inelastic spin parameters would be very useful to clarify the nature of the structure in $\Delta\sigma_L$ and $\Delta\sigma_T$ near 1.2 GeV/c.

There have been a number of theoretical papers which have attempted to explain the behavior of $\Delta\sigma_T$ and/or $\Delta\sigma_L$ or which have criticized the dibaryon resonance interpretation.[15-18] These models require a large contribution to $\Delta\sigma_L$ from the elastic channel

$$\Delta\sigma_L^{el} = -2 \int C_{LL}(d\sigma/d\Omega)d\Omega$$

for $P_{LAB} \lesssim 2$ GeV/c. Kroll[14] estimates $\Delta\sigma_L^{el}$ to be -6 to -8 mb at 1.47 GeV/c from the measured $C_{LL}$ data[11] and from results at $\theta=0°$ from the dispersion analysis. This estimate is inconsistent with the values required by some of the inelastic models.

Criticisms of the dibaryon interpretation of the structures near 1.2 and 1.5 GeV/c (Refs. 18-21 and other references cited) have been raised before the $\Delta\sigma_T$, $\Delta\sigma_L$ and/or $C_{LL}$ results were published. These results may invalidate some of the criticisms. In particular, the $C_{LL}$ data contradict the predictions of the "modified phase shift analysis" of Ref. 18.

The interpretation of dibaryon resonances is described in terms of SU(6) in Ref. 22, in terms of the MIT bag model in Refs. 23,24, in terms of the $\pi$NN system with the $\pi$ forming a $\Delta$ with both nucleons in Ref. 25, and in a quark-cluster model in Ref. 26.

C.  Structure Near 2.0 GeV/c

The peak in $\Delta\sigma_T$ near 2.0 GeV/c has been confirmed and more accurately defined by recent measurements of the Rice University group.[4] Although there is no apparent structure in $\Delta\sigma_L$, there is a hint of a shoulder occurring around 2.1 GeV/c in $\sigma^{tot}(\rightleftarrows)$, and structure is again seen in $C_{LL}$ (Fig. 4). As discussed before, the behavior of $\Delta\sigma_T$ and $\sigma^{tot}(\rightleftarrows)$ is characteristic of spin-singlet partial waves ($^1D_2$, $^1G_4$, $^1I_6$, ...).

Grein and Kroll[13] find resonance-like structure in the spin-singlet partial waves near 2.0 GeV/c. The Argand diagram for the amplitude ($-P_{cm}^2 F_2/P_{LAB}$) is shown in Fig. 5b. Recall that the imaginary part of the amplitude $F_2$ is proportional to $\Delta\sigma_T$, and $\Delta\sigma_T$ contains no contribution from spin-triplet J=L partial waves. The mass and width derived by Grein and Kroll from the results in Fig. 5b are

Mass ~ 2390 MeV
Width ~ 100 MeV

A spin of 4 is conjectured, which gives an elasticity of $\lesssim 0.1$. If the structure in $\Delta\sigma_T$ near 2.0 GeV/c is caused by a spin-singlet partial wave, a Legendre coefficient analysis of $Pd\sigma/d\Omega$ (which contains no contributions from S=0 partial waves) would not help to determine the resonance quantum numbers. However, J=4 or the $^1G_4$ partial wave is consistent with the behavior of the $C_{LL}$ data, since the $^1G_4$ contribution to $C_{LL}$ should vanish at $\theta_{c.m.} = 70.1°$. The data indicate a considerably stronger variation of $C_{LL}$ with momentum at 90° than at 73°.

The Rice group collected data on the elastic scattering parameter

$$C_{NN} = [d\sigma/dt\,(\uparrow\uparrow) - d\sigma/dt\,(\uparrow\downarrow)]/[d\sigma/dt\,(\uparrow\uparrow) + d\sigma/dt\,(\uparrow\downarrow)]$$

at several momenta near 2.0 GeV/c. Their preliminary results[4] and data at other energies[27,28] indicate a substantial variation of $C_{NN}$ with angle and energy. Their new results should considerably assist phase shift analyses near 2.0 GeV/c; however additional data are probably required to completely understand the nature of the structure in $\Delta\sigma_T$.

D. <u>High Energy Behavior of $\Delta\sigma_L$</u>

Various predictions of the high energy behavior of $\Delta\sigma_L$ have been made, all assuming dominance of an $A_1$-like (unnatural parity) exchange. One such prediction by Berger, Sorenson and Irving[29] is shown in Fig. 6. They predict

$$\Delta\sigma_L \sim (P_{LAB})^{\alpha-1}$$

with $\alpha = -0.19$. Grein and Kroll[13] assumed a similar behavior with $\alpha = -0.15$ as input to their dispersion relation calculation for the $F_3$ amplitude. Finally, Stacey[30] gives a more complicated expression which is numerically very close to the values of Grein and Kroll at ZGS energies. All three predictions give $\Delta\sigma_L \sim -0.5$ mb at $P_{LAB} = 11.75$ GeV/c, which agrees with the preliminary experimental value (Fig. 1). The predictions differ by almost a factor of 2 at 100 GeV/c. Stacey also gives a model independent way to obtain the pure unnatural parity exchange amplitudes from the $\Delta\sigma_L$ and $\Delta\sigma_T$ values. He concludes that the $\Delta\sigma_L$ values near 5 GeV/c contain a substantial ($\sim$30%) natural parity exchange contribution.

2. <u>Structure in the I=0 NN System</u>

The second topic to be covered deals with the I=0 system. The shape of the spin-averaged total cross section is again smooth, just as in the pp case (see Fig. 9). Data from several experimental groups are shown,[31-33] and it can be seen that systematic errors are sizeable. This is also the case for the $\Delta\sigma_L$ (I=0) results. The rise in the spin-averaged total cross section occurs at a somewhat higher energy than in the pp case. Thresholds for the various inelastic channels are the same as in the I=1 case except that the $N\Delta$ channel, which was suspected to be the cause of the structure in $\sigma^{Tot}(pp)$ and $\Delta\sigma_L(pp)$ near 1.5 GeV/c, cannot occur in I=0. In particular, threshold effects are not the cause of structure observed near 1.5 GeV/c in the I=0 channel.

Before describing the data, consider a few relevant experimental details. Unfortunately, there are experimental problems at the ZGS with the use of either a polarized neutron or longitudinally polarized deuteron beam incident on a polarized target. It is also impossible to make a target of free neutrons, that is neutrons not bound in some nucleus. Therefore, the experiment used a beam of polarized protons from the Argonne ZGS and precessed the proton spin to the longitudinal direction. The polarization direction was reversed on alternate pulses, which greatly assisted in reducing systematic errors. The target material was partially deuterated ethylene glycol. Thus the target neutrons are not free, but are bound inside a deuteron. Most of the experimental runs had either the deuterons polarized or <u>both</u> the free protons and the deuterons polarized.

The number of incident protons was counted with three scintillators in the beam. Rejection of particles that would miss the target was performed with various beam veto counters. A set of circular scintillators, located about 2m downstream of the target, recorded the number of charged particles as a function of the solid angle subtended. The number of noninteracting beam particles transmitted through the target was determined for the two beam spin directions by extrapolating the counts in the circular scintillators to zero solid angle. The difference in pure spin total cross sections, $\Delta\sigma_L$, was then computed.

In order to extract the I=0 total cross section difference $\Delta\sigma_L(I=0)$, a number of subtractions and corrections were required. First, for runs in which both free target protons and deuterons were polarized, $\Delta\sigma_L(pp)$ was subtracted to give the directly measured quantity $\Delta\sigma_L(pd)$. These values agreed well with $\Delta\sigma_L(pd)$ for runs where only the deuterons in the target were polarized. Alternately, both $\Delta\sigma_L(pp)$ and $\Delta\sigma_L(pd)$ could be extracted using all the runs and the known target polarizations for protons and deuterons. The values of $\Delta\sigma_L(pp)$ from this method of analysis are compared to previous measurements in Fig. 6. Most of the new measurements were performed at energies where $\Delta\sigma_L(pp)$ data already existed. As can be seen, the agreement is quite good, which gives added confidence in the $\Delta\sigma_L(pd)$ results.

The preliminary results of $\Delta\sigma_L(pd)$ are given in Fig. 7 along with the spin-averaged total cross section $\sigma^{Tot}(pd)$. These data are quite smooth and featureless compared to the pp case. Errors shown are statistical only. At present, systematic errors are estimated to be ± 10-15%, due mainly to the target polarization uncertainty. These errors may be reduced after we perform a more refined data analysis. Note that $\sigma^{Tot}(pd)$ shows more structure than $\Delta\sigma_L(pd)$.

We have had to rely on some theory to extract $\Delta\sigma_L(pn)$ from the $\Delta\sigma_L(pd)$ results. In the simplest approximation

$$\Delta\sigma_L(pd) \approx \Delta\sigma_L(pp) + \Delta\sigma_L(pn) \quad .$$

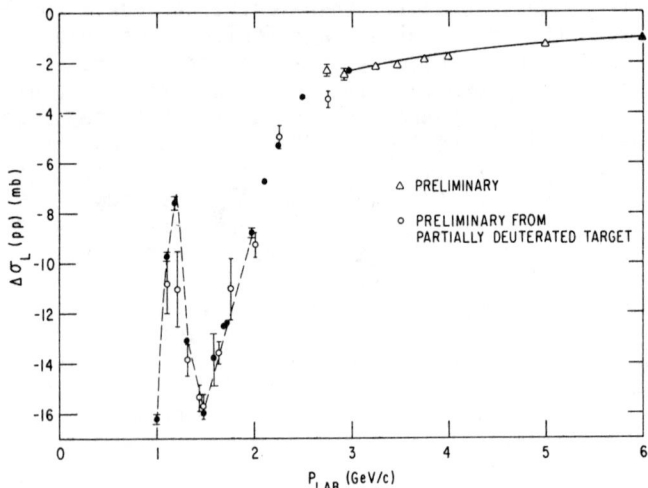

Fig. 6.  $\Delta\sigma_L(pp)$ vs. $P_{LAB}$.  The solid circles are data from Refs. 5,6,8 and the open triangles are preliminary results using a polarized proton target.  The open circles are preliminary $\Delta\sigma_L(pp)$ values measured with a partially deuterated target at the same time as the $\Delta\sigma_L(pd)$ results.  The dashed curve at low energy is to guide the eye, and the solid curve at high energy is a prediction from Ref. 29.

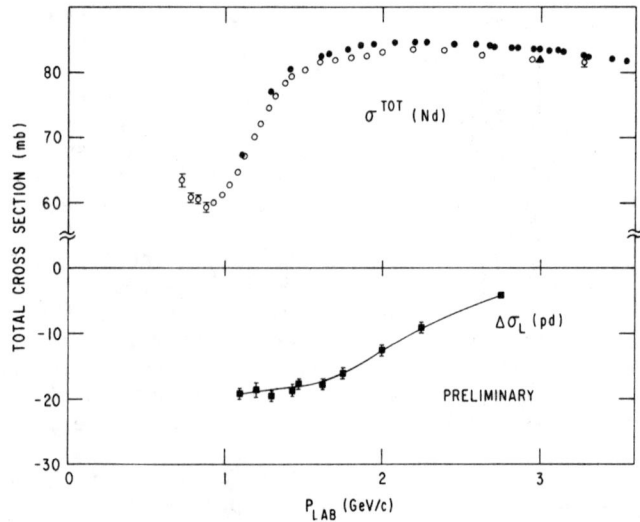

Fig. 7.  pd Total Cross Sections.  The values of $\sigma^{Tot}(pd)$ are from Refs. 31-33.  All errors shown are statistical only.  Systematic errors for the $\Delta\sigma_L(pd)$ results are ± 10-15%.

However, deviations from this simple approximation occur because the neutron can be shadowed by the proton when the incident particle strikes the deuteron, and also because the neutron is moving inside the deuteron; termed Fermi-motion. C. Sorensen[34] derived a formula relating $\Delta\sigma_L(pn)$ with the quantities $\Delta\sigma_L(pd)$, $\Delta\sigma_L(pp)$, $\sigma^{Tot}(pn)$, $\sigma^{Tot}(pp)$ and the real parts of the forward pp and pn amplitudes within the framework of Glauber theory (see also Ref. 35). Since the I=0 forward amplitudes are poorly known, especially the real parts, an approximation was made to extract $\Delta\sigma_L(pn)$. The real parts of all forward amplitudes were assumed to be zero for the data shown. (More refined Glauber corrections will be made in the future.) Simple corrections were applied for the D-state of the deuteron wave function. In order to account for Fermi-motion effects, the $\Delta\sigma_L(pp)$ values were averaged for the momentum distribution of the proton inside the deuteron. The results are shown in Fig. 8. The features in both $\sigma^{Tot}(pn)$ and $\Delta\sigma_L(pn)$ are smeared by Fermi-motion of the neutron inside the deuteron. Statistical errors are shown; however systematic errors from the corrections and target polarization uncertainty are larger than the statistical errors.

Finally, $\Delta\sigma_L(I=0) = 2\Delta\sigma_L(pn) - \Delta\sigma_L(pp)$ is shown in Fig. 9. Recall from Fig. 7 that $\Delta\sigma_L(pd)$ is smooth. Therefore, the peak in $\Delta\sigma_L(I=0)$ and $\Delta\sigma_L(pn)$ at 1.5 GeV/c arises from the subtraction of $\Delta\sigma_L(pp)$. Fermi-motion effects cannot smear $\Delta\sigma_L(pp)$ sufficiently to make the peak in $\Delta\sigma_L(I=0)$ disappear, because the smear is typically ± 100 MeV/c in $P_{LAB}$. It is also believed that the refined analysis of the shadowing or Glauber corrections, especially taking the real parts of the forward amplitudes into account, and the D-state corrections will change the values of $\Delta\sigma_L(I=0)$ in a smooth and slowly varying manner. Therefore, it appears that there is a striking peak in $\Delta\sigma_L(I=0)$ near 1.5 GeV/c. As mentioned previously, this structure cannot be caused by the $N\Delta$ channel, since it cannot contribute to I=0.

If the I=0 total cross sections are expressed in terms of partial waves, then

$$\Delta\sigma_T \sim \text{Singlet} - \text{Triplet } (J=L\pm 1, \text{ odd})$$

$$\sigma^{Tot}(\rightleftarrows) \sim \text{Singlet} + \text{Triplet } (J=L\pm 1, \text{ odd})$$

$$\sigma^{Tot}(\rightleftarrows) \sim \text{Triplet } (J=L, \text{ even}) + \text{Triplet'} (J=L\pm 1, \text{ odd})$$

in the same notation as that used for the I=1 case. A peak in $\Delta\sigma_L(I=0)$ would be characteristic of a spin-singlet partial wave or waves, $^1P_1$ and/or $^1F_3$. Unfortunately, additional information is lacking at this time. Polarization data for I=0 do not cover wide angular ranges, and thus a Legendre coefficient analysis is not possible. Furthermore, $Pd\sigma/d\Omega$ only contains contributions from spin-triplet terms. If the structure in $\Delta\sigma_L(I=0)$ is caused by spin-singlet partial waves, $Pd\sigma/d\Omega$ would not assist the identification of quantum numbers. Contributions from spin-singlet terms

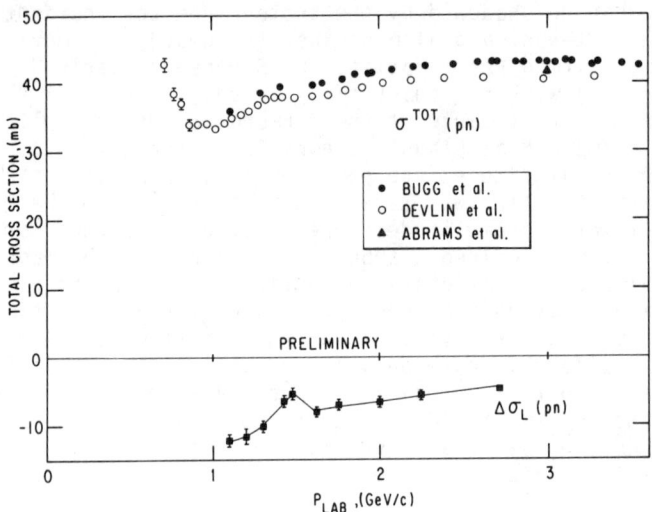

Fig. 8. pn Total Cross Sections. The values of $\sigma^{Tot}(pn)$ are from Refs. 31-33. Errors on the $\Delta\sigma_L(pn)$ data are statistical only. Simple corrections for shadowing, for the D-state of the deuteron and for Fermi-motion of the proton have been applied (see the text). All cross sections are smoothed by Fermi-motion of the neutron in the deuteron target. The curve is a guide to the eye.

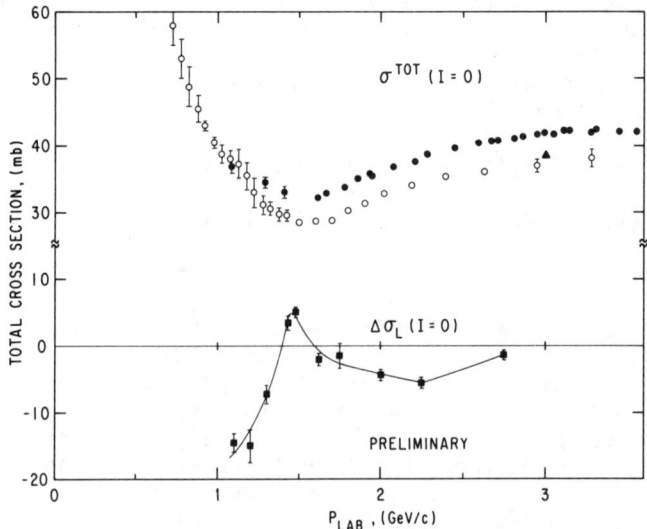

Fig. 9. I=0 Total Cross Sections. (See the caption for Fig. 8.)

are isolated in the quantity $\Delta\sigma_T + \sigma^{Tot}(\rightleftarrows)$; unfortunately $\Delta\sigma_T(I=0)$ has not been measured. Although it appears there is remarkable structure in $\Delta\sigma_L(I=0)$, the interpretation is unclear until additional data are available in the I=0 system.

## REFERENCES

1. E. F. Parker et al., Phys. Rev. Lett., 31, 783 (1973).
2. W. deBoer et al., Phys. Rev. Lett., 34, 558 (1975).
3. E. K. Biegert et al., Phys. Lett., 73B, 235 (1978).
4. T. Mulera, these proceedings.
5. I. P. Auer et al., Phys. Lett., 67B, 113 (1977).
6. I. P. Auer et al., Phys. Lett., 70B, 475 (1977).
7. K. Hidaka et al., Phys. Lett., 70B, 479 (1977).
8. I. P. Auer et al., Phys. Rev. Lett., 41, 354 (1978).
9. E. Colton et al., Nucl. Instrum Methods, 151, 85 (1978).
10. M. G. Albrow et al., Nucl. Phys., B23, 445 (1970).
11. I. P. Auer et al., Phys. Rev. Lett., 41, 1436 (1978).
12. K. Hidaka, Argonne National Laboratory preprint, ANL-HEP-CP-78-15.
13. W. Grein and P. Kroll, Nucl. Phys., B137, 173 (1978).
14. P. Kroll, University of Wuppertal preprint, WU B-78-13. The data used in this analysis were not corrected for Coulomb-nuclear interference effects.
15. G. L. Kane and G. H. Thomas, Phys. Rev., D13, 2944 (1976); and E. L. Berger, P. Pirila, G. H. Thomas, Argonne National Laboratory preprint ANL-HEP-75-72 (unpublished).
16. D. W. Joynson, J. of Phys., G2, L65 (1976).
17. W. M. Kloet, R. R. Silbar, R. Aaron, R. D. Amado, Phys. Rev. Lett., 39, 1643 (1977).
18. M. Arik and P. G. Williams, Nucl. Phys., B136, 425 (1978).
19. S. Minami, Phys. Lett., 74B, 120 (1978).
20. H. Suzuki, Prog. Theor. Phys., 54, 143 (1975).
21. D. D. Brayshaw, Phys. Rev. Lett., 37, 1329 (1976).
22. F. J. Dyson and N. H. Xuong, Phys. Rev. Lett., 13, 815 (1964).
23. R. L. Jaffe, Phys. Rev. Lett., 38, 195 (1977).
24. A. T. M. Aerts, P. J. G. Mulders, J. J. deSwart, Phys. Rev., D17, 260 (1978).
25. T. Ueda, Phys. Lett., 74B, 123 (1978).
26. D. B. Lichtenberg, E. Predazzi, D. H. Weingarten, J. G. Wills, Phys. Rev., D18, 2569 (1978).
27. D. Miller et al., Phys. Rev., D16, 2016 (1977).
28. A. Lin et al., Phys. Lett., 74B, 273 (1978).
29. E. L. Berger, A. C. Irving, C. Sorensen, Phys. Rev., D17, 2971 (1978).
30. R. Stacey, Westfield College, University of London preprint (submitted to J. of Phys. G.).
31. D. V. Bugg et al., Phys. Rev., 146, 980 (1966).
32. R. J. Abrams et al., Phys. Rev., D1, 2477 (1970); See also K. F. Riley, Phys. Rev. D1, 2481 (1970).
33. T. J. Devlin et al., Phys. Rev., D8, 136 (1973).

34. C. Sorensen, Argonne National Laboratory preprint, ANL-HEP-PR-78-26, to be published in Phys. Rev. D.
35. G. Alberi, M. Bleszynski, T. Jaroszewicz, S. Santos, contribution to this conference.

# pp Phase Shifts and Dibaryon Resonances

Norio Hoshizaki

Department of Nuclear Engineering
Kyoto University, Kyoto

## §1. Introduction

In these two years, dibaryon resonances became a subject of great interest that was previously inaccessible.[1)-5)] I shall report results of a pp phase shift analysis from Kyoto for $p_{lab}=1.1 - 3$ GeV/c, and discuss on the basis of them the possibilities of at least two dibaryon resonances: $B^2$(2.17 GeV) and $B^2$(2.22 GeV). Important structures of some experimental data will be explained on the basis of these two resonances.[6)]

## §2. pp Phase shifts at $p_{lab}=1.1 - 3$ GeV/c

We first summarize results of our phase shift analysis. The input data are as follows: At the forward angle we were able to use a complete set of data in practically the whole momentum range considered. They are $\sigma_{tot}=\frac{2\pi}{k}\text{Im}(\phi_1+\phi_3)_{0°}$, $\Delta\sigma_L=\frac{4\pi}{k}\text{Im}(\phi_1-\phi_3)_{0°}$, $\Delta\sigma_T=-\frac{4\pi}{k}\text{Im}\phi_2(0°)$, Re/Im ratio of the forward amplitude $(\phi_1+\phi_3)_{0°}$, $\text{Re}(\phi_1-\phi_3)_{0°}$ and $\text{Re}\phi_2(0°)$, in addition to $\sigma_{el}$ and $\sigma_{inel}$. Here, I have inputted as "experimental data" all three real forward amplitudes derived from the dispersion analysis by Grein and Kroll.[7] At nonforward angles, the employed data are d$\sigma$/dt, P, all or some of the spin-spin correlation parameters $C_{kp}$, $C_{nn}$, $C_{\ell\ell}$ and $C_{qkn}$, and the triple scattering parameters D, $D_t$, R, $R_t$, A, $A_t$, R' and $R_t$' from Dubna, Berkeley, Saclay, CERN, Argonne, Vancouver, Zurich, etc.[8-20] The number of kinds and points of data at each momentum is summarized in Table I.

Table I. Review of the experimental data used in the analysis.[8-20] The $\chi^2$ values of the phase shift solution are shown for reference.

| $P_L$ (GeV/c) | 1.1 | 1.2 | 1.25 | 1.3 | 1.38 | 1.5 | 1.6 | 1.7 | 1.9 | 2.0 | 2.5 | 3.0 |
|---|---|---|---|---|---|---|---|---|---|---|---|---|
| Nr. Kinds of data | 17 | 22 | 20 | 19 | 13 | 14 | 13 | 13 | 11 | 12 | 11 | 15 |
| Nr. data points | 150 | 157 | 101 | 163 | 130 | 137 | 107 | 116 | 116 | 90 | 78 | 116 |
| $\chi^2$ | 170.2 | 141.8 | 76.5 | 113.6 | 102.5 | 119.7 | 83.6 | 108.4 | 102.4 | 69.3 | 44.6 | 74.9*  74.2** |

* sol. a;  ** sol. a'.

There are rouphly 20 kinds at each momentum for $p_{lab} \lesssim 1.3$ GeV/c and 13 kinds for $p_{lab} > 1.3$ GeV/c. These data have been $\chi^2$-fitted at each momentum. The number of data points at each momentum is about 130 and number of free parameters is about 20. The $\chi^2$ value at each momentum is about 100.

The phase shifts for the higher partial waves with J>6 and L>6 for $p_{lab} \leq 2.5$ GeV/c and J>8 and L>8 at 3 GeV/c have been assumed to be represented by the OPE contribution. For absorptions in the higher angular momentum states in which resonances do not seem to exist, the OPE model by Amaldi et al. has been taken into account for $p_{lab} \approx 1.5-2$ GeV/c.[21)] In order to obtain a physically reasonable phase shift solution, continuity of the phase shifts as functions of energy has been required for an acceptable solution. This has been realized by examining the continuity of the phase shifts from the elastic region, i.e., below 1 GeV/c towards the inelastic region above 1 GeV/c step by step with the momentum interval of $\Delta p_{lab} \sim 0.1$ GeV/c insofar as the experimental data allow it.

Our outputs are summarized in Figs. 1—3. The solution is represented as functions of $p_{lab}$ and $\sqrt{s}$, the two-nucleon c.m. total energy. Recent phase shift results below 1 GeV/c are also shown in comparison. I would like to note that the phase shifts for the high partial waves with $J \geq 2$ are rather well determined. It is likely that they are unique at least up to 1.3 GeV/c, although no explicit attempt has been made here to investigate uniqueness of the present solution.

Clearly we have a $^3F_3$ resonance around 1.4 GeV/c and a possible $^1D_2$ resonance around 1.25 GeV/c. The Argand diagram of these waves is shown in Fig. 4. We also see some

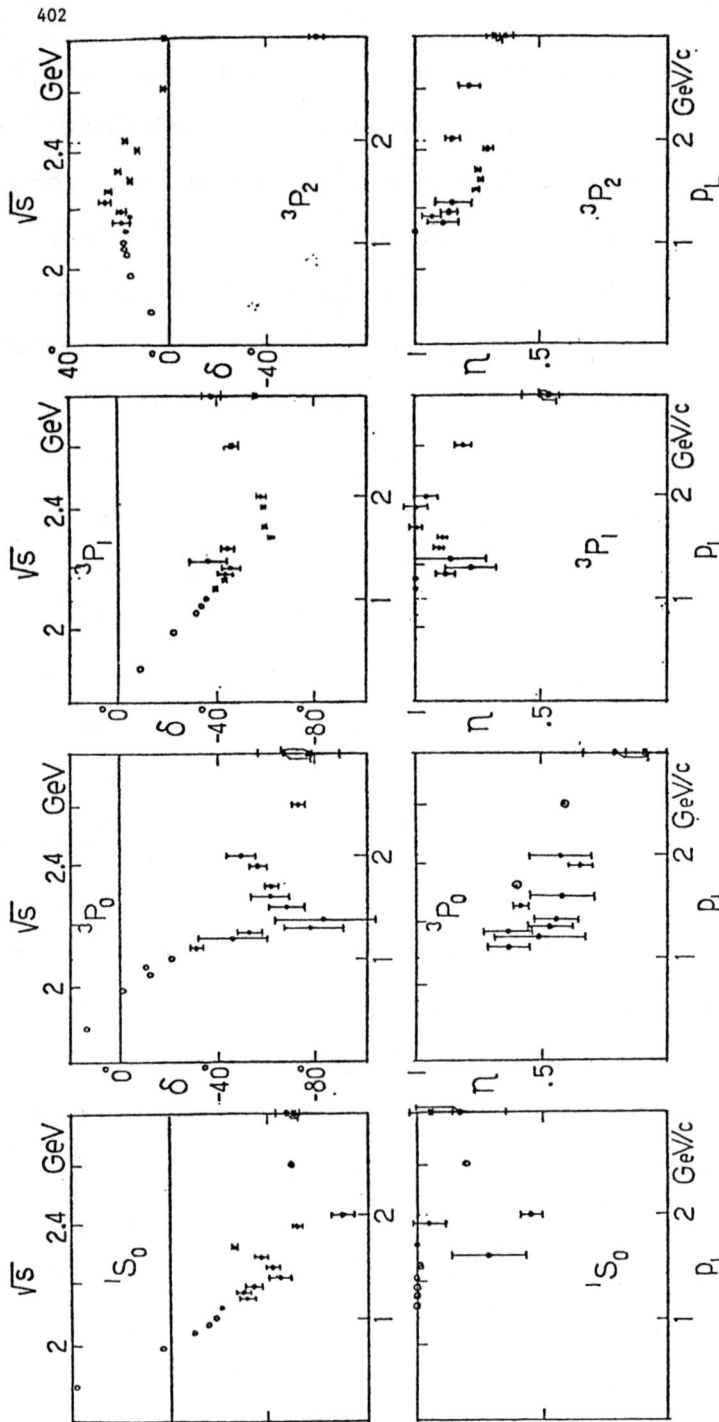

Fig. 1. Plots of the phase shift solution versus $p_{lab}$ (lower abscissa) and $\sqrt{s}$ (upper abscissa). Sol.a (sol.a') at 3 GeV/c is shown by black circles (squares). Fixed phase shifts are shown by double circles. Recent low-energy solutions (white circles) are given in comparison.[6],[22]

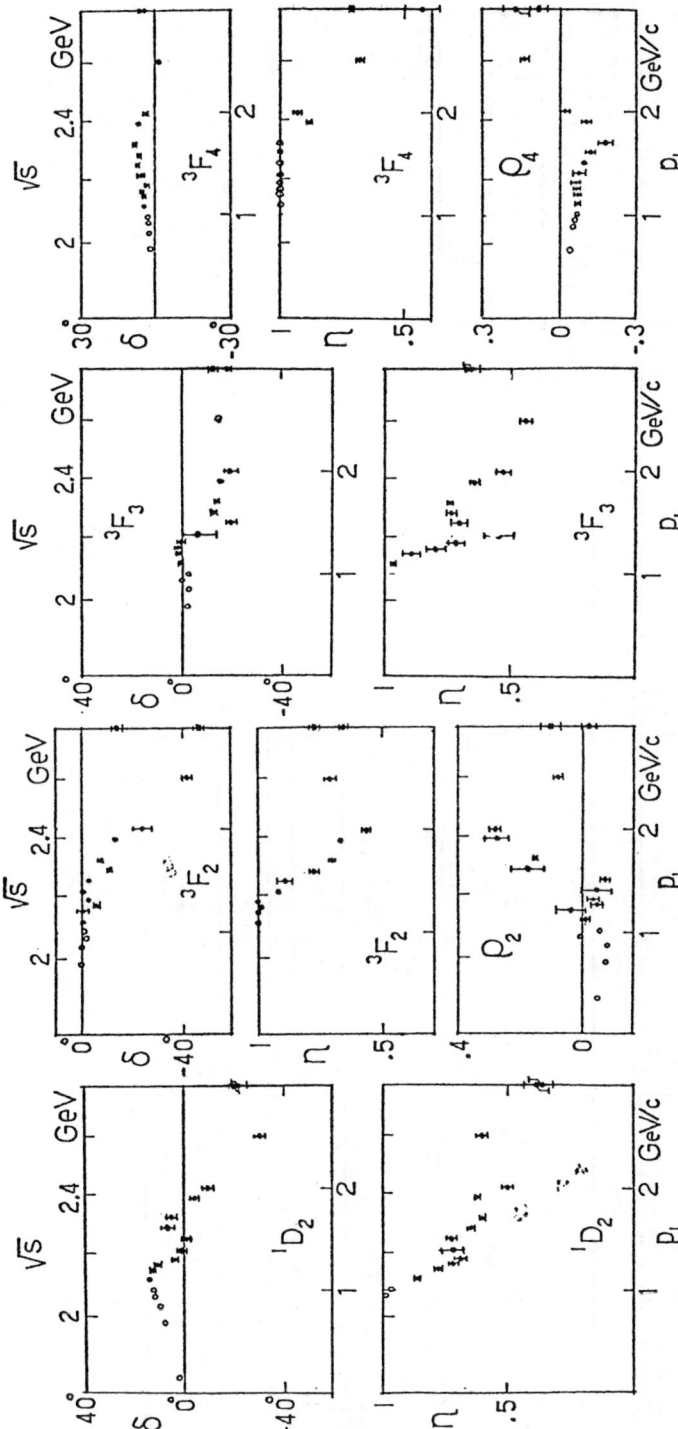

Fig. 2. Plots of the phase shift solution versus $p_{lab}$ and $\sqrt{s}$. Symbols are the same as in Fig. 1.

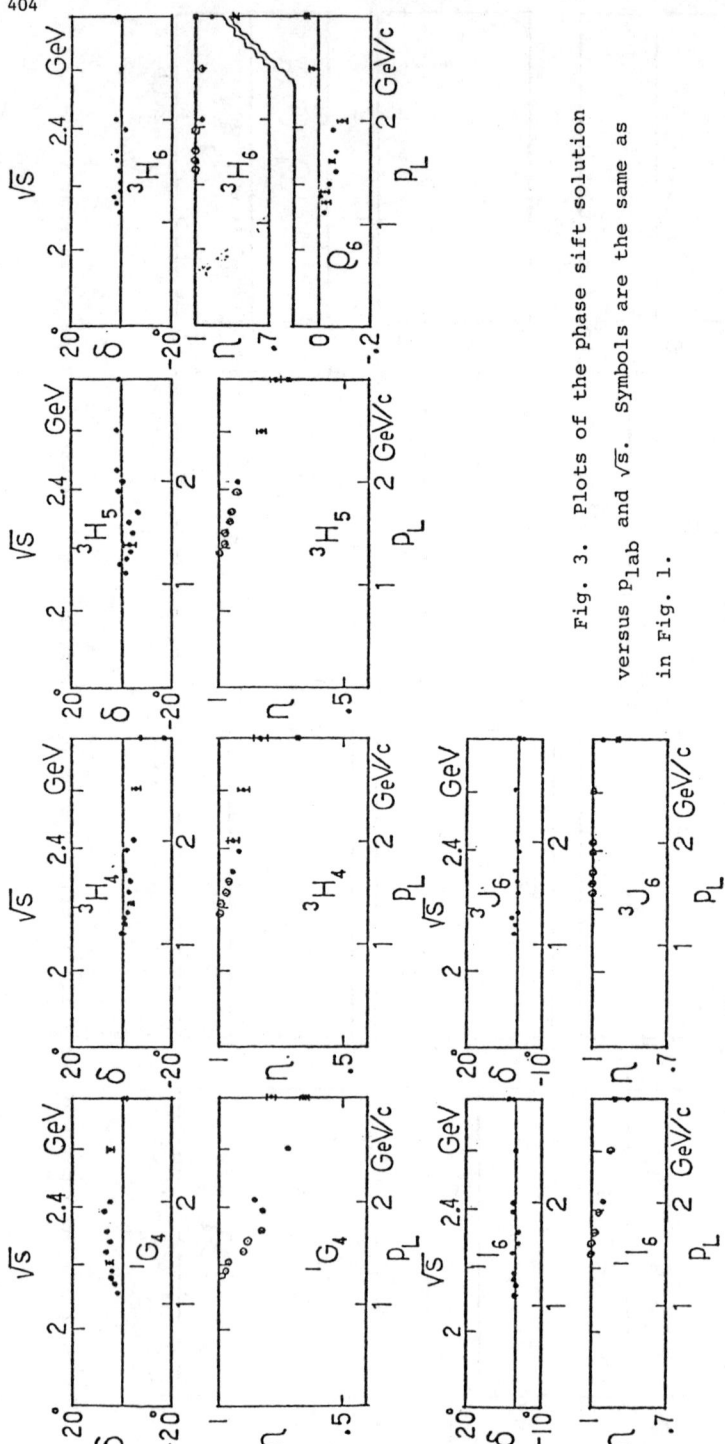

Fig. 3. Plots of the phase shift solution versus $p_{lab}$ and $\sqrt{s}$. Symbols are the same as in Fig. 1.

structures of the $^1S_0$ phase shift around 2 GeV/c and all three of the triplet P phase shifts in the range of $p_{lab} \sim 1.3 - 1.7$ GeV/c, although they do not trace any clear resonant behavior yet. Error bars are still large.

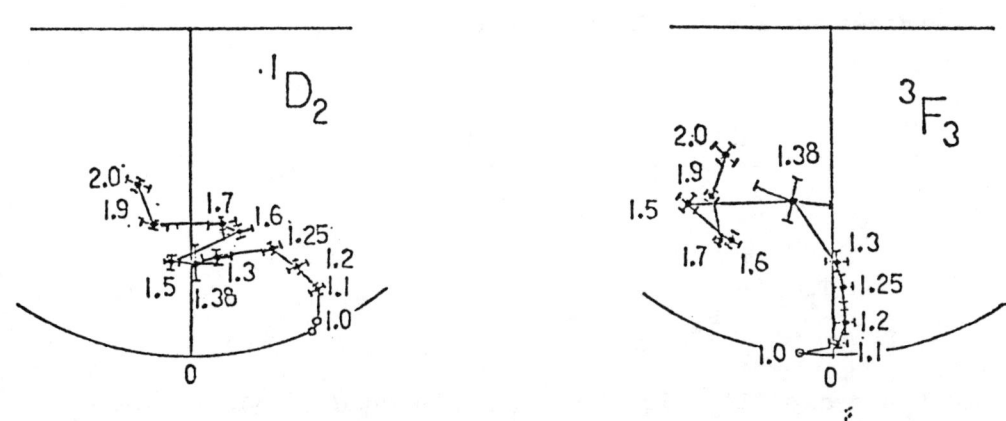

Fig. 4. Argand plots of the $^3F_3$ and $^1D_2$ amplitudes. Each $p_{lab}$ value at which the phase shifts have been obtained is attached in units of GeV/c.

## §3. The $^3F_3$(2.22 GeV) and $^1D_2$(2.17 GeV) diprotons

We now estimate the resonance parameters of the $^3F_3$ diproton. This can be done by first separating out the background amplitude $T_\infty$ in the $^3F_3$ wave from the true $^3F_3$ amplitude T of the present solution. The resonant part $T_R$ in T is then parametrized as a Breit-Wigner resonance:

$$T = \frac{\eta(^3F_3)e^{2i\delta(^3F_3)} - 1}{2i} = S_\infty T_R + T_\infty, \quad (3.1)$$

$$T_R = \frac{T - T_\infty}{S_\infty} = \frac{x}{\varepsilon - i}, \quad (3.2)$$

where $S_\infty = \eta_\infty \exp(2i\delta_\infty)$ is the $^3F_3$-background S-matrix element and $T_\infty = (S_\infty - 1)/2i$. x is the elasticity: $x = \Gamma_{el}/\Gamma$, with $\Gamma_{el}$ being the elastic partial width and $\Gamma$, the total width. $\varepsilon$ is given by $\varepsilon = 2(M_R - E)/\Gamma$, where $E(=\sqrt{s})$ is the two-nucleon total energy and $M_R$, the mass of the $^3F_3$ diproton.

Practically, the background phases $\delta_\infty$ and $\eta_\infty$ are assumed to take values as given in Fig. 5 by dashed curves, although they are ideally to be determined together with the resonance parameters by least-squares-fitting to the phase shift solution. The resonant amplitude is obtained from Eq.(3.2) as is shown in Fig. 6. This gives the $^3F_3$ resonance parameters as

$$M_R \approx 2.22 \text{ GeV}, \quad \Gamma_{E=M_R} \sim 100\text{–}150 \text{ MeV}, \quad x_{E=M_R} \sim 0.2. \quad (3.3)$$

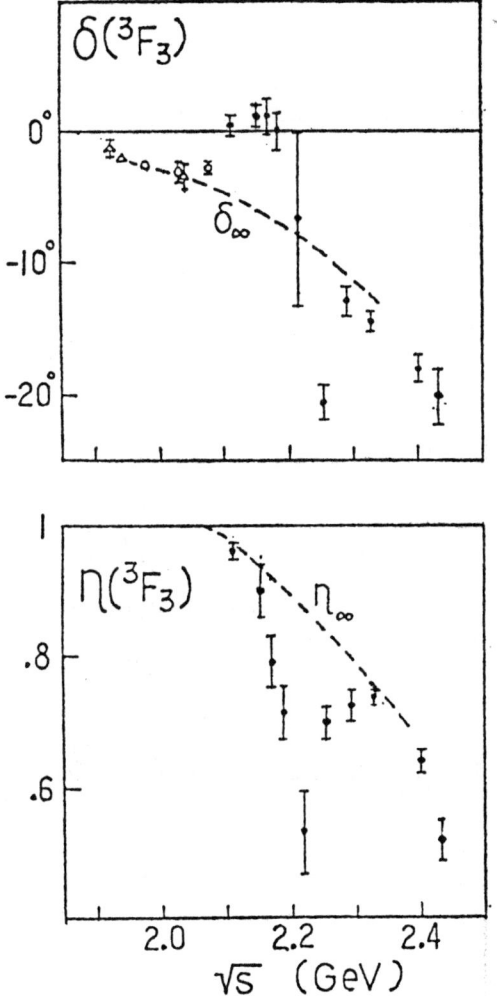

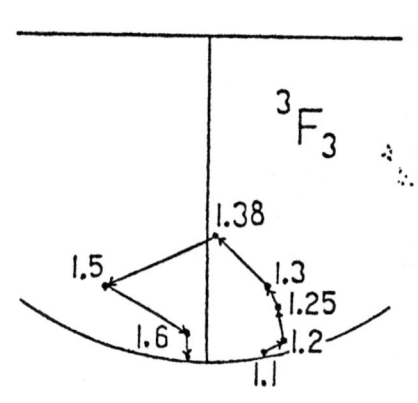

Fig. 6. Argand diagram of the $^3F_3$ resonant amplitude $T_R$. Numbers are $p_{lab}$ in GeV/c.

Fig. 5. Plots of $\delta(^3F_3)$ and $\eta(^3F_3)$ versus $\sqrt{s}$. Dashed curves represent background parts in the $^1D_2$-wave. Low-energy phase shifts are shown in comparison (white circles).[16),22),23)]

In the same way, the $^1D_2$ resonance parameters are estimated to be

$$M_R \approx 2.175 \text{ GeV}, \quad \Gamma_{E=M_R} \sim 50\text{--}100 \text{ MeV}, \quad x_{E=M_R} \sim 0.1. \tag{3.4}$$

The values of the x indicate that both these resonances are highly inelastic.

Unlike the $^3F_3$(2.22 GeV) resonance, the $^1D_2$(2.17 GeV) diproton is quite near the threshold for the pp→NΔ process which is fed from the $^1D_2$ pp-state. So, there is a threshold effect, i.e., a cusp. In order to take this effect into account, we try to perform a Breit-Wigner fit to the $^1D_2$ phase shift of the phase shift solution near the resonance energy. It is assumed that the $^1D_2$(2.17 GeV) diproton decays into the reaction channel r via the NΔ S-state with momentum q. The momentum dependence of the Glashow-Rosenfeld form is used for $\Gamma_{el}$ and $\Gamma_r$:

$$\Gamma_{el} = \gamma_{el} \left| \frac{p^2}{p^2 + X^2} \right|^2 \frac{p}{M_R}, \tag{3.5}$$

$$\Gamma_r = \gamma_r \frac{q}{M_R}, \tag{3.6}$$

where X is related to the size of the interaction.

A cusp effect is naturally taken into account by letting q imaginary (+i|q|) below the NΔ-threshold. The cusp in the present case may not be sharp due to the fact that the Δ-particle itself is a resonance. It is expected to get rounded with a width of the order of the width $\Gamma_\Delta$ of the $\Delta$-particle in the channel r. We considered this effect by letting q complex, as evident from the energy-conservation law for pp → NΔ: $E_N+E_\Delta=E$ with $E_N=(q^2+M_N^2)^{1/2}$ and $E_\Delta=(q^2+(M-i\Gamma_\Delta/2)^2)^{1/2}$. Note that Im(q)>Re(q) below the threshold, Im(q)<Re(q) above the threshold and q≠0 at the threshold, washing out the sharpness of the cusp at the threshold. The reaction width $\Gamma_r$ and hence the total width $\Gamma$ is now complex by Eq.(2.5), whose imaginary part gives rise to a level shift of the resonance mass. In Eq.(2.2), ε and x are now given by

$$\varepsilon = 2(M_R-E)/\text{Re}\Gamma, \quad M_R=M_R^0+\tfrac{1}{2}\text{Im}\Gamma, \quad x=\Gamma_{el}/\text{Re}\Gamma. \quad (3.7)$$

A Breit-Wigner curve is drawn in Fig. 7 by solid lines obtained from the following parameters:

$$\begin{cases} \gamma_{el} \approx 0.05 \text{ GeV}, & \gamma_r \approx 0.9 \text{ GeV}, \\ X \approx 0.28 \text{ GeV}, & M_R^0 \approx 2.145 \text{ GeV}. \end{cases} \quad (3.8)$$

We note that $(\text{Im}\Gamma/2)_{E=M_R} \approx 30$ MeV, $(\text{Re}\Gamma)_{E=M_R} \approx 90$ MeV. The resonant amplitude $T_R$ got in this way is plotted on the Argand plane as in Fig. 8.

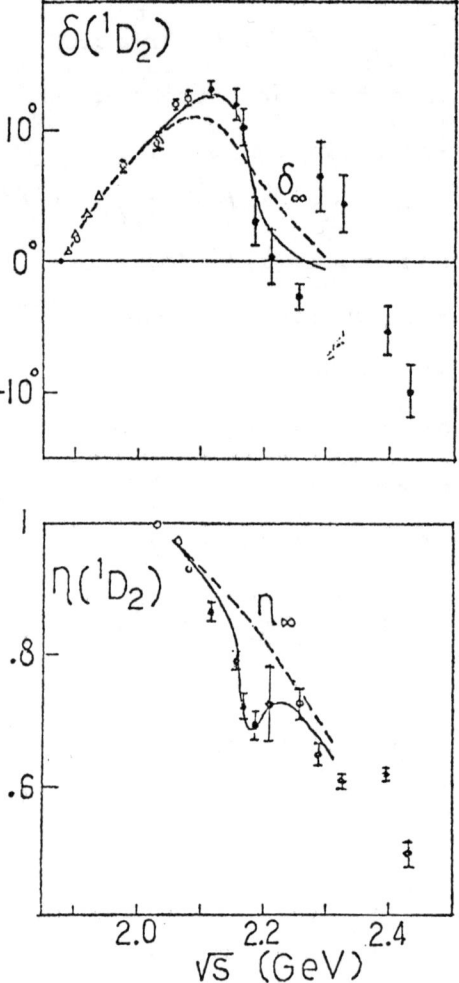

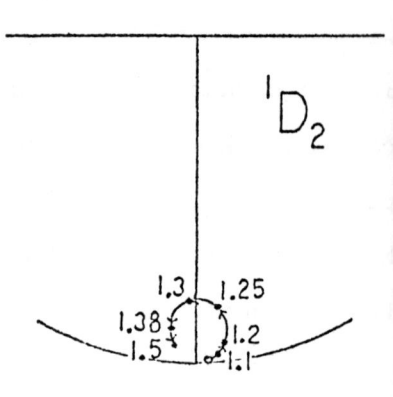

Fig. 8. Argand diagram of the $^1D_2$ resonant amplitude $T_R$. Numbers are $p_{lab}$ in GeV.

Fig. 7. Plots of $\delta(^1D_2)$ and $\eta(^1D_2)$ versus $\sqrt{s}$. Dashed curves represent background parts in the $^1D_2$-wave. Low-energy phase shifts are shown in comparison.[16),23)] Solid curves are the Breit-Wigner fit whose parameters are given by Eq.(3.8).

## §4. Interpretation of experimental data

There exist a variety of structures in the experimental data. I would like to explain some of them on the basis of the reported two resonances. Fig. 9 shows the data on $\Delta\sigma_L$, $\Delta\sigma_T$, $\sigma_{tot}$ and the Re/Im ratio of the forward amplitude $\phi_1+\phi_3$ as functions of $p_{lab}$. According to the present phase shift solution, the peak of the $\Delta\sigma_L$ data corresponds to $B^2$(2.17 GeV, $^1D_2$) and the valley, to $B^2$(2.22 GeV, $^3F_3$). A small peak in $\Delta\sigma_T$ is due to $B^2$(2.17 GeV, $^1D_2$). No structure is seen around 2.22 GeV, in agreement with the fact that the $\Delta\sigma_T$ does not contain the $^3F_3$ amplitude. These structures of $\Delta\sigma_L$ and $\Delta\sigma_T$ should be compared with the smoothly rising behavior of $\sigma_{tot}$. Spin effects are significant.

We next interpret the Re/Im ratio of the forward amplitude $\phi_1+\phi_3$, which rapidly falls off and even changes its sign from positive to negative at $p_{lab} \sim 1.4$ GeV/c. This important feature has been known long before but with no reasonable explanation.[24] According to the present phase-shift solution, this can be essentially explained in terms of the existence of two resonances: the $^1D_2$(2.17 GeV) and $^3F_3$(2.22 GeV). We emphasize that such a feature as rapid fall-off with sign change is characteristic of a resonant real amplitude ($\frac{1}{2}\eta\sin2\delta$) near a resonance, if

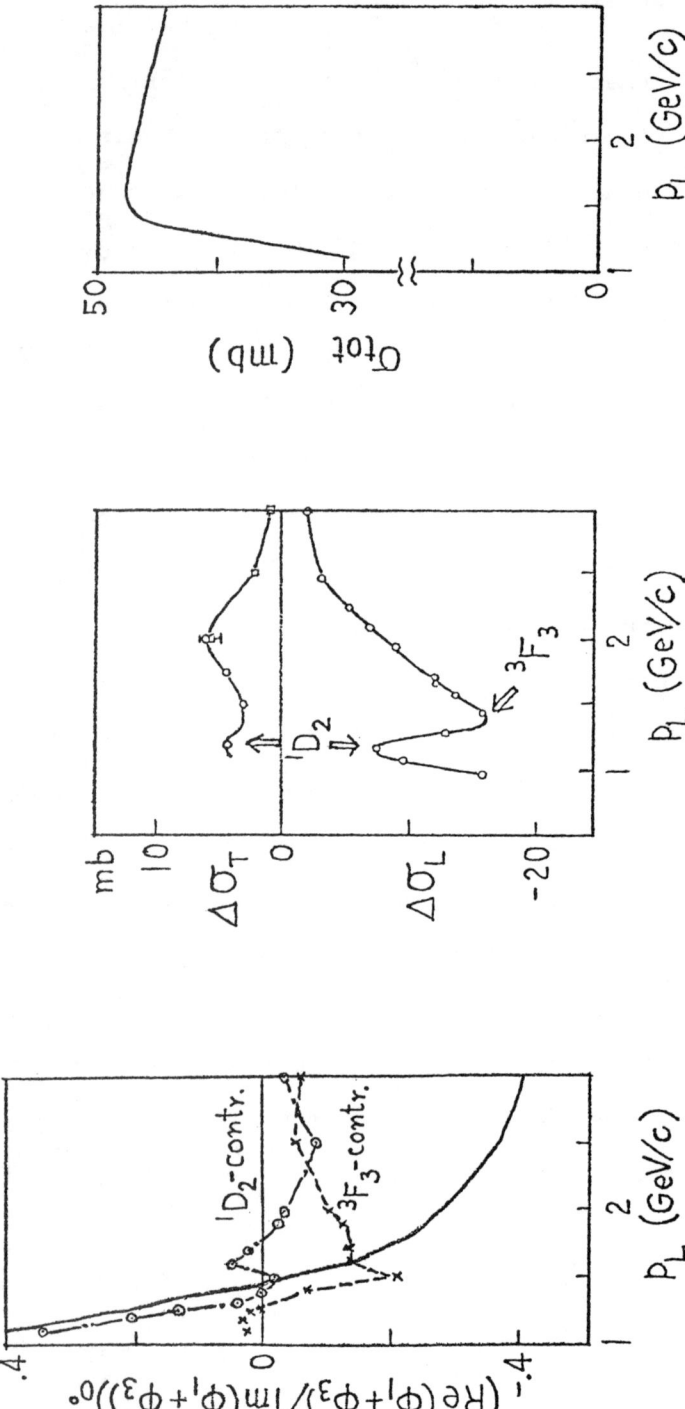

Fig. 9. Plots of $\Delta\sigma_L$, $\Delta\sigma_T$, $\sigma_{tot}$ and Re/Im of $(\phi_1+\phi_3)_{0°}$. Solid lines represent experimental data.

no attention is paid to a background effect. The background phase shift $\delta_\infty$ is positive for the $^1D_2$ and negative for the $^3F_3$ wave, and hence two resonances of the $^1D_2$ (2.17 GeV) and $^3F_3$ (2.22 GeV) are surely needed for an explanation of the data.

Next interpreted is the differential cross section $d\sigma/dt$ as represented in Fig. 10 against $p_{lab}$. Here the scattering angle $\theta$ is fixed to 10°, 20°,..., 90°. A significant peak is seen at $p_{lab} \sim 1.5$ GeV/c for small $\theta$, which decreases as $\theta$ goes up and vanishes for $\theta \gtrsim 50°$. We find a clear correspondence between this remarkable structure of the peak and the $^3F_3$ diproton. The variation of the peak as a function of $\theta$ may be characteristic of $P_2(\cos\theta)$ if $d\sigma/dt$ is expanded as $d\sigma/dt = \sum_{\ell=0,2,..} a_\ell(E) P_\ell(\cos\theta)$. It is difficult to explain this feature of the peak in terms of a $^3P$-wave resonance, because if it exists the $P_0(\cos\theta)$-term would also have a sizable structure which is against experimental facts. It is to be interpreted as a $^3P$-$^3F$ interference with the $^3P$-amplitudes serving as background amplitudes and a $^3F$ one, as a resonance. We note that the $^1D_2$ (2.17 GeV) diproton contributes little to the peak. In this way, the structure of $d\sigma/dt$ at $\sim 1.5$ GeV/c is strongly in favor of the asignment of $^3F_3$ to $B^2$ (2.22 GeV) diproton.

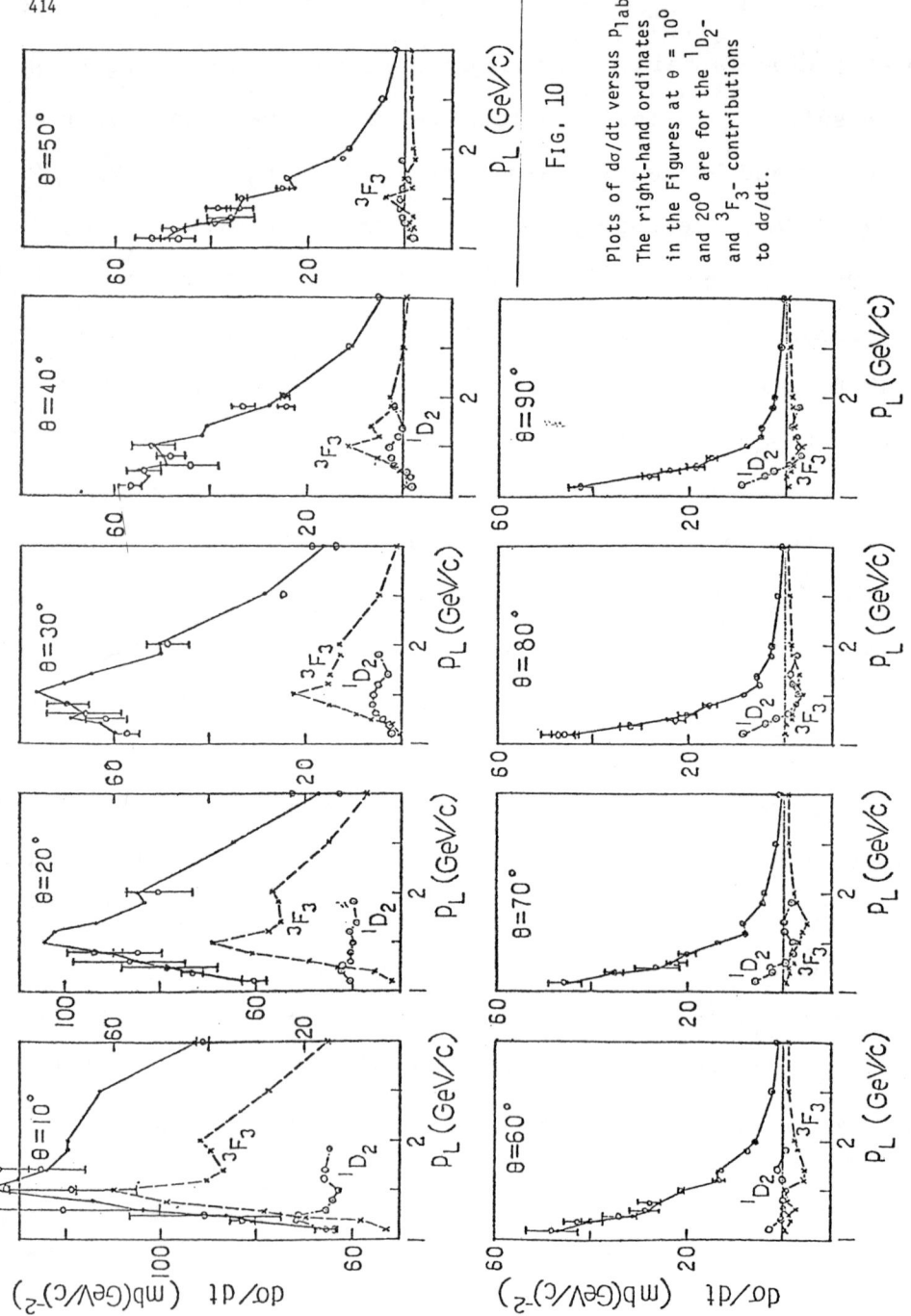

FIG. 10

Plots of $d\sigma/dt$ versus $p_{lab}$. The right-hand ordinates in the Figures at $\theta = 10°$ and $20°$ are for the $^1D_2$- and $^3F_3$- contributions to $d\sigma/dt$.

The polarization data also have a peak around 1.4 GeV/c. The peak varies from sharp to broad form as the scattering angle increases, and this variation of the peak form is also a manifestation of the $^3F_3$(2.22 GeV) diproton. We note that the singlet waves have nothing to do with the polarization.

Lastly mentioned are the spin-spin correlation parameters $C_{nn}$ and $C_{\ell\ell}$. There is a steep rise of the $C_{nn}(90°)$ data from $P_{lab} \sim 1.2$ GeV/c, which is a reflection of the structure in the singlet amplitude. This feature of the data remains for $\theta=80°$ and tends to disappear as $\theta$ goes down, in agreement with the behavior of the $^1D_2$-contribution to the $C_{nn}$. We note that the recent $C_{nn}(90°)$ data are strongly in favor of $B^2(2.17$ GeV, $^1D_2)$.[25] The recent Argonne data on the $C_{\ell\ell}(70°-90°)$ also have this feature. Here appears a dip at $\sim 1.2$ GeV/c for $\theta=90°$ and this dip tends to vanish as $\theta$ becomes smaller, in correspondence with the behavior of the $^1D_2$-contribution or of the singlet amplitude.

## §5. Discussion and conclusion

So far, I have discussed two structures in the pp system, namely the $^3F_3$(2.22 GeV) as a strong candidate for resonance and another possible one, the $^1D_2$(2.17 GeV). Let me discuss more about the $^1D_2$(2.17 GeV), since there has been a controversy whether this state is a resonance or an NΔ threshold effect like a nonresonant cusp or a final state interaction with an intermediate Δ.[26],[27] As was shown in Fig. 4, the $^1D_2$ amplitude traces a half-circle in a counter-clockwise direction on the Argand plane as $P_{lab}$ goes up from 1.0 to 1.5 GeV/c. This is much larger than the 90° left-turn of amplitude peculiar to the pure cusp state. If we subtract a background effect as was done in §3, the rotation of the resonant amplitude $T_R$ marks a three-quarter circle when $P_{lab}$ increases from 1.0 to 1.5 GeV/c, a three times larger than the 90° left-turn in question, as is seen in Fig. 8. Hence it is difficult to consider the discussed $^1D_2$ state as a rounded cusp state without a resonance interpretation. We have a resonance effect in addition to the cusp one. It is a problem beyond the range of the present talk whether a final state interaction in an inelastic channel can also explain the reported $^1D_2$ and $^3F_3$ diprotons in the elastic channel. All that we have shown here is that they can be Breit-Wigner resonances at least in the pp elastic channel, with the resonant behavior induced by the strong coupling to inelastic channels.

A hint of the third candidate for diproton resonances may be seen in the peak of the $\Delta\sigma_T$ data at 2 GeV/c ($\sqrt{s}\sim 2.43$ GeV) in Fig. 9. A dispersion analysis has suggested a resonance in the singlet state.[4] In our preliminary consideration, the anomalous behavior of the $^1S_0$ phase shift around 2 GeV/c shown in Fig. 1 corresponds to this peak. But error bars of the phase shifts are still large. Even more, ambiguities arise from the fact that this peak lies almost at the threshold for pp→NN*(1.47 GeV). Here also we encounter the problem of threshold effects.

There might be more diproton states in the $^3P$ wave as predicted by quark models or channel-coupling calculations.[28)-32)] A slight suggestion for their existence may be seen in Fig. 1 for the $^3P$ phase shifts, although it is not significant at present.

In conclusion we note that it is an interesting question whether a dibaryon resonance as reported here is similar to the deuteron or it is a single hadron in a six quark state. We mention that the $^1D_2$(2.17 GeV) and $^3F_3$(2.22 GeV) correspond to 0.8fm and 1.1fm of the impact parameter between the incident protons. These values are consistent with the size of the ordinary baryon resonances.

## References

1) I. P. Auer et al., Phys.Lett. __67B__ (1977), 113; __70B__ (1977), 475

2) N. Hoshizaki, Prog. Theor. Phys. __58__ (1977), 716.

3) K. Hidaka et al., Phys. Lett. __70B__ (1977), 479.

4) W. Grein and P. Kroll, Nucl. Phys. __B137__ (1978), 173.

5) T. Kamae et al., Phys. Rev. Lett. __38__ (1977), 468; 471.

6) N. Hoshizaki, NEAP-18, 19 (1978).

7) Sol. 3 recommended by P. Kroll, Private communication.

8) J. Bystricky et al., CEA-N-1947(E) (1972).

9) W. de Boer et al., Phys. Rev. Lett. __34__ (1975), 558.

10) G. W. Abshire et al., ANL_HEP-CP-75-73(1975) 22.

11) D.M. Miller et al., Phys. Rev. Lett. __36__ (1976), 763.

12) D. Aebischer et al., Phys. Rev. __D13__ (1976), 2478.

13) I.P. Auer et al., Phys. Lett. __70B__ (1977), 475.

14) D. Besset et al., preprint (1977)

15) D. Besset et al., preprint for Zurich Conf. (1977).

16) D. Axen et al., Lett. Nuov. Cim. __20__ (1977), 151.

17) P. Kroll, loc. cit.

18) I. P. Auer et al., ANL-HEP-PR-78-13 (1978).

19) Ed. K. Biegert et al., Phys. Lett. __73B__ (1978), 235.

20) A. Yokosawa, ANL-HEP-CP-78-11 (1978); ANL-group, private communications.

21) U. Amaldi, Jr. et. al., Nuov. Cim. __47__ (1967), 85.
    S. Francaviglia, Prog. Theor. Phys. __39__ (1968), 676.

22) N. Tamura et al., J. Phys. Soc. Japan Suppl. 44 (1978), 289.

23) R. A. Arndt and M. H. MacGregor, Phys. Rev. 141 (1966), 873.

24) W. Grein, Nucl. Phys. B131 (1977), 255. Early papers are cited therein.

25) A. Lin et al., Phys. Lett. 74B (1978), 273.

26) R. A. Arndt, Phys. Rev. 165 (1968), 1834.
    G. L. Kane and G. H. Thomas, Phys. Rev. D13 (1976), 2944.
    S. Furuichi and H. Suzuki, Prog. Theor. Phys. 57 (1977), 1803 and early papers cited therein.

27) M. A. Abolins et al., Pys. Rev. Lett. 15 (1965), 125.
    B. Werner et al., Phys. Rev. 188 (1969), 2023.
    G. Cozzika, CEA-N-1720 (1974).
    W. Grein and P. Kroll, Nucl. Phys. B137 (1978), 173.
    P. Kroll, WU B-78-13 (1978).

28) P. J. G. Mulders et al., Phys. Rev. Lett. 40 (1978), 1543.

29) M. Matsuda and M. Yonezawa, Prog. Theor. Phys. 58 (1977), 1327.

30) T. Ueda, Prog. Theor. Phys. 59 (1978), 657; Phys. Lett. 74B (1978), 123; OUAM 78-6-5.

31) V.G. Neudatchin, Yu. F. Smirnov and R. Tamagaki, Prog. Theor. Phys. 58 (1977), 1072.

32) O. Hara, S. Ishida and S. Y. Tsai, Prog. Theor. Phys. 58 (1977), 1325; S. Ishida and M. Oda, Prog. Theor. Phys. 59 (1978), 959; 60 (1978), No. 3, to be published.

## Proton-Proton Elastic Scattering Spin Correlative Parameter

## $A_{nn}(\theta)$ at 643 and 796 MeV.

H. B. Willard, M. W. McNaughton, P. R. Bevington, E. Winkelmann,
H. W. Baer*, and F. Cverna*, Case Western Reserve University,
E. P. Chamberlin, J. J. Jarmer, N. S. P. King, J. E. Simmons,
Los Alamos Scientific Laboratory, H. Willmes, University of Idaho,
and M. A. Schardt, Arizona State University.

The spin correlation parameter $A_{nn}(\theta)$ has been measured for proton-proton elastic scattering with the beam and target spins polarized normal to the scattering plane. In our experiment the polarized beam at LAMPF was produced by a Lamb shift source, generally with a transverse (normal) polarization of about 85% as monitored by our EPB polarimeter which was calibrated by the quench ratio technique. The direction of the beam polarization was reversed 180° on a 3 minute cycle. The polarized target constructed at P-Division at Los Alamos Scientific Laboratory produced a transverse (normal) polarization of 80-85% by $^3$He-$^4$He cooling to 0.4° K in a 25 kG conventional magnetic field and optical pumping with 70 Ghz microwaves. The target material is propanedial doped with chromium with a volume of about 7 cm$^3$. The direction of target polarization was reversed on a cycle of several hours.

Forward and conjugate protons scattered left and right by the polarized target were detected in coincidence by our system of MWPC's and fast digital readout system. Coplanarity and kinematic restrictions on the azimuthal and inclusive angles greatly reduces background from target material other than hydrogen. In addition MWPC telescopes on the conjugate arms further reduced background from the thin stainless steel walls of the cryogenic target system. The strong magnetic field altered the trajectories of forward and conjugate protons and careful attention to the kinematics was required particularly at the back angles.

Four measurements with different spin orientations and left and right detector arms give eight independent sets of data at each angle (and energy). Since the beam polarization is measured independently by the polarimeter (secondary calibration) this allows us to determine $A_y(\theta)$, $A_{nn}(\theta)$, and $P_T$ while averaging out instrumental asymmetries. The values of $P_T$ can be compared to the independent NMR measurements and $A_y(\theta)$ with the previously obtained analyzing power measurements.

* now at Los Alamos Scientific Laboratory

The spin correlation parameter $A_{nn}(\theta)$ had been obtained at 643 and 796 MeV. Results for 796 MeV are shown in Figure 1. The precision of these new data is about 5 to 10 times better than previous measurements in this energy range as indicated in Figure 2 where $A_{nn}(90°)$ for selected data 1,2 are plotted as a function of proton kinetic energy (laboratory) from 14 MeV. to 5 GeV. It should be noted that the 90° c.m. value of this spin correlation parameter gives the ratio of the singlet $\sigma_s(90°)$ to triplet $\sigma_t(90°)$ scattering cross sections:

$$\frac{\sigma_s(90°)}{\sigma_t(90°)} = \frac{1 - A_{nn}(90°)}{1 + A_{nn}(90°)}$$

Thus a value of $A_{nn}(90°) = -1$ corresponds to pure singlet scattering (as observed up to about 20 MeV) and $A_{nn}(90°) = +1$ corresponds to pure triplet scattering. In the medium energy range we observe $A_{nn}(90°) \approx 2/3$ indicating triplet dominance. The rapid decrease in $A_{nn}(90°)$ immediately above the medium energy range observed by Lin et al.[2] indicates that singlet and triplet scattering become more nearly equal at high energies. The reason for the sharp transition had not been found. However the observation[3] of resonance like structures just below this transition region may well be related.

Energy dependent phase shift analyses are in progress to fit the $\sigma(\theta)$, $A_y(\theta)$ and $A_{nn}(\theta)$ data at 643 and 796 MeV.

1. J. Bystricky et al., Centre d'Etude Nucléaires de Saclay, Report No. CEA-N-1547, (1972) unpublished.

2. A. Lin, J. R. O'Fallon, K. G. Ratner, P. F. Schultz, K. Abe, D. G. Crabb, R. C. Fernow, A. D. Krisch, A. J. Salthouse, B. Sandler, and K. M. Terwillinger, Phys. Lett. 74B, 273 (1978).

3. I. P. Auer, E. Colton, H. Halbern, D. Hill, H. Spinka, G. Theodosion, D. Underwood, Y. Watanabe, and A. Yogosawa, Phys. Rev. Lett. 41, 354 (1978).

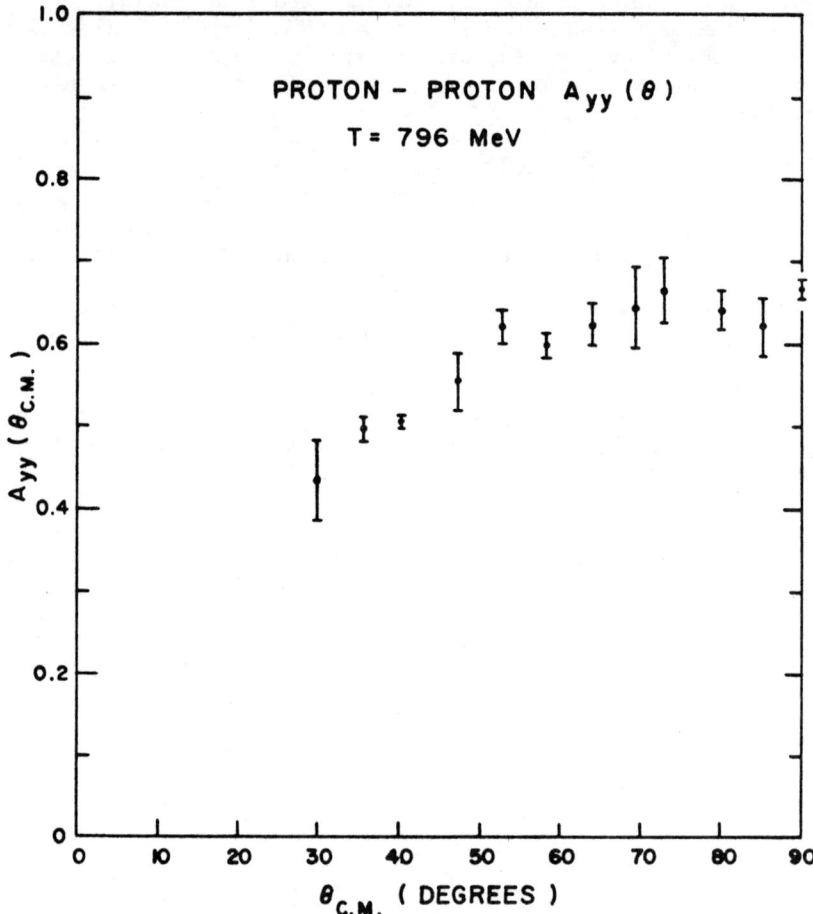

Figure 1. Proton-proton elastic scattering spin-correlation parameter $A_{nn}(\theta)$ at 796 MeV.

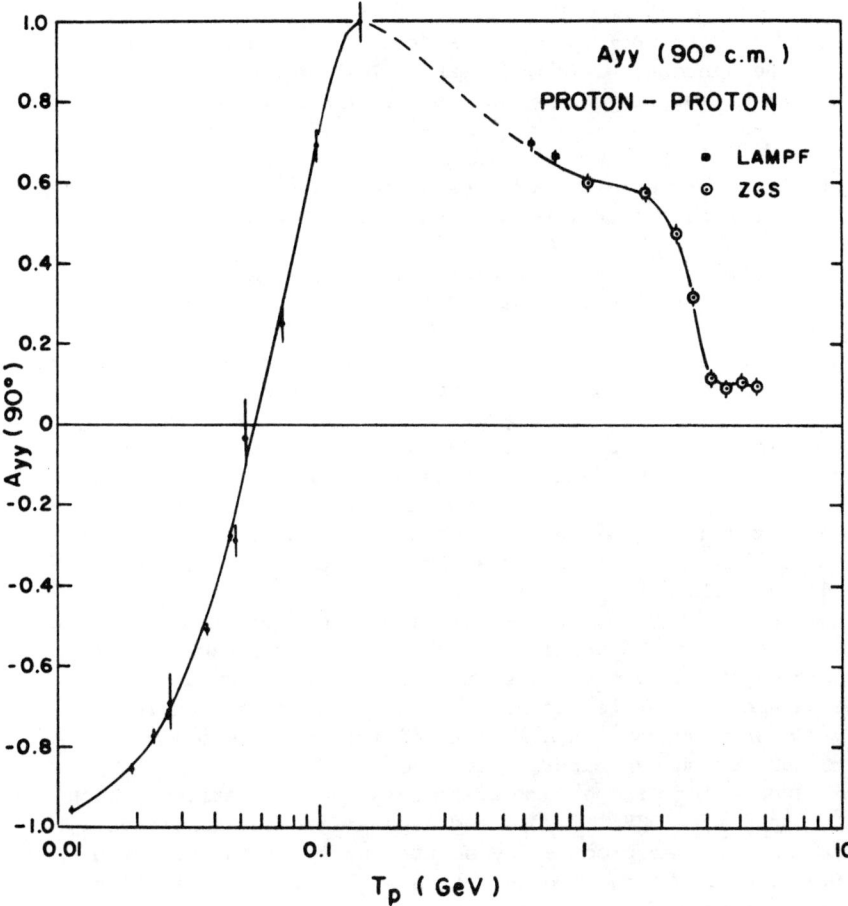

Figure 2. Proton-proton elastic scattering spin-correlation parameter $A_{nn}(90°)$ as a function of proton kinetic energy.

## MEASUREMENTS OF THE POLARIZATION A AND THE SPIN-CORRELATION $A_{nn}$***
## IN ELASTIC p-p SCATTERING BETWEEN 400 AND 600 MeV

D. Besset, Q.H. Do, B. Favier, R. Hausammann, E. Heer, R. Hess,
C. Lechanoine, W. Leo, D. Rapin, D.W. Werren*
DPNC, University of Geneva, Geneva, Switzerland

Ch. Weddigen
Institut für Experimentelle Kernphysik, der Universität Karlsruhe,
Kernforschungszentrum Karlsruhe, West Germany

J. Cameron**, S. Jaccard, S. Mango
SIN - 5234 Villigen, Switzerland

In recent years measurements of complex polarization parameters in elastic p-p scattering at intermediate energies have extended the success of the phase shift analysis (PSA) approach in describing the p-p scattering matrix up to about 500 MeV. Above this energy, however, there is still no definite agreement on the correct phase shift values. A survey of the data in this region, furthermore, shows a large gap between 400 and 600 MeV containing almost no spin-correlation data - the bulk of the existing data here being concentrated around 400 and 600 MeV.

In light of this and as a first step in our program towards a complete experimental determination of the scattering matrix at medium energies [1], we have measured the spin correlation and polarization parameters for large angle p-p scattering at 7 energies between 400 and 600 MeV. A new data acquisition method was also put into use for these measurements.

The experiment was performed at SIN using a polarized proton beam with $|\vec{p}_B|$ = (41.65 ± 0.43)% and a polarized butanol target. Beam polarization was produced by scattering the main 590 MeV unpolarized beam from a thin Be target and selecting the elastically scattered portion at 8°. The energy of the beam was then lowered using copper degraders which produced no observable depolarization effects. The polarization orientation was changed by using a superconducting solenoid located just before the experimental apparatus to rotate $\vec{p}_B$. Any angle including a 180° reversal could be reached.

The polarized target [2] consisted of a small thin-walled (.1 mm) copper cylinder, 19 mm in diameter and 20 mm high filled with 1 mm diameter butanol droplets immersed in liquid $^3$He. The cylinder was kept at a temperature of 0.5°K by $^3$He pumping and placed in a 25kG magnetic field. Vertical polarization of the free butanol protons

---

\* Present address: Landis & Gyr, Zug, Switzerland
\*\* On sabbatical leave from Univ. of Alberta, Edmonton, Canada
\*\*\* The notation used is in accordance with the Ann Arbor convention

was produced by dynamic nuclear orientation and monitored using an NMR technique. The maximum attainable polarization, $\vec{p}_T$, was ~ 65 % although the values used during the experiment were somewhat lower (~ 50 %).

The scattered and recoil protons which emerged from the target were detected by two telescopes each consisting of three X-Y plane multi-wire proportional chambers. The two telescopes were mounted on movable platforms which could be rotated about the central vertical axis of the target. For this experiment, the telescopes were positioned so as to allow a simultaneous measurement of elastic pp events between 30° and 90° center of mass scattering angle.

Events which satisfied pre-set electronic requirements were constructed on-line by a novel event reconstruction program [3]. This program, based on a least squares linearization of the reconstruction equations [4], enabled a complete determination of the relevant scattering parameters i.e. scattering angles, vertices, etc.., at a rate up to ~300 events/sec. This limit was actually due, in fact, to the speed of the hardware information transfers rather than to the program itself. During reconstruction, the program rejected those events which failed to meet the correct kinematical characteristics for elastic scattering and other relevant criteria. Surviving events were plotted as a set of histograms and stored on magnetic computer tape in this form. Data acquisition by this method was extremely efficient: it drastically reduced the amount of tape and tape handling required and greatly facilitated the final analysis.

At each energy, data were taken for six different beam-target configurations, i.e. $P_B$ up-down, $P_T$ up-down and zero. Background information was also collected at these energies using a "dummy" target made of Carbon and $^3$He set in the same proportion as that for butanol. In the analysis, the data were corrected for background contamination, chamber inefficiencies, system acceptance and bin centering.

Final results for $A_{nn}$ at seven energies are shown in Fig. 1 and 2 along with the PSA predictions [5] at these energies. In general a good agreement was found except at 557 and 578 MeV, where our measured $A_{nn}$ values were somewhat higher. Indeed when our measurements were admitted into the PSA program, the predictions at 578 MeV were significantly modified (see dashed curve). In addition, they also produced a factor of two reduction in the PSA errors. In the case of the polarization A, which comes out as a byproduct of our data, no discrepancy with the PSA values was found at any energy (see example Fig.2). This experiment, as well, has increased by several fold the amount of $A_{nn}$ and A measurements between 400 and 600 MeV, thereby remedying the deficiency in this region.

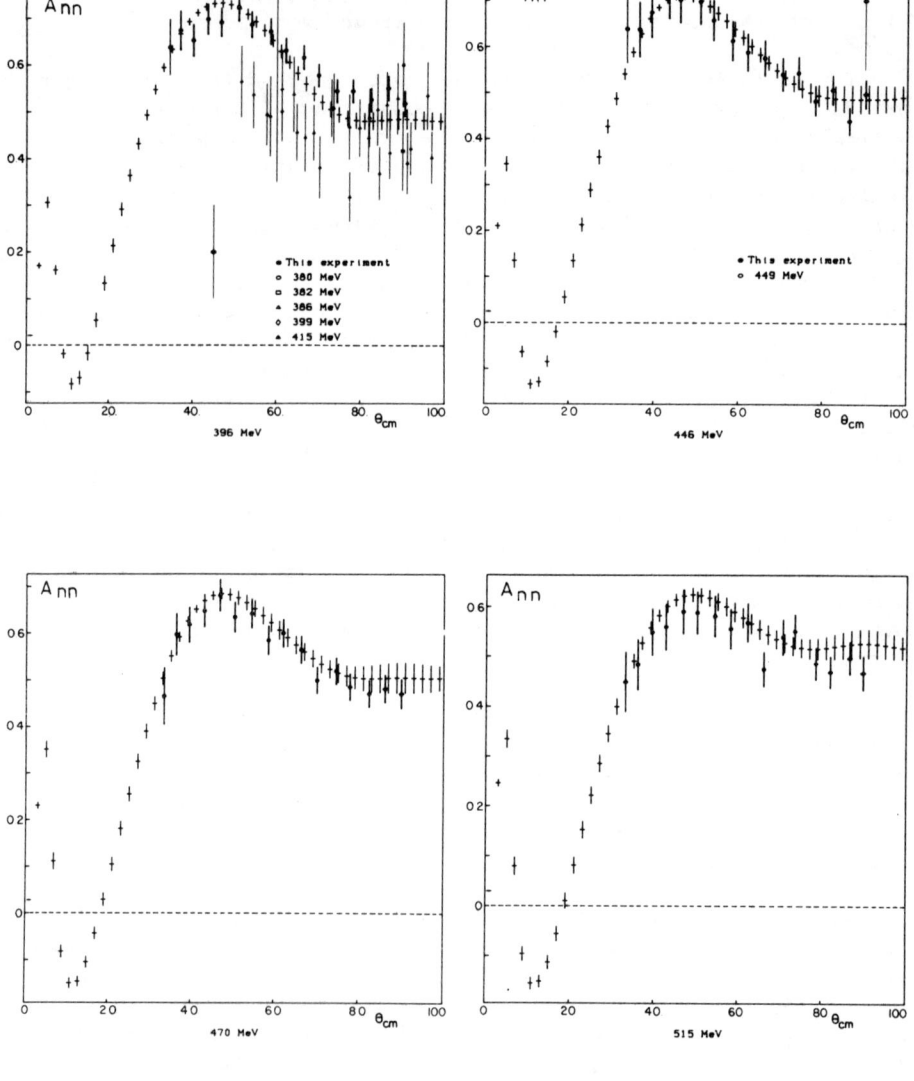

Fig. 1 $A_{nn}$

References

1. D. Besset et al., Nucl. Instr. and Meth. <u>148</u>, 129 (1978).
2. SIN Annual Report 1974.
3. D. Besset, Ph. D. thesis, 1978, unpublished.
4. For a general discussion of linearization technique, see for example Peter W.M. John, Statistical Design and Analysis of Experiments, The MacMillan Co., New York, 1971.
5. J. Bystricky and F. Lehar, 4th International Symposium on Polarization Phenomena in Nuclear Reactions, Zurich, August 1975, Birkhauser Verlag, W. Gruebler and V. Koenig ed.

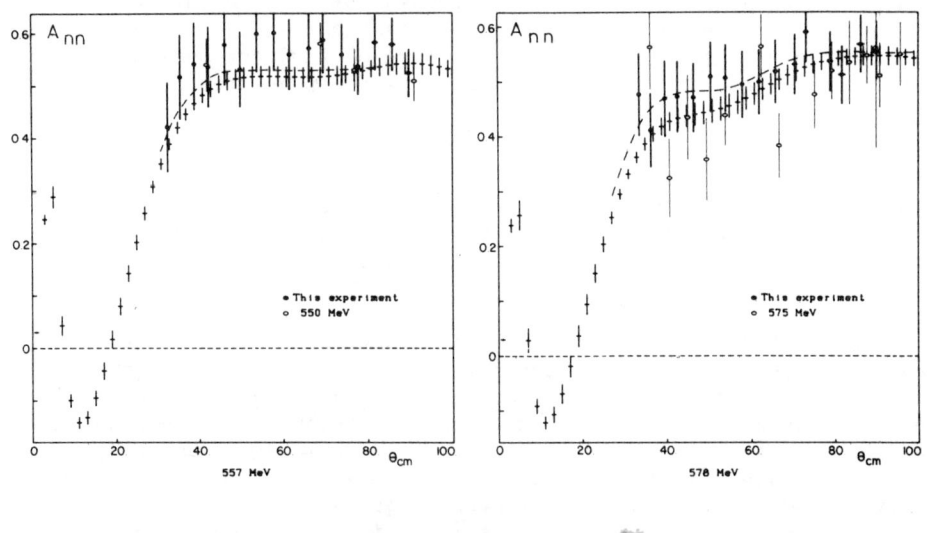

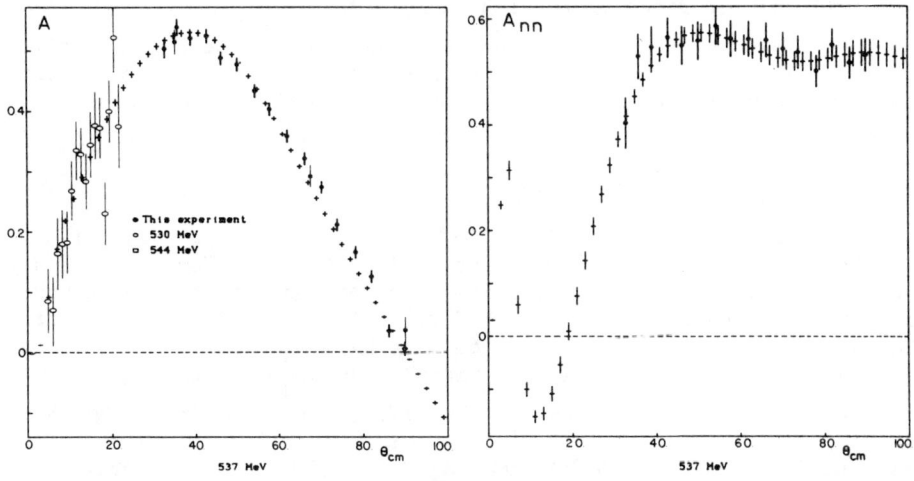

Fig.2 $A_{nn}$ and A

MEASUREMENT OF THE ENERGY DEPENDENCE OF $\Delta\sigma_T$ AND $A_{nn}$
FOR pp SCATTERING IN THE 1-3 GeV/C REGION*

T. A. Mulera[§]
Rice University, Houston, TX 77001

ABSTRACT

The polarized proton beam at the Argonne ZGS was used in conjunction with an N-type polarized proton target to extend previous measurements of $\Delta\sigma_T$ [1,2] in the region of incident momentum 1-3 GeV/c. In addition, a double arm spectrometer was used to measure the spin-spin asymmetry parameter, $A_{nn}$, for pp elastic scattering over complete angular distributions in this range of incident momentum. Preliminary results of these measurements are presented.

INTRODUCTION

Previous measurements [1,2,3] of the spin dependence of the pp total cross section for both transverse and longitudinal initial state spin orientations between laboratory momenta of 1 and 3 GeV/c have revealed an energy dependent structure whose interpretation remains unclear.

Attempts[4-8] have been made, without a great deal of success, to explain these structures as the effects of the crossing of various inelastic production thresholds. The Yokosawa Group has proposed[9] an explanation requiring the existence of various diproton resonances. Arik and Williams[10] take exception to this explanation and claim that the phase shifts between 1 and 2 GeV/c provide no evidence for a dibaryon resonance. Both authors agree that the picture would be clarified by measurement of $A_{nn}$, $A_{ss}$, and $A_{\ell\ell}$ in this range of laboratory momentum. In the hopes on contributing to this clarification, we have made further measurements of $\Delta\sigma_T$ and measurements of $A_{nn}$ over the full angular range in the 1-3 GeV/c region.

Preliminary results of these measurements are presented in this report. The preliminary nature of these results should be emphasized as only the most superficial off-line analysis of the data has been performed.

EXPERIMENT

Both measurements utilized the ZGS polarized proton beam with the beam polarization in the vertical direction and alternated in sign on a pulse to pulse basis. The beam was focused onto the new Argonne N-type polarized proton target (PPT VI). This target is a "standard" $He^4$-$He^3$ refrigerator cooling a 7 cm long by 2 cm diameter cavity containing ethanediol doped with chromium paramagnetic complexes. The

*Work supported by U.S. Dept. of Energy, Contract No.EY-76-S-05-5096.

free hydrogen protons in the ethanediol are dynamically polarized by a 70 GHz microwave system and this polarization is measured by a 106.5 MHz NMR system.

The measurement of $\Delta\sigma_T$ was carried out by the standard "good geometry" attenuation technique. The experimental layout did not differ significantly from that described for our previous measurements.[2]

For the measurement of $A_{nn}$, elastic scattering from the free hydrogen was detected by the two-arm spectrometer shown schematically in Fig.1. Both spectrometer arms were easily moveable about a pivot point under the PPT. The incident beam intensity was monitored by the scintillation telescopes labeled L, R, PL, PR which viewed polyethylene targets upstream of the PPT. (Beam intensities while measuring elastic scattering were typically 10-20 times higher than while measuring $\Delta\sigma_T$ making direct counting of the beam impractical.) The sums L+R and PL+PR were taken to compensate for effects due to the polarization state of the beam.

The forward scattered proton was momentum analyzed and its scattering angle measured by the multiwire proportional chambers (MWPC), P3-P6 and a 20X42 C-magnet. The angle of the recoil particle was measured by MWPCs P1 and P2. Scintillation counters S1-S4 provided the fast electronic trigger for MWPC readout. Measurement of momentum and angle of the forward particle and angle of the recoil particle combined with measurement of the time of flight between, and pulse height in, scintillation counters allowed clean separation of elastic scattering from free hydrogen from the scattering from the carbon and oxygen in the target material.

After obtaining the normalized event rate, N(ij), from the number of elastic scatters in each of the four initial spin states (i,j=beam, target), the spin-spin asymmetry parameter $A_{nn}$ was obtained as

$$A_{nn} = \frac{N(\uparrow\uparrow)+N(\downarrow\downarrow)-N(\uparrow\downarrow)-N(\downarrow\uparrow)}{P_B P_T \sum_{ij} N(ij)}$$

where $P_B$ and $P_T$ are the polarizations of the beam and target respectively.

## RESULTS

Preliminary results of the new measurements of $\Delta\sigma_T$ are shown in Fig.2. along with previous results.[1,2] The previously observed dip at 1.5 GeV/c and the maximum at 2.0 GeV/c would appear to be confirmed with no new structure making itself evident.

Some preliminary results of the measurement of $A_{nn}$ are shown in Fig.3. These values are on-line estimates and are extremely crude. Nevertheless, the data appear to be in good agreement with other measurements[11,12] in this range of incident momentum except possibly near $\Theta_{cm}=90°$. Although these data are extremely preliminary, one can see a suggestion that the angular dependence of $A_{nn}$ undergoes a qualitative change around 1.5 GeV/c. Below 1.5 GeV/c, $A_{nn}$ shows a maximum at 90° CM while above 1.5 GeV/c the data peak at somewhat smaller angles.

Off line analysis of our data should clarify the situation by providing measurements at additional values of incident momentum (especially in the vicinity of 1.5 GeV/c) and reducing our relative errors to less than ±2-3% and allowing the comparison of these results to the predictions of phase-shift calculations.

## ACKNOWLEDGMENTS

§The other physicists involved were D. Bell, J. A. Buchanan, M. C. Calkin, J. M. Clement, W. H. Dragoset, M. Furić, J. H. Hoftiezer, K. Johns, D. M. Judd, J. D. Lesikar, H. E. Miettinen, G. S. Mutchler G. P. Pepin, G. C. Phillips, J. B. Roberts, and S. E. Turpin, all from the T. W. Bonner Nuclear Laboratories, Rice University, Houston, Texas.

## REFERENCES

1. W. deBoer et al., Phys.Rev.Lett. 34, 558 (1975).
2. Ed.K. Biegert et al., Phys.Lett. 73B, 235 (1978).
3. I. P. Auer et al., Phys.Lett. 67B, 113 (1977).
4. S. Mandlestam, Proc.Roy.Soc. A244, 491 (1958).
5. G. L. Kane and G. H. Thomas, Argonne Report, ANL-HEP-PR-75-56 (1975).
6. E. L. Berger, P. Pirila and G. H. Thomas, Argonne Report, ANL-HEP-PR-75-72 (1975).
7. G. N. Epstein and D. O. Riska, Michigan State Preprint (1976) unpublished.
8. W. M. Kloet, R. R. Silbar, R. Aaron and R. D. Amado, LASL Report LA-UR-77-2321 (1977) and to be published.
9. H. Hidaka et al., Phys.Lett. 70B, 479 (1977).
10. M. Arik and P. G. Williams, Nucl.Phys. B136, 475 (1978).
11. D. Miller et al., Phys.Rev.Lett. 36, 763 (1976).
12. Presentation of J. Simmons at LAMPF Nucleon-Nucleon Workshop, 27 July 1978 [unpublished].

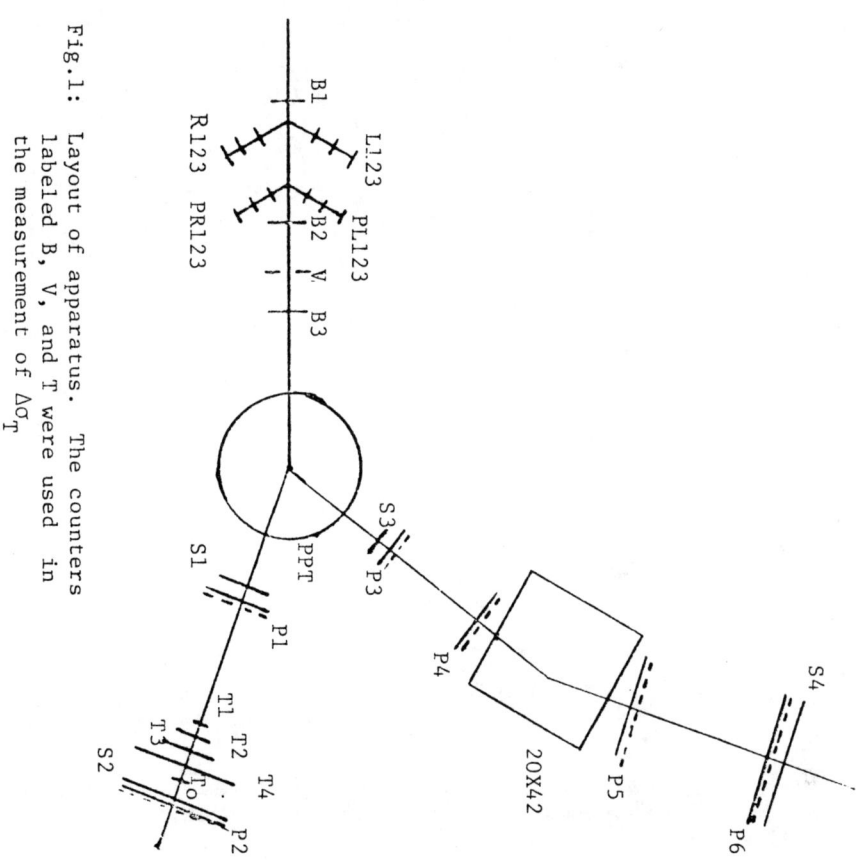

Fig.1: Layout of apparatus. The counters labeled B, V, and T were used in the measurement of $\Delta\sigma_T$

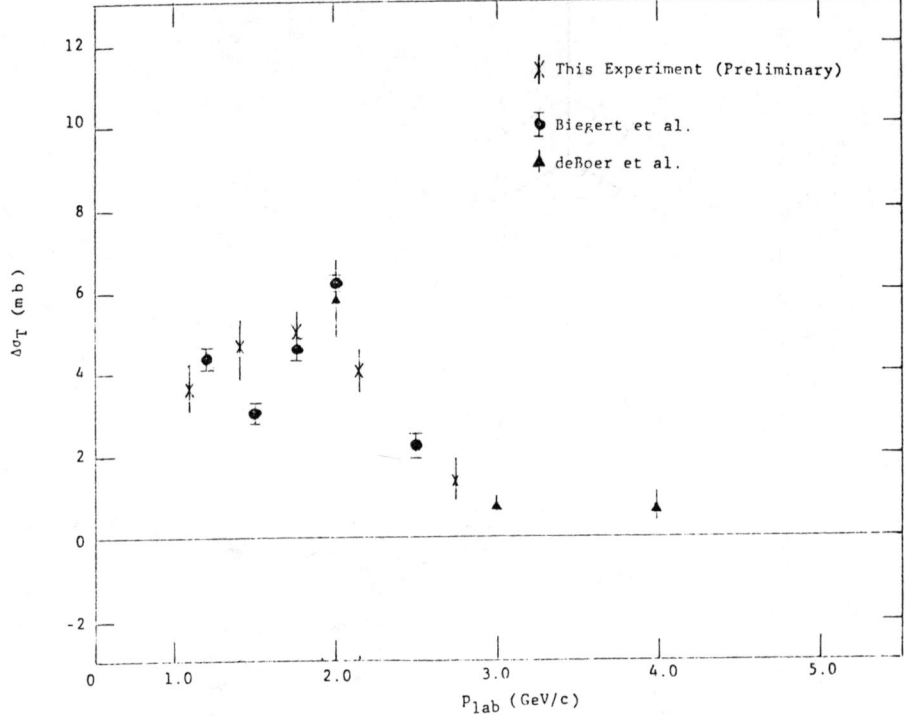

Fig. 2 The proton proton transverse total cross section difference as a function of incident momentum.

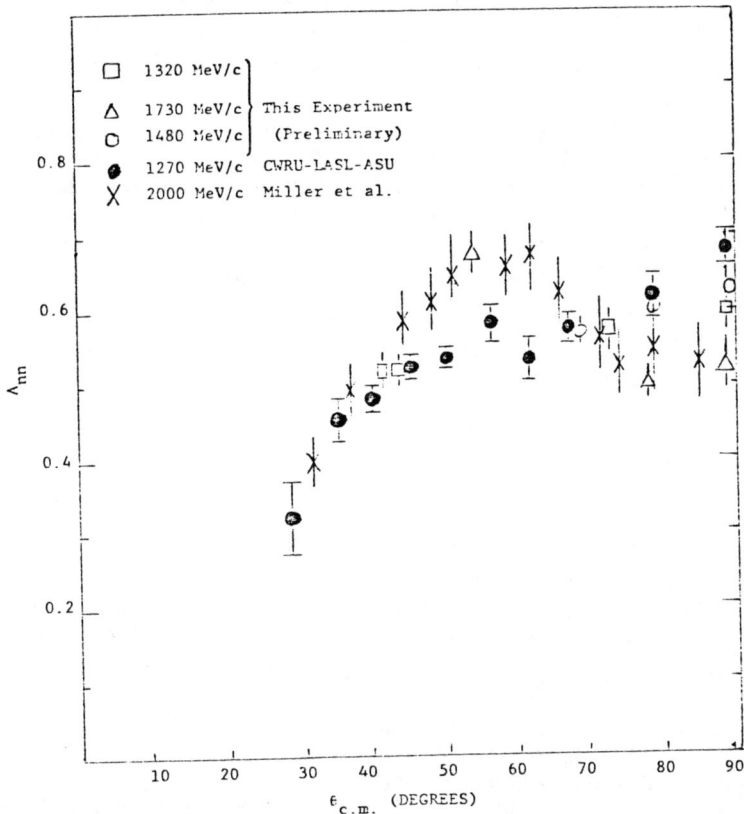

Fig. 3 Measurements of $A_{nn}$ in proton proton elastic scattering.

Chapter 6   Hadronic Interactions

## Polarized Target Experiments at CERN

M. Fidecaro
CERN - European Organization for Nuclear Research
1211 Genève 23, Switzerland

In recent years the study of spin effects in elementary particle physics at CERN has followed different patterns. Here I consider only experiments designed for the use of polarized targets, leaving aside other experiments from which interesting information on the spin of the constituents comes out. No polarized hadron beam is in operation.

At CERN polarized targets are used in the GeV to the few hundred GeV range in two lines of research, namely:

i) in two body or quasi two body processes, and there are six experiments of this type. At the lower energies one aims at constructing unambiguous sets of reaction amplitudes. At the intermediate energies one is still able to compare experimental data with prediction of models, in a range of energy where spin retains selection power for the choice of models. As the energy becomes higher the question to answer is primarily whether spin effects are important at all.

ii) in experiments where the polarized nucleon is probed in an inclusive way via either hadronic or electromagnetic interactions. Here the presence of bounds (unpolarized) nucleons in the target is more disturbing. There are two experiments of this type.

What follows is an index of the CERN experiments which make use of a polarized target. The additional information I gave at the time of my talk was reported in extenso in the parallel sessions directly by the authors, and is published in these proceedings, or elsewhere. I include also experiments not any more on the floor, but for which the analysis was either under way or had been recently completed at the time of the conference. Of the eight experiments mentioned, six refer to the intermediate energy programme with the 25 GeV Proton-Syncrotron (PS), and two make use of the high energy beams available at the 450 GeV Super Proton-Synchrotron (SPS), one in the West Area (WA) and the other one in the North Area (NA).

The layout of these experiments, together with the complete list of the authors and the reference to the original proposal can be found in the CERN experiment book, published yearly. The targets, except for Exp. PS 160, were developed by the CERN Polarized Target Group (T. Niinikoski et al.).

1. Experiment PS 134, CERN-Zürich (ETH)-Helsinki-London (Imperial College) collaboration [1].

    Beam: 5 GeV/c negative pions ($q_{11}$).
    Target: propanediol, 15 cm long, 1.7 cm diameter, frozen spin technique, 90 % polarization in a holding field of $\sim$ 10 kG with a decay smaller than 1 % per day, and of 4 % per day, respectively,

for positive and negative sign, 25 kG polarizing field, dilution refrigerator [2].

Aim: P, A, R parameters in the reaction $\pi^-p \to K^0\Lambda^0$ are measured up to $-t \sim 1.5$ GeV$^2$ so that the amplitudes can be constructed in the corresponding angular range. The $\Lambda$'s associated to the forward produced $K^0$'s are observed inside a large magnet filled with a system of optical spark chambers having a wide angular acceptance. The $\Lambda$ polarization is measured by taking advantage of the $\Lambda$ analysing power. About 3.5 % of the pictures show both the $K^0$ and the $\Lambda^0$. The rather delicate analysis is in progress and the final statistics should include $\sim 10^5$ $\Lambda$'s.

2. Experiment PS 136, CERN-München (MPI) Collaboration [3].

   Beam: 17.2 GeV/c negative pions ($p_{14}$).
   Target: butanol, 10 cm long, 1.5 cm diameter, 70 % polarization, He$^3$ refrigerator [4].

   Aim: the moments of the ($\pi^+\pi^-$) angular distribution are measured in the reaction $\pi^-p \to \pi^+\pi^-n$. By combining these moments with the polarization independent moments from a previous experiment (in hydrogen) the production amplitudes can be determined except for one relative phase between two sets of amplitudes (and the over-all phase). The ($\pi^+\pi^-$) four-momentum is determined up to $-t = 0.2$ GeV$^2$ and m = 1.8 GeV by means of a magnetic spectrometer with magnetostrictive readout wire chambers. The dipion distribution shows a large up-down asymmetry pointing to the importance of the $A_1$ type terms in the exchange amplitudes.

3. Experiment PS 137, CEN-Saclay [5].

   Beam: 6 and 12 GeV/c positive kaons ($p_{16}$).
   Target: deuterated propanediol, 12 cm long, 1.7 cm diameter, 40 % neutron polarization, dilution refrigerator [6].

   Aim: the P parameter for $K^+n \to K^0p$ is measured up to $-t \sim 1$ GeV$^2$. The charged particles in the final state are detected in a MWPC spectrometer. In order to construct the corresponding amplitudes one has to take into account the information from other experiments. The polarization values from this experiment together with those for $\pi^-p \to \pi^0n$, $\pi^-p \to \eta n$, $K^-p \to K^-p$ measured at comparable energies, fulfil the SU(3) sum rule.

4. Experiment PS 160, Edinburgh-RHEL-London (Westfield College) collaboration, in preparation.

   Beam: 1.5 to 2.6 GeV/c positive pions ($q_{13}$).
   Target: propanediol, 12 cm long, 1.5 cm diameter, frozen spin technique, dilution refrigerator (similar to the one for experiment PS 134).

   Aim: A and R parameters for the reaction $\pi^+p \to K^+\Sigma^+$ in the 1930-2400 MeV c.m. region are determined by measuring the final state charged particles in a large acceptance magnet equipped with a set of capacitative readout spark chambers. The amplitudes will be constructed by using the hydrogen data already available from the same set up run at NIMROD. The purpose of the experiment is to obtain information on the SU(6) structure of high mass Δ* resonances.

5. Experiment PS 141, CERN-LAPP-Oxford collaboration [7].

   Beam: 24 GeV/c protons ($c_9$).

   Target: propanediol, 4.3 cm long, 1.4 cm diameter, 70 % polarization, $He^3$ evaporation refrigerator [4].

   Aim: i) measurement of the P parameter for proton-proton elastic scattering, $0.2 \leq -t \leq 5$ GeV$^2$, in a two arm spectrometer, equipped with scintillators and Cerenkov counters;

   ii) determination of the A asymmetry in pp → $\pi^0$+anything, by means of lead glass detectors.
   The data for the elastic channel, which are now final, indicate a double bump structure between 1 and 3 GeV$^2$ followed by a negative dip at ∼ 4 GeV$^2$, and they do not seem to fulfil the present theoretical predictions, at least beyond 1.5 GeV$^2$.

6. Experiment PS 156, CERN-LAPP-Oxford collaboration [7].

   Beam: 24 GeV/c protons ($c_9$).

   Target: i) deuterated propanediol, 4.3 cm long, 1.4 cm diameter, 20 % polarization, $He^3$ evaporation refrigerator [4];
   ii) deuterated propanediol, 3.5 cm long, 1.4 cm diameter, 35 % polarization, dilution refrigerator [8].

   Aim: measurement of the P parameter in proton-neutron elastic scattering in the t interval 0.1 - 1 GeV$^2$, by combining a MWPC spectrometer (proton) and a time-of-flight detector (neutron). The preliminary data are almost mirror symmetric to the elastic proton data and show a trend different from the one at lower energy.

7. Experiment WA6, CERN-Padua-Trieste-Vienna collaboration [9].

   Beam: 40 to 150 GeV/c protons and pions (H3).

   Target: propanediol, 15 cm long, 2 cm diameter, 90 % polarization, $He^3$ evaporation refrigerator [4].

   Aim: the P parameter is determined in forward proton-proton and pion-proton elastic scattering, in a two arm MWPC spectrometer, in the range $0.3 \leq -t \leq 3$ GeV$^2$. Proton data from a first run at 150 GeV/c, already published, tend to indicate a polarization somewhat larger than expected, for $-t \lesssim 1$ GeV$^2$. More proton data, about a factor ten in statistics, have been collected at 150 GeV/c to cover the t-region where at higher energies the dip in the cross-section develops. The new data are being analysed, as well as the proton and pion data at 100 GeV/c, up to $-t = 2$ GeV$^2$.

8. Experiments NA2 and NA9, European Muon Collaboration.

Beam: 280 GeV/c muons, ∿ 90 % polarized (M2).

Target: at present the experiment is run with an unpolarized hydrogen target. A butanol* polarized target is being constructed. This target will be 1 m long, 5 cm diameter, cooled with a dilution refrigerator. It will be placed inside a 25 kG superconducting solenoid parallel to the beam direction.

Aim: Determination of the nucleon structure functions by detecting the forward muons in a magnetic spectrometer equipped with hodoscopes and drift chambers. The polarized target will allow the determination of the spin dependent part by measuring the a-symmetry $A = (d\sigma\uparrow\uparrow - d\sigma\uparrow\downarrow)/(d\sigma\uparrow\uparrow + d\sigma\uparrow\downarrow)$ in a region $q^2 \lesssim 120$ GeV$^2$, $40 \lesssim \nu \lesssim 170$ GeV, and the investigation of the role of the constituents, as far as the spin is concerned, in a region of very large $q^2$.

## REFERENCES

1. P. Astbury, private communication.
2. T. Niinikoski and F. Udo, Nuclear Instrum. Methods, <u>134</u>, 219 (1976)
3. - H. Becker et al, Measurement and analysis of the reaction $\pi^- p \to \rho^0 n$ on a polarized target, submitted to Nuclear Phys. B.
   - H. Becker et al, A model independent partial wave analysis of the $\pi^+\pi^-$ system produced at low four-momentum transfer in the reaction $\pi^- p_\uparrow \to \pi\pi n$ at 17.2 GeV/c, submitted to Nuclear Phys. B
   - G. Lutz and K. Rybicki, MPI report MPI-PAE/Exp. EL. 75 (October 1978).
   - See also K. Rybicki communication at this conference.
4. J. Vermeulen, Proc. 2nd Int. Conf. on polarized targets (ed. G. Shapiro), Berkeley 1971, p. 69.
5. - M. Fujisaki et al, Polarization measurements in $K^+ n$ charge exchange at 6 and 12 GeV/c, D.Ph.P.E. 78-09;
   - M. Fujisaki et al, Construction of tensor exchange amplitudes in $\overline{K}N$ and KN CEX reactions, D.Ph.P.E. 78-12;
   - M. Babou et al, Apparatus for two body scattering experiments at 6 and 12 GeV/c with the CERN polarized deuteron target, D.Ph.P.E. 78-11, submitted to Nuclear Instrum.Methods;
   - M. Fujisaki et al, Polarization measurements in $\pi^+ p$ and $K^+ p$ elastic scattering at 6 and 12 GeV/c with the CERN polarized deuteron target, D.Ph.P.E. 78-14;
   - See also L. Van Rossum and M. Svec communications at this conference.
6. T. Niinikoski, Proc. Symposium on High Energy Physics with Polarized Beams and Targets (ed. M. Marshak), 1976, AIP Conf. Proc. 35, p. 458.
7. - J. Antille et al, Measurements of polarization in elastic pp and pn scattering and of asymmetry in inclusive $pp\uparrow \to \pi^0$+anything

---

* The development of new doping agents is under study and could lead to the use of different material than butanol.

    at 24 GeV/c, submitted to this conference;
    - See also P. Kyberd, communication at this conference.
8. T. Niinikoski and M. Rieubland, Advances in Refrigeration at the lowest temperatures, I.I.R., Zürich 1978, p. 181.
9. G. Fidecaro et al, Phys. Letters 36B, 369 (1978).

# POLARIZATION MEASUREMENTS IN
# $\pi^{\pm}$p AND pp ELASTIC SCATTERING AT 100 AND 300 GEV/C

A. Jonckheere, P. F. M. Koehler
Fermi National Accelerator Laboratory, Batavia, IL 60510

I. P. Auer, D. Hill, B. Sandler,[*] D. Underwood, A. Yokosawa
Argonne National Laboratory, Argonne, IL 60439

J. Snyder, M. E. Zeller
Yale University, New Haven, CT 06520

R. V. Kline, M. E. Law, F. M. Pipkin
Harvard University, Cambridge, MA 02138

W. Johnson
Suffolk University, Boston, MA 02114

W. Brückner,[†] O. Chamberlain, H. Steiner, G. Shapiro
Lawrence Berkeley Laboratory, Berkeley, CA 94720

## ABSTRACT

Polarization measurements in $\pi^{\pm}$p elastic scattering at 100 GeV/c and pp elastic scattering at 100 and 300 GeV/c are presented. These data are compared to other experiments and to two models.

## INTRODUCTION

We present the results of polarization measurements in $\pi^{\pm}$p elastic scattering at 100 GeV/c and pp elastic scattering at 100 and 300 GeV/c. A double arm spectrometer employing multiwire proportional chambers fully measured both the scattered and the recoil particles. The apparatus, Fig. 1, has been described in detail several times,[1,2] so only its principal features will be discussed here. It was designed to use up to $10^8$ particles per second, incident flux. Thus it was built with no detectors directly in the beam. Momentum analysis was performed in each arm, with 1.7%

---

[*]Present address: University of Michigan, Ann Arbor, MI 48104.
[†]Present address: CERN, Geneva, Switzerland.

ISSN: 0094-243X/79/510439-06$1.50 Copyright 1979 American Institute of Physics

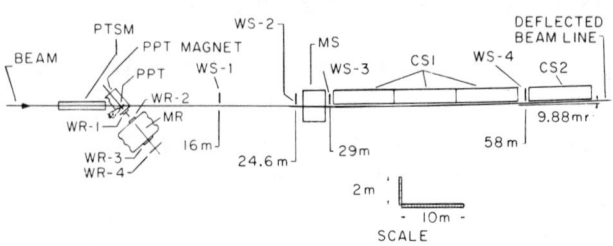

Fig. 1: E61 apparatus, 100 GeV/c layout. At 300 GeV/c the length of the forward arm was doubled. MS and MR are large superconducting analysis magnets, WS and WR are multiwire proportional chambers, and CS1 and CS2 are threshold Cerenkov counters.

resolution at 500 MeV/c in the recoil arm and 1% in the forward arm at 100 GeV/c. The overall acceptance was very flat, being 4-5% of $2\pi$ over a four momentum transfer (t) range from -0.2 to -1.8 $(GeV/c)^2$ with some acceptance out to t = -1.95 $(GeV/c)^2$.

The 3.5 mr beam of the Meson Department at Fermilab produced typically $2 \times 10^7$ ($5 \times 10^7$) particles per pulse, with a spot $2 \times 2$ ($1 \times 1$) cm$^2$ in size and a divergence of $\pm 0.2$ ($\pm 0.1$) mr at 100 (300) GeV/c. This small divergence allowed the separation of interactions on free protons from those on bound protons which have Fermi motion. The $2 \times 2 \times 8.2$ cm$^3$ ethylene glycol target presented a free proton thickness of 0.580 gm/cm$^2$ of a total target thickness of 6.47 gm/cm$^2$, and was polarized to typically 75-85%.

Since there were no detectors in the beam, the beam flux was monitored indirectly. The primary monitors were two triple scintillation counter telescopes viewing the target at 100 mr from the forward direction, perpendicular to the scattering plane, and coincidences between veto counters mounted above and below the target.

The data presented here were analyzed using a four momentum transfer (t) balance technique[1,2] coupled with coplanarity constraints. This technique involved calculating three independent t values, with errors, from directly measured quantities. An average t and a $\chi^2$ were calculated from these. In addition a normalized coplanarity ($\Delta\phi/\delta\phi$) was calculated. $\Delta\phi$ is the coplanarity angle and $\delta\phi$ is its error. Samples of these two quantities are shown in Fig. 2. The calculated polarization of the background (high $\chi^2$ or high $\Delta\phi/\delta\phi$) was consistent with zero,

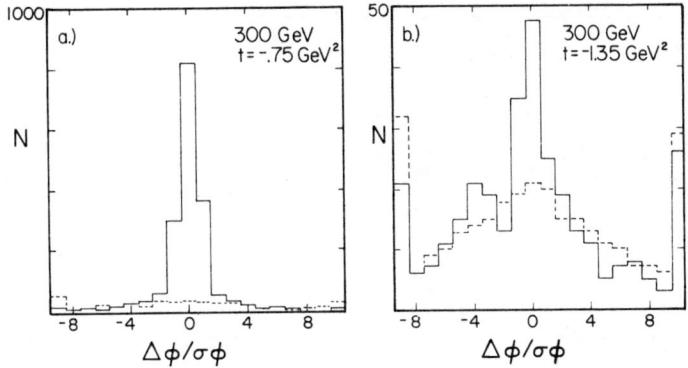

Fig. 2: Normalized coplanarity for two t bins. The solid line is for events with $\chi^2 < 2$, the dotted line is $\chi^2 > 10$, normalized to the tails of the low $\chi^2$ plots.

except in the 300 GeV/c pp data. This data showed a polarization of 0.005 ± 0.002, nearly independent of t and independent of any data selection procedure used.

## PION DATA

The pion data at 100 GeV/c shows no great surprises. The measured angular distributions are shown in Fig. 3, compared to the data of Akerlof et al.[3] Also shown is a curve from a version of the Kane and Seidl[4] strong absorption model. Our data has been

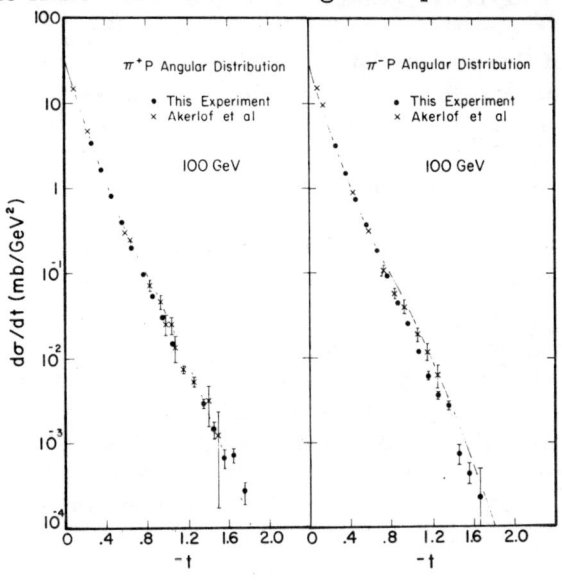

Fig. 3: $\pi^{\pm}p$ angular distributions normalized to Akerlof et al.[3]

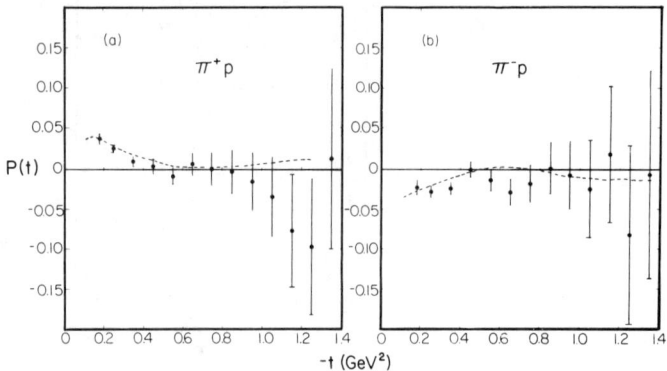

Fig. 4: $\pi^\pm p$ polarization parameter at 100 GeV/c. Dashed curves are lower energy data scaled as explained in the text.

normalized to the Akerlof data at -t = 0.6 $(GeV/c)^2$. Figure 4 shows the polarization results. The curves are a compendium of lower energy data[5-7] scaled by $s^a$, where a = $\alpha_\rho - \alpha_p$ = -0.5 + 0.9 t, and s is the square of the center of mass energy. This energy dependence is predicted by simple Regge pole exchange models dominated by non-spin flip pomeron exchange and spin flip rho ($\rho$) exchange. These models also predict the observed shapes and near mirror image nature of the $\pi^+ - \pi^-$ polarizations.

## PROTON DATA

The interpretation of the proton data is not as straightforward as the pion data. The angular distributions of Fig. 5 demonstrate the onset of a dip structure just above 100 GeV/c. This structure is well established at 300 GeV/c. These data have again been normalized to that of Akerlof et al.[3] at 100 GeV/c and to Kwak et al.[8] at 300 GeV/c. The modified Kane and Seidl model curves are shown. The 100 GeV/c data represents a total of 370 K events while the more precise 300 GeV/c data contains 520 K elastic events.

The polarization parameter shows equally interesting behavior. At energies up to 24 GeV/c[5,6] the data show smooth positive polarizations with a double zero near -t = 0.8 $(GeV/c)^2$. At 45 GeV/c[7] this has changed to a single zero near -t = 0.5 $(GeV/c)^2$ with rather large, negative values above this. This behavior persists at 100 and 300 GeV/c as is pictured in Fig. 6. The more

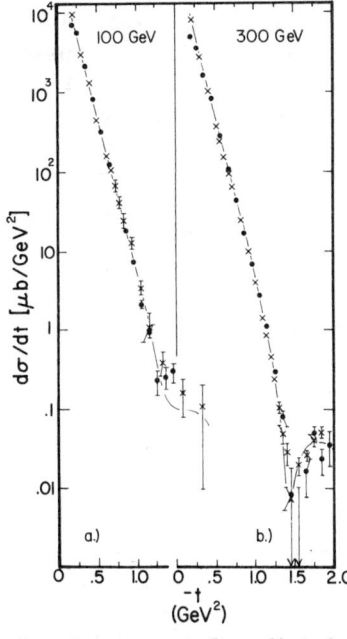

Fig. 5: pp angular distributions. Crosses are data of Akerlof et al. at 100 GeV and Kwak et al. at 300 GeV.

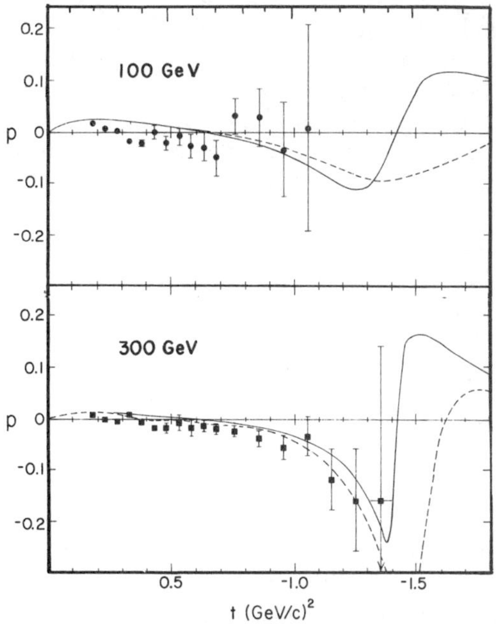

Fig. 6: pp polarization data with model curves explained in the text.

precise 300 GeV/c data also indicate that the polarization may go to, or through, zero again in the region of the dip. The data of Fidecaro et al.[9] at 150 GeV/c indicate the same type of behavior, with even larger negative polarizations near -t = 1.0 (GeV/c)$^2$.

Several models have been proposed to explain high energy polarization data. Curves from two of these are drawn on Fig. 6. The solid curve is a prediction of the Kane and Seidl strong absorption model with parameters refit by one of the authors (J.S.). The fit used elastic scattering data at 50, 100, 200, 300 and 2050 GeV/c[3,8,10] and polarization data up to 45 GeV/c.[5-7,11] The details of this model and the refit of the parameterization are given in detail in Refs. 2 and 4. This model generally fits the data quite well above 10 GeV/c, except that it predicts too small a polarization at 150 GeV/c.

The dashed curves are the predictions of the model of Bourrely, Soffer and Wu[12] which explains the polarization in terms of the impact picture and matter currents in the proton. This model also fits our data very well but predicts polarizations which are too small at large -t at both 45 and 150 GeV/c. This is not entirely surprising however, since our preliminary 100 GeV/c data was used, in part, to determine the model's parameters. This data was and is rather poor statistically in this region. This model does however show a rather strong energy dependence in the high t region which qualitatively describes the data. At low energies the model predicts small, positive polarizations which go to zero at ~40 GeV/c, rapidly become large and negative at the onset of the dip and then become smaller at larger energies. Both of these models predict large polarizations in the dip region with a zero near the dip. Unfortunately none of the data at present is precise enough to determine the polarization behavior at and beyond the dip in the elastic cross section.

## CONCLUSION

Data have been presented on $\pi^{\pm}p$ polarizations at 100 GeV/c and on pp polarizations at 100 and 300 GeV/c. The pion data agrees well with predictions of simple Regge pole exchange models. The pp data indicate a much more complex situation. Although several models fit our data reasonably well, the two investigated here do not agree with the data at 150 GeV/c, [9] predicting polarizations which are too small just at this energy.

## REFERENCES

1. I. P. Auer et al., Phys. Rev. Lett. 39, 313 (1977), J. H. Snyder et al., Phys. Rev. Lett. 41, 781 (1978).
2. J. H. Snyder, Ph.D. Thesis, Yale University, 1978 (unpublished).
3. C. W. Akerlof et al., Phys. Rev. D 14, 2864 (1976).
4. G. Kane and A. Seidl, Rev. Mod. Phys. 48, 309 (1976).
5. M. Borghini et al., Phys. Lett. 36B, 493 (1971).
6. D. G. Crabb et al., Nucl. Phys. B121, 231 (1977).
7. A. Gaidot et al., Phys. Lett. 57B, 389 (1975), A. Gaidot et al., Phys. Lett. 61B, 103 (1976).
8. N. Kwak et al., Phys. Lett. 58B, 233 (1974).
9. G. Fidecaro et al., Phys. Lett. 76B, 369 (1978).
10. Fermilab Single Arm Spectrometer Group, Phys. Rev. Lett. 35, 1195 (1975).
11. M. Borghini et al., Phys. Lett. 36B, 501 (1971).
12. C. Bourrely et al., A New Impact Picture for Low and High Energy Proton-Proton Elastic Scattering, to be published.

# IMPLICATIONS OF NUCLEON-NUCLEON SPIN-POLARISATION MEASUREMENTS

A. C. Irving*
University of Liverpool, Liverpool, England

E. L. Berger and C. Sorensen
Argonne National Laboratory, Argonne, Illinois

## ABSTRACT

We interpret the available data on polarised nucleon-nucleon elastic scattering. By comparing these with the simplest exchange model predictions we can identify features of particular interest such as low-lying $A_1$-like and isoscalar exchanges, and a helicity-flip Pomeron component. Our maximum-simplicity Regge model is intended to facilitate interpretation of forthcoming pp amplitude analysis results.

## EXCHANGE STRUCTURE OF pp SCATTERING

The elastic reaction $pp \to pp$ appears to be an excellent candidate for a study of hadron dynamics at low momentum transfer — it looks simple, highly symmetric and is particularly well-measured. In an exchange context, however, it has an embarassingly rich structure. The 5 amplitudes combinations $N_n$, $U_n$ (N, U stand for natural, unnatural parity exchange respectively and n is total s-channel helicity flip) can have contributions from almost every known Regge exchange (see Table I). According to symmetry arguments and coupling systematics established from factorisation studies of many processes, these exchanges are expected to couple in a distinctive way as shown in Table I.

Table I  Exchanges in pp Elastic Scattering

| Amplitude | Dominant Contributions | Suppressed Contributions |
|---|---|---|
| $N_0$ | $\mathbb{P} + f + \omega$ | $\rho + A_2$ |
| $N_2$ | $\rho + A_2$ | $\mathbb{P} + f + \omega$ |
| $N_1$ | $\rho + A_2 / \mathbb{P} + f + \omega$ | |
| $U_0$ | $A_1 + Z$ | $D + Z_0$ |
| $U_2$ | $\pi + B$ | $\eta + H$ |

The complex numbers which are the end product of pp amplitude analyses[1] will remain somewhat sterile quantities unless there exist

*Presented by A. C. Irving

model predictions of good theoretical pedigree, with which to compare them. We have constructed such a model[2] using the simplest possible amplitude structure satisfying SU(3), exchange degeneracy (EXD) and factorisation constraints[3] together with an f-dominated Pomeron amplitude. Details of this highly simplified, and hence predictive, model and of the relation between amplitudes and observables may be found in Ref.2. The basic model amplitudes (at 6 GeV/c and $t = -0.3$ GeV$^2$) are similar to those shown in Fig.1 except that $U_0$ ($A_1 + Z$ exchange), in common with other components having no Pomeron contribution, is purely real. In particular, the sign of each component is a theoretical prediction since the process is an elastic one.

As an example, the $A_1 + Z$ amplitude is predicted[2] by identification of the Feynman diagram

$$U_0(A_1) = \frac{g^2_{A_1 pp}}{t - m_A^2} (\bar{u}\gamma_\mu \gamma_5 u)(\bar{u}\gamma^\mu \gamma_5 u) \qquad (1)$$

with the Regge pole expression. The coupling $G_{A_1 pp}$ is estimated using current algebra and axial vector dominance of the weak form factor. Using $A_1 - Z$ EXD, one obtains a real and negative prediction for $U_0$.

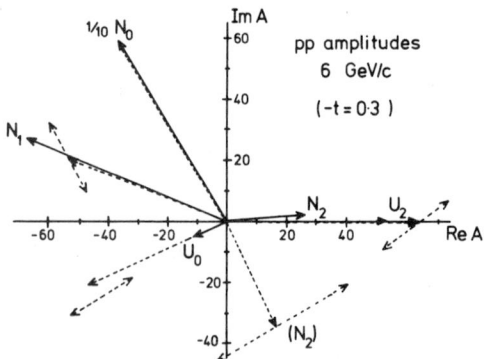

Fig.1 Argand plot of pp amplitudes. Solid vectors are the model of Ref.2. Dashed vectors are the tentative amplitude analysis results (and associated error estimates) of Ref.1.

The only non-real contribution in the basic model is due to Pomeron exchange whose flip and non-flip couplings are proportional to those of the ω and f.

## COMPARISON WITH SPIN POLARISATION DATA

Of the many interesting features of the polarisation data in np and pp scattering, the following have particularly direct implications for exchange models.

A. The sizeable value of $\Delta\sigma_{tot}^L (\propto \text{Im } U_0)$ measured at 6 GeV/c[4] implies a non-EXD $A_1$-like exchange. To account for this, we have put in an ad hoc imaginary contribution with the energy dependence of the $A_1$ $(\alpha_{A_1}(0) = -.19)$[2]. The agreement of our 12 GeV/c prediction with preliminary data[5] for $\Delta\sigma_{tot}^L$ (Fig.2) shows this to be a reasonable approximation. Our description of the amplitude $U_0$ is reinforced by the measurement $C_{LL}$ but will be most strongly tested by the triple scattering measurement $H_{LSN}$.[2]

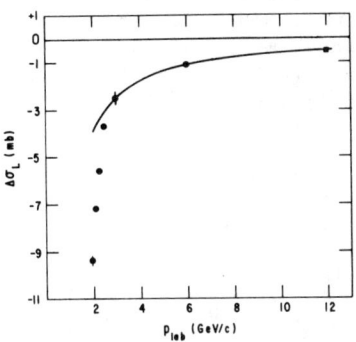

Fig.2. A comparison of model[2] and data[4] for the longitudinally polarised total cross-section difference $\Delta\sigma_{tot}^L$(pp). The preliminary data point (■) at 12 GeV/c is from Ref.5.

B. The isospin zero contribution to the nucleon-nucleon polarisation (P(pp)+P(pn)) shows an anomalously rapid energy dependence[6] which may be interpreted[7] as scalar $\varepsilon$ and/or $\omega'$ exchange (see Fig.3).

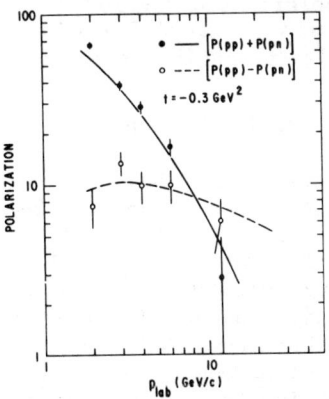

Fig.3. The isoscalar (●) and isovector (○) components of the NN elastic polarisation.[6] The model curves are from Ref.2.

In fact, a pole extrapolation estimate using knowledge of the on-shell coupling constant gives a good account of the magnitude of this low-energy effect as well as the correct sign (Fig.3). This exchange component also appears to play an important role in producing the highly energy dependent effects seen in $C_{NN}$.[2]

C. The negative polarisation measured in pp scattering for $|t| \gtrsim 0.4$ and $p_{LAB} \gtrsim 50$ GeV/c (and similar results in np scattering at lower energies) suggests a diffractive (i.e. mildly energy dependent) helicity flip amplitude component[8]. This has the sign predicted by eikonal models of the Pomeron with f-dominated couplings.[9] The implications for our simple model are that the real part of our helicity-flip Pomeron component must be reduced to reproduce this behavior. Fig.4 shows recent P(pp) data at 100 GeV/c[11] which illustrate this effect. Regge models with no helicity-flip pomeron predict very small but <u>positive</u> polarisation at 100 GeV/c.

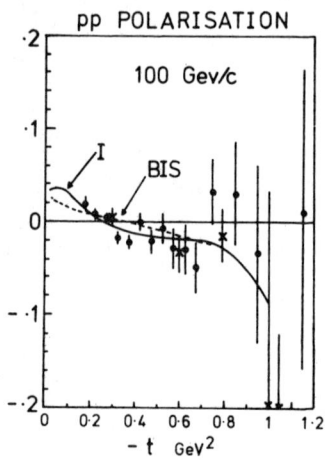

Fig.4. The pp elastic polarisation at 100 GeV/c (● Snyder et al, X Corcoran et al).[11] The curves marked I and BIS are predictions of Refs. 8 and 2 respectively and include electromagnetic corrections.

## AMPLITUDE ANALYSIS

Fig.1 gives an example of our basic pp amplitude predictions modified as described in A, B and C above. The dashed argand vectors are the preliminary amplitude analysis results of Ref.[1] rotated to have our model phase for $N_0$ (not experimentally measurable for $t \neq 0$) and scaled to have the same magnitude for $N_0$ (trivially given by $\sqrt{\frac{d\sigma}{dt}}$). Since our model gives a good overall description of all available spin polarisation measurements (single, double and triple correlation) some degree of agreement is to be expected. The discrepancy in $U_0$ and $N_2$ is easily traced to the approximations made in the preliminary data analyses[10]

$$C_{LL} \sim U_0^{\parallel} \qquad (2a)$$

$$C_{NN} \sim N_2^{\|} \tag{2b}$$

$$H_{NSS} \sim N_2^{\perp} \tag{2c}$$

where $\|$ and $\perp$ refer to the direction of $N_0$ in the Argand plot. In our model, these appear to be bad approximations due to the large contributions from $Re(N_2 U_2^*)$, $|N_1|^2$ and $Im(N_0 N_1^*)$ to 2a, 2b and 2c respectively, in this t-range (near -0.3).

In any case, model amplitudes are vital in motivating, testing and interpreting pp amplitude analyses.

## NON-ASYMPTOTIC CONTRIBUTIONS

The s-channel helicity components $N_0$, $N_1$, $N_2$, $U_0$ and $U_2$ are well-known to correspond to definite t-channel parity at leading order in 1/s only. Since many of the interesting exchange contributions in pp appear at low s only (e.g. "$\varepsilon$" and "$A_1$" exchange) it is essential to know what the exact parity content is. A detailed kinematical analysis[12] gives the results[2] shown in Table II.

Table II  t-Channel Parity in s-Channel Helicity Amplitudes

| Amplitude | Leading Contributions | 1/s Contributions |
|---|---|---|
| $N_0 = \frac{1}{2}(\phi_1 + \phi_3)$ | $s^{\alpha_N}$ | $s^{\alpha_U} \times 0(1/s)$ |
| $N_1 = \phi_5$ | $s^{\alpha_N}$ | $s^{\alpha_U} \times 0(1/s)$ |
| $N_2 = \frac{1}{2}(\phi_4 - \phi_2)$ | $s^{\alpha_N}$ | $s^{\alpha_U} \times 0(1/s)$ |
| $U_0 = \frac{1}{2}(\phi_1 - \phi_3)$ | $s^{\alpha_U}(A_1)$ | $N_0(-t/2s)$* |
| $U_2 = \frac{1}{2}(\phi_4 + \phi_2)$ | $s^{\alpha_U}(\pi)$ | none |

*The factor -t/s comes from assuming factorising contributions to $N_0$. The factor of 1/2 results if one neglects their contribution to $N_1$.

We can safely neglect the $s^{\alpha_U - 1}$ contributions to "natural parity" combinations, but the natural parity (Pomeron) contamination is potentially important for $t \neq 0$ (i.e. in $C_{LL}$)[2] and even at $t = 0$ (i.e. in $\Delta\sigma_{tot}^L$) if conspiratorial solutions are admitted. Stacey[13] has studied the latter possibility in detail.

It has been pointed out[12,13] that the $A_1$ or Z contribution to $U_0$ at $t = 0$ must vanish unless daughter contributions are present to satisfy an analyticity requirement on their 1/s "wrong naturality" contributions. Since there are many other theoretical reasons for the existence of daughter Reggeon contributions one need not be unduly worried by this curiosity.

## OUTLOOK

Considerable light could be shed on the isospin decomposition of the pp amplitudes (Table I) by a selection of np elastic scattering spin measurements.[10] Our model suggest that $\Delta\sigma_{tot}^L$, $C_{NN}$ $C_{SS}$ and $H_{LSN}$ would be particularly valuable, as would $C_{xx}$ (x = L,S,N) and $D_{NN}$ measurements of np charge exchange in which spin correlation effects should be especially large.[2]

In the early seventies, amplitude analysis of $\pi N \to \pi N$ dealt an almost fatal blow to Regge exchange models. From the chaotic remains of these models some battered ideas on hadronic amplitude structure have struggled back into the daylight. Will the NN amplitude results be the coup de grace or will some totally new insight emerge? For my part, I expect these Regge ideas will still be around long after elephants have learned to fly and the $A_1$ has achieved respectability.

## REFERENCES

1. A. Yokosawa, Proc. Second Int. Conf. on the Nucleon-Nucleon Interaction, Vancouver (1977).
2. E. L. Berger, A. C. Irving and C. Sorensen, Phys Rev. D17, 2971 (1978).
3. A. C. Irving and R. P. Worden, Phys. Reports 34C, 117 (1977).
4. I. P. Auer et al., Phys. Lett. 70B, 475 (1977).
5. I. P. Auer et al., contribution to the XIX Int. Conf. on High Energy Physics, Tokyo, 1978.
6. R. Diebold et al, Phys. Rev. Lett. 35, 632 (1975); S. L. Kramer, Phys. Rev. D17, 1709 (1978).
7. J. Dash and H. Navelet, Phys. Rev. D13, 1940 (1976).
8. A. C. Irving, Nucl. Phys. B101, 263 (1975).
9. A. Martin and H. Navelet, J. Phys. G, 4, 647 (1978).
10. G. H. Thomas, Proc. Second Int. Conf. on the Nucleon-Nucleon Interaction, Vancouver (1977).
11. J. H. Snyder et al., Phys. Rev. Lett. 41, 781 (1978); M. D. Corcoran et al., Phys. Rev. Lett. 40, 1113 (1978).
12. E. Leader and R. Slansky, Phys. Rev. 148, 1491 (1966); E. Leader ibid 166, 1599 (1968).
13. R. Stacey, J. Phys. G. (to appear).

# INTERPRETATION OF THE SPIN STRUCTURE IN NUCLEON-NUCLEON SCATTERING

P. Kroll[*]
Dept. of Physics, Univ. of Wuppertal, Germany

## INTRODUCTION

We have shown that the rapid variation and change of shape with energy of the polarizing power P and the spin-correlation parameter $A_{NN}$, can be understood in terms of the properties of the dominant helicity non-flip (i.e. diffractive) amplitude, which is known from fixed-t dispersion relations. Experimental dips and zeros appear as the energy increases because of the <u>rapid variation</u> with energy of terms proportional to the real part of the non-flip amplitude. These terms are either ignored or quite <u>incorrectly</u> described in standard dynamical models.

## DOMINANCE OF THE NON-FLIP AMPLITUDE

By far the most dominant amplitude in elastic NN scattering is the spin-averaged amplitude

$$\phi_+ = 1/2(\phi_1 + \phi_3) \tag{1}$$

where $\phi_i$ are the usual s-channel helicity amplitudes. At medium and high energies and not too large $|t|$ values the NN differential cross-sections can be well approximated by $\phi_+$. This fact allows to calculate $\phi_+ = |\phi_+| \exp(i\delta_+)$ from the differential cross sections with the aid of a phase-modulus dispersion relation[1].

The main features of $\phi_+$ are (compare fig.1)
i) while $\text{Im}\phi_+$ always dominates at small $|t|$, it is $\text{Re}\phi_+$ that dominates at large $|t|$ for medium energies.
ii) $\text{Im}\phi_+$ has a zero at about $-1.1$ to $-1.5$ GeV$^2$ at all energies.
iii) The ratio $\text{Re}\phi_+/\text{Im}\phi_+$ drops rapidly as the energy increases at all $|t|$ values, except the region very close to the zero of $\text{Im}\phi_+$. For very high energies ($\gtrsim 250$ GeV) the ratio starts to increase again.

Owing to the method used in ref.1, the accuracy in the determination of $\delta_+$ diminishes at lower energies, but should suffice for our present purpose of obtaining a qualitative understanding of the structure of P and $A_{NN}$.

---

[*] This work has been done in collaboration with E. Leader and W. von Schlippe, Westfield College, London NW 3 7ST.

## STRUCTURE OF THE POLARIZING POWER

The dominance of $\phi_+$ linearizes the relations between oberservables and helicity amplitudes. Thus we have for the sum ($\Sigma P$) and the difference ($\Delta P$) of pp and pn elastic polarizations

$$\frac{1}{2}(P(pp)+(-1)^I P(pn)) = 2\cos\delta_+ \text{Im}\phi_5^{(I)} - 2\sin\delta_+ \text{Re}\phi_5^{(I)} \qquad (2)$$

assuming $\phi_+(pp) \simeq \phi_+(pn)$.

$\phi_5^{(1)}$ is parametrized by a simple $\rho$-$A_2$ Regge pole model

$$\phi_5^{(1)} = \sqrt{-t}\, p_L^{\alpha-1}\{\alpha\beta_{A_2}(t)(i-\cot\pi\alpha/2) - \beta_\rho(t)(i+\tan\pi\alpha/2)\} \qquad (3)$$

where we have taken $\alpha_{A_2}(t) = \alpha_\rho(t) = \alpha(t) = 0.48+0.93t+0.21t^2$. The parametrization differs slightly from recent parametrizations by a missing factor of $\alpha$ in front of the $\rho$-contribution. The origin of the $\alpha$ factor lies in a belief in <u>strong</u> $\rho$-$A_2$ exchange degeneracy, which, as we demonstrated in an earlier paper[2], definitely does not hold. This parametrization is forced upon us by

i) $P\frac{d\sigma}{dt}(\bar{p}p\to n\bar{n}) + P\frac{d\sigma}{dt}(pn\to np) = \pi\text{-cut}\cdot\text{Im}A_2 \simeq 0$

ii) $P\frac{d\sigma}{dt}(\bar{p}p\to n\bar{n}) - P\frac{d\sigma}{dt}(pn\to np) = \pi\text{-cut}\cdot\text{Im}\,\rho$ large

iii) $\Delta P(\text{elastic}) \simeq -2\sin\delta_+ \text{Re}\phi_5^{(1)} \simeq 0$

near $\alpha = 0$ at 24 GeV.

The isospin zero amplitude is parametrized by a $\omega$-$f$-$\varepsilon$ Regge pole model

$$\phi_5^{(0)} = \sqrt{-t}\{\alpha\beta_f(t) p_L^{\alpha-1}(i-\cot\pi\alpha/2) - \alpha\beta_\omega(t) p_L^{\alpha-1}(i+\tan\pi\alpha/2)$$
$$+ \alpha_\varepsilon(\alpha_\varepsilon+2)\beta_\varepsilon(t) p_L^{\alpha_\varepsilon-1}(i-\cot\alpha_\varepsilon/2)\} \qquad (4)$$

For simplicity we used the same trajectory for $\alpha_f$ and $\alpha_\omega$ as for $\rho$-$A_2$ whereas the $\varepsilon$-trajectory was left free (result of the fit $\alpha_\varepsilon = 0.02+0.89$). Parametrizing all residues as simple exponentials, we fitted all pp and pn polarization data from 6 to 100GeV (FNAL data[3]) and $(t) \leq 2\text{GeV}^2$. In fig.2 we show some of our results. Bearing in mind the properties of $\delta_+$ and that of the parametrization of $\phi_5$, the behaviour of the two terms making up $\Sigma P(\Delta P)$ is easily understood. At medium energies the terms proportional to $\cos\delta_+$ <u>mask</u> the asymptotic shape which of course is built up by the terms proportional to $\sin\delta_+$.

The fit agrees reasonable well with the data. The only problem arises from the SPS data[4] at 15oGeV. If we include these data instead of the FNAL data, the parametrization of $\phi_5$ has unavoidably to be modified by an additional Pomeron flip term. Our interpretation of the polarization data, however, remains unaltered.

## THE STRUCTURE OF $A_{NN}$

The dominance of $\phi_+$ allows to write

$$A_{NN} = \cos\delta_+ \text{Re}(\phi_2-\phi_4) + \sin\delta_+ \text{Im}(\phi_2-\phi_4) \qquad (5)$$

We have explained in detail in ref.2 that $\phi_2-\phi_4$ can be described by simple I=1 exchanges but that a "new" signature odd contribution besides $\rho, A_2, \pi, \pi\text{-cut}$ is needed in order to understand the behaviour of $A_{NN}$. Labeling this "object" c and using the Williams prescription[5] for the $\pi$-cut, we take

$$\phi_2-\phi_4 = -\beta_\pi(t) p_L^{-1} - (1+\alpha_c)(i+\tan\pi\alpha_c/2)\beta_c(t) p_L^{\alpha_c-1} \qquad (6)$$
$$+ t\beta_{2A_2}(t) p_L^{\alpha-1} \alpha(i-\cot\pi\alpha/2) - t\beta_{2\rho}(t) p_L^{\alpha-1}(i+\tan\pi\alpha/2)$$

With this parametrization we are able to fit the $A_{NN}$ data. The interpretation of the rapid variation with energy of $A_{NN}$ is similar to that of the polarizing power.

## REFERENCES

1. W.Grein, R.Guigas and P.Kroll, Nucl.Phys.B89,(1975)93.

2. P.Kroll, E.Leader and W.von Schlippe, Journal of Phys. G4(1978)1oo3.

3. J.H.Snyder et al., Phys.Ref.Lett.41(1978)781.

4. G.Fidecaro et al., Phys.Lett.76B(1978)369.

5. P.K.Williams, Phys.Rev.181(1969)1963.

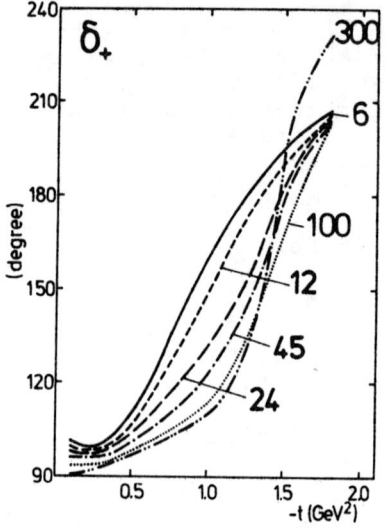

Fig.1: The phase $\delta_+(s,t)$ of the dominant spin-averaged amplitude at several energies as taken from ref.1.

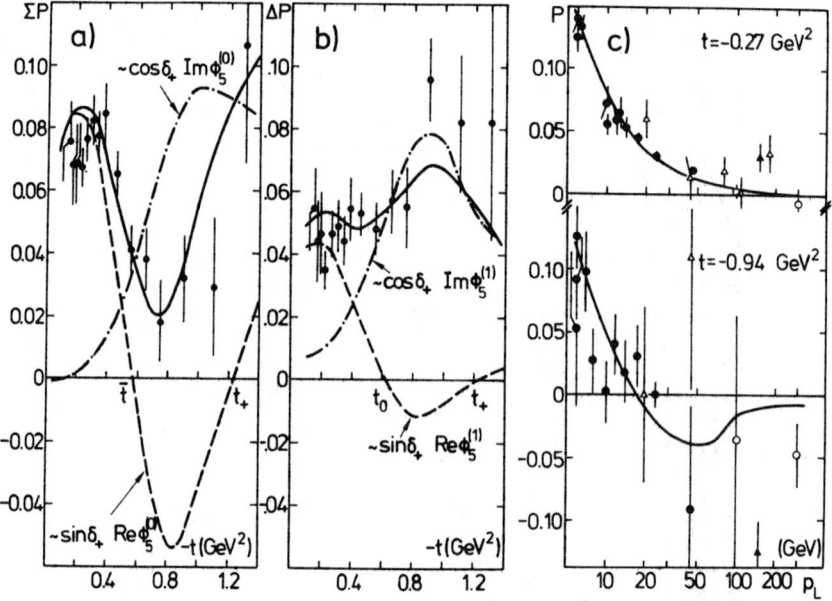

Fig.2: $\Sigma P$ (a) and $\Delta P$ (b) data at 6 GeV compared to our fit (solid lines). The decomposition according to eq.(2) is also shown. In c, the fit is compared to P(pp) at t= -0.27 and -0.94 GeV².

POLARIZATION IN P-N ELASTIC SCATTERING AT 24 GEV/C

D. G. Crabb, P. Kyberd and G. L. Salmon
Nuclear Physics Laboratory, Oxford University, Oxford, U.K.

J. Antille and M. Werlen
CERN, Geneva, Switzerland

K. Kuroda, A. Michalowicz and D. Perret-Gallix
LAPP, Annecy-le-Vieux, France

(Presented by P. Kyberd)

## ABSTRACT

Data are presented on the analyzing power A for p-n elastic scattering at 24 GeV/c in the range $0.1 \leqslant -t \leqslant 0.8$ $(GeV/c)^2$.

## INTRODUCTION

The apparatus used in the measurement is shown in Fig. 1 A 24 GeV/c extracted beam from the CERN PS was focussed onto a polarized deuterium target; the intensity was $< 5 \times 10^8$ ppp with a spot size of $5 \times 6$ mm$^2$ sufficient to slightly overmatch the NMR coil in the target and ensure a correct measurement of target polarization even after radiation damage. The target was deuterated propanediol doped with $C_r^V$ complexes giving initial deuteron polarizations of about 25% in a He$^3$ evaporation refrigerator. When this was replaced by a He$^3$/He$^4$ dilution refrigerator initial polarizations of 35% were obtained.

The apparatus for detecting the scattered particles was fairly standard. The recoil neutron was detected in a 3 x 3 array of plastic scintillator blocks 15x15x27 cm$^3$ the charged particles and γ-rays being vetoed by a large wall of scintillation counters. The forward proton was bent by 3° in the spectrometer magnet and detected by scintillation counter, hodoscopes and MWPC's. The elastic signal was enhanced by operating the Cerenkov counter at pressure that rejected protons with momenta below that of the elastically scattered ones.

For the range $0.1 \geqslant -t \leqslant 0.8$ $(GeV/c)^2$ it was sufficient to operate with a reduced acceptance using small trigger counters and simply measuring the neutron time of flight. Above $-t = 0.8$ $(GeV/c)^2$ the complete detection system was used.

To monitor the beam or target we used a) a thin plate ionization chamber b) two three counter telescopes directed at the target and in the plane of polarization and 3) a magnetic spectrometer to detect 300 MeV/c protons scattered backwards from unpolarized nucleons in the target. All were consistent to the 0.1% level.

For the low t region presented here the neutron time of flight showed a prominent elastic signal over a random background. Dummy target measurements were made using a normal propanediol target and quasi-elastic scatters also produced a peak which varied from 20% to 40% of the elastic peak.

The analysing power A for p-n elastic scattering is shown in Fig. 2 together with a previous measurement of pp elastic scattering at 24 GeV/c[1]. The two curves are not inconsistent with mirror symmetry.

## REFERENCES

1. D. G. Crabb et al. Nuclear Physics B121, 231 (1977).

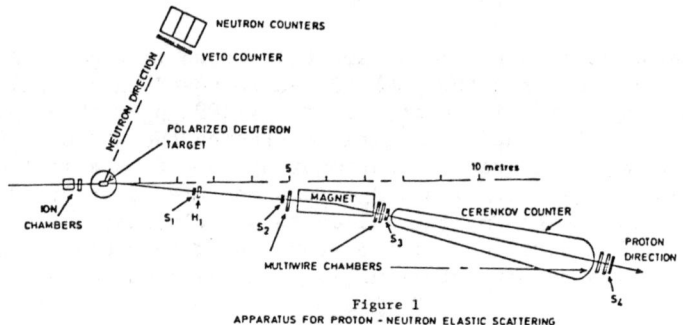

Figure 1
APPARATUS FOR PROTON - NEUTRON ELASTIC SCATTERING

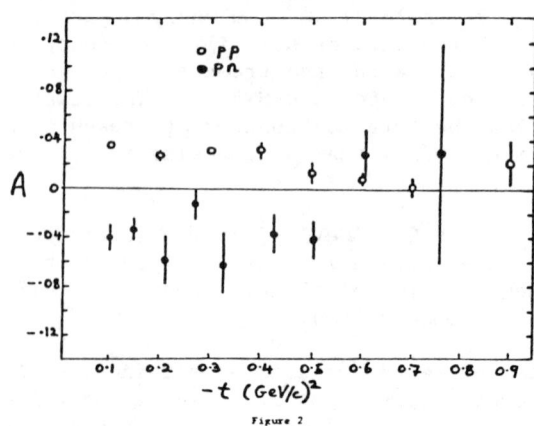

Figure 2

INCLUSIVE $\pi^-$ ASYMMETRIES NEAR X=0*

S. C. Ems, S. W. Gray, D. R. Rust
Indiana University, Bloomington, Indiana 47405

R. D. Klem and Y. L. Cho
Argonne National Laboratory, Argonne, Illinois 60439

ABSTRACT

New results from the Argonne ZGS on the scattering asymmetries of inclusively produced pions are presented. The preliminary data is for the reaction $p \uparrow p \to \pi^- + X$ near X=0 for a Pt range extending past 1.0 GeV/c.

INTRODUCTION

We selected this particular sample from our data taken this April, May, and June for quick preliminary analysis because we perceived a growing interest in X=0 inclusive pion asymmetries as a possible test of calculations based on a constituent picture. Earlier in this conference J. Soffer presented the results of a calculation he made with C. Bourrely limiting the size of a possible X=0 asymmetry through the mechanism of quark-quark elastic scattering[1]. Babcock, Monsay, and Sivers have published some QCD based calculations of two spin asymmetries for inclusively produced pions at X=0[2,3]. Recently Ranft and Ranft have made similar calculations[4]. Sivers and Monsay are now calculating quark-gluon contributions to X=0 inclusive asymmetries.

EXPERIMENTAL EQUIPMENT

This experiment was part of a program of inclusive asymmetry measurements carried out with the polarized beam at the Argonne Zero Gradient Synchrotron. The experimental area is shown in Figure 1. An 11.75 GeV/c extracted proton beam of $\sim 10^8$/burst was brought down Beam 5. The spot size was $\sim$.5 inch. The beam position and size were monitored by SPIC's (Segmented Ion Proportional Chambers) 1m and 5m upstream of the target. The spot size was 8 mm fwhm with a divergence ~1 MR fwhm. The beam intensity was monitored by a scintillator intercepting the entire beam; charge integrating electronics eliminated rate effects. A three counter telescope viewing a target in the beam from below and an ionization chamber were also used. The beam polarization, averaging about 65%, varied only slightly during these runs. The polarized source alternated the beam polarization direction from spill to spill. The standard ZGS monitors checked the polarization at 50 MeV and at each front porch or flattop. The hydrogen target used for this part of the experiment was a 3 inch diameter cylinder

*Work supported in part by the U.S. Department of Energy

perpendicular to the beam. The spectrometer moved on a steel track about a pivot under the target. Most of the data presented was taken at a lab angle of ~22 degrees, corresponding to 90 degrees in the center of mass.

The spectrometer used sets of drift chambers and proportional chambers on both sides of a septum magnet. A scintillator 75 cm from the target and a second scintillator 280 cm from the first defined the fast logic trigger and provided the start and stop signals for the time of flight system (resolution < 800 ps fwhm). There were two pairs of XY MWPC's in front of the magnet and a single XY pair downstream of the magnet. The wires in the X chambers were connected to a trigger processor. Four X drift chambers formed two high resolution telescopes with the magnet in between. A large aperture Freon 12 threshold Cerenkov counter sat behind the last chambers but in front of the final scintillator. A smaller scintillator was added to the trigger to guarantee that the triggering particle passed through the useful aperture of the Cerenkov counter. The spectrometer had an acceptance ~3 degrees in theta and a solid angle of 1.5 milli-steradians.

An important part of the experiment was the ability to suppress triggers from high rate kinematic regions and allow rarer events to be accumulated. The trigger processor or "Bend Detector" used the latched MWPC information to limit the angular acceptance and to select the momentum range. Events outside the chosen kinematic range were rejected in less than 3 micro seconds. This allowed reasonable livetimes on all runs. The Bend Detector acceptance rate ranged from 3% to 30% for these runs; the corresponding livetimes ranged from 99% to 70%.

## THE DATA

The seven data points were from a corresponding set of seven runs analyzed separately. The acceptance ranged from $|X| < .01$ and $\Delta Pt = \pm 25\%$ for the lowest Pt point to $|X| < .05$ and $\Delta Pt = +50\% -10\%$. Each point was plotted at the average Pt for that run. Corrections were made for the relative integrated beam for each incident beam polarization. The largest correction was 1.3% to the highest Pt point; most points were corrected by less than 0.5%. The relative livetime for each beam polarization was also monitored and used as a correction. The largest correction was 3% to the largest Pt point; all other points were corrected by less than 0.5%. No corrections were made for accidentals which were less than 2% for all points. Target empty runs were made at each point. The asymmetries from the full and empty runs were combined with the appropriate weights to extract the hydrogen asymmetry. The largest target empty rate was 40% at the highest Pt point. The lower Pt points had target empty fractions ranging down to less than 25%. The errors reflect the statistical error on both the full and empty runs. The preliminary asymmetry, shown in Figure 2, is small and slightly negative. The final data sample will include several times as many events, obtained primarily by extending the X range to ±0.2 . A slightly larger sample of $\pi^+$ data was also taken.

## MODEL IMPLICATIONS

The "QCD" calculations of the type mentioned earlier have a common approach to studying the reaction. Each of the incident particles is pictured as emitting a quark or gluon; these constituents interact with each other; and then, in the final state, one of the constituents dresses itself as the detected particle in the inclusive reaction. The calculation or model focuses on this constituent-constituent scattering (quark-quark or quark-gluon).

The approach of Bourrely and Soffer was to assume the inclusive process was dominated by quark-quark elastic scattering and that the polarization was 1 (in quark-quark elastic scattering). This allowed them to place a limit ~40% on the inclusive asymmetry near Pt=1 GeV/c (for s=200). Using the Kuti-Weiskopf model they obtained a lower limit of ~20% at Pt=1 GeV/c. Our result of ~5% does not let us put a limit on the polarization in quark-quark elastic scattering because the quark-gluon contribution was not included.

The Argonne group went a step further and calculated the individual diagrams contributing to the constituent scattering process. For instance, for the quark-quark scattering there would be one gluon exchange, two gluon exchange, and so forth. All the first order diagrams contribute nothing to the inclusive asymmetry. Kane, Pumplin, and Remko[5] have found that the two gluon exchange contribution in quark-quark scattering gives essentially no asymmetry. Sivers and Monsay[6] are now calculating the two gluon exchange contribution to quark-gluon scattering, which they feel may be the dominant contribution to the inclusive asymmetry. We are hoping they will have a calculation to compare with our more extensive full data set soon.

## ACKNOWLEDGEMENTS

We would like to acknowledge technicians B. Martin and P. Smith of Indiana for their sizeable efforts in putting the experiment together. We also wish to thank H. Neal, H. Ogren, and R. Polvado for many helpful discussions and assistance in running shifts. The ANL Experimental Area Group, in particular engineers A. Passi and W. Siljander, were extremely helpful in getting the experiment running.

## REFERENCES

1. C. Bourrely and J. Soffer, Phys. Lett. 71B, 330 (1977)
2. J. Babcock, E. Monsay, and D. Sivers, Phys. Rev. Lett. 40, 1161 (1978)
3. K. Hidaka, E. Monsay, and D. Sivers, ANL Preprint

ANL-HEP-PR-78-47
4. J. Ranft and G. Ranft, Phys. Lett. **77B**, 309 (1978)
5. G. L. Kane, J. Pumplin, and W. Repko, Univ. of Michigan Preprint UM-HE-78-29
6. Private Communication

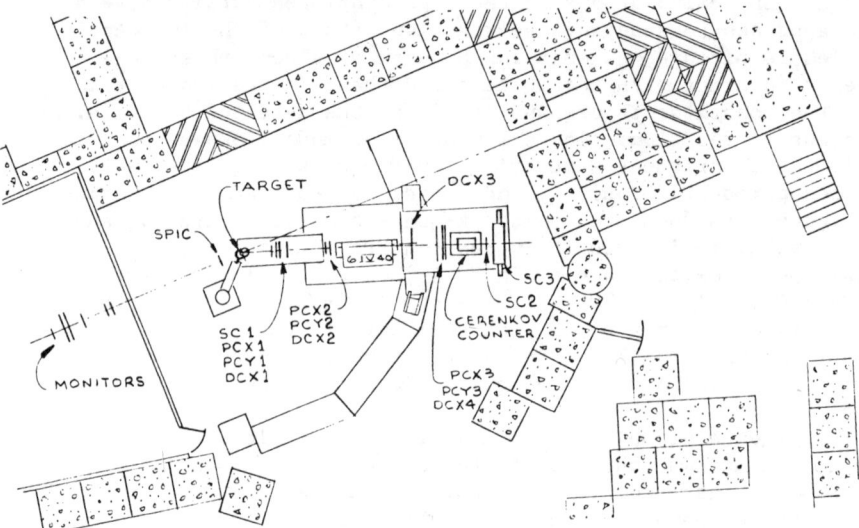

Fig. 1 Experimental Area Plan View

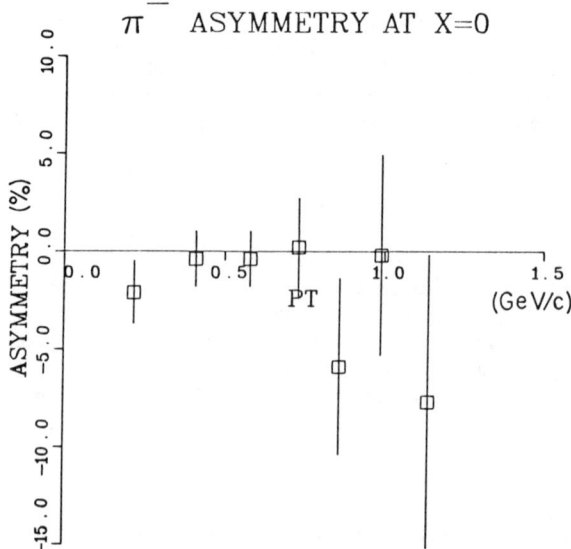

Fig. 2 Inclusive $\pi^-$ Asymmetry at X=0 Produced by 12 GeV/c Polarized Beam.

A MODEL INDEPENDENT PARTIAL WAVE ANALYSIS OF THE $\pi^+\pi^-$ SYSTEM

PRODUCED AT LOW FOUR-MOMENTUM TRANSFER IN THE REACTION

$\pi^- p_\uparrow \to \pi^+\pi^- n$ AT 17.2 GeV/c

H. Becker, G. Blanar, W. Blum, M. Cerrada, V. Chabaud, J. De Groot,
J. Dietl, J. Gallivan, L. Görlich, B. Gottschalk, G. Grayer,
G. Hentschel, B. Hyams, E. Lorenz, G. Lütjens, G. Lutz, W. Männer,
B. Niczyporuk, D. Notz, T. Papadopoulous, R. Richter, K. Rybicki,
U. Stierlin, B. Stringfellow and A. Zalewska

CERN-Cracow-Munich Collaboration

The CERN-Munich experiment [1] yielding ~300 000 events of the reaction $\pi^- p \to \pi^+\pi^- n$ at 17.2 GeV/c originated a series of investigations of high mass $\pi^+\pi^-$ partial waves. These were the studies of the CERN-Munich group [2,3], Estabrooks and Martin [4], Froggatt and Petersen [5] and, more recently, of Martin and Pennington [6]. In the absence of polarization data some physical assumptions had to be introduced. These were the following:

a) s-channel nucleon spin flip dominance in the unnatural spin-parity amplitudes;

b) phase coherence of the unnatural spin-parity exchange amplitudes with helicity m = 0 and m = 1;

c) equality of moduli of m = 1 natural and unnatural exchange amplitudes (this assumption is well supported by the vanishing of $M \geq 2$ t-channel moments).

In addition the above studies were based on an energy-dependent analysis assuming the Breit-Wigner (plus possible background) shape of amplitudes or some parametrization of the Barrelet zeros [1]. The latter were also used for the search for ambiguities. In most recent works [5,6] analyticity was imposed on solutions. There was a general consensus about two solutions below 1450 MeV and four above this mass. The $\pi^0\pi^0$ data at 40 GeV/c [8] seemed to exclude two of the latter but even then two solutions α and β (using the notation of Ref.[6]) were left. These solutions differ in the resonance structure of the lower waves (S,P,D) above 1450 MeV.

In the present work we use the results of the experiment [1] and our polarized target experiment which yielded ∼1 200 000 events on butanol ($C_4H_9OH$) also at 17.2 GeV/c. About $\frac{1}{3}$ of them correspond to collisons with free protons (the hydrogen nuclei) with the average polarization P = 68%. Only these protons contribute to the polarization-dependent part of the angular distribution. We combine this part with a polarization-independent one from the experiment [1] since the latter is heavily contaminated by interactions with bound protons in the polarized target experiment. This combination allows a completely model independent partial wave analsis avoiding any *a priori* physical assumptions.

## VARIABLES, MOMENTS, AMPLITUDES AND RELATIONS BETWEEN THEM

At a given energy our reaction can be described by the following five variables:

- $m_{\pi\pi}$ — effective mass of the $\pi$-$\pi$ system
- $t$ — four momentum transfer to the nucleon
- $\psi$ — polarization angle (the angle between a normal to the production plane and the polarization direction)
- $\theta$ — decay angles of the $\pi^-$ in the $\pi^+\pi^-$ rest system
- $\phi$ (here the Gottfried-Jackson system)

Due to parity conservation the angular distribution I is of the general form

$$I(m_{\pi\pi},t,\theta,\phi,\psi) = \sum_{L,M} t_M^L(m_{\pi\pi},t) \, \mathrm{Re}\, Y_M^L(\cos\theta,\phi) + P_t \cos\psi \sum_{L,M} p_M^L(m_{\pi\pi},t)$$

$$\mathrm{Re}\, Y_M^L(\cos\theta,\phi) + P_t \sin\psi \sum_{L,M} r_M^L(m_{\pi\pi},t) \, \mathrm{Im}\, Y_M^L(\cos\theta,\phi)$$

where

$P_t$ is the polarization component perpendicular to the beam direction (in our case this is the full proton polarization P), and $Y_M^L(\cos\theta,\phi)$ are the spherical harmonic functions.

The full set of moments $t_M^L(m_{\pi\pi},t)$, $p_M^L(m_{\pi\pi},t)$ and $r_M^L(m_{\pi\pi},t)$ shown in fig.1 represents the complete basis of our partial wave analysis. This is an energy-dependent analysis performed in 40 MeV mass bins from 580 MeV to 1780 MeV and a single t bin between 0.01 $GeV^2/c^2$ and 0.20 $GeV^2/c^2$. In this bin the t dependence of moments is fairly weak.

The moments are related to the nucleon transversity (spin component perpendicular to the production plane) amplitudes by the following formulae (for the explicit form see Refs. [9-11]):

$$t_M^L = \sum_{j,k} c_{jk}^{LM} \text{Re}(^Ug_j\, ^Ug_k^* + {^Uh_j}\, ^Uh_k^* + {^Ng_j}\, ^Ng_k^* + {^Nh_j}\, ^Nh_k^*)$$

$$p_M^L = \sum_{j,k} c_{jk}^{LM} \text{Re}(^Ug_j\, ^Ug_k^* - {^Uh_j}\, ^Uh_k^* - {^Ng_j}\, ^Ng_k^* + {^Nh_j}\, ^Nh_k^*)$$

$$r_M^L = \sum_{j,k} c_{jk}^{LM} \text{Re}(^Ug_j\, ^Ug_k^* - {^Uh_j}\, ^Uh_k^* + {^Ng_j}\, ^Ng_k^* - {^Nh_j}\, ^Nh_k^*)$$

where

g(h) are the transversity down (up) amplitudes;

U(N) are the unnatural (natural) spin parity exchange;

j or k stands for the dipion spin $\ell$ and helicity m indices;

$c_{jk}^{LM}$ contains the Clebsch-Gordan coefficients etc.

A $\pi$-exchange dominance is equivalent to the equality $^Ug_j \equiv {^Uh_j}$ for any j. This in turn leads to the vanishing of the $p_M^L$ and $r_M^L$ moments. Experimentally only $r_M^L = 0$ (see fig.1) and this is due to the weakness of the natural exchange at low t. On the contrary the $p_0^L$ and $p_1^L$ moments differ significantly from zero. This immediately rules out the assumption a).

Since the $M \geq 2$ moments vanish we use only amplitudes with dipion helicit m $\leq$ 1. As it has been found empirically the $\ell \geq 2$ ($\ell \geq 3$) amplitudes can be neglected below 900 MeV (1380 MeV). In each mass bin various starting points were used always independently of the results for the neighbouring bins. The details of the fit are described in Refs. [9-11].

The fits are generally very good and they produce unique solutions apart from the mass regions 900-1100 MeV and ~1500 MeV. Nevertheless even here one solution is usually favoured by demanding the resonance shape for the leading waves ($P_0$ and/or $D_0$ and/or $F_0$).

## INTENSITIES OF THE PARTIAL WAVES

As we have seen, the relations between moments and amplitudes do not contain the phase between the g and h amplitudes. This missing information could only be supplied by the measurement of the polarization of the recoil neutron. In the absence of this information we can determine moduli and relative phases inside each set of transversity amplitudes. This allows us to calculate an important quantity

$$|L_m|^2 = |g_m^\ell|^2 + |h_m^\ell|^2$$

which is an intensity of each partial wave characterized by the dipion spin $\ell$ and helicity m. These intensities are shown in fig.2. Each point represents the partial wave cross section in a 40 MeV mass bin and a t range of $0.01$ GeV$^2$/c$^2$ < $|t|$ < $0.20$ GeV$^2$/c$^2$. If there is more than one solution the full circle shows the solution which is more consistent with the tails of the leading resonances $\rho(770)$, f(1270) and g(1690) which dominate the $P_0, D_0$ and $F_0$ waves, respectively. In the S wave above the $\rho(770)$ resonance the full circles represent the solution more similar to the "down" solution generally favoured in the previous studies.

The results of the Breit-Wigner fits to the leading partial waves are compared in Table I with the relevant Particle Data Group [12] values (in brackets)

Table I

| Resonance | $\rho$ | f | g | $\rho'$ |
|---|---|---|---|---|
| Mass(MeV) | 776.1 ± 2.6 (776.0 ± 3.0) | 1273.8 $^{+2.8}_{-2.7}$ (1271.0 ± 5.0) | 1716 $^{+58}_{-29}$ (1688 ± 20) | 1598 $^{+24}_{-22}$ (1600) |
| Width(MeV) | 161.8 $^{+7.6}_{-7.2}$ (155.0 ± 3.0) | 183.2 $^{+8.3}_{-7.9}$ (180.0 ± 20) | 325 $^{+\infty}_{-119}$ (180 ± 30) | 175 $^{+98}_{-53}$ (300) |
| Elasticity(%) | 100 (input) (100) | 85.7 ± 1.6 (80.3 ± 0.3) | 25.9 $^{+1.8}_{-1.9}$ (24.0 ± 5.0) | 28.7 $^{+4.3}_{-4.2}$ (25.0 ± 10) |
| $\chi^2$/NDF | 8.1/6 | 11.8/13 | 20/5 | 5.4/5 |
| Mass range | 580-980 | 980-1660 | 1420-1780 | 1420-1780 |

The consistency with the previous results is fairly good (the large errors for the g resonance are due to the lack of high mass bins). Since we fit the partial waves obtained from two high statistics experiments in a model independent way our results may well represent the best parameters for the $\rho(770)$ and f(1270) resonances.

Now let us turn to the lower partial waves in the mass region above 1400 MeV. As we have already mentioned, it is here that one can distinguish between various solutions obtained in the previous studies. Fig.3 shows that the α solution of Martin and Pennington [6] is strongly disfavoured while β and especially β' are consistent with our results. The distinction between β and β' is of minor importance since according to Ref.[6] they are two versions of the same solution. Since the ρ'(1600) resonance is an essential feature of this solution we have tentatively fitted the $P_0$ wave at high mass by the Breit-Wigner formula. The results are also shown in Table I.

In the lower mass region the S wave is certainly very complicated showing a first maximum at 800 MeV, an ambiguity at (800-900) MeV, a dip at 1000 MeV and a broad bump abruptly terminated at 1500 MeV. The behaviour of the S wave phase (see Ref.[10]) is consistent with the narrow $S^*(980)$ and broad ε(1200). However, the rapid fall of the S wave intensity may indicate the presence of a high mass object interfering with the tail of the ε(1200) resonance.

## NON-PION EXCHANGE

It has been widely assumed that at low four-momentum transfer pion exchange is by far the dominating mechanism of our reaction. The polarized traget experiment enables us to investigate the other exchanges in a model-indepdent way. Namely, we can determine both the exact amount of the m = 1 waves and the deviation of the m = 0 waves from pure pion exchange.

A so-called "Poor Man's Absorption Model" of Williams [13] predicts the following relation between the unnatural exchange amplitudes:

$$\frac{|L_U|}{|L_0|} = \frac{C_A^L}{m_{\pi\pi}} \sqrt{L(L+1)}$$

where $|L_U|^2 = |^U g_1^L|^2 + |^U h_1^L|^2$ and $C_A^L$ in an absorption constant. Thus

$$\frac{|P_U|}{\sqrt{2}|P_0|} = \frac{|D_U|}{\sqrt{6}|D_0|} = \frac{|F_U|}{\sqrt{12}|F_0|}$$

It has been found in previous studies [2-4] that $C_A^L$ is independent of L but decreases with the increasing π-π mass. In fig.4 we show the ratio

$|L_U|/(|L_0|\sqrt{L(L+1)})$ as a function of $m_{\pi\pi}$ for our results. It is immediately seen that this function is a fairly complicated one. The $m_{\pi\pi}^{-1}$ dependence (normalized in the ρ region) is obviously too weak thus confirming the above-mentioned trend of the absorption constant. On the other hand the analysis of Estabrooks and Martin [4] clearly underestimated the amount of the m = 1 unnatural spin-parity exchange especially above 1400 MeV. Consequently the natural exchange amplitudes were also underestimated due to assumption c).

Now let us discuss the leading m = 0 waves. As it has been already mentioned the pure π-exchange corresponds to $|g_0^\ell| = |h_0^\ell|$. Fig.5 shows that this is not the case in a wide range of $m_{\pi\pi}$ for S, $P_0$ and $D_0$ waves. The deviation of the $|g_0^\ell|/|h_0^\ell|$ ratio from unity clearly demonstrates the presence of a significant s-channel spin no-flip component (an exchange of an object with the quantum numbers of the $A_1$). We cannot determine this component exactly without the (unknown) phase between the g and h amplitudes. However we can find its lower limit by exploring the difference between $|g_0^\ell|$ and $|h_0^\ell|$. This yields 20% for the ρ(770) and 15% for the f(1270).

Also considering the m = 1 amplitudes discussed above we can generally state that the pion exchange dominance is weaker than was thought previously. This exchange, however, becomes more dominant with increasing mass.

## THE BARRELET ZEROS OF TRANSVERSITY AMPLITUDES

The π-π scattering amplitude for $L_{max}$ = 3 can be written as follows:

$$A(m_{\pi\pi}, z = \cos\theta) = S + \sqrt{3}\, P_0 z + \sqrt{5}\, D_0\left(\frac{3}{2}z^2 - \frac{1}{2}\right) + \sqrt{7}\, F_0\left(\frac{5}{2}z^3 - \frac{1}{2}\right)z = c\prod_{i=1}^{3}(z - z_i)$$

where c and $z_i$ are complex functions of $m_{\pi\pi}$. From the experimental π-π cross section we only determine the modulus of the amplitude

$$\frac{d\sigma_{\pi\pi}}{dt} \propto |A(m_{\pi\pi},z)|^2 = |c|^2 \prod_{i=1}^{3}\left[(z - z_i)(z - z_i^*)\right] = |c|^2 \prod_{i=1}^{3}\left[(z - \text{Re}\,z_i)^2 + |\text{Im}\,z_i|^2\right]$$

The Barrelet ambiguity is just the uncertainty of the sign of the imaginary parts of the complex zeros since the π-π cross section is invariant under $z_i \rightleftarrows z_i^*$. This manifold ambiguity is fortunately reduced since each new zero must enter the physical region with a large negative imaginary part.

In fig.6 we show the Barrelet zeros for the (better determined) h amplitude (the plot for the g amplitude is very similar, only fluctuations are larger). The imaginary parts fluctuate around zero, apart from the initial trend at low mass mentioned above. There are some deviations (Im $z_{1,2} \neq 0$) around 1500 MeV and they produce large ambiguities but otherwise there is a unique solution. This is in striking contrast to the results of the old studies (the curves are after Estabrooks and Martin [4]). Our result, however, explains why we have obtained unique solutions using various independent starting points for the fit in each mass bin independently.

## REFERENCES

[1] G. Grayer et al., Nuclear Phys. B75 (1974) 189.

[2] B. Hyams et al., Nuclear Phys. B64 (1973) 134.

[3] B. Hyams et al., Nuclear Phys. B100 (1975) 205.

[4] P. Estabrooks and A.D. Martin, Nuclear Phys. B95 (1975) 322.

[5] C.D. Froggatt and J.L. Petersen, Nuclear Phys. B129 (1977) 89.

[6] A.D. Martin and M.R. Pennington, CERN preprint TH-2353.

[7] E. Barrelet, Nuovo Cimento 8A (1972) 331.

[8] V.D. Apel et al., Phys. Letters B57 (1975) 398.

[9] H. Becker et al., Nuclear Phys. (submitted for publication).

[10] H. Becker et al., Nuclear Phys. (submitted for publication) - this is a full version of the present work.

[11] G. Lutz and K. Rybicki, MPI report MPI-PAE/Exp.El.75 (October 1978).

[12] Particle Data Group, Phys. Letters B75 (1978) No.1.

[13] P.K. Williams, Phys. Rev. D1 (1970) 312.

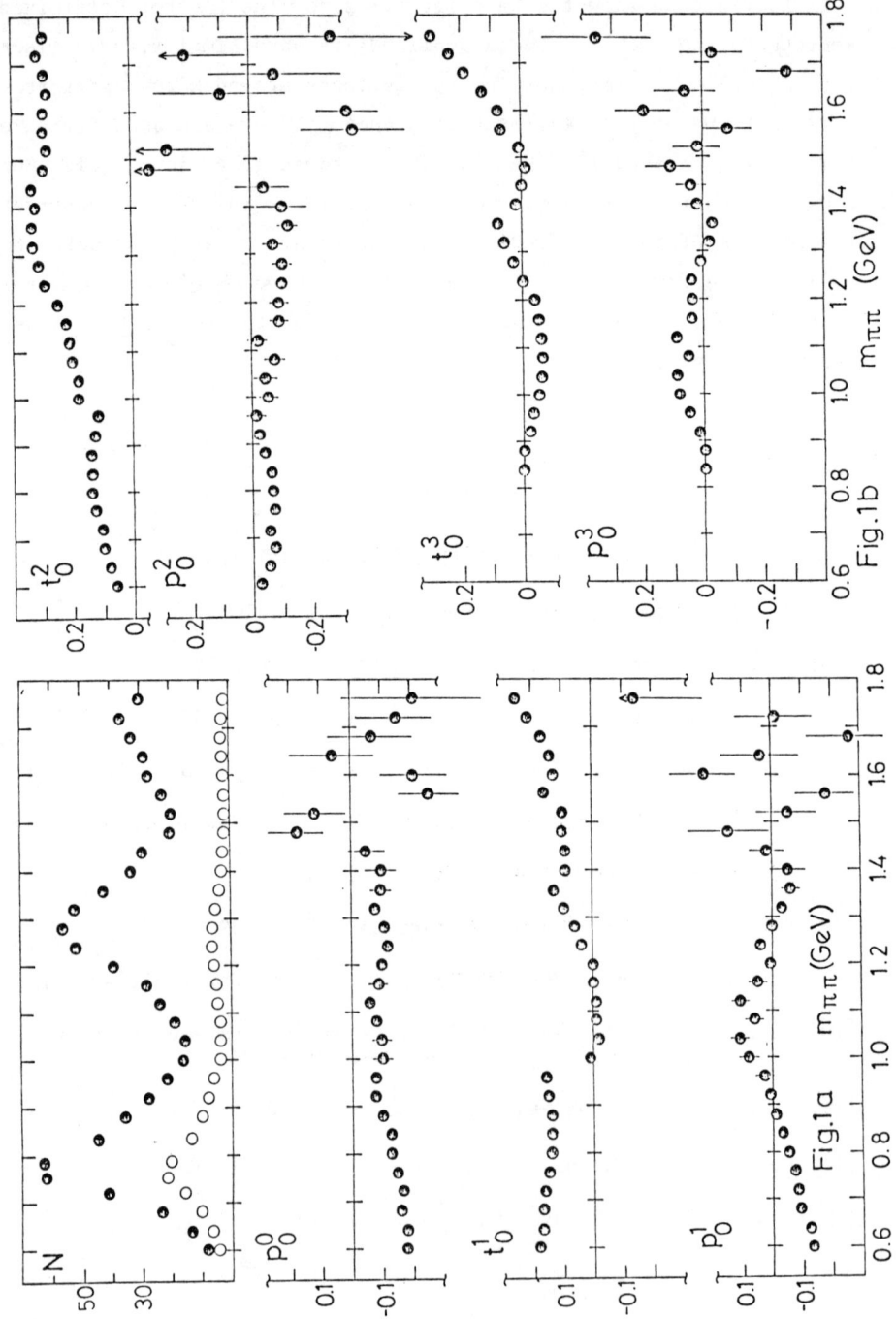

Fig.1a, Fig.1b

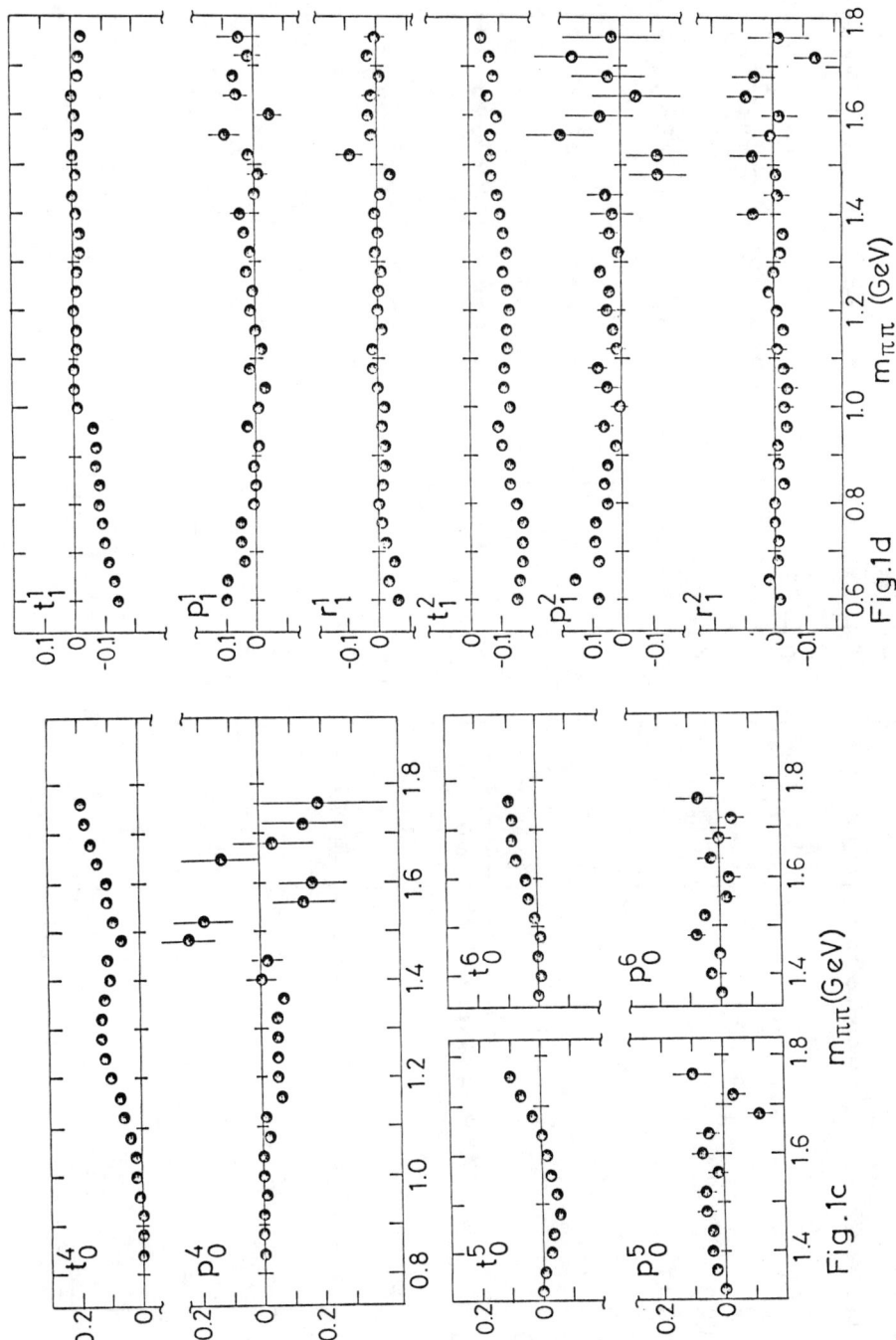

Fig.1c

Fig.1d

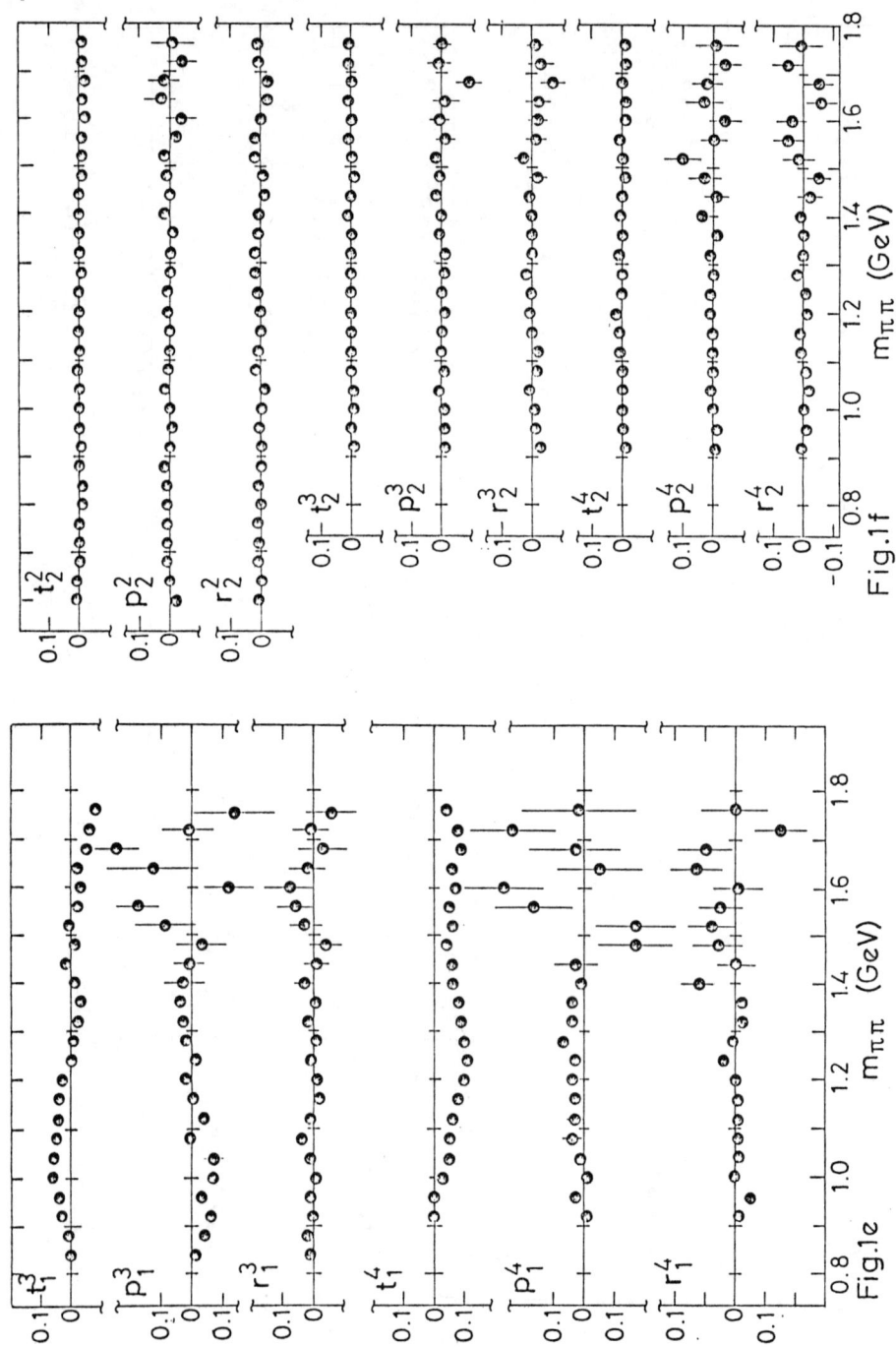

Fig.1e Fig.1f

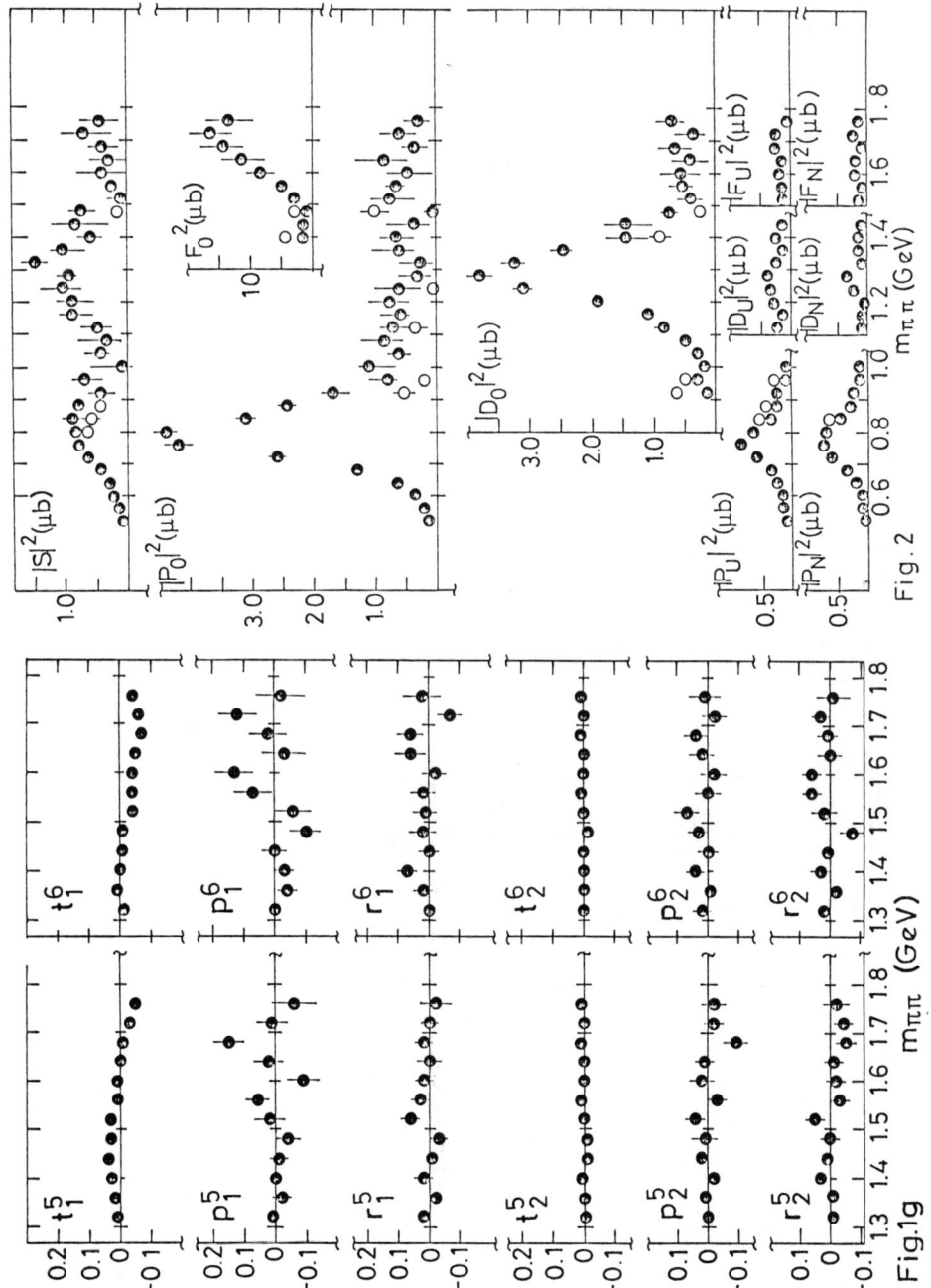

Fig.1g  Fig.2

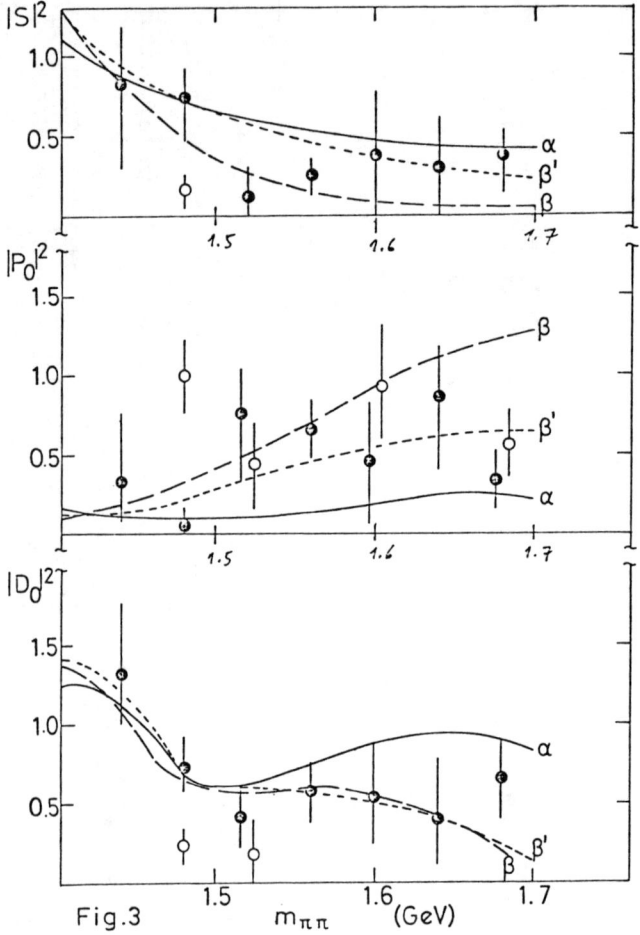

Fig. 3

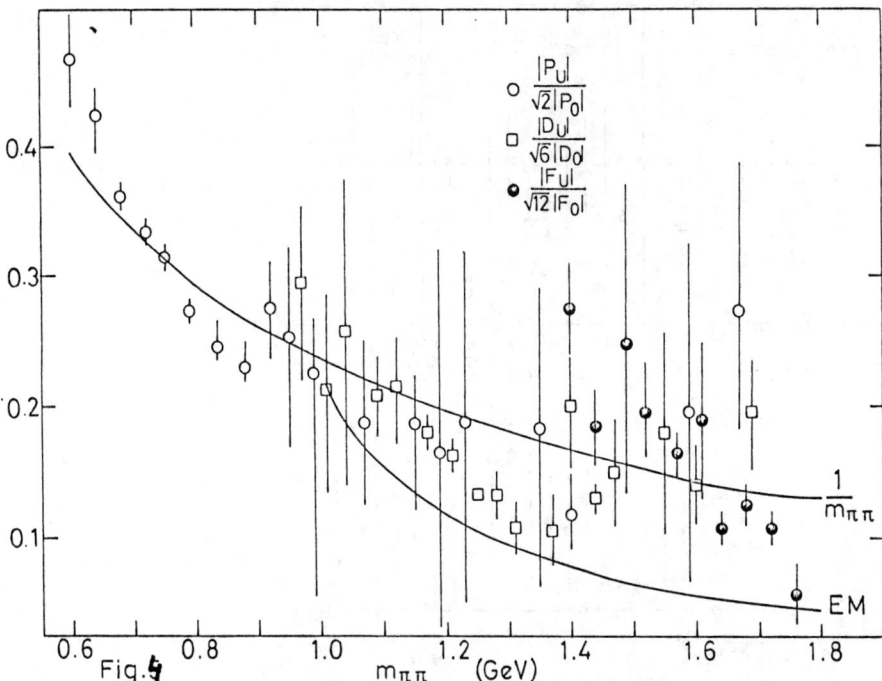

Fig. 4

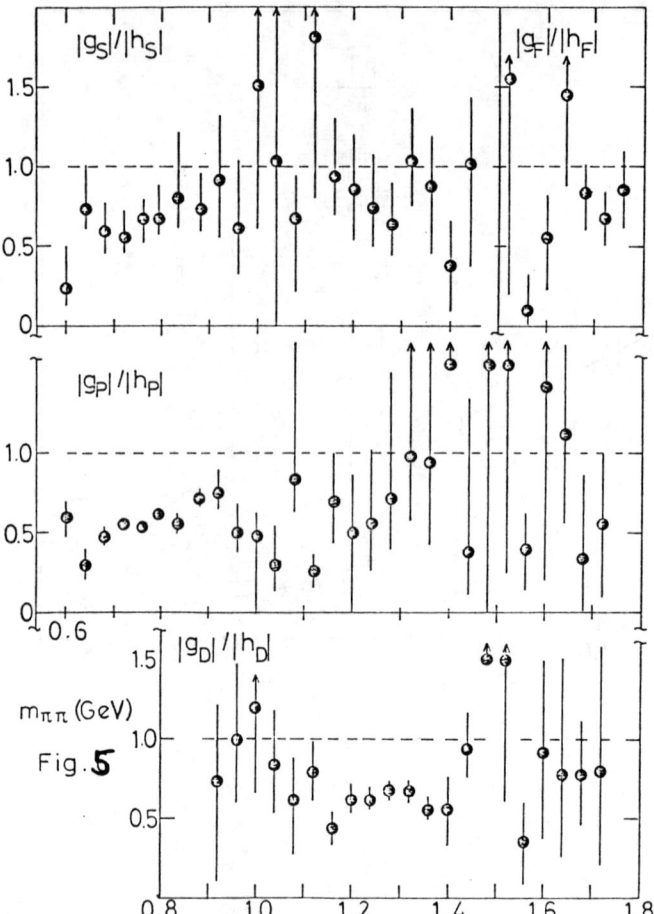

Fig. 5

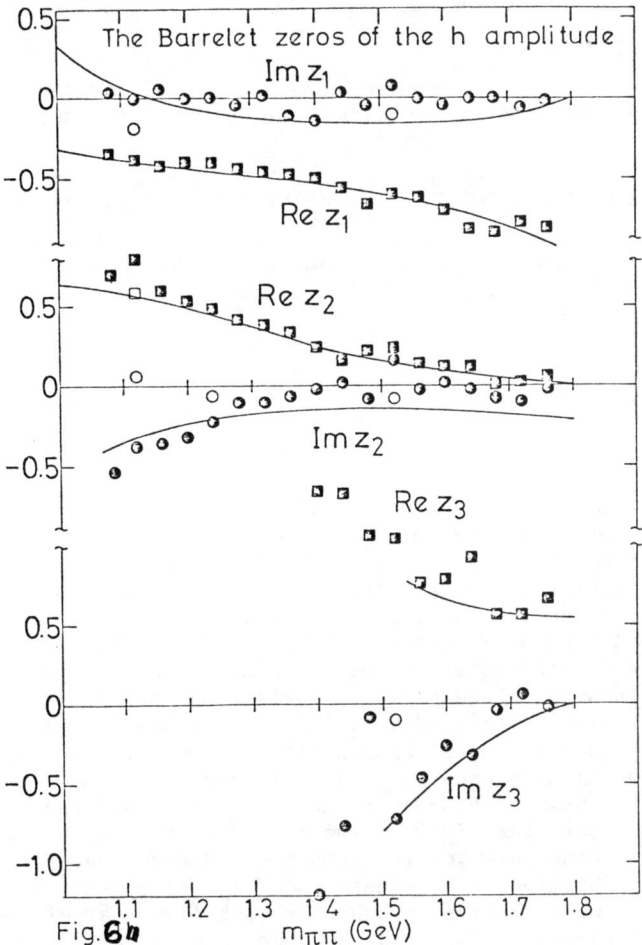

Fig. 6b

## POLARIZATION MEASUREMENTS AND A NARROW ε(750)

J. T. Donohue
Argonne National Laboratory, Argonne, IL 60439
Univ. de Bordeaux I, Domaine du Haut-Vigneau, F-33170 Gradignan, France

Y. Leroyer
Univ. de Bordeaux I, Domaine du Haut-Vigneau, F-33170 Gradignan, France

Low energy $\pi$-$\pi$ scattering has been studied mainly via the detailed analysis of the reactions

$$\pi^- p \to \pi^- \pi^+ n \qquad\qquad A$$

and

$$\pi^+ p \to \pi^+ \pi^- \Delta^{++}. \qquad\qquad B$$

Over the years the quality of the experimental data has steadily improved, and we have seen at this conference the latest results of the CERN-MPI experiment,[1] observing reaction A with a polarized target. The set of observables measured is sufficient, in principle, to determine all amplitudes up to independent arbitrary phases for those with recoil neutron transversity up and down, respectively. The most striking result of this experiment is that polarization effects are substantial, especially indicating the presence of nucleon-non-flip helicity amplitudes, called $A_1$ exchange. Since the most thorough phenomenological analyses[2,3] of $\pi$-$\pi$ explicitly assumed the absence of $A_1$ exchange, it is conceivable that their conclusions concerning the S-wave phase shift may be wrong. All analyses show that at the ρ-mass, the $I = 0$ S-wave phase-shift is approximately $\pi/2$. The remaining question is whether the S-wave phase-shift varies slowly as a function of mass near $m_{\pi\pi} = m_\rho$, or whether it varies rapidly, showing a resonant behavior much like that of the P-wave phase-shift. Although the narrow resonance interpretation was once popular and still has some advocates, the recent consensus favors a broad ε at a high mass.

It is clear that two major differences between a narrow and broad ε are: 1) The narrow ε will have a smaller cross section, integrated over the $\pi\pi$ mass interval around the ρ. 2) The relative phase of S- and P-wave amplitudes will vary appreciably across the ρ-mass region for the broad ε, but only slightly for the narrow ε.

In a recent detailed analysis of reaction B,[4,5] we have found that the integrated S-wave cross section is substantially smaller than previous analyses have suggested.[6] The essential feature of our approach is to use the joint-angular decay distribution for the $\pi^+\pi^-\Delta^{++}$ system, rather than just the $\pi\pi$ decay moments. In principle, the full set of joint-decay moments provides enough information to determine all amplitudes, up to some generalized phases. Unfortunately, our methods cannot be applied directly to the raw data of

ISSN: 0094-243X/79/510476-02$1.50 Copyright 1979 American Institute of Phys

Ref.4, since certain necessary pasitivity conditions on the joint-density-matrix are not satisfied. However, if one modifies the data (within errors) so as to satisfy pasitivity, one finds that the S-wave fraction, integrated over the $\rho$ mass region, is about 30% smaller than that found by the analysis of Ref.6. In order for a small integrated S-wave to produce the large S-P interference terms observed, the S-wave must be closely in phase with the P-wave, hence it must resonate with a width of order 200 MeV. In consequence little or no mass-dependence of the moments is expected. The narrow $\varepsilon$ we find is also consistent, within errors, with the findings of Ref.1, provided one chooses the smaller S-wave solution in the $\rho$-mass region. Again the mass-dependence of observables is predicted to be small throughout the $\rho$-region. Finally, we remark that a recent analysis of reaction B at 3.75 GeV/c,[7] which again fitted the joint-decay moments as a function of dipion mass, concluded that the data were consistent with a narrow $\varepsilon$ resonance having a mass and width comparable to the $\rho$.

The importance of detailed polarization measurements in reducing the theoretical abmiguities present in such analyses cannot be over-emphasized. Reliance on simplifying phenomenological assumptions often biases the results of the analysis, and leads to false conclusions.

## REFERENCES

1. CERN-MPI Collaboration, presented to this conference by K. Rybicki.
2. P. Estabrooks et al., Proceedings of the AIP Conference on Scattering, Tallahassee (1973).
3. B. Hyams et al., Nucl. Phys. B64 134 (1973).
4. J. F. Owens et al., Nucl. Phys. B112, 514 (1976).
5. J. T. Donohue and Y. Leroyer, preprint U. of Bordeaux, 1978.
6. S. D. Protopopescu et al., Phys. Rev. D7, 1279 (1973).
7. N. Gelfand et al., Nucl. Phys. B 138, 365 (1978).

## POLARIZATION MEASUREMENTS IN $K^+N$ CHARGE EXCHANGE AT 6 AND 12 GeV/c

M. Fujisaki, M. Babou, J. Bystricky, G. Cozzika,
T. Dobrowolski*, Y. Ducros, A. Gaidot, C.F. Hwang**,
A. Itano, F. Khantine-Langlois, F. Lehar, A. de Lesquen***,
J.C. Raoul, L. van Rossum and G. Souchere
Département de Physique des Particules Elémentaires,
CEN-Saclay, France.

### ABSTRACT

The polarization in $K^+n\uparrow \to K^0p$ has been measured at 6 and 12 GeV/c in the interval $0.1 < |t| < 1.0$ using a polarized deuteron target. The results are compared to predictions from SU(3), EXD and line reversal, and from various phenomenological models.

The polarization in $K^+p\uparrow \to K^+p$ measured simultaneously with high statistics is compared to previous results on free polarized protons.

---

Comparison of the reactions $K^-p \to \bar{K}^0n$ and $K^+n \to K^0p$ is essential for studying vector and tensor exchange components in meson-nucleon scattering. Measurements of the differential cross sections exist for both reactions, but the polarization parameter has been measured only for $K^-p \to \bar{K}^0n$ at 8 GeV/c [1].

The present experiment measures the polarization in $K^+n \to K^0p$ at 6 and 12 GeV/c for t ranging from -0.1 to -1.0 $(GeV/c)^2$. It was made possible by recent progress in the development of polarized deuteron targets by the CERN Polarized Target Group. The experiment contributes to the determination of the amplitudes dominated by $\rho$ and $A_2$ exchange. The results can be compared to predictions from SU(3) invariance, from exchange degeneracy and from line reversal invariance and to the predictions from various phenomenological models for kaon-nucleon scattering. Finally, the measurement is necessary for obtaining a complete

* Present address : Brookhaven National Laboratory, Upton, LI., NY 11973.

** Present address : Los Alamos Scientific Laboratory, P.O. Box 1663, NM 87544.

*** Temporary address : KEK, National Laboratory for High Energy Physics, Oho-machi, Tsukuba-gun, Ibaraki-ken, 300-32, Japan.

ISSN: 0094-243X/79/510478-13$1.50 Copyright 1979 American Institute of Phy

set of observables from which the amplitudes can eventually be constructed without phenomenological hypothesis.

Methods to use the $K^+n \to K^0p$ polarization data for constructing the tensor exchange amplitudes from an incomplete set of data are presented in separate papers [2,3]. Descriptions of the polarized deuteron target [4] and of the experimental apparatus [5] (fig. 1) are published elswhere.

The experiment was carried out in 1976-1977 at the CERN-PS in an unseparated beam produced from an external target in the slow extracted proton beam. Kaons, identified by three gas Cherenkov counters, represent 2.7 percent of the beam particles at 6 GeV/c and 1.7 percent at 12 GeV/c. The beam intensity is about $1.3 \times 10^6$ particles per burst at both energies. At 6 GeV/c, because of the increased multiple scattering in the Cherenkov counters, about 30 percent of the particles miss the target. The effective kaon intensity is thus approximately the same at both energies.

The polarized target [6] uses a $^3He-^4He$ dilution refrigerator [7]. It is made of 18 gr of 92 percent deuterated propanediol in a cylindrical volume of 120 mm length and 17 mm diameter. The polarization of the neutrons in the deuterons is close to 40 percent, corresponding to a free proton polarization exceeding 97 percent. This figure illustrates the performance of the cryostat.

The $K^0$ emitted forward is detected through its decay mode $K^0_S \to \pi^+\pi^-$. The momenta of both the scattered and the recoil particles can thus be measured by recording charged tracks deflected in magnetic fields. Knowing also the incident momentum vector, the Fermi momentum of the target nucleon can be reconstructed. The target momentum distribution is different for events on polarized neutrons in deuterons and for target background events on quasi-free neutrons in the heavier nuclei. Discrimination against target background is thus possible by appropriate kinematical cuts [8].

The pion pairs are measured in the forward spectrometer consisting in five MWPC's and a large aperture magnet. The track segments before and after the magnet are measured by two pairs of chambers, with a lever arm of 100 cm at 6 GeV/c and 200 cm at 12 GeV/c. The first chamber of the spectrometer, at 80 cm and 140 cm, respectively, for 6 and 12 GeV/c, defines the end of the $K^0_S$ decay region. The beginning of the decay region is defined by an anticoincidence counter 4 cm after the end of the target. The target magnet is used to measure the momentum of the recoil protons by recording the tracks in scintillator hodoscopes close to the target, and in MWPC's at the edge of the magnetic field and in a plane 100 cm farther out in the field free region. The deflection of the recoil protons by the target magnet leads to an asymmetric acceptance for scatterings to the left and to the right. However, measuring simultaneously on either side the change of rate due to reversal of the target polarization $P_t$ considerably reduces the influence of possible errors in the relative normalization of runs with opposite sign of $P_t$.

The data acquisition is triggered by requiring an incident $K^+$, no charged particles 4 cm after the end of the target, two charged particles in the hodoscopes at the end of the spectrometer and one charged particle in the one of the recoil arms consistent with the position of the forward particles. The trigger is vetoed if additional charged or neutral particles are detected in one of the triggering hodoscopes or in one of the lead-scintillator sandwich anticoincidence counters surrounding the target. The PDP11/45 to CAMAC interface [9,10] transfers all relevant information on the event to the computer. The acquisition capacity being much higher than required for the rare events $K^+n \to K^0p$ occuring less than once per burst, a multiple trigger logic allows to take data simultaneously on several other reactions, in particular $K^+p$ and $\pi^+p$ elastic scattering and $K^{0*}(890)$ production by $K^+$ or $\rho^0$ production by $\pi^+$ on neutrons.

The primary data tapes are treated off-line by a CDC 7600 program providing fast return additional quality controls, and sorting the events according to the type of trigger, leading to a small number of secondary tapes for the rare triggers. Tracks of pions pairs in the spectrometer are reconstructed from the points in the wire planes by the principal component method [11]. The HWHM mass resolution on the $K_s^0$ mass is 9 MeV and the average value agrees with the world average figure within 0.4 percent. The recoil proton momentum is measured with an error of 3 to 4 percent. The global fit at the scattering and decay vertices yields the value of the target momentum $P_F$ with a most probable error of 30 MeV/c, as shown by reconstructing events on liquid hydrogen (fig. 2a). The distributions shown on figure 2 are obtained by analysing events $K^+p \to K^+p$ which are available with high statistics. The difference between the experimental target momentum distribution for the polarized target and for the carbon target (fig. 2b) is the same as obtained from liquid deuterium (fig. 2c), after subtracting the small hydrogen contribution in the polarized target. The liquid deuterium distribution is also well fitted by Monte Carlo events generated with a distribution of the form [12]

$$\frac{P_F^2}{(P_F^2 + \alpha^2)^2 (P_F^2 + \beta^2)^2}$$ with $\alpha = 45$ MeV/c and $\beta = 270$ MeV/c

and reconstructed taking into account the experimental resolution. The excess of real events at $P_F > 300$ MeV/c is not significant since it could be due to a small contamination by inelastic events from liquid deuterium. The deuteron distribution peaks at 60 MeV/c and the carbon or oxygen distribution at about 140 MeV/c. The deuteron/(carbon+oxygen) ratio for the polarized target is enhanced by a factor 2.3 by the final cuts on the module and the transverse components of the reconstructed target momentum. This is the average enhancement for all momentum transfers in the reaction $K^+p \to K^+p$. For $K^+n \to K^0p$ the determination of the enhancement is less precise because the background from inelastic events on deuterons complicates the calculation. The remaining target background is taken into account by a maximum likelihood method assigning to each event a probability to belong to the background, depending on the value of the reconstructed Fermi momentum distribution of the target neutron. This method leads to a systematic uncertainty of $\left|\frac{\Delta P}{P}\right| < 0.1$. The resulting carbon normalization coefficient is checked for consistency between left and right scatterings and for compatibility with normalization by the beam telescope monitor and the measured weights of the polarized and the carbon targets.

For relative normalization of runs with opposite sign of the target polarization the beam telescope ratio is checked against the ratios given by the beam Cherenkov signatures, by the strobe gate rates and by two target monitor telescopes in the vertical plane containing the target polarization, all ratios being corrected for counting rate effects and for specific incidents in the operation of each of these monitors. These different normalization procedures all agree within $\leqslant 3 \times 10^{-3}$. The resulting possible systematic error is $|\Delta P| \leqslant 1 \times 10^{-2}$ for the final results calculated from all events left and right.

The contamination of $K^+n \to K^0p$ by inelastic events on deuterons is $5 \pm 3$ percent as determined by a detailed Monte-Carlo method using as input all information on the various inelastic channels from Bubble Chamber experiments, and fitting the observed kinematical distributions by a weighted sum of the inelastic channels. An upper bound for the average polarization of this background, $|P_b| \leqslant 0.5$, is obtained by analysing the background events, mostly $K^+n \to K^0 \Delta^+$ with $\Delta^+ \to p\pi^0$, in neighbouring kinematical regions. The resulting possible systematic error on $P(K^+n \to K^0p)$ is therefore $|\Delta P| \leqslant (2.5 \pm 1.5) \times 10^{-2}$.

The measurement of the target deuteron polarization $P_D$ is based on several calibration methods [13]. It has an estimated systematic error of $|\Delta P_D/P_D| = 5 \times 10^{-2}$ for the average values. The neutron polarization $P_n$ $(0.91\pm0.015)P_D$ is calculated assuming $(6 \pm 1)$ percent D-wave [12] contribution to the deuteron wave function.

Possible errors due to changes in the detection or reconstruction efficiency are minimized by reversing the sign of the target polarization at the end of each data tape, i.e. approximately 200 times. Separate data analyses of several successive running periods of approximately six weeks all give the same result within statistics.

The results for the polarization parameter P in $K^+ n \to K^0 p$ at 6 and 12 GeV/c as function of the four momentum transfer t [14] are given in tables 1 and 2 and in figures 3 and 4. Only the statistical errors are given. None of the several systematic errors exceeds $|\Delta P| = 0.01$ or $|\frac{\Delta P}{P}| = 0.1$.

The results at 6 GeV/c are presented on figure 3 in bins which are only slightly larger than the experimental resolution. The polarization reaches a maximum of $P(t) \simeq 0.50 \pm 0.15$ at $t \simeq 0.35$ $(GeV/c)^2$ and decreases in the region $0.4 < |t| < 0.6$ where the absolute value of $P(K^- p \to \bar{K}^0 n)$ is still increasing.

An alternate way to extract the information on $P(t)$ contained in the data is the maximum likelihood method described in reference [34]. This methods searches for a function $P(t)$ most likely to be presented by all of the events. Lacking a prescription for the analytical expression of $P(t)$ we have assumed, for simplicity, $P(t)$ to be given by a polynomial in t of unknown order N, multiplied by $\sqrt{-t}$ to insure the correct behaviour at $t = 0$. Experience shows that the polynomial expansion can be used only in a relatively small region of the variable. We therefore apply the method separately to two regions of t. For the events with $|t| < 0.5$ the likelihood plateau is reached for $N = 6$. The most likely function $P(t) = \sqrt{-t} \sum_{i=0}^{6} a_i t^i$ is represented by the solid curve on figure 3.
The same method applied to the events with $0.25 < |t| < 0.75$ reaches constant likelihood at $N = 5$. The resulting function $P(t)$ for this region is shown by the dotted curve. The solutions for the two sets of events connect smoothly in the region of overlap. The numerical results of this analysis are given table 2. This table also gives the results obtained with $N = 7$ and $N = 6$ for the two regions, respectively, showing that no significant change occurs when using the next higher order polynomials. The maximum likelihood analysis leads to the same conclusions about the t-structure of the polarization as the presentation of the data in narrow bins, close to the instrumental resolution. The present data don't exclude the possibility of negative values of P at $|t| < 0.2$.

The results at 12 GeV/c shown on figure 4, are obtained with large statistical errors and provide less information about the t-structure. For comparison, figure 4 also shows the results at 6 GeV/c with the same binning. The average polarization in the interval $0.1 < |t| < 1.0$

is $0.23 \pm 0.17$ at 12 GeV/c, as compared to $0.23 \pm 0.06$ at 6 GeV/c. The statistics correspond to about 2500 hours of running at either energy.

Our data confirm that predictions of strong EXD are not satisfied. The deviations from the mirror symmetry of $P(K^+n \to K^0p)$ and $P(K^-p \to \bar{K}^0n)$ which is predicted by weak EXD, suggest EXD violating contributions in the $K^{\pm}N$ CEX amplitudes at these energies.

In figure 5a we compare our data with predictions of the Barger-Cline SU(3) sum rule [15] assuming $\eta = \eta_8$. When $\eta - \eta'$ mixing is introduced with a mixing parameter $X = 1.18$ [16] the predictions for $P(K^+n \to K^0p)$ are not significantly modified. Agreement is good up to $t \approx -0.35$, but for $|t| \gtrsim 0.5$ the sum rule predicts too high polarization by more than two standard deviations for the average value of P in the interval $0.5 \leqslant |t| \leqslant 1.0$. The sum rule for $d\sigma/dt$ is more sensitive to the value of X and is strongly violated for $X = 1.18$. We remark that the same assumptions about $\eta - \eta'$ mixing also lead to strong disagreement with the transition rates of $\psi \to \pi^0\gamma$, $\eta\gamma$ and $\eta'\gamma$ [17].

Figure 5b shows the bounds for $P(K^+n \to K^0p)$ at 4 GeV/c by Martin, Michael and Phillips [18] obtained from hypercharge exchange data assuming that vector and tensor exchanges in charge and hypercharge exchange reactions are related by SU(3) octet symmetry with common F/D ratios in hypercharge exchange. Our results are consistent with the bounds up to $t \approx -0.35$, but not with the predictions by the same authors, based on additional estimates of the (F/D) ratios for non-flip and flip amplitudes and SU(3) symmetry breaking.

In figures 6a,b our data at 6 GeV/c are compared with the predictions representative for the new generation of weak [19-23], strong [24,25], and effective [26,27] Regge cut models. There is consistency for sign and order of magnitude between the model predictions and the data for $|t| \lesssim 0.4$, although our data do not exclude the possibility of negative polarizations at $|t| < 0.15$. Most models predict $P(K^+n \to K^0p)$ too large for $|t| > 0.4$. Exceptions are solution B of [19], and [24] which both involve cuts in the tensor exchange s-channel helicity single-flip amplitude.

Figure 6c shows predictions of analyses based on fixed-t dispersion relations with FESR constraints [28-31]. These analyses combine elastic and CEX KN reactions except [28] which is restricted to KN CEX. The most recent analysis [31] includes a careful treatment of the vacuum exchange amplitude and uses reliable low-energy phase shifts.

Figure 6d compares our data at 12 GeV/c with available predictions near to this energy. The energy dependence is most marked in one of the two versions of the strong cut model [24].

From comparisons in figure 6 we conclude that for $|t| \lesssim 0.4$ models with weaker cuts [19,22,31] give somewhat better description. However models with stronger cuts are less restrictive and could adapt to our data. Best overall agreement at 6 and 12 GeV/c appears to be with analysis [31] which predicts a maximum of $P \approx 0.5$ in the region of $-t \approx 0.3$, and weak energy dependence.

In the same experiment the polarization in $K^+p \to K^+p$ (table 3 and fig. 7) and $\pi^+p \to \pi^+p$ at 6 and 12 GeV/c [14] was measured with high statistics by scattering on protons in the polarized deuterons. The results are in good agreement with those obtained by scattering on free polarized protons [32]. The comparison for $K^+p \to K^+p$ on figure 7 shows no evidence for the left-right asymmetry of the cross sections on polarized protons in deuterons to be modified measurably by spin dependent final state interaction between the recoil and the spectator nucleon. Such effects would be expected to be strongest at small t-values.

The authors wish to acknowledge many stimulating discussions with M. Svec on the phenomenology of kaon - nucleon scattering.

## REFERENCES

1. W. Beusch et al., Phys. Letters 46B, 477 (1973).
2. M. Fujisaki et al., "Construction of Tensor Exchange Amplitudes in KN and $\overline{KN}$ CEX Reactions", XIX International Conference on High Energy Physics, August 23-30, 1978, Tokyo, Japan.
3. M. Fujisaki et al., "Construction of Tensor Exchange Amplitudes in KN and $\overline{KN}$ CEX Reactions", to be submitted to Nuclear Physics.
4. M. Borghini et al., "A polarized Deuteron Target", to be published in Nuclear Instr. Meth.
5. M. Babou, Thèse de Doctorat 3ème Cycle, University of Paris VII, June 1977.
   M. Babou et al., "Apparatus for two Body Scattering Experiments at 6 and 12 GeV/c with the CERN Polarized Deuteron Target", to be published in Nuclear Instr. Meth.
6. T.O. Niinikoski, "Polarized Targets at CERN", Symposium on High Energy Physics with Polarized Beams and Targets, August 1976, Argonne, p. 458.
7. T.O. Niinikoski and F. Udo, Nuclear Instr. Meth. 134, 219 (1976).
8. M. Babou et al., "Progress Report on the Measurement of Polarization in $K^+$ Neutron Charge Exchange", Symposium on High Energy Physics with Polarized Beams and Targets, August 1976, Argonne, p. 208.
9. J. Pascual and J.C. Raoul, Saclay Internal Report Note CEA-N-2023, April 1978.

10. B. Ollivier, Saclay Internal Report, Journées d'Information Electronique, CEN-S, SES, April 1975.
11. H. Wind, CERN Internal Report CERN 72-11 (1972).
    M. Hansroul et al., CERN Internal Report CERN DD/73/31 (1973).
12. A. Fridman, Fortschritte der Physik 23, 243 (1975).
13. K. Guckelsberger and F. Udo, Nuclear Instr. Meth. 137, 415 (1976).
14. M. Fujisaki, Thèse de Doctorat d'Université, Université Paris-Sud, Orsay, July 1978.
    M. Fujisaki et al., "Polarization Measurements in $K^+n$ Charge Exchange at 6 and 12 GeV/c",
    M. Fujisaki et al., "Addendum to Polarization Measurements in $K^+n$ Charge Exchange at 6 and 12 GeV/c", Contributed papers to the XIX International Conference on High Energy Physics, August 23-30, Tokyo, Japan. Submitted to Physics Letters.
15. V. Barger and D. Cline, Phys. Rev. 156, 1522 (1967).
16. T. Inami et al., Tokyo Preprint UT-Komaba 76-5 (1976).
17. T.F. Walsh, Preprint DESY 76/13 (1976).
18. A. Martin et al., Nucl. Phys. B43, 13 (1972).
19. G.A. Ringland et al., Nucl. Phys. B44, 395 (1972).
20. B. Sadoulet, Nucl. Phys. B53, 135 (1973).
21. G. Girardi et al., Nucl. Phys. B47, 445 (1972).
22. G. Girardi et al., Nucl. Phys. B69, 107 (1974).
23. S.E. Egli et al., Phys. Rev. D9, 1365 (1974).
24. B.J. Hartley and G.L. Kane, Nucl. Phys. B57, 157 (1973).
25. G.L. Kane and A. Seidl, Rev. Mod. Phys. 48, 309 (1976).
26. J.P. Ader et al., Il Nuovo Cimento 27A, 385 (1975).
27. M. Navelet and P.R. Stevens, Nucl. Phys. B117, 475 (1977).
28. E.N. Argyres et al., Phys. Rev. D10, 2095 (1974).
29. J. Dronkers and P. Kroll, Nucl. Phys. B82, 130 (1974).
30. M.D. Lyberg, Nucl. Phys. B89, 142 (1974).
31. B.W. Joynson and B.R. Martin, Preprint University College London (1977).
32. M. Borghini et al., Phys. Letters 31B, 405 (1970).
    M. Borghini et al., Phys. Letters 36B, 497 (1971).
    M. Borghini et al., CERN Internal Report, January 15 (1972).
33. R. Diebold et al., Phys. Rev. Letters 32, 904 (1974).
    J.J. Phelan et al., Phys. Letters B61, 483 (1976).
    P. Sonderegger et al., Phys. Letters 20, 75 (1966).
    M.H. Schaevitz et al., Phys. Rev. Letters 36, 5 (1976).
    P. Bonamy et al., Nucl. Phys. B52, 392 (1973).
    D. Hill et al., Phys. Rev. Letters 30, 239 (1973).
    W. Beusch et al., Phys. Letters 46B, 477 (1973).

34. J. Bystricky, A. de Lesquen, F. Lehar, "Presentation of Experimental Data by a Function", in 1st European Conference on Computational Physics, CERN, 1972.

TABLE CAPTIONS

Table 1 : $P(K^+n \to K^0p)$. Results at 6 and 12 GeV/c (figs. 3, 4).

Table 2 : Maximum likelihood polynomials

$$P(t) = \sqrt{-t} \sum_{i=0}^{N} a_i t^i$$

fitted to two samples of events at 6 GeV/c (fig. 3). The value of -2S measures the quality of the fit.

Table 3 : $P(K^+p \to K^+p)$. Results at 6 and 12 GeV/c (figs. 7ab,).

FIGURE CAPTIONS

Fig. 1 : Schematic view of the apparatus at 6 GeV/c.

Fig. 2 : Distributions of reconstructed momentum of the target nucleon in the laboratory frame.
  a) Events on liquid hydrogen, empty target effect subtracted (H-V).
  b) - Events on the polarized target (A), solid histogram.
   - Events on the carbon target (B), dashed histogram. The normalization is described in the text. Both histograms include the same empty target effect.
   - The difference (A) - (B) is shown by the shaded histogram.
  c) - Events on liquid deuterium, empty target effect subtracted, solid histogram.
   - Deuteron effect from the polarized target (A) - $\alpha$(B) - $\gamma$(H-V) as described in the text, dashed histogram.
   - Monte-Carlo simulation of deuteron events generated with HULTHEN wave function, curve. The three distributions are normalized for $0 < P_F \leq 0.2$ GeV/c.

Fig. 3 : Results for $P(K^+n \to K^0p)$ at 6 GeV/c in narrow bins. The curves represent maximum likelihood functions explained in the text.

Fig. 4 : Results for $P(K^+n \to K^0p)$ at 12 GeV/c (crosses) and at 6 GeV/c (points) in wide bins.

Fig. 5 : Results for $P(K^+n \to K^0p)$ at 6 GeV/c and comparison with SU(3) predictions or bounds.
  a) Predictions from SU(3) Polarization Sum Rule calculated from the data [33] with :

o $\frac{d\sigma}{dt}$ ($K^{\pm}N$ CEX) DIEBOLD and P($\pi^-$p CEX) CERN
x ............. PHELAN and ........... CERN
Δ ............. DIEBOLD and ........... ARGONNE
+ ............. PHELAN and ........... ARGONNE

The dotted lines show an estimate of the overall error.

b) Bounds from SU(3) and EXD, and predictions from additional assumptions (open circles) at 4 GeV/c [18].

Fig. 6 : Results for P($K^+n \to K^0p$) at 6 and 12 GeV/c and comparison with model predictions.
a) 6 GeV/c, weak cut models
——— R1, R2 6 GeV/c [19]
- - - - - G1 5 GeV/c [21]  - - - - - G2 5 GeV/c [22]
—.— S 5.5 GeV/c [20].
b) 6 GeV/c, strong and effective cut models
——— K1 6 GeV/c [24]  ——— K2 8 GeV/c [25]
- - - - - NS 8 GeV/c [27]  —.— A 8 GeV/c [26]
c) 6 GeV/c, models using fixed -t dispersion relations
——— ACH 8 GeV/c [28]  - - - - - L 5 GeV/c [30]
—x— DK 8 GeV/c [29]  —.— JM 6 GeV/c [31]
d) 12 GeV/c, model predictions
——— K1 12 GeV/c [24]  ——— K2 15 GeV/c [25]
- - - - - G2 10 GeV/c [22]  —.— JM 15 GeV/c [31]

Fig. 7 : Results for P($K^+p \to K^+p$) at 6 and 12 GeV/c and comparison with previous experiments on a polarized proton target.
a) Results at 6 GeV/c of the present experiment, full dots, and of reference [32], open circles, at the same energy.
b) Results at 12 GeV/c of the present experiment, full dots, and of reference [32] at 10 GeV/c, open circles, and at 14 GeV/c, crosses.

| INCIDENT MOMENTUM | -t (GeV/c)$^2$ LIMITS | | AVERAGE | NB. OF EVENTS | P ± ΔP $K^+n \to K^0p$ |
|---|---|---|---|---|---|
| | Narrow bins : | | | | |
| | 0.09 | 0.13 | 0.113 | 247 | 0.05 ± 0.59 |
| | 0.13 | 0.18 | 0.158 | 814 | -0.40 ± 0.24 |
| | 0.18 | 0.23 | 0.206 | 1471 | 0.21 ± 0.16 |
| | 0.23 | 0.28 | 0.255 | 1656 | 0.26 ± 0.16 |
| | 0.28 | 0.33 | 0.305 | 1691 | 0.45 ± 0.17 |
| | 0.33 | 0.38 | 0.354 | 1574 | 0.52 ± 0.18 |
| | 0.38 | 0.43 | 0.404 | 1312 | 0.38 ± 0.21 |
| | 0.43 | 0.48 | 0.454 | 1049 | 0.21 ± 0.25 |
| | 0.48 | 0.53 | 0.503 | 770 | 0.01 ± 0.30 |
| 6 GeV/c | 0.53 | 0.60 | 0.564 | 810 | -0.02 ± 0.29 |
| | 0.60 | 0.70 | 0.646 | 747 | 0.34 ± 0.30 |
| | 0.70 | 0.80 | 0.746 | 429 | 0.01 ± 0.42 |
| | 0.80 | 1.00 | 0.879 | 413 | -0.12 ± 0.43 |
| | 1.00 | 1.50 | 1.158 | 223 | 0.38 ± 0.58 |
| | Wide bins : | | | | |
| | 0.10 | 0.27 | 0.206 | 3823 | 0.10 ± 0.11 |
| | 0.27 | 0.37 | 0.319 | 3296 | 0.46 ± 0.12 |
| | 0.37 | 0.50 | 0.428 | 3002 | 0.37 ± 0.14 |
| | 0.50 | 1.00 | 0.652 | 2819 | -0.03 ± 0.16 |
| | Average : | | | | |
| | 0.10 | 1.00 | 0.384 | 12940 | 0.23 ± 0.06 |
| | 0.10 | 0.27 | 0.204 | 718 | 0.02 ± 0.28 |
| 12 GeV/c | 0.27 | 0.37 | 0.318 | 658 | 0.32 ± 0.32 |
| | 0.37 | 0.50 | 0.426 | 531 | 0.28 ± 0.35 |
| | 0.50 | 1.00 | 0.636 | 473 | 0.44 ± 0.42 |
| | 0.10 | 1.00 | 0.371 | 2380 | 0.23 ± 0.17 |

Table 1

| -t (GeV/c)$^2$ | P ± ΔP 10164 events $0.09 < \|t\| < 0.5$ N=6, -2S=27.39 (N=7, -2S=27.42) | P ± ΔP 9163 events $0.25 < \|t\| < 0.75$ N=5, -2S=23.9 (N=6, -2S=24.3) |
|---|---|---|
| 0.15 | -0.06 (-0.08) ± 0.20 (0.23) | |
| 0.20 | 0.05 (0.06) ± 0.14 (0.15) | |
| 0.25 | 0.21 (0.21) ± 0.13 (0.13) | |
| 0.30 | 0.44 (0.43) ± 0.12 (0.13) | 0.41 (0.46) ± 0.13 (0.15) |
| 0.35 | 0.56 (0.56) ± 0.13 (0.13) | 0.62 (0.63) ± 0.13 (0.13) |
| 0.40 | 0.39 (0.40) ± 0.14 (0.16) | 0.46 (0.41) ± 0.12 (0.14) |
| 0.45 | | 0.17 (0.14) ± 0.14 (0.15) |
| 0.50 | | -0.03 (-0.01) ± 0.16 (0.16) |
| 0.55 | | -0.04 (+0.03) ± 0.17 (0.20) |

Table 2

| INCIDENT MOMENTUM | -t (GeV/c)$^2$ LIMITS | AVERAGE | NB. OF EVENTS x 10$^3$ | P ± ΔP $K^+p \to K^+p$ | |
|---|---|---|---|---|---|
| 6 GeV/c | 0.10  0.20 | 0.164 | 104.9 | 0.234 | ± 0.012 |
|  | 0.20  0.25 | 0.226 | 101.6 | 0.246 | 0.012 |
|  | 0.25  0.30 | 0.275 | 113.0 | 0.238 | 0.012 |
|  | 0.30  0.35 | 0.324 | 100.0 | 0.240 | 0.014 |
|  | 0.35  0.40 | 0.374 | 80.5 | 0.268 | 0.016 |
|  | 0.40  0.50 | 0.447 | 130.8 | 0.248 | 0.013 |
|  | 0.50  0.60 | 0.547 | 91.9 | 0.234 | 0.017 |
|  | 0.60  0.70 | 0.646 | 57.9 | 0.229 | 0.021 |
|  | 0.70  0.90 | 0.783 | 55.2 | 0.192 | 0.022 |
|  | 0.90  1.20 | 1.015 | 24.1 | 0.255 | 0.034 |
|  | 1.20  1.50 | 1.315 | 5.6 | 0.066 | 0.070 |
|  | 1.50  2.00 | 1.642 | 1.1 | 0.220 | 0.171 |
| 12 GeV/c | 0.10  0.20 | 0.166 | 44.8 | 0.172 | ± 0.021 |
|  | 0.20  0.25 | 0.226 | 50.0 | 0.170 | 0.018 |
|  | 0.25  0.30 | 0.275 | 55.0 | 0.197 | 0.018 |
|  | 0.30  0.35 | 0.325 | 53.2 | 0.156 | 0.019 |
|  | 0.35  0.40 | 0.374 | 47.4 | 0.167 | 0.021 |
|  | 0.40  0.50 | 0.447 | 73.4 | 0.133 | 0.017 |
|  | 0.50  0.60 | 0.546 | 46.5 | 0.117 | 0.023 |
|  | 0.60  0.70 | 0.646 | 27.4 | 0.188 | 0.031 |
|  | 0.70  0.90 | 0.784 | 27.0 | 0.101 | 0.032 |
|  | 0.90  1.20 | 1.015 | 12.6 | 0.088 | 0.052 |
|  | 1.20  1.50 | 1.312 | 2.8 | -0.001 | 0.122 |
|  | 1.50  2.00 | 1.649 | 0.7 | -0.211 | 0.208 |

Table 3

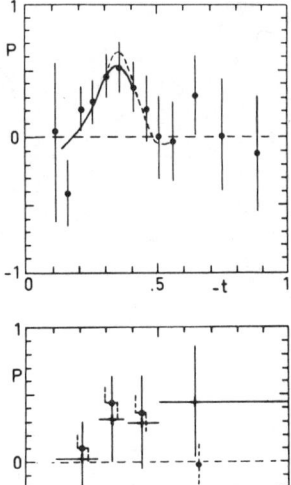

Fig. 3

Fig. 4

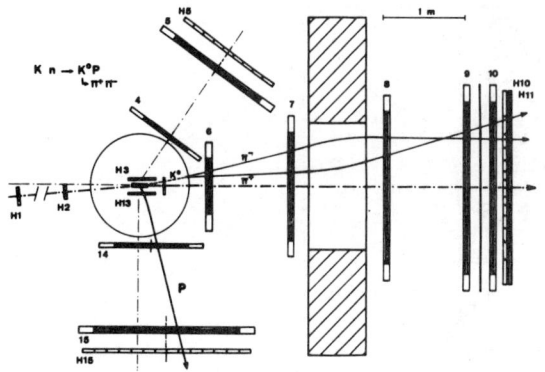

Fig. 1

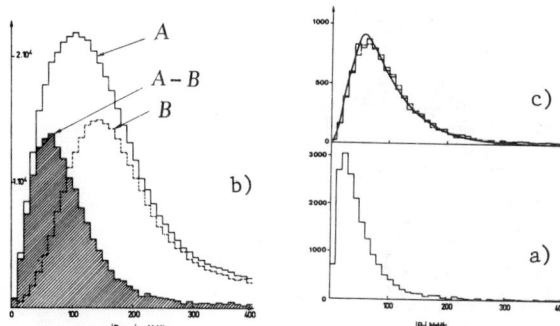

Fig. 2

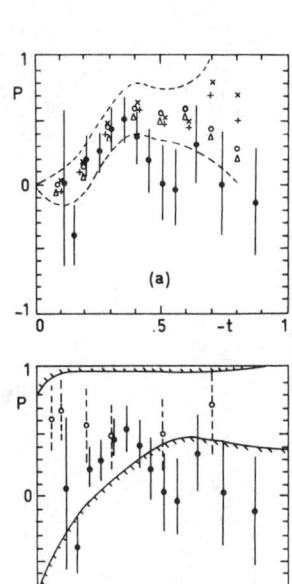

Fig. 5

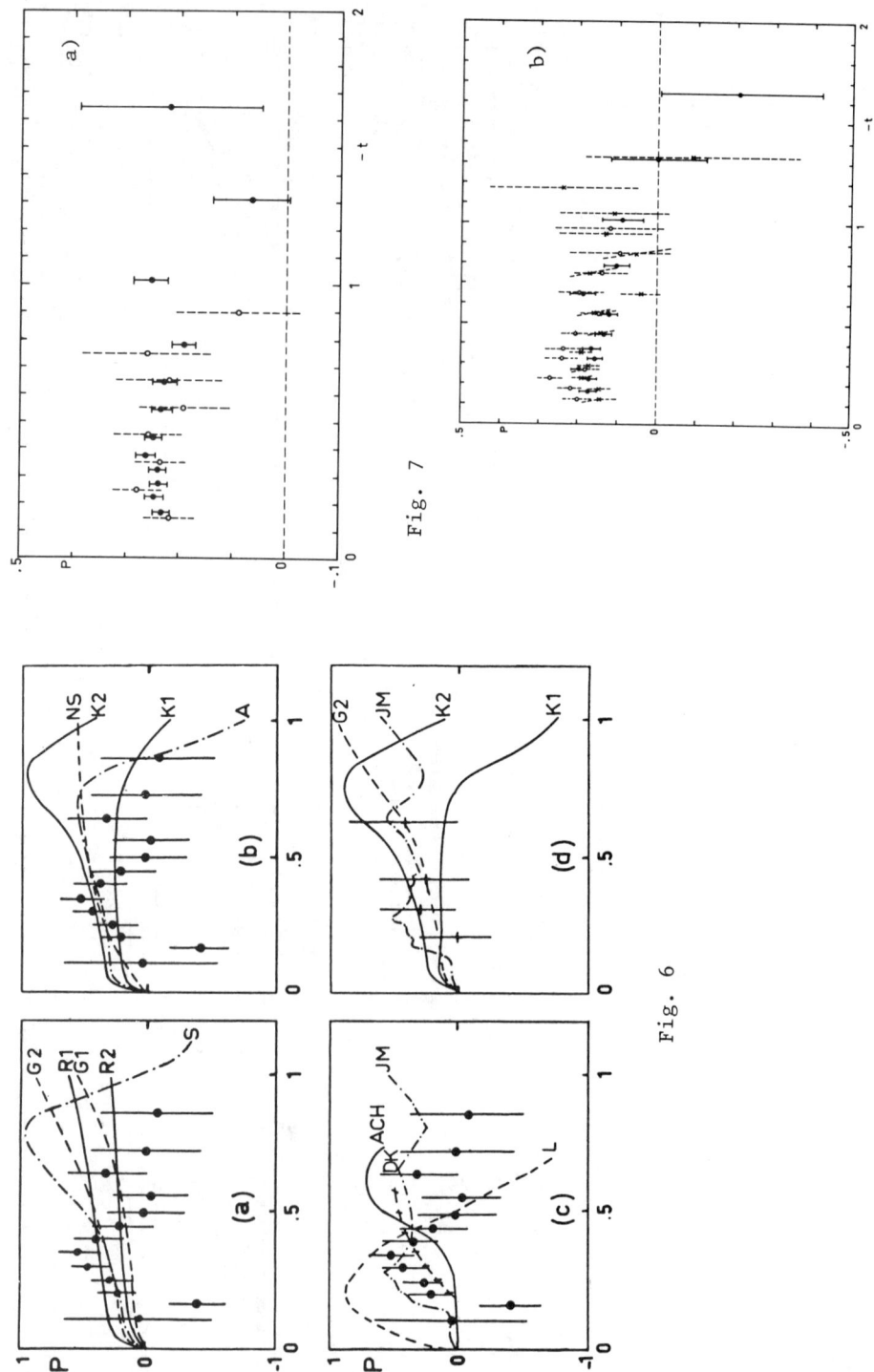

Fig. 6

Fig. 7

# TENSOR EXCHANGE AMPLITUDES IN $K^{\pm}N$ CHARGE EXCHANGE REACTIONS

M. Svec*
Mc Gill University, Montreal, PQ, Canada

## ABSTRACT

Tensor ($A_2$) exchange amplitudes in $K^{\pm}N$ charge exchange (CEX) are constructed from the $K^{\pm}N$ CEX data supplemented by information on the vector($\rho$) exchange amplitudes from $\pi N$ scattering. We observe new features in the t-structure of $A_2$ exchange amplitudes which contradict the t-dependence anticipated by most of the Regge models. The results also provide evidence for violation of weak exchange degeneracy.

---

Recent measurement [1] of polarization in $K^+n \to K^0p$ at 6 GeV/c closes a set of measured observables in $K^{\pm}N$ CEX reactions which allows to construct the tensor exchange amplitudes directly from data. The s-channel helicity non flip (n = 0) and flip (n = 1) amplitudes are resolved into tensor ($A_2$) and vector ($\rho$) exchange components $T_n$ and $V_n$, n = 0,1. In our notation

$$\sigma_+ = d\sigma/dt\ (K^+n \to K^0p) = |T_0-V_0|^2 + |T_1-V_1|^2 \quad (1a)$$

$$P_+\sigma_+ = P(d\sigma/dt)(K^+n \to K^0p) = 2\ \mathrm{Im}\left[(T_0-V_0)(T_1-V_1)^*\right] \quad (1b)$$

$$\sigma_- = d\sigma/dt\ (K^-p \to \bar{K}^0n) = |T_0+V_0|^2 + |T_1+V_1|^2 \quad (1c)$$

$$P_-\sigma_- = P(d\sigma/dt)(K^-p \to \bar{K}^0n) = 2\ \mathrm{Im}\left[(T_0+V_0)(T_1+V_1)^*\right] \quad (1d)$$

The set of equations (1) can be solved [2] for the amplitudes $T_0$ and $T_1$ in terms of the vector amplitudes $V_0$ and $V_1$, $\sigma_+$ and $P_+$. The differential cross sections (dcs) $\sigma_+$ were measured at ANL [3] and polarization $P_-$ at CERN [4]. We assume that $\rho$-exchange amplitudes in $K^{\pm}N$ CEX are related by SU(3) to $\rho$-exchange amplitudes in $\pi N$ scattering which at 6 GeV/c are well determined from the complete set of $\pi N$ scattering data [5, 6]. This means that in fact we combine the $K^{\pm}N$ CEX data (1) with

---

* The co-authors of this work are M. Fujisaki, J. Bystricky, G. Cozzika, A. Itano, F. Lehar, A. de Lesquen and L. van Rossum
Département de Physique des Particules Elémentaires, CEN-Saclay, France

data from πN scattering to form a complete set from which both $V_n$ and $T_n$ exchange amplitudes are reconstructed.

In our analysis we examined at each value of the four momentum transfer t the response of solutions for $T_n$ to uncertainties in all input. This method allows to identify those input observables to which the structure of the amplitudes is particularly sensitive. Input dcs $\sigma_+$ and polarizations $P_+$ are shown in reference [2]. Three different $\sigma_+$ inputs were finally used corresponding to the experimental uncertainties. For $P_+$ and $P_-$ we introduced an estimate of the maximum (H), mean (A) and minimum (L) values of polarization at a given t. The tensor amplitudes were then calculated at each value of t for the 27 combinations of possible "experimental data". To examine the dependence of solutions on the small differences between published vector amplitudes we have used results of five different analyses [5-9].

Figure 1 shows the range of solutions for the 27 input combinations, with the vector amplitudes of Barger and Phillips [7]. The solutions for the mean experimental values are also indicated. We note that Im $T_0$ has two pronounced dips our double zeros at $t \simeq -0.16$ and $t \simeq -0.45$ $(GeV/c)^2$. Im $T_1$ has two distinct zeros at $t \simeq -0.45$ and $t \simeq -0.70$. Re $T_0$ shows a dip at $t \simeq -0.20$ and rapid change of sign at $t \simeq -0.45$. Re $T_1$ shows a shoulder at $t \simeq -0.45$ and a change of sign at $t \simeq -0.65$. The t-structure of Im $T_0$ and Re $T_0$ for $|t| \lesssim 0.40$ is most sensitive to $P_+$ (Fig. 2). The slope of $\sigma_+$ for $|t| \gtrsim 0.40$ affects the t-structure of Im $T_0$ and Im $T_1$ for $|t| \gtrsim 0.50$ while Re $T_n$, n = 0,1 are not sensitive [2]. Im $T_0$ and Im $T_1$ are remarkably insensitive to $P_-$ for $|t| \lesssim 0.40$ [2].

We tried to reproduce the $K^{\pm}N$ CEX data with different "test" amplitudes $T_0$ with no dip-bump structure at $|t| < 0.4$. The best of the smooth amplitudes fits the $P_+$ data in this region with a $\chi^2$ 3.3 times larger than the amplitude reconstructed with the mean experimental values. Amplitudes without dip-bump structure in Im $T_0$ yield polarizations $P_+(t)$ with slopes smaller than indicated by the data points or by the experimental maximum likelihood function [1].

To examine weak exchange degeneracy (EXD) we calculated phase-effective Regge trajectories $\alpha_n^V$ and $\alpha_n^T$, n = 0,1 defined by

$$\text{Re } V_n = \tan(\pi/2 \; \alpha_n^V) \text{ Im } V_n$$
$$\text{Re } T_n = -\cot(\pi/2 \; \alpha_n^T) \text{ Im } T_n$$

In figure 3 we notice that $\alpha_0^V - \alpha_0^T \simeq 0.10$ for $t \simeq 0$ and $\alpha_1^V - \alpha_1^T \simeq 0.10$ for $0.05 \lesssim |t| \lesssim 0.10$. We conclude that weak EXD is violated by $\Delta\alpha \simeq 0.10$ in regions of t where either amplitude dominates.

Reaction $\pi^-p \to \eta n$ is dominated by pure $A_2$ exchange. Using SU(3) without $\eta$ mixing we predict $d\sigma/dt(\pi^-p \to \eta n)$ in satisfactory agreement [2]. Our predictions for the polarization $P(\pi^-p \to \eta n)$ are shown in figure 4. We notice that the rapid rise of $P_+$ polarization for $0.10 \lesssim |t| \lesssim 0.40$ correlates with the change of sign of $P(\pi^-p \to \eta n)$. This observation of SU(3) invariance is supported by the correct prediction of $P_+$ by the Barger-Cline sum rule for $|t| \lesssim 0.40$ [1].

The current Regge models with weak, strong and effective Regge cuts typically predict smooth and nonperipheral $T_0$. They have also in common a "$J_1$" t-structure of Im $T_1$ and Re $T_1$ nonvanishing for $|t| \lesssim 1.0$. They predict $P_+ > 0$ with little t-dependence for $0.10 \lesssim |t| \lesssim 0.40$ and $P(\pi^-p \to \eta n) > 0$ for $|t| \lesssim 0.30$ in conflict with data [10].

The $A_2$ exchange amplitudes [11] constructed from $\pi^-p \to \pi^0 n$, $\eta n$ and $K^-p \to \bar{K}^0 n$ data are in good agreement with our results (fig. 12 of ref. [2]). Also, the recent analysis [12] of all previous kaon-nucleon data using dispersion relations has predicted a t-dependence of $P_+$ for $|t| \lesssim 0.40$ similar to the observed one [1] and hence supports our conclusion about Im $T_0$.

The measurements of the polarization in pion-nucleon CEX [10] led to revisions of Regge models in 1972-1976. The presents status of $A_2$ exchange illustrates once more the importance of direct experimental determination of the t-structure of amplitudes in two-body scattering.

## REFERENCES

1. M. Fujisaki et al., "Measurements of Polarization in $K^+n$ Charge Exchange at 6 and 12 GeV/c". To be published in Physics Letters.
2. M. Fujisaki et al., "Construction of Tensor Exchange Amplitudes in KN and $\bar{K}N$ Charge Exchange Reactions". Submitted to Nuclear Physics.
3. R. Diebold et al., Phys. Rev. Letters 32, 904 (1974). J.J. Phelan et al., Phys. Letters B61, 483 (1976).
4. M. Beusch et al., Phys. Letters 46B, 477 (1973).
5. R.L. Kelly, Phys. Letters B39, 635 (1972).
6. G. Cozzika et al., Phys. Letters B40, 281 (1972).
7. V. Barger, R.J.N. Phillips, Phys. Letters 53B, 195 (1974).
8. M. Svec, Lettere al Nuovo Cimento 18, 45 (1977).
9. H. Navelet, P. R. Stevens, Nucl. Phys. B118, 475 (1977)
10. P. Bonamy et al., Nucl. Phys. B52, 392 (1973). D. Hill et al., Phys. Rev. Letters 30, 239 (1973).
11. G. Girardi, H. Navelet, Nucl. Phys. B83, 377 (1974).
12. D.W. Joynson, B.R. Martin, Nucl. Phys. B134, 83 (1978).
13. O.I. Dahl et al., Phys. Rev. Letters 37, 80 (1976).

## FIGURE CAPTIONS

Fig. 1 : Tensor exchange amplitudes at 6 GeV/c. The vertical bars represent the range covered by the 27 input combinations. Solutions for the mean experimental values $P_+(A)$, $P_-(A)$ and $\sigma_+(A)$ are represented by horizontal bars. The input vector amplitudes [7] are shown by open circles.

Fig. 2 : Dependence of tensor amplitudes on the $P_+$ polarization input with $\sigma_+(A)$ and $P_-(A)$.
$\cdot\ P_+(A)$, $+\ P_+(H)$, $\times\ P_+(L)$.

Fig. 3 : Phase-effective trajectories $\alpha_0^T$ and $\alpha_1^T$ of $A_2$-exchange ($\sigma_+(A)$, $P_-(A)$ and $P_+(A)$) and $\alpha_0^V$ and $\alpha_1^V$ of $\rho$-exchange [7]. For comparison we show $\alpha_{eff}$ obtained from $d\sigma/dt\ (\pi^- p \to \eta n)$ between 20 and 200 GeV/c [13].

Fig. 4 : Predictions for $P(\pi^- p \to \eta n)$ at 6 GeV/c with $\sigma_+(A)$. The solid line corresponds to $P_+(A)$, $P_-(A)$ input. The dashed lines are upper und lower bounds corresponding to $P_+(H)$, $P_-(H)$ and $P_+(L)$, $P_-(L)$ respectively. Data are from reference [10].

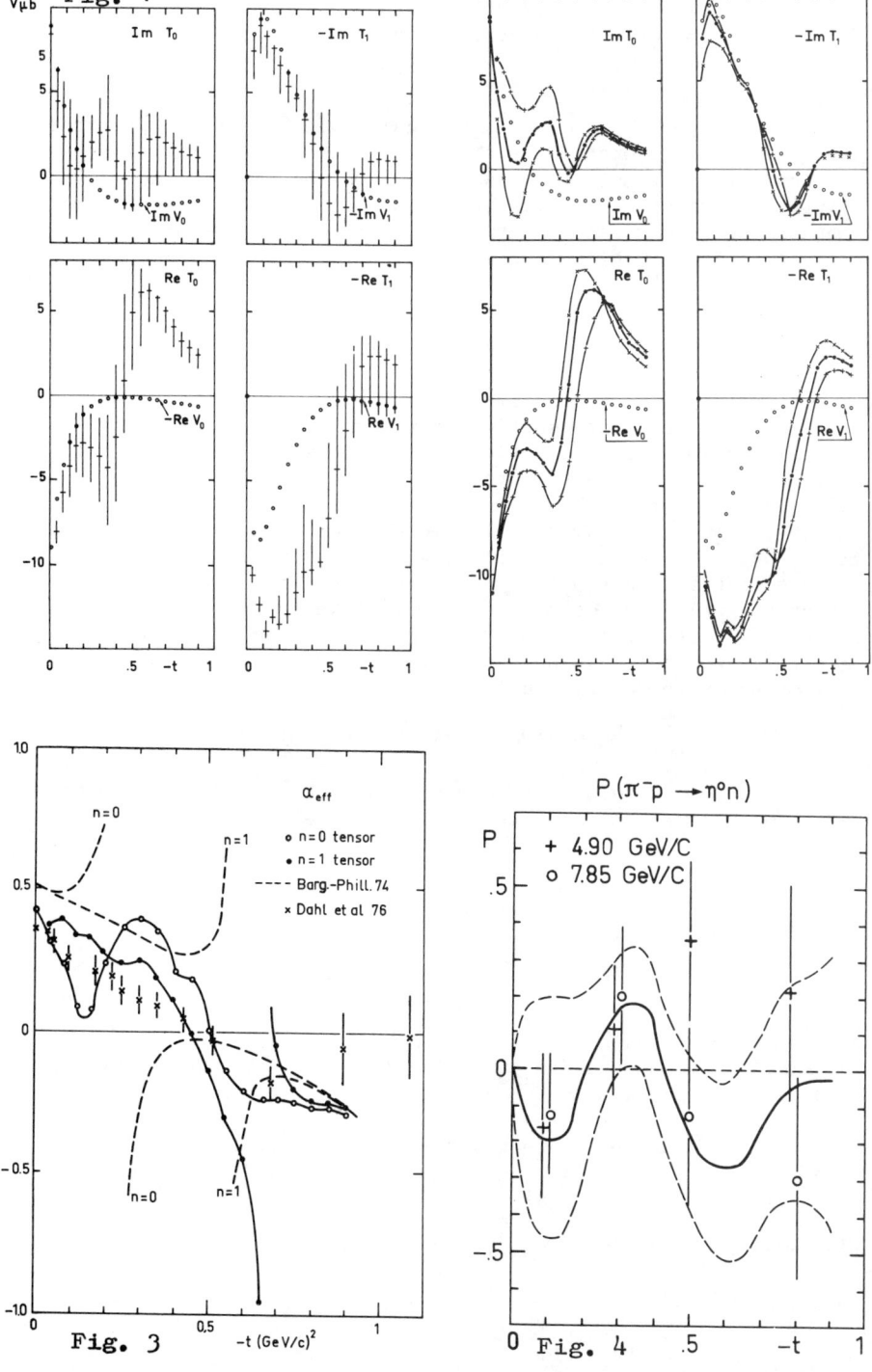

## ASYMPOTIC THEOREMS AND POLARIZATION PHENOMENA

Nguyen Van Hieu
Institute of Physics, Nghia Do Tu Lien Hanoi Vietnam

### ABSTRACT

In this report we give a critical review of the theoretical studies of the asymptotic relations between the polarizations and the asymmetry parameters in two-body crossing elastic and inelastic scattering processes at high energies. In particular we emphasize what asympototic relations are derived only from the analytical properties of the scattering amplitudes without the assumption on the T-invariance of strong interactions.

### INTRODUCTION

After the discovering of the asympototic equality between the total interaction cross-sections for particles and antiparticles by Pomeranchuk and the asymptotic equality between the differential cross-sections of two-body crossing scattering processes by Van Hove and Logunov et al.,[1] the asymptotic relations for the polarizations and the asymmetry parameters in two-body crossing processes were also studied by Logunov, Bilenki, Ryndin and myself. Since at the present time the possibility to check experimentally these asymptotic relations is coming I would like to use this opportunity to pay your attention to them and to review them critically. In particular we shall discuss the role of the T-invariance in the derivation of these relations.

### CROSSING SYMMETRY RELATIONS

First let us follow the derivation of the crossing symmetry relations between the amplitudes of three related scattering processes

$$a + b \rightarrow a' + b', \qquad (I)$$

$$\bar{a} + b' \rightarrow \bar{a}' + b, \qquad (II)$$

$$\bar{a}' + b \rightarrow \bar{a} + b'. \qquad (III)$$

For simplicity we consider in detail the case when the particles a and a' are spinless.

Denote by $q, p, q', p'$ the 4-momenta of particles $a, b, a', b'$, resp., in the process (I). For its matrix element we have the expression

ISSN: 0094-243X/79/510496-05$1.50 Copyright 1979 American Institute of Physics

$$\bar{u}_{b'}(p') M^I(p,q;p',q') u_b(p) = \tag{1}$$

$$\int d^4x\, e^{i\frac{q+q'}{2}x} \langle b'(p') | \frac{\delta j_a^+(\frac{x}{2})}{\delta \varphi_{a'}^+(-\frac{x}{2})} | b(p) \rangle , \tag{2}$$

$$j_a^+(x) = i \frac{\delta S}{\delta \varphi_a(x)} S^+ ,$$

where $\varphi_a$ and $\varphi_{a'}$ are the field operators for the particles a and a'. Due to the analytical properties of the amplitude (1) we can make the substitution $q \to -q'$, $q' \to -q$, and have

$$\bar{u}_b(p') M^I(p,-q';p',-q) u_b(p) =$$

$$\left[ \int d^4x\, e^{i\frac{q+q'}{2}x} \langle b(p) | \frac{\delta j_a(\frac{x}{2})}{\delta \varphi_{a'}(-\frac{x}{2})} | b'(p') \rangle \right]^+ . \tag{3}$$

On the other hand the matrix element of the process (II) equals

$$\bar{u}_b(p) M^{II}(p',q';p,q) u_{b'}(p') =$$

$$\int d^4x\, e^{i\frac{q+q'}{2}x} \langle b(p) | \frac{\delta j_a(\frac{x}{2})}{\delta \varphi_{a'}(-\frac{x}{2})} | b'(p') \rangle , \tag{4}$$

where $q'$, $p'$, $q$, $p$ are the 4-momenta of the particles $\bar{a}$, $\bar{b}$, $\bar{a}'$, $\bar{b}$, resp., in this process. We conclude that without any assumption on the T-invariance of the strong interaction as well as on the parity conservation or the charge conjugation invariance we can derive the crossing symmetry relations between the amplitudes of the processes (I) and (II) from their analyticity:

$$M^I(p,-q';p',-q) = \gamma_4 \left[ M^{II}(p',q';p,q) \right]^+ \gamma_4 . \tag{5}$$

For the scattering processes of two spin $\frac{1}{2}$ particles we have also similar crossing symmetry relations.

Note that if the processes (I) and (II) are the elastic scattering ones, then both they can be the observable processes, for example

$$\pi^+ + p \to \pi^+ + p \quad \text{and} \quad \pi^- + p \to \pi^- + p,$$

$$p + p \to p + p \quad \text{and} \quad \bar{p} + p \to \bar{p} + p.$$

However, if (I) is the following inelastic process

$$\pi^+ + p \to K^+ + \Sigma^+,$$

for example, then (II) will be an unobservable one

$$\pi^- + \Sigma^+ \to K^- + p.$$

In this case we must assume the T-invariance of strong interactions in order to establish the crossing symmetry relations between the amplitude of two observable processes (I) and (III)

$$\pi^+ + p \to K^+ + \Sigma^+ \quad \text{and} \quad K^- + p \to \pi^- + \Sigma^+ \tag{6}$$

## MESON-BARYON AND BARYON-BARYON SCATTERING

Now we consider the polarization phenomena in the scattering processes (I), (II), (III), at infinitely high energies when the particles a and a' are the pseudoscalar mesons and the particles $b$ and $b'$ are the baryons. Denote by $P^I$, $P^{II}$, $P^{III}$ the polarizations of the particles $b'$, $b$, $b'$, in the final states of the processes (I), (II), (III), resp., with the unpolarized targets, and by $\eta^I$, $\eta^{II}$, $\eta^{III}$, the symmetry parameters of the cross sections of these processes with the polarized targets. From the analytical properties of the scattering amplitudes and the above mentioned crossing symmetry relations we can derive following asymptotic relations between the polarizations and the asymmetry parameters without any assumption on the invariance with respect to the discrete symmetry transformations T,C,P:

$$P^I \approx -\eta^{II}, \quad P^{II} \approx -\eta^I. \tag{7}$$

In particular, the relations (7) hold for the particle and antiparticle elastic scattering processes of the form

$$\pi^+ + p \to \pi^+ + p \quad \text{and} \quad \pi^- + p \to \pi^- + p, \tag{8}$$
$$K^+ + p \to K^+ + p \quad \text{and} \quad K^- + p \to K^- + p.$$

If we make the additional assumption on the T-invariance of the strong interactions, then we obtain new relations

$$P^{I} \approx -P^{III}, \quad \eta^{I} \approx -\eta^{III} \tag{9}$$

They hold not only for the elastic scattering processes of the form (8), but also for the inelastic processes of the form (6).

In the case of the baryon-baryon scattering we should make the distinction between the polarizations $P_a^J$, $J = I, II, III$ of the particles a', ā', ā and the polarizations $P_b^J$, $J = I, II, III$ of the particles b, b' in the final states of the processes I,II,III with unpolarized beams and unpolarized targets, and also between the asymmetry parameters $\eta_a^J$ and $\eta_b^J$ in these processes with the polarized beams (particles a, ā, ā') or the polarized targets (particles b, b'). From the analytical properties of the scattering amplitudes and the above mentioned crossing symmetry relations we can derive following asymptotic relations without the assumption on the T,C,P - invariances:

$$P_b^{I} \approx -\eta_b^{II}, \quad P_b^{II} \approx -\eta_b^{I}, \tag{10}$$

and

$$P_a^{I} \approx -P_a^{II}, \quad \eta_a^{I} \approx -\eta_a^{II} \tag{11}$$

These relations hold, in particluar, for the elastic scattering of some baryon and of its antiparticle on the proton, e.g.

$$p + p \to p + p \quad \text{and} \quad \bar{p} + p \to \bar{p} + p,$$
$$\Sigma^+ + p \to \Sigma^+ + p \quad \text{and} \quad \overline{\Sigma^+} + p \to \overline{\Sigma^+} + p. \quad (12)$$

If we assume the T-invariance, we obtain also new relations

$$P_\ell^{I} \approx - P_\ell^{III} \quad , \quad \eta_\ell^{I} \approx - \eta_\ell^{III} \qquad (13)$$

and

$$P_a^{I} \approx - \eta_a^{III} \quad , \quad P_a^{III} \approx - \eta_a^{I} \qquad (14)$$

### ACKNOWLEDGEMENT

In conclusion I would like to express my appreciation to Professor G. Thomas and th Organizing Committee for your kind invitation and your hospitality.

### REFERENCES

1. Nguyen Van Hieu, Rapporteur Talk at the Internation Conference on High Energy Physics, Kien, 1970, and therein references.

# COMMENT ON $Z_1^*$(1800) AND SPIN-ROTATION PARAMETER MEASUREMENTS

R. L. Kelly
Lawrence Berkeley Laboratory, Berkeley, CA   94720

## ABSTRACT

The $K^+p \to K^+p$ partial wave analysis of Arndt et al. finds a $J^P=3/2^+$ $Z_1^*$ resonance with M=1797 MeV and $\Gamma$=220 MeV. This disagrees with the analysis of Cutkosky et al. which found no evidence for this $Z_1^*$. It is pointed out that this disagreement can probably be resolved by measurements of the $K^+p$ elastic spin-rotation angle in the backward hemisphere near $p_{lab}$=1 GeV/c.

The possible existence of a $J^P=3/2^+$ $Z_1^*$ with a mass of 1800-1900 MeV has been a controversial feature of $K^+p$ elastic partial wave analyses for some time. The 1975 combined I=0 and 1 analysis of Martin[1] has been found[2] to contain a $J^P=3/2^+$ pole in the $K^+p$ channel at 1820-134i MeV, while the 1976 I=1 analysis of Cutkosky et al.[3] found no evidence for such a pole. The most recent $K^+p$ elastic analysis is that of Arndt et al.[4] which again finds evidence for a $J^P=3/2^+$ pole at 1797-110i MeV. The predictions of Cutkosky et al. are consistent with the newer data used by Arndt et al., so we have here two solutions, one resonant and one non-resonant, which can not be distinguished by current data. This situation is reviewed in more detail in Ref. 5. The purpose of this comment is to point out a spin-rotation measurement which could probably resolve the ambiguity.

It is useful to discuss spin-rotation measurements in meson-nucleon scattering in terms of the Wolfenstein rotation angle, $\beta$, defined in Fig. 1(a), rather than the unnecessarily complicated A and R parameters. On recoil from a target polarized in the scattering plane, the c.m. frame polarization is rotated through an angle $\beta$ and multiplied by a factor which depends only on the ordinary polarization parameter, P. The magnitude of the recoil polarization will thus be known from the results of transversley polarized target experiments, and the task of the spin rotation experiment is to measure its orientation, i.e., $\beta$. The most sensitive way of measuring the orientation would be to search for the direction in the scattering plane along which the recoil polarization has a vanishing component. Such a null experiment would be quite insensitive to uncertainties in the analyzing power of the second-scattering analyzer, but might also require unrealistically flexible experimental equipment to cover a reasonable range of scattering angles and lab momenta. A less demanding approach is illustrated in Fig. 1(b), where a moderately accurate prediction of $\beta$ is used to choose a fixed direction along which the recoil polarization has a small component.

Predictions for $\beta$ from Arndt et al. and Cutkosky et al. are shown in Fig. 2. There are large differences, particularly in the backward hemisphere. The uncertainties in these predictions are generally small compared to the differences between them in the energy range shown. This alone does not guarantee that an accurate spin-rotation measurement would resolve the $Z_1^*$ ambiguity. Monte Carlo studies of the sensitivity of the resonance to β-data of typical distribution and accuracy could be useful in evaluating the effectiveness of various experimental designs. However, it is possible that there are some surprises in store here, for example both sets of predictions in Fig. 2 might be wrong, so detailed Monte Carlo studies should also not be given undue weight.

In the lab momentum range 700-1300 MeV/c, a proton scattered through 150° in the center-of-mass system emerges from the target with a lab recoil angle of about 13° and a lab recoil momentum about 140 MeV/c larger than the beam momentum. It should be fairly easy to design polarizing coils which would not obstruct the flight of a recoil proton within 13° of the beam direction. The high momentum of the proton, however, presents a problem. The analysing power of carbon is known accurately only up to about 600 MeV/c[7]. It would be useful, for many experimental applications, to extend this range during the remaining year of operation of the ZGS polarized beam. Another possibility would be the use of a hydrogen analyzer, for which the analyzing power is known in this momentum range. In any case, the sensitivity of the final results to uncertainties in the analyzing power are minimized by choosing the experimental geometry to approximate that of a null experiment.

## REFERENCES

1. B. R. Martin, Nucl. Phys. B94, 413 (1975).
2. B. R. Martin, in Proceedings of the Topical Conference on Baryon Resonances (Oxford, 1976), edited by R. T. Ross and D. H. Saxon, pg. 409.
3. R. E. Cutkosky et al., Nucl. Phys. B102, 139 (1976).
4. R. A. Arndt et al., Virginia Polytechnic Institute and State University preprint, 1978.
5. R. L. Kelly, "KN Partial-Wave Analysis and $Z^*$ Resonances", LBL-7976, 1978.
6. R. L. Kelly et al., Phys. Rev. D 10, 2309 (1974).
7. G. Waters et al., Nucl. Instr. and Meth. 153, 401 (1978).

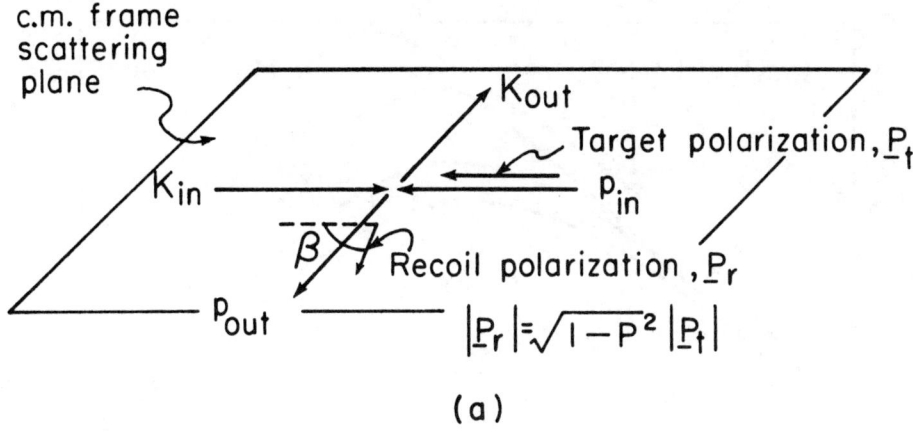

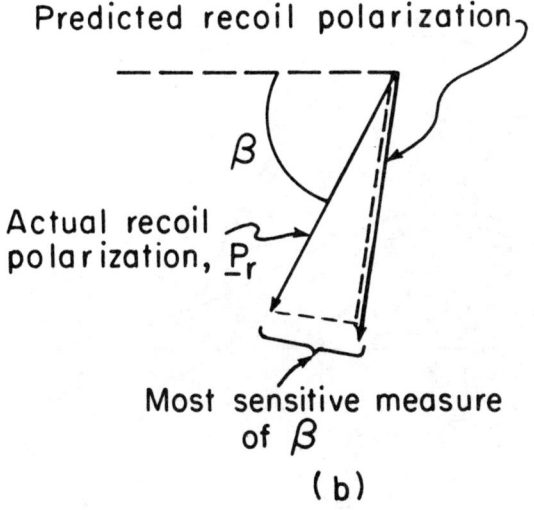

Fig. 1. (a) Definition of the Wolfenstein spin rotation angle, $\beta$. The nucleon polarization in the c.m. frame scattering plane rotates through an angle $\beta$ as it recoils from the target. (b) Components of $\underline{P}_r$ with respect to its predicted value are shown as dashed lines. The component perpendicular to the predicted value is the most sensitive measure of $\beta$.

XBL 788-10097

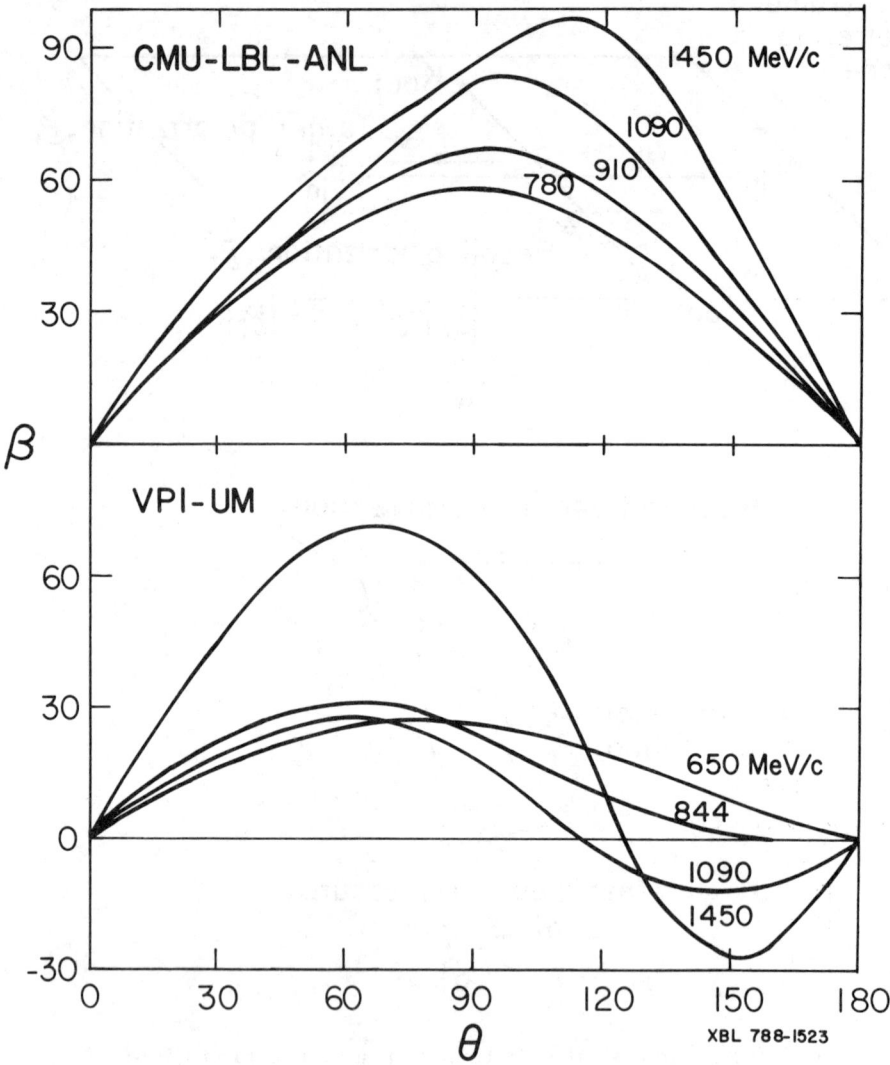

Fig. 2. Comparison of the spin-rotation angle predictions of Arndt et al.[4] (lower plot) and Cutkosky et al.[3] (upper plot).

## HIGH-$P_T$ THEORY IN A SPIN*

Dennis Sivers
Argonne National Laboratory, Argonne, IL 60439

### ABSTRACT

The role of measurements using polarized hadron beams or polarized targets is discussed. Future experiments can be compared with explicit calculations to provide tests of QCD.

---

*"...Round and round I go, In a spin, lovin' the spin I'm in,..." That Ol' Black Magic, Harold Arlin and Johnny Murcer, 1943.

---

There has been substantial recent development of the theoretical ideas relevant to the description of large-transverse-momentum production processes. One effect of this growth in understanding has been to push the possible role of experiments involving polarized beams and polarized targets closer to the firing line of confrontation with fundamental theory. If the obstacles associated with the design and execution of these experiments can be overcome it may very well turn out that many of the decisive confrontations between quantum chromodynamics (QCD) and measurement in the next few years will take place in the arena provided by polarized beams and targets.

### Inclusive Reactions

Two years ago the state of the art of large-$p_T$ phenomenology could be characterized by the existence of two types of ad hoc models, the Field-Feynman black box model[1] and the constituent-interchange-model (CIM) of Blankenbecler, Brodsky and Gunion.[2] Both approaches gave an adequate description of "most" of the data and many experimental proposals were written and experiments designed to decide between them--largely on the basis of multiparticle correlations. Polarized beam or target experiments were not too relevant to this issue. However, the situation was aesthetically unattractive since the models apparently lacked contact with fundamental QCD. Recently, a new phenomenological approach has emerged--firmly based on the foundations of QCD perturbation theory--which co-opts many of the successes of the older ad hoc models. This has the immediate effect of putting measurements of large-transverse-momentum hadron production on a comparable footing with $ep$, $\nu p$, $e^+e^-$ collisions and the production of high mass lepton pairs, $pp \to (\ell^+\ell^-)X$, as tests of QCD perturbation theory. Let's examine the basic elements of this new approach in order to get some idea of that role that measurements involving particles with spin can play.

The starting point for our discussion is the hard-scattering diagram of Fig.1. This leads to the simple expression for the spin-averaged inclusive cross section

$$E d\sigma/d^3p (AB \to CX) \simeq \sum_{ab \to cd} \int dx_a dx_b G_{a/A}(x_a) G_{b/B}(x_b)$$

$$\times \frac{1}{z_c} D_c^C(z_c) \frac{1}{\pi} \frac{d\hat{\sigma}}{d\hat{t}} (ab \to cd) \quad (1)$$

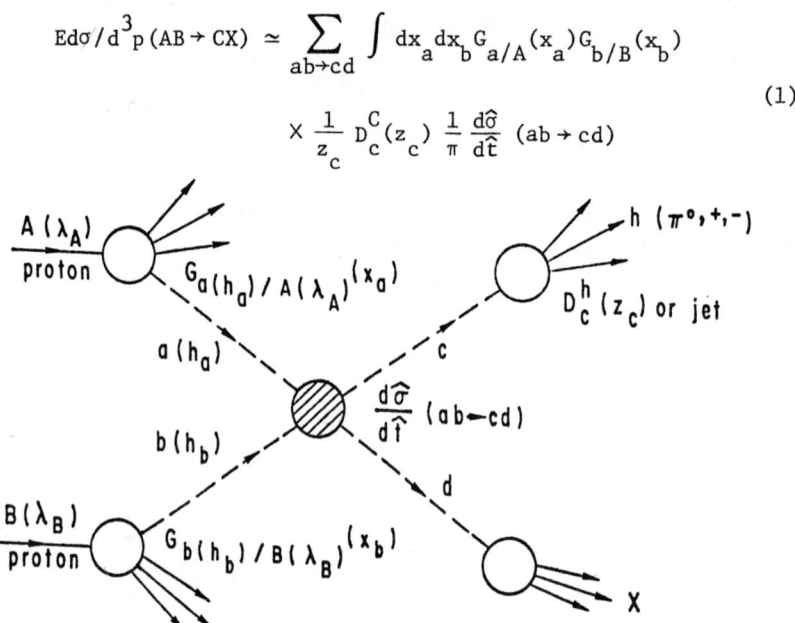

Fig.1. Hard-scattering model.

where $G_{a/A}(x_a)$ is the probability of finding constituent a in hadron a with fraction $x_a$ of the longitudinal momentum and $D_C^c(z_c)$ is the probability that the outgoing constituent produces a hadron with fraction $z_c$ of its momentum. Included in the internal hard scattering are all fundamental $2 \to 2$ processes involving quarks and gluons.[3,4]

Because QCD is a renormalizable theory we can include the effects of higher-order perturbative corrections such as those indicated schematically below.

It is now beginning to be understood that QCD factorizes so that the IR singularities in (higher-order) diagrams appropriate to inclusive processes can be included in the hard-scattering model[5] by introducing the momentum-dependent distribution functions

$$G_{a/A}(x,k_T^2) \to G_{a/A}(x,k_T^2,\hat{t})$$
$$D_c^C(z,k_T^2) \to D_c^C(z,k_T^2,\hat{t}) \qquad (2)$$

and the momentum-dependent effective coupling in the constituent hard scattering cross section

$$d\hat{\sigma}(g^2,\hat{s},\hat{t}) \to d\hat{\sigma}(g^2(\hat{t}),\hat{s},\hat{t}). \qquad (3)$$

We will assume in the discussion below that these modifications to the simple version of the hard-scattering model have been included even though we will not always show them explicitly in the equations. For simplicity, we will often suppress the $k_T$-fluctuations in the distribution functions as well.

It is important to observe that the partially-coherent hard processes such as $q\pi \to q\pi$ which appear in the CIM have not disappeared. They constitute a specific type of subasymptotic contribution to the cross section. In the language of the operator product expansion they are analogous to the "higher twist operators" which appear in leptoproduction. These subasymptotic corrections can be more important in purely hadronic reactions where a hadron is probed by another hadron instead of by an electron or neutrino. In hadronic collisions local color fluctuations can be significant since a virtual $q\bar{q}$ pair in an $SU_3^{color}$ singlet inside a proton can act vigorously as a "meson". The CIM terms therefore can give important (although nonleading) contributions which would not appear in leptoproduction. Certain internal processes such as $qV \to q\pi$ (where V is a vector gluon) and $q\pi \to q\pi$ are enhanced by a trigger bias effect since the outgoing $\pi$ can carry all of the available momentum in the hard collision. The main problem with including these CIM terms is that the normalization of the subasymptotic corrections is intrinsically more difficult than for the dominant, fundamental QCD processes. Efforts to normalize these effects have been made but they involve many assumptions.[2]

In the discussion which follows we will assume that we are considering a kinematic regime where CIM contributions are unimportant. This same requirement will assure that controversies associated with the precise way in which $k_T$ fluctuations are included in the distribution functions will be unimportant. We will therefore be able to concentrate on possible spin-dependent effects in the theory.

## The Prediction $PEd\sigma/d^3p(pp\uparrow \to CX) \to 0$

It is often claimed that the hard-scattering model predicts that single spin asymmetries (commonly called polarizations) vanish at large transverse momentum.[6] This statement needs to be reconsidered in the QCD framework outlined earlier. Since we are now dealing with a specific theory we can ask more detailed questions and it should be possible to make a more specific prediction. In what follows we will use the convention that ↑ and ↓ refer to spins in a transversity basis while + and − label helicities.

After some manipulation we can write the initial spin asymmetry for inclusive large-$p_T$ production normal to the spin direction[7]

$$A_N E d\sigma/d^3p (A\uparrow B \to CX) \cong \sum_{ab \to cd} \int dx_a dx_b \, \Delta^T G_{a/A}(x_a) G_{b/B}(x_b)$$

$$D_c^C(z) \frac{1}{z} \left[ \hat{A}_N \frac{d\hat{\sigma}}{d\hat{t}} (ab \to cX) \right]$$

(4)

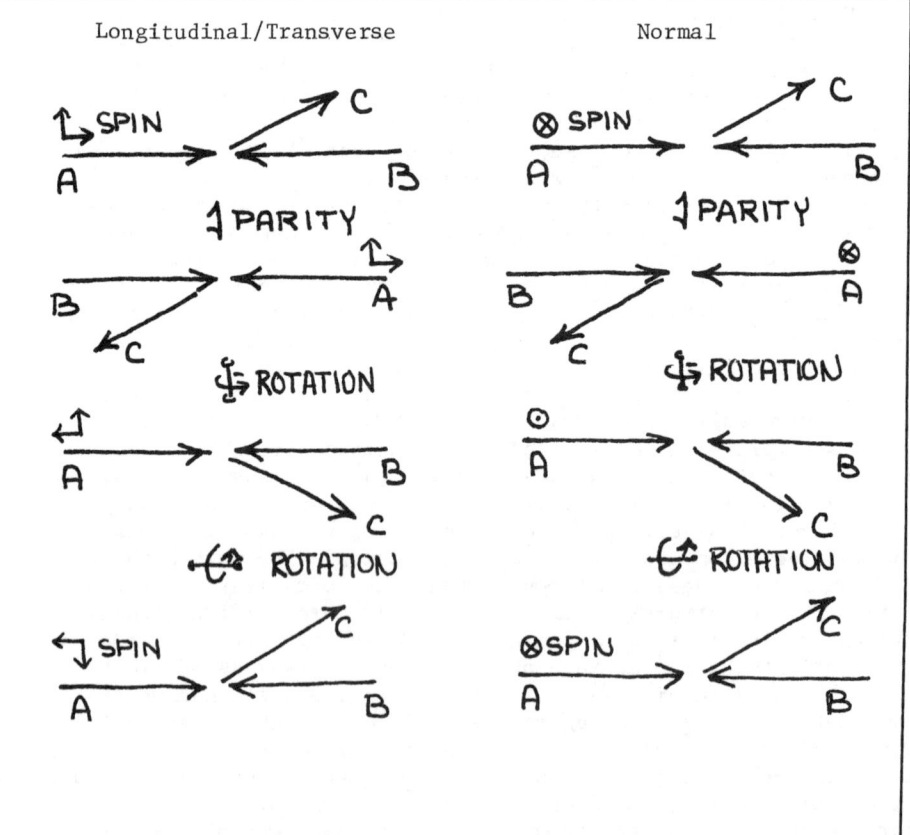

Fig.2. Single spin asymmetries involving longitudinal spins or transverse spins in the scattering plane vanish by the parity invariance of the strong interactions while those with spins normal to the plane are allowed unless they conflict with dynamical principle such as the factorization of the dynamics at large-$p_T$.

in a manner similar to the hard-scattering equation for the spin-averaged cross section

$$\Delta^T G_{a/A}(x_a) = G_{a\uparrow/A\uparrow}(x_a) - G_{a\downarrow/A\uparrow}(x_a) \qquad (5)$$

while $\hat{A}_N$ is the single-spin asymmetry for the internal hard process. This internal asymmetry should be calculable in perturbation theory--not from lowest-order perturbation diagrams, which leads to real amplitudes so that polarizations or initial spin asymmetries vanish--but from higher order diagrams.

We can write[8] the asymmetry in terms of the imaginary part of a helicity-flip amplitude where we are taking the discontinuity of a Mueller diagram indicated schematically below.

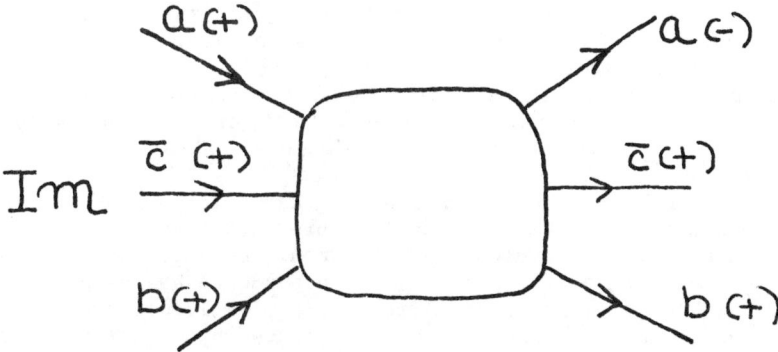

Now for processes where $a$ is a (light) quark (such as $qq \to q$) we can use quark helicity conservation to show to leading non-vanishing order

$$\text{Im}\langle q(+)X | q(-)X \rangle \propto m_q/\sqrt{\hat{s}} \qquad (6)$$

where $\hat{s}$ is some large invariant in the internal hard-scattering process. Because we believe that the $u$ and $d$ quarks can be considered very light,[9]

$$m^u = m_o^u - \frac{m_o^u g_o^2}{2\pi^2} \left( \frac{1}{1 + 2(m_o^u/Q_o)^2} \right) \ln(Q/Q_o) + O(g_o^4),$$
$$m_o^u \lesssim 5 \text{ MeV} \qquad (7)$$

helicity conservation will lead to very small polarizations. The implications of this fact for final quark polarizations have been pointed out by Kane, Pumplin and Repko.[10] In what follows we will neglect effects proportional to $m_q/\sqrt{s}$.

It is important to note that this does not mean we can conclude that QCD predicts $A_N \equiv 0$. Let's look at asymmetries involving gluons. The arguments which we have reviewed for quarks simply do not apply.

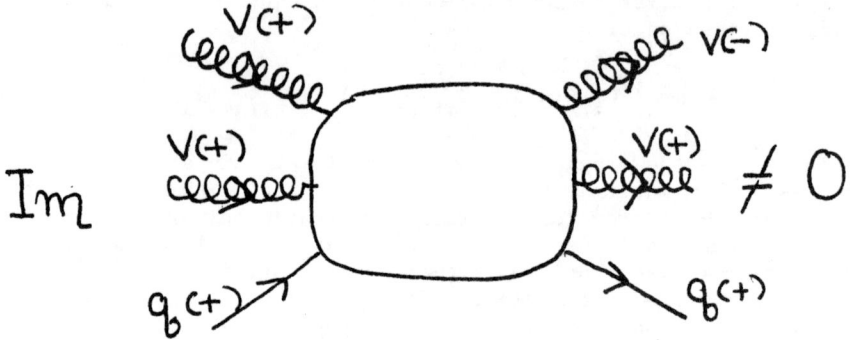

Under the quite reasonable assumptions that gluons (as well as quarks) remember the proton's spin we can get a small, but nonzero asymmetry. The asymmetry does vanish at very large $p_T$ using the asymptotic freedom argument that $\alpha_s \to 0$ so that higher order corrections are suppressed, but this is only a logarithmic effect. Precision experiments to measure this type of asymmetry should produce results which can be compared with calculations. These calculations are in progress.[11]

I should mention also one indication that a large $p_T$ polarization measurement should be nonzero. Measurements involving polarized beams incident on heavy nuclear targets should be very interesting in light of the fact that nontrivial A dependence has been observed[12] in large $p_T$ production from nuclei. Multiple scattering explanations of this enhancement could be tested by polarized beam experiments and give insight into the space-time development of interactions.

### Spin-Spin Asymmetries

Unlike the single-spin asymmetries which test the higher order terms in the perturbation expansion, there are sizable two-spin asymmetries in the fundamental $2 \to 2$ cross sections of QCD calculated to lowest order. A list of these cross sections involving quarks and gluons in a helicity basis is shown in Table 1.[13] We can write the longitudinal spin-spin asymmetry for large-$p_T$ production using the hard scattering model

$$A_{LL} \frac{E d\sigma(AB \to CX)}{d^3 p} \simeq \sum_{ab \to cd} \int dx_a dx_b$$

$$\Delta^L G_{a/A}(x_a) \Delta^L G_{b/B}(x_b) D_{C/c}(z_c) \frac{1}{z_c}$$

$$\frac{1}{\pi} a_{LL} d\hat{\sigma}/d\hat{t} \qquad (8)$$

where

$$\Delta^L G_{a/A}(x) = G_{a(+)/A(+)}(x) - G_{a(-)/A(+)}(x) \quad (9)$$

and $\hat{a}_{LL}$ is the constituent $2 \to 2$ asymmetry.

TABLE OF QCD CROSS SECTIONS* FOR DEFINITE HELICITY STATES

| PROCESS $ab \to cd$ | $d\sigma/dt\,(a(+)b(+) \to cd)$ | $d\sigma/dt\,(a(+)b(-) \to cd)$ |
|---|---|---|
| $q_\alpha q_\beta \to q_\alpha q_\beta$ | $\left(\frac{8}{9}\right)\left[\frac{s^2}{t^2} + \delta_{\alpha\beta}\left(\frac{s^2}{u^2} - \frac{2}{3}\frac{s^2}{tu}\right)\right]$ | $\left(\frac{8}{9}\right)\left[\frac{u^2}{t^2} + \delta_{\alpha\beta}\frac{t^2}{u^2}\right]$ |
| $q_\alpha \bar{q}_\beta \to q_\delta \bar{q}_\gamma$ | $\left(\frac{8}{9}\right)\left[\delta_{\alpha\delta}\delta_{\beta\gamma}\frac{s^2}{t^2}\right]$ | $\left(\frac{8}{9}\right)\left[\delta_{\alpha\delta}\delta_{\beta\gamma}\frac{u^2}{t^2}\right.$ $+ \delta_{\alpha\beta}\delta_{\delta\gamma}\frac{(t^2+u^2)}{s^2}$ $\left. - \frac{2}{3}\delta_{\alpha\gamma}\delta_{\alpha\beta}\delta_{\delta\gamma}\frac{u^2}{st}\right]$ |
| $qV \to qV$ | $\left[\frac{2s^2}{t^2} - \frac{8}{9}\frac{s^2}{us}\right]$ | $\left[\frac{2u^2}{t^2} - \frac{8}{9}\frac{u^2}{us}\right]$ |
| $\bar{q}V \to \bar{q}V$ | $\left[\frac{2s^2}{t^2} - \frac{8}{9}\frac{s^2}{us}\right]$ | $\left[\frac{2u^2}{t^2} - \frac{8}{9}\frac{u^2}{us}\right]$ |
| $VV \to q\bar{q}$ | 0 | $\left[\frac{1}{3}\frac{(u^2+t^2)}{ut} - \frac{3}{4}\frac{(t^2+u^2)}{s^2}\right]$ |
| $q\bar{q} \to VV$ | 0 | $\left[\frac{64}{27}\frac{t^2+u^2}{ut} - \frac{16}{3}\frac{t^2+u^2}{s^2}\right]$ |
| $VV \to VV$ | $\left(\frac{9}{2}\right)\left[\frac{2s^2}{ut} - \frac{su}{t^2} - \frac{st}{u^2}\right]$ | $\left(\frac{9}{2}\right)\left[6 - \frac{2s^2}{ut} - \frac{su}{t^2} - \frac{st}{u^2}\right.$ $\left. - \frac{2ut}{s^2}\right]$ |

*All cross sections contain a common factor $\pi\alpha_s^2/s^2$.

The distribution functions $\Delta^L G_{q/p}(x)$ and $\Delta^L G_{V/p}(x)$ which parametrize how quarks and gluons carry spin information in a longitudinally polarized proton. We do not understand these functions very well although data on electroproduction spin-spin asymmetries are beginning to become available.[14] We currently have to resort to building theoretical models within the constraints of the data. Several models are discussed in Ref.15 where they are used to calculate the asymmetries for $pp \to \pi^0 X$ shown in Fig.3. Because the internal processes $qq \to qq$, $qV \to qV$ and $VV \to VV$ which should dominate large $p_T$ production all have positive $\hat{a}_{LL}$ asymmetries, the reasonable hypothesis that quarks and gluons carry some of the proton's spin leads to a positive production asymmetry. A large subasymptotic (CIM) contribution to $\pi$ production from $q\pi \to q\pi$ can diminish this asymmetry for $\pi$ production but should not change the production substantially for jets. Because a wide range of values for $A_{LL}$ are allowed between the different candidate models for distribution functions it cannot now be said that QCD makes a prediction for this quantity. However, improved data on lepton scattering spin asymmetries and on spin-spin asymmetries in lepton pair production[16], will substantially remove this ambiguity. The functions $\Delta^L G_{a/p}(x)$ should be measurable in different experiments so that a complete set of spin-spin asymmetry measurements will be able to provide a substantive test of QCD.

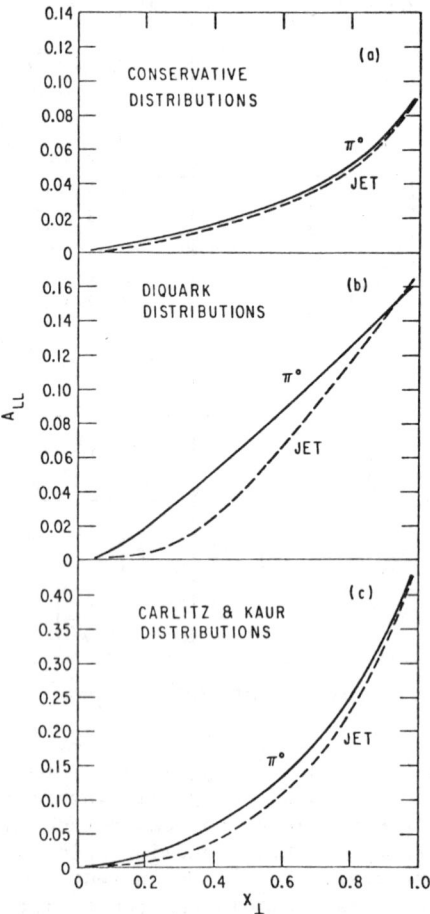

Fig.3. Longitudinal spin-spin asymmetries for the production of large $p_T$ $\pi^o$'s and jets. The three different models for the distribution functions $\Delta^L G_{a/p}(x)$ are explained in Ref.15. The existence of positive constituent asymmetries in the underlying perturbation series should be testable with precision experiments using longitudinally polarized beams and targets.

An interesting spin-flavor correlation can be observed in our models for the distribution functions based on usual quark-model ideas. This is shown in Fig.4 where the asymmetry for $pp \to \pi^+ X$ is compared with $pp \to \pi^- X$ in each of the models for the distribution function.

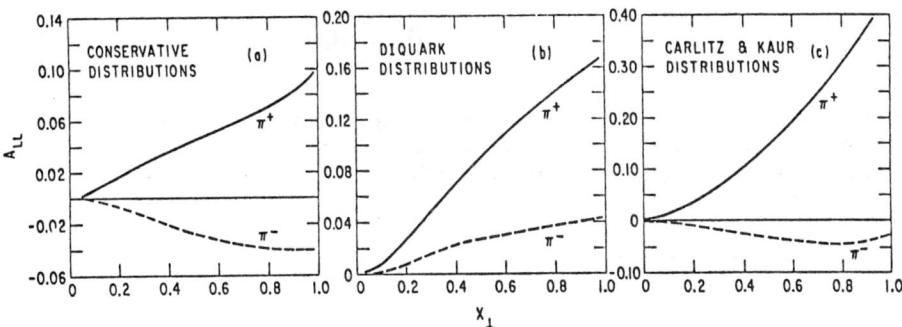

Fig.4. Longitudinal spin-spin asymmetries for $pp \to \pi^\pm$ reflect the fact that our models for spinning distribution functions have $\Delta^L G_{u/p}(x) > 0$ and $\Delta^L G_{d/p}(x) \leq 0$. More details can be found in Ref.15.

Inclusive spin-spin asymmetries with transversely polarized beam and targets can also be calculated. We have an expression

$$A_{NN} \frac{Ed\sigma(AB \to CX)}{d^3p} \simeq \int dx_a dx_b \Delta^T G_{a/A}(x_a) \Delta^T G_{b/B}(x_b)$$
$$\frac{1}{z_c} D_{C/c}(z_c) \frac{1}{\pi} \hat{a}_{NN} d\hat{\sigma}/d\hat{t}_{ab \to cd} \quad (10)$$

analogous to Eq.(8) with $\Delta^T G_{a/A}(x)$ defined in Eq.5 and $\hat{a}_{NN}$ is the transverse constituent asymmetry. If low-order perturbation theory and the hard-scattering model are to be believed, transverse spin-spin asymmetries should be significantly smaller than longitudinal ones. There are two ingredients in the argument which leads to this conclusion.[17] In the kinematic regime where quark-quark scattering dominates the cross section, $x_T \gtrsim 0.4$, we are dealing with smaller fundamental asymmetries. At $\theta_{cm} = 90°$ we have

$$\hat{a}_{NN}(q_\alpha q_\beta \to q_\alpha q_\beta) = \delta_{\alpha\beta}(-\frac{1}{11} + \frac{2}{11}\cos^2\phi) \quad (11)$$

where $\phi$ is the angle between the spin-direction and the production plane. This is to be compared with large values (of order 1/2) for $\hat{a}_{LL}$ in the same region. In addition there is good reason to believe that the values of $\Delta^T G_{q/p}(x)$ may be much smaller at large x than those for $\Delta^L G_{q/p}(x)$. One way of picturing this is to imagine that at large x a quark has a definite helicity, even if it is in a transversely polarized proton. A simplified version of a sum rule due to Wandzura and Wilczek[18] based on this picture can be written

$$\Delta^T u(x) = \int_x^1 \frac{dy}{y} \Delta^L u(y) \qquad (12)$$

to relate the transverse and longitudinal asymmetry distributions for up quarks. This leads to a predicted transverse asymmetry in electroproduction as shown in Fig.5.

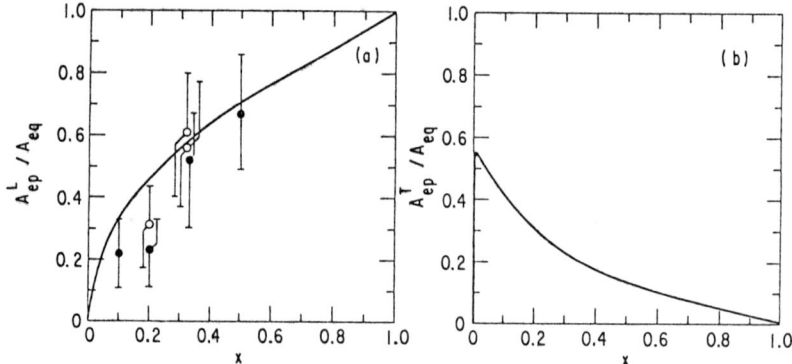

Fig.5. Asymmetries for polarized ep scattering; (a) longitudinal polarization (data are from Ref 19); (b) transverse polarization.

Combining this result with (10) and (11), we get small values for $A_{NN}$

$$A_{NN}(pp \to \pi X) \lesssim 10^{-3} \qquad (x_T \gtrsim 0.4)$$

At this level higher-order or nonperturbative effects can be much more important than those we've calculated. The rule $A_{NN} < A_{LL}$ seems likely to be true in some average sense even if we include other effects. More discussion can be found in Ref.17.

## Spin Observables in Large-Angle Elastic Scattering

We can now turn our attention to a problem which requires a different theoretical starting point. The elastic scattering of composite systems is intrinsically sensitive to the long-range forces which assemble a collection of constituents into a specific state. Even for large angles, we cannot be as careless with the infra-red behavior of QCD in discussing elastic pp scattering as we can in inclusive processes. At the present time we have to rely more on theoretical models and less on the rigorous application of theoretical ideas. Currently the most successful and consistent model approach to the scattering of composite systems through large angles is the CIM.[2]

The application of the CIM to elastic processes involves a set of theoretical questions distinct from those involved in using specific subasymptotic diagrams in inclusive production. The CIM gives a specific realization of the constituent counting rules[20] which predict $d\sigma/dt\,(pp \to pp) \sim s^{-10}$ and $d\sigma/dt\,(\pi p \to \pi p) \sim s^{-8}$. The model has also been successful in explaining the angular behavior of cross sections near 90°.[2,21] Following these successes it should prove interesting to examine the model's predictions for large-angle spin observables.

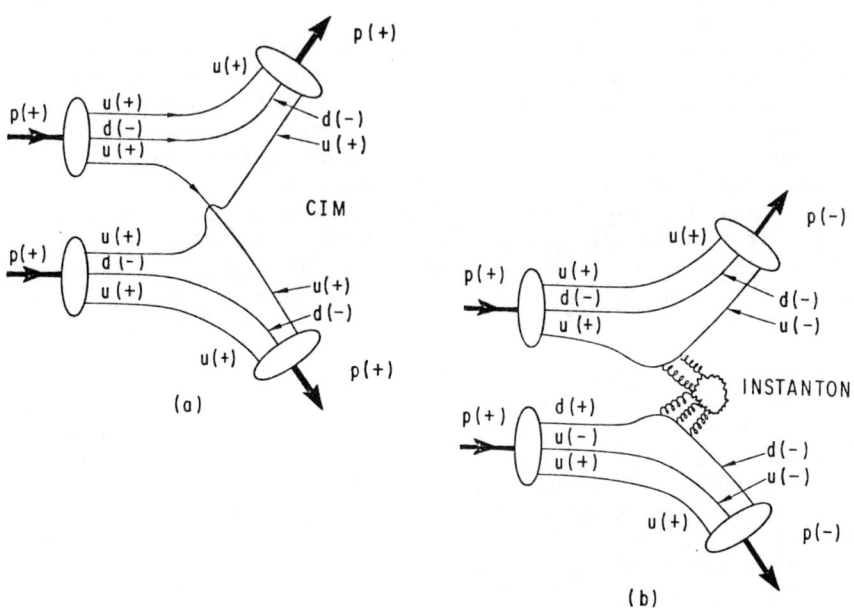

Fig.6. Constituent picture of large angle elastic scattering.
    (a) Constituent interchange with quark helicity conservation.
    (b) Instanton generated quark-quark scattering with double helicity flip.

A typical CIM diagram for $pp \to pp$ is shown in Fig.6(a). If we make the assumptions

1. The masses of the u and d quarks in the nucleon are neglected. The helicity of each quark is therefore preserved by the perturbative (gluon-exchange) interactions which connect the quarks.

2. After summing over all gluon configurations and averaging over color, the magnitude and angular dependence of a given CIM diagram do not depend on the helicities of the exchanged quarks.

3. The helicity amplitudes for $pp \to pp$ and $pn \to pn$ are obtained by projecting the possible quark-interchange amplitudes on the $SU_6$ wave functions for the proton and neutron.

With these assumptions we can calculate the Jacob-Wick helicity amplitudes and predict[21,22]

$$A_{nn} = 1/3$$

$$A_{ss} = A_{\ell\ell} = -1/3 \qquad (13)$$

$$A_{s\ell} = 0$$

approximately, independent of $\theta$. These predictions are shown compared with data[23,24,25] in Fig.8.

The assumptions are very strong. We have a clue that we are neglecting something important since the theory with two massless quarks we are using displays an $SU_2 \otimes SU_2 \otimes U_1^B \otimes U_1^A$ symmetry. The $U_1^A$ symmetry, which is evidently not observed in nature, needs to be broken. Our current understanding of how this symmetry is broken involves vacuum fluctuations called instantons.[26] These instantons can generate an effective interaction between the light u and d quarks. This interaction has the following properties:[27,22]

1. It is present only when the initial quarks have opposite flavor (i.e., one u and one d quark) and the same helicities. It flips the helicity of both the quarks.

2. It generates an NN double-flip amplitude $\langle ++ | -- \rangle$ indicated in the diagram in Fig.6(b).

The magnitude of this quark helicity-flipping amplitude is not presently known. It is an effect which would not be there in perturbative QCD and which can have important consequences for NN spin-spin observables. Some estimates for $A_{nn}$, $A_{ss}$ and $A_{\ell\ell}$ combining a small admixture of the instanton-generated double-flip amplitude are shown in Fig.8.

It is obviously very interesting to use the data to test for quark helicity conservation. At a CM angle of 90° in $pp \to pp$ we have the model independent constraint

$$1 - A_{nn}(\pi/2) + A_{ss}(\pi/2) + A_{\ell\ell}(\pi/2) = 0 \qquad (14)$$

In the absence of quark helicity flip we have the additional result

$$A_{nn} = -A_{ss} \qquad (15)$$

In view of the interesting structure observed in $A_{nn}$ near 90° at $P_{LAB} = 12$ GeV/c it is important to measure either $A_{ss}$ or $A_{\ell\ell}$ in the same kinematic regime. If $A_{\ell\ell}$ is measured, the constraint (14) could then be used to test (15).

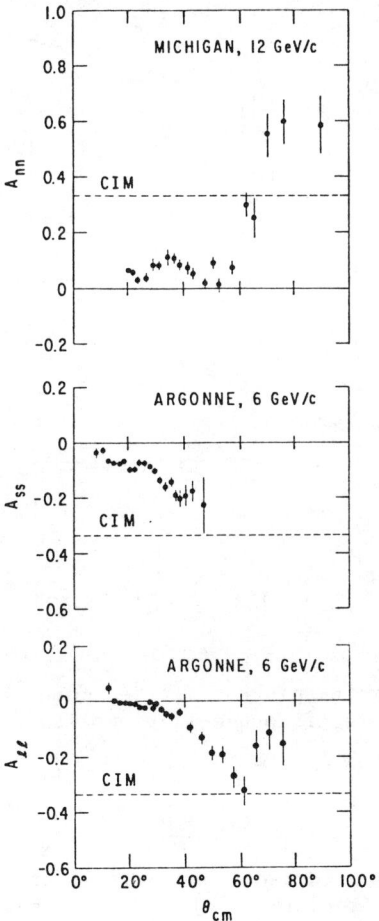

Fig. 7. Proton-proton elastic scattering spin-spin asymmetries calculated in the CIM as explained in the text and compared with measurements done at the Argonne ZGS.

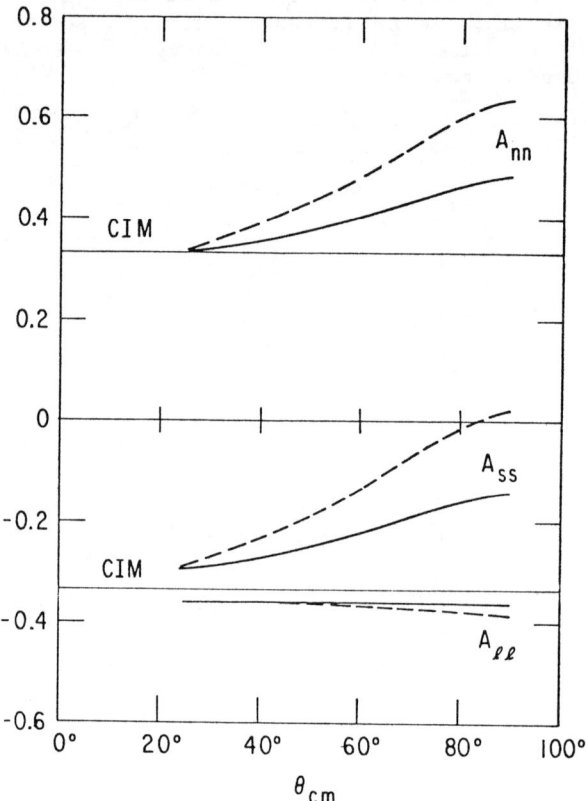

Fig.8. Proton-proton spin-spin asymmetries found by combining a small admixture of double-flip amplitude with CIM amplitudes.

## Testing QCD

QCD is no longer considered a one experiment theory. Although the first quantitative predictions of the perturbative theory involved scaling violations in deep inelastic scattering we are beginning to learn how to apply the same techniques in other processes, including the large-$p_T$ production of hadrons. There now seems to be no one dramatic experiment which will provide irrefutable confirmation of the theory but many experiments can play a role. Experiments with polarized hadron beams or polarized targets can play a particularly important role if they can be done with the precision necessary to confront theoretical calculations.

Because we are not given beams of polarized quarks and gluons, the interpretation of experiments necessarily involves the distribution functions $G_{a/A}^T(x)$ or $G_{a/A}^L(x)$ which tell how hadrons transmit spin information to the fundamental constituents. We need to do a large number of spin-dependent experiments such as deep inelastic lepton scattering, production of massive lepton pairs, production of new massive flavors such as charm in addition to the large-$p_T$ production of hadrons and jets in order to determine these distribution functions. The determination of these functions will allow us to isolate the portion of the dynamics sensitive to the form of the QCD perturbation expansion.

At this conference, we have heard optimistic projections concerning future high-energy polarized proton beams at the Brookhaven AGS, at Fermilab and at ISABELLE. These projections may provide the incentive for theorists to do more spin-dependent calculations involving QCD. Those predictions discussed here may prove to be only a small sample of the spin effects in QCD.

## ACKNOWLEDGMENTS

Many of the ideas in this presentation can be traced to discussions with my collaborators John Babcock, Frank Close, Glennys Farrar, Steve Gottlieb, Keisho Hidaka, Evelyn Monsay and Jerry Thomas. I have also benefited immensely from conversations with Stan Brodsky, Rick Field, Harry Lipkin and Jeff Owens. I'd like to thank Cristian Sorensen for reading the manuscript and offering his comments.

## REFERENCES

1. R. D. Field and R. P. Feynman, Phys. Rev. D15, 2590 (1977).
2. R. Blankenbecler, S. Brodsky and J. Gunion, SLAC-PUB-2057 (1978). D. Sivers, S. J. Brodsky and R. Blankenbecler, Physics Reports 23C, 1 C (1976).
3. R. D. Field, Phys. Rev. Lett. 40, 1161 (1978); J. F. Owens, E. Reya and M. Glück, Phys. Rev. D (1978); A. P. Contogourio, R. Gaskell and A. Nicolaidis, Phys. Rev. D17, 839 (1978).
4. R. Cutler and D. Sivers, Phys. Rev. D17, 196 (1978). B. Combridge, J. Kripfganz and J. Ranft, Phys. Lett. 70B, 234 (1977).
5. C. Sachrajda, Phys. Lett. 73B, 185 (1978); R. K. Ellis, H. Georgi, M. Machacek, H. D. Politzer and G. G. Ross, Phys. Lett. (to be published); Y. L. Dokshitsev, D. I. D'Yakanov and S, I. Troyau, Proc. XIII Leningrad Winter School; S. Libby and G. Sterman, SUNY preprints ITP-SB-78-41 and ITP-SB-78-42.
6. R. Field, Proc. of the Symposium on Experiments using enriched antiproton, polarized proton and polarized antiproton beams at Fermilab Energies, (Editor A. Yokosawa, June 10, 1977 ANL-HEP-CP-77-45).
7. We are assuming a stronger form of factorization than has been proved in the papers in Ref.5.
8. The conventions for inclusive spin asymmetries are given in R. D. Field, Workshop on Physics with Polarized Targets, Brookhaven National Laboratory, 1974, ed. J. S. Russ.
9. H. Georgi and H. D. Politzer, Phys. Rev. D14, 1829 (1976).

10. G. Kane, J. Pumplin and W. Repko, University of Michigan preprint UM HE 73-29.
11. Work in progress with Evelyn Monsay.
12. L. Kluberg et al, Phys. Rev. Lett. 38, 670 (1977); U. Becker et al, Phys. Rev. Lett. 37, 1731 (1976); R. McCarthy et al, Phys. Rev. Lett. 40, 213 (1978).
13. J. Babcock, E. Monsay and D. Sivers, Phys. Rev. Lett

14. M. J. Algard et al, Phys. Rev. Lett. 37, 1258 and 1261 (1971), ibid 41, 70 (1978).
15. J. Babcock, E. Monsay, D. Sivers, Argonne preprint ANL-HEP-PR-78-39.
16. F. Close and D. Sivers, Phys. Rev. Lett. 39, 1116 (1977); J. Soper, U. Oregon preprint and presentation at this conference.
17. K. Hidaka, E. Monsay and D. Sivers, Argonne preprint ANL-HEP-PR-78-47.
18. S. Wandzura and F. Wilczek, Phys. Lett. 72B, 195 (1977); S. Wandzura, Nucl. Phys. B122, 412 (1977).
19. M. J. Alguard et al., Phys. Rev. Lett. 37, 1258 and 1261 (1976) and Phys. Rev. Lett. 41, 70 (1976).
20. S. Brodsky and G. Farrar, Phys. Rev. Letts. 31, 1153 (1973) and Phys. Rev. D11, 1309 (1975), V. A. Matveev, R. M. Muradyan and A. N. Tavkhelidze, Lett. al Nuovo Cumento 7, 719 (1973). See, B. Pire, Nucl. Phys. B114, 11 (1976).
21. S. J. Brodsky, C. Carlson and H. Lipkin, private communication.
22. G. Farrar, S. Gottlieb, D. Sivers, G. Thomas, Argonne preprint ANL-HEP-PR-78-43.
23. Data on $A_{ss}$ from I. P. Auer et al, Phys. Rev. Lett. 37, 1727 (1976) and private communication from H. Spinka.
24. I. P. Auer et al, Phys. Lett. 70B, 475 (1977). Preliminary data submitted 1978 Int'l. Conf. on High Energy Physics (Tokyo) ANL-HEP-CP-78-37 and private communication from H. Spinka.
25. D. G. Crabb, et al, UM-HE-78-36 and private communication from D. G. Crabb.
26. G. 't Hooft, Phys. Rev. D14, 3432 (1976).
27. R. Carlitz, Phys. Rev. D17, 3225 (1978) and R. Carlitz and C. Lee, Phys. Rev. D17, 3238 (1978).

# SPIN MEASUREMENTS IN HADRONIC HIGH MOMENTUM TRANSFER SCATTERING*

Kent M. Terwilliger
University of Michigan, Ann Arbor, Michigan 43109

## ABSTRACT

The results of recent experiments investigating spin effects in hadronic high momentum transfer scattering are reviewed. There is evidence for very large spin dependences at high momentum transfer in measurements involving both the polarization of a single particle and the polarizations of two particles.

## INTRODUCTION

I will review recent spin measurements in hadronic reactions, concentrating on the higher momentum transfer range, with $P_\perp \geq 1$ GeV/c .[1] Momentum transfers of ~1 GeV/c are not exactly high but for some reactions are the highest that are available.

One general characteristic of high momentum transfer hadronic processes is the low cross sections, cross sections which fall continuously with increasing momentum transfer. A complete spin amplitude analysis of a particular reaction requires the measurement of final state as well as initial state polarizations. This final state polarization determination usually involves a rescattering, decreasing the event rate by a factor of 100 below the unrescattered processes. With present polarized beam and targets, carrying out the necessarily large set of initial and final state polarization measurements is only feasible at modest momentum transfers in a straightforward reaction such as p-p scattering. However, at higher momentum transfers the low rates preclude such a set of measurements and a complete spin amplitude analysis. So, we do what we can. What we can measure are spin effects involving a single particle and, with a polarized proton beam and target, two particles, enabling us to determine p-p cross sections in pure initial spin states.

Single particle spin effects in the high momentum transfer region have been investigated in a wide variety

---

*Work supported by the United States Department of Energy

of hadronic reactions. They have been studied in inclusive processes, as in the production of polarized $\Lambda^\circ$ and protons using unpolarized proton beams, or, using polarized beams or targets, to create asymmetries in reaction products such as pions. They have been studied to modest momentum transfers in the exclusive process $p_\uparrow + p \to \Delta^{++} + n$ and to substantial momentum transfer in elastic p-n scattering. The p-$p_\uparrow$ elastic scattering process is of course the most intensively studied with single particle asymmetry measurements now up to 300 GeV/c beam momentum. The 12 GeV/c polarized proton beam at the ZGS has made possible the investigation of two particle p-p spin effects, with measurements now to $P_\perp^2 = 5 \ (\text{GeV/c})^2$.

## SPIN PARAMETERS

I will now discuss the spin parameters which are measured in the various reactions. Fig. 1 shows a portion of the spin parameter convention agreed on at the Ann Arbor Polarization Workshop in the fall of 1977.[2] It should be noted that there are historical differences from this convention - and certainly no unanimous agreement to use it - so in this talk there will be considerable switching of notation. In the laboratory $\vec{N}$ is normal to the scattering plane, $\vec{L}$ along the beam, and $\vec{S} = \vec{N} \times \vec{L}$ in the scattering plane. Small letters denote the directions in the CM; they are similarly defined. For the two-spin beam-target processes considered today, the lab and CM directions are identical.

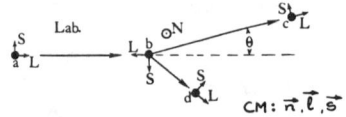

Fig. 1. Spin parameter convention. See Ref. 2.

Fig. 2 shows the type of experiments we will consider and the Wolfenstein spin parameters which are determined (again, using the Ann Arbor Convention). For one polarized particle, beam or target or one of the

EXPERIMENTS

I  Asymmetries with one polarized particle

| REACTION | | DETERMINE WOLFENSTEIN PARAMETERS $(i,j;k,\ell)$ |
|---|---|---|
| Inclusives: | | |
| $p$ + nucleus $\rightarrow \Lambda_\uparrow + X$ | | $P = (o,o;n,o)$ |
| | $p_\uparrow^+ + X$ | $P = (o,o;n,o)$ |
| $\left.\begin{array}{l}p_\uparrow + p \\ p' + p_\uparrow\end{array}\right\} \rightarrow \binom{p}{\pi} + X$ | | $A = (n,o;o,o)$ $A = (o,n;o,o)$ |
| Exclusives: | | |
| $p_\uparrow + p \rightarrow \Delta^{++} + n$ | | $A = (n,o;o,o)$ |
| Elastics: | | |
| $\left.\begin{array}{l}p_\uparrow + n \\ n' + p_\uparrow\end{array}\right\} \rightarrow p + n$ | | $A = (n,o;o,o)$ $A = (o,n;o,o)$ |
| $\pi + p_\uparrow \rightarrow \pi + p$ | | $A = (o,n;o,o)$ |
| $\left.\begin{array}{l}p_\uparrow + p \\ p' + p_\uparrow\end{array}\right\} \rightarrow p + p$ | | $A = (n,o;o,o)$ $A = (o,n;o,o)$ |
| $p + p \rightarrow p_\uparrow + p$ | | $P = A = (o,o;n,o)$ |

II  Correlations with two polarized particles

| | | |
|---|---|---|
| $p_\uparrow + p_\uparrow \rightarrow p+p$ | | $A_{nn}(C_{nn}) = (n,n;o,o)$ |
| $p_\rightarrow + p_\rightarrow \rightarrow p+p$ | | $A_{\ell\ell}(C_{\ell\ell}) = (\ell,\ell;o,o)$ |
| $p_\odot + p_\rightarrow \rightarrow p+p$ | | $A_{s\ell}(C_{s\ell}) = (s,\ell;o,o)$ |

Measure as

$$A = \frac{1}{P_{B(T)}} \frac{N(\uparrow) - N(\downarrow)}{N(\uparrow) + N(\downarrow)} \qquad A_{nn} = \frac{1}{P_B P_T} \frac{N(\uparrow\uparrow) + N(\downarrow\downarrow) - N(\uparrow\downarrow) - N(\downarrow\uparrow)}{N(\uparrow\uparrow) + N(\downarrow\downarrow) + N(\uparrow\downarrow) + N(\downarrow\uparrow)}$$

Fig. 2

reaction products, the measurement is always of an asymmetry - for example to determine the asymmetry parameter for vertical polarization, $A=(n,0,0,0)$, one measures the difference between the number of events with up and down beam polarization divided by the sum. Since the beam polarization is never 100% one must divide by it to get the pure state asymmetry. Equivalently, from rotational invariance, one could measure the number of left scatters minus right scatters over the sum and divide by the beam polarization. To determine a polarization parameter, as for the $\Lambda^\circ$, $P=(0,0,n,0)$, one measures the asymmetry in the $\Lambda^\circ$ decay and divides by the known $\Lambda^\circ$ asymmetry parameter. These asymmetry and polarization parameters are perhaps easiest to visualize as the possible measure of a spin-orbit coupling, in analogy with atomic and nuclear physics.

For two polarized particles one measures correlations - determines the dependence of the event rate on the product of the two particles' polarizations. So $A_{nn}$ is the difference between the event rate with parallel and antiparallel polarizations divided by the sum, the result then divided by the product of the beam and target polarizations. $A_{nn}$ is then a measure of the correlation of the particle spins - possibly due to a direct spin-spin interaction. Fig. 3 shows the p-p two-spin differential cross section for $\vec{n}$ polarization in terms of the Wolfenstein parameters A and $A_{nn}$, the beam and target polarizations and

p-p TWO-SPIN CROSS SECTION RELATIONS
( given for $\vec{n}$-polarization)

$$\frac{d\sigma}{dt} = \left\langle\frac{d\sigma}{dt}\right\rangle \left[ 1 + (P_B + P_T)A + P_B P_T A_{nn} \right]$$

Pure state cross sections:

$$\left.\frac{d\sigma}{dt}\right)_{\uparrow\uparrow} = \left.\frac{d\sigma}{dt}\right)_{\uparrow\uparrow \rightarrow \infty} = \left\langle\frac{d\sigma}{dt}\right\rangle \left[ 1 + 2A + A_{nn} \right]$$

$$\left.\frac{d\sigma}{dt}\right)_{\downarrow\downarrow} = \left.\frac{d\sigma}{dt}\right)_{\downarrow\downarrow \rightarrow \infty} = \left\langle\frac{d\sigma}{dt}\right\rangle \left[ 1 - 2A + A_{nn} \right]$$

$$\left.\frac{d\sigma}{dt}\right)_{\uparrow\downarrow} = \left.\frac{d\sigma}{dt}\right)_{\downarrow\uparrow} = \left\langle\frac{d\sigma}{dt}\right\rangle \left[ 1 - A_{nn} \right]$$

Fig. 3

the spin averaged cross section. This equation is equivalent to the previous relations for A and $A_{nn}$.[3] Also shown are the pure spin state parallel and antiparallel cross sections directly obtained from the above relation. It will be of use later to note that since A = 0 at 90° CM in pp scattering, because of particle identity and rotational invariance, the spin-up and spin-down parallel cross sections become equal there.

## SPIN EFFECTS WITH ONE POLARIZED PARTICLE INCLUSIVE REACTIONS

I will now discuss experiments involving one polarized particle, the measurement of a single asymmetry, and will begin with inclusive experiments. I will first consider reactions in which the incident beam is unpolarized and the polarization of a scattered particle is measured. One of the most exciting experiments showing high momentum transfer spin effects is that of the collaboration involving Michigan-Rutgers-Wisconsin and others, whose recent results were presented at this conference by Heller.[4] This group was the first to demonstrate the existence of large spin effects at very high energies - at 300 GeV at Fermilab.[5] They had observed the polarization of the inclusively produced $\Lambda°$ to increase with momentum transfer to over 20% at a $P_\perp \sim 1.6$ GeV/c. Very strikingly, their results were consistent with the same $P_\perp$ dependence seen at 24 GeV - implying an energy independent process. Their present experiment extends the energy to 400 GeV and $P_\perp \sim 2.1$ GeV/c. Their new data and the previous 300 GeV data are shown in Fig. 4. It is evident that the 400 GeV data tracks the 300 GeV - with a possible flattening at the higher $P_\perp$. One should

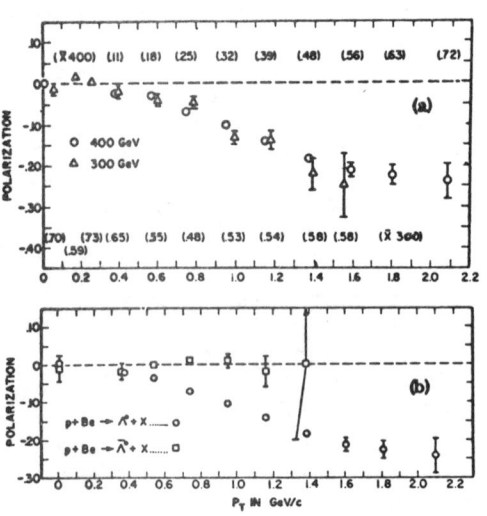

Fig. 4. (a) $\Lambda°$ polarization as a function of $P_T$. The number in parentheses is the average value of x for that point. (b) $\Lambda°$ and $\bar{\Lambda}°$ polarizations. The polarization is defined as positive along $\vec{n} = (\vec{k}_p \times \vec{k}_\lambda)/|\vec{k}_p \times \vec{k}_\lambda|$. From Ref. 4.

note that the observed $\Lambda°$ may be partially from $\Sigma°$ decay, with a consequent dilution of the measured polarization. The initially produced $\Lambda°$ may have a considerably larger polarization. They also measured $\bar{\Lambda}°$ polarization - consistent with zero, as indicated on the figure. Their conference presentation also showed evidence for a significant $\Xi°$ polarization, ~10%. In addition they looked for an indication of proton polarization through a two step process producing $\Lambda°$ but saw none. Why such strong $\Lambda°$ polarization, but nothing for $\bar{\Lambda}°$? A possibly pertinent difference is that $\Lambda°$ are leading particles, carrying the valence quarks, the $\bar{\Lambda}°$ are not. Wicklund commented that perhaps the mechanism might be a low energy resonance production process. In the theory panel presentation, Kane said that in a reasonable QCD model, at high $P_\perp$ and energy, the $\Lambda°$ polarization should disappear. The group intends to test this next year at a $P_\perp$ of ~6 GeV/c.

The lack of proton polarization in a high transverse momentum inclusive process was also observed (at the 10% level) in a recent Brookhaven experiment, Carroll et al.[6], with 28 GeV/c protons on platinum. They measured the polarization of 7 GeV/c scattered protons at a $P_\perp$ =1.46GeV/c. Since this corresponds to nearly 90° in the cms the p-p process would not be expected to give an asymmetry, but there is no such restriction on p-n scattering.

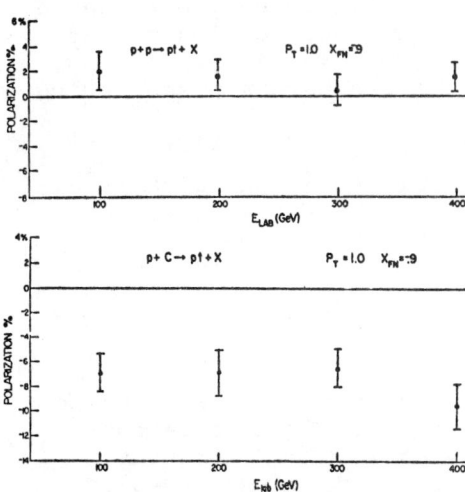

Fig. 5. Polarization of inclusively produced protons as a function of proton beam energy. Top curve: Hydrogen jet target; bottom: Carbon target. From Ref. 7.

Another experiment investigating the polarization of inclusively produced protons was carried out at Fermilab by the Indiana University group: their results were presented at the Tokyo Conference, Baranko et al.[7], and at this conference by Polvado. This group uses a carbon scattering polarimeter to analyze the polarization of the recoil proton. Their primary targets were an internal hydrogen gas jet and a rotating carbon target. Some of their results are shown in Fig. 5. Looking at large x at a $P_\perp$ =1 GeV/c on hydrogen they see only a 2% level proton polarization. On carbon they observe -3% polarization -- they attribute the difference to rescattering of the recoil proton by the carbon nuclei in the target, creating a proton polarization going into their polarimeter much larger than that from the primary reaction.

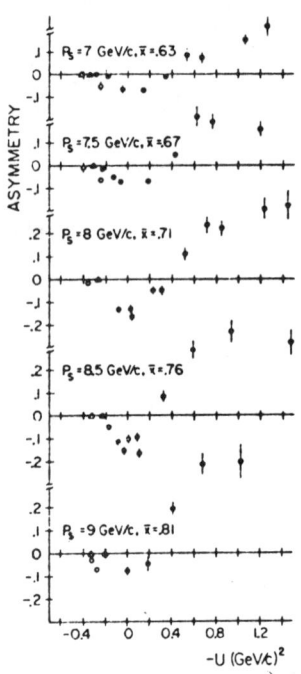

Figure 6. Asymmetry for inclusive $\pi^+$ production in $p_\uparrow p$ collisions at 11.8 GeV/c (solid circles) and 6 GeV/c (open circles). $P_s$ is the scattered momentum at 11.8 GeV/c. From Ref. 8.

A second type of inclusive experiment involves the production of asymmetries in inclusively produced pions or protons with the use of a polarized beam or target. Some earlier work on inclusive pion production using the ZGS polarized proton beam was by an Argonne, Minnesota, Rice collaboration, Klem et al.[3]. Some of their results are shown in Fig. 6. They saw a large asymmetry in $\pi^+$ production at their largest momentum transfer. They interpreted their large x (forward production) results in terms of a baryon-exchange model and concluded they were essentially looking at backward $\pi$-p scattering.

A CERN, Oxford, Orsay collaboration, Dick et al.[9], was the first to use a polarized proton target to study asymmetries in inclusive reactions. Their original work was on inclusive $\pi^\pm$ production in $\pi^\pm p$ scattering at 8 GeV/c, giving evidence for mirror symmetry. In more recent work, Aschman et al.[10], they used 7.9 GeV/c protons from Nimrod at Rutherford Lab to study asymmetries in inclusive proton and pion production in $pp_\uparrow$ scattering. Some of their results are shown in Fig. 7. The proton and $\pi^+$ asymmetries are non-zero, at the ~5% level, with the $\pi^+$ asymmetry possibly large at their highest $P_\perp$.

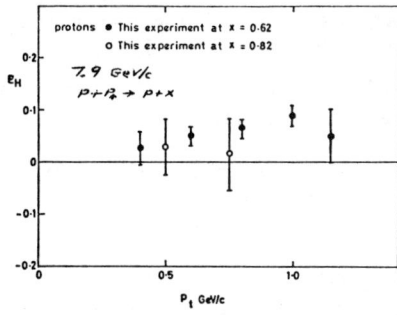

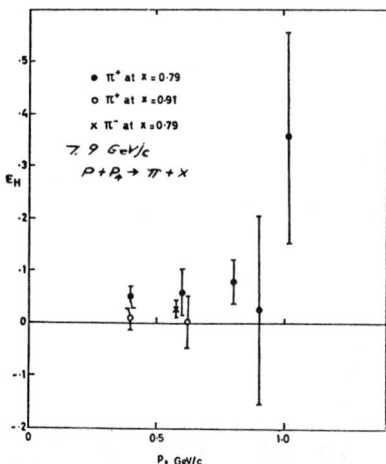

Fig. 7. Asymmetries in inclusive proton and pion production in 7.9 GeV/c pp$_\uparrow$ collisions, Ref. 10.

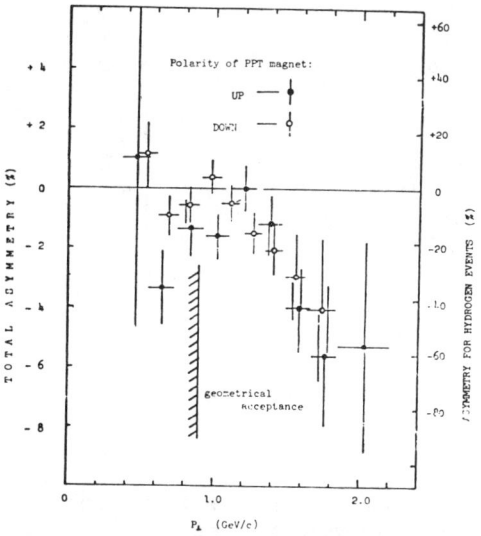

Fig. 8. Asymmetry in inclusive $\pi^0$ ($2\gamma$) production in 24 GeV/c pp$_\uparrow$ collisions, central region, $x \simeq 0$. Left scale, raw measured asymmetry; right, hydrogen asymmetry after estimating the target dilution. Preliminary results, Ref. 11.

The group also studied inclusive $\pi^0$ ($2\gamma$) production at 24 GeV/c at CERN. Their very preliminary results were presented at Tokyo, Antille et al.[11], and at this conference by M. Fidecaro and are shown in Fig. 8. The raw asymmetry is indicated on the left hand scale. This asymmetry is of course strongly diluted by the non-polarized material in their target. They estimated the dilution factor from previous experiments to obtain a measure of the pure proton asymmetry, shown using the right hand scale, obtaining a dramatic -60% at $P_\perp \sim 2$ GeV/c.

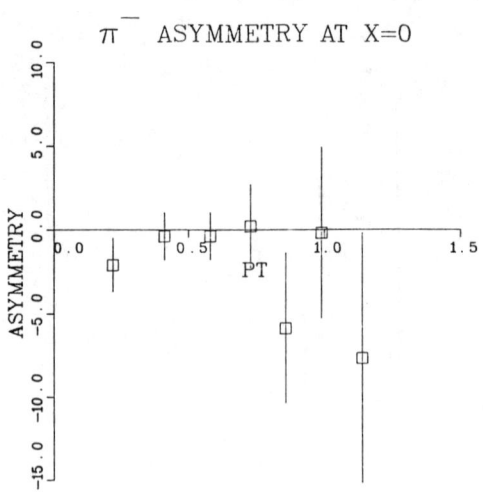

Fig. 9. Inclusive $\pi^-$ asymmetry as a function of momentum transfer, 12 GeV/c $p_\uparrow p$ collisions, $x \cong 0$ region.

This is very exciting, the largest inclusive asymmetry seen at high $P_\perp$, if it holds up when they complete their analysis.

The asymmetry in inclusive produced $\pi^-$ at x = 0 has been studied at 6 and 12 GeV/c here at the ZGS by an Argonne-Indiana collaboration. Some preliminary results presented at this conference by Gray are shown in Fig. 9. They are mostly at low momentum transfer and the asymmetry is small, at the few percent level. They hope to use their final data to test QCD calculations.

## EXCLUSIVE, QUASI ELASTIC REACTION

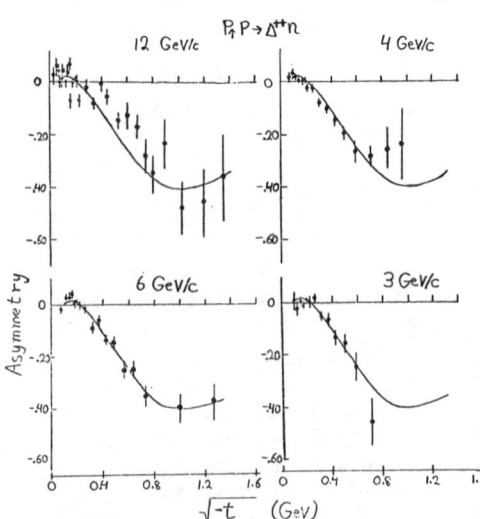

Fig. 10. Asymmetry in $p_\uparrow p \to \Delta^{++} + n$. 12 GeV/c data are preliminary. Ref. 12.

I will now switch to a quasielastic experiment carried out by the Argonne Effective Mass Spectrometer (EMS) group, Wicklund et al.[12] They studied the reaction

$p_\uparrow p \to \Delta^{++} + n$, 3 to 12 GeV/c, but only to modest momentum transfer. Their measured asymmetries are shown in Fig. 10. The 12 GeV/c results are preliminary and based on 20% of their data sample. The hand drawn curve is common to all energies. The asymmetries are large and consistent with energy independence. The group compares their results with a quark-vector meson production model, with qualitative agreement.

## ELASTIC SCATTERING

A comparison of the asymmetries in $p_\uparrow n$ and $p_\uparrow p$ elastic scattering was first done at at the ZGS at low momentum transfer in the 2-6 GeV/c range by the Argonne EMS group, Diebold et al.[13] This type of comparison has been extended to much higher momentum transfers in a ZGS experiment by the Minnesota group, Marshak et al.[14] presented at Tokyo. They have data at 2, 3, and 6 GeV/c - on the latter run covering practically the full angular range. Fig. 11 shows their 6 GeV/c data. I have shrunk their p-p plot to align the t scales. 90° CM corresponds to $|t| = 4.8$, and $P_\perp^2 = 2.4 (\text{GeV/c})^2$. It is clear that the $p_\uparrow n$ asymmetry is completely unlike the $p_\uparrow p$ past the $|t| = 1.5 (\text{GeV/c})^2$ region, going dramatically negative to -.3 at 90° CM and going further negative in the backward hemisphere. These appear to be the highest elastic symmetries observed at high transverse momentum.

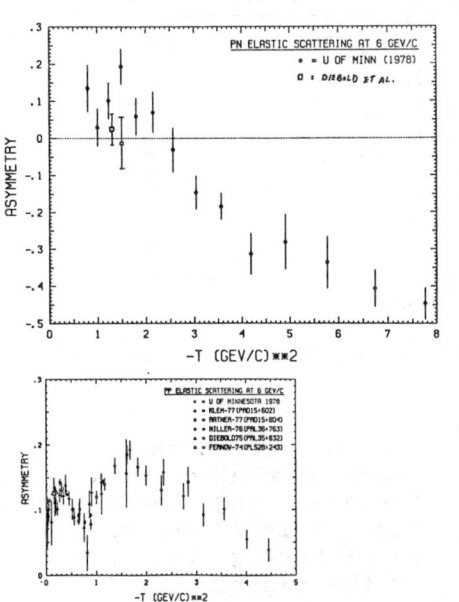

Fig. 11. Asymmetry in $p_\uparrow n$ and $p_\uparrow p$ elastic scattering at 6 GeV/c. Ref. 14

Turning now to the higher energies, asymmetry measurements at substantial transverse momenta were made in a recent experiment by the CERN-Oxford-Annecy collaboration, Crabb et al.[15], $pp_\uparrow$ elastic scattering at 24 GeV/c. Their results are shown in Fig. 12. Their data for $|t| < .9$ were published earlier.[15] The pp asymmetry parameter generally appears to hold in the few percent range, with evidence for a significant negative value, $-(10 \rightarrow 15\%)$ or so at $|t| \sim 3.5 (\text{GeV/c})^2$.

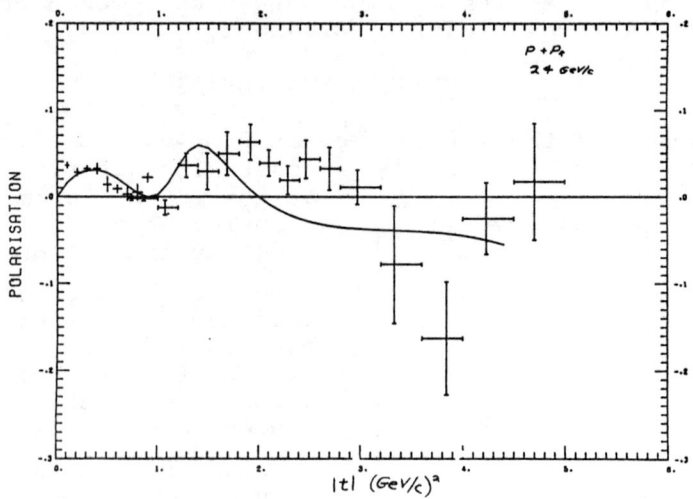

Fig. 12. Asymmetry parameter in 24 GeV/c pp↑ elastic scattering. Ref. 15.

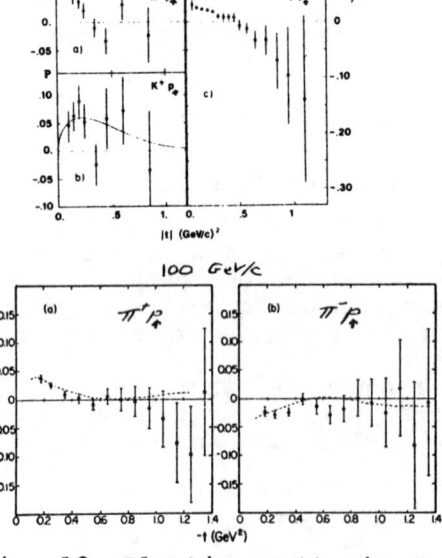

Fig. 13. Elastic scattering asymmetry parameters. Top, 45 GeV/c, Ref. 16. Bottom, 100 GeV/c, Ref. 17

At even higher energies the momentum transfer range becomes more limited and the error bars quite large at the experimenter's maximum t values. The top of Fig. 13 shows the results of the earlier 45 GeV/c experiment of Gaidot et al.[16] done at Serpukhov. The error bars are large but there is an indication that the asymmetry parameter has become larger in the $|t| = 1(\text{GeV/c})^2$ range than at 24 GeV/c. The 100 GeV/c elastic $\pi^{\pm}p$ asymmetry parameter data shown on the bottom of Fig. 13 were obtained at Fermilab by the large scale collaboration, Auer et al.[17], and presented

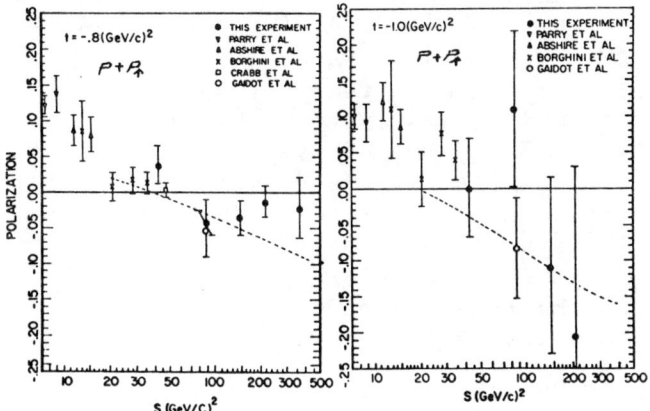

Fig. 14. Polarization parameter in pp elastic scattering as a function of S. Ref. 18. Dashed curve, Pumplin-Kane model, Ref. 19.

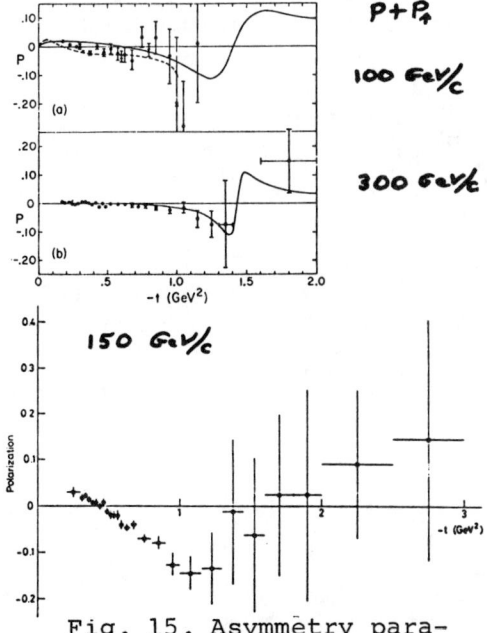

Fig. 15. Asymmetry parameter in $pp_\uparrow$ elastic scattering. Top, Ref. 20; theoretical curve, updated Kane-Seidl model, Ref. 21; bottom, Ref. 22.

here by Jonkheere. The asymmetry parameter values appear to be low, under 10%, in the $|t| \sim 1.2 (\text{GeV/c})^2$ range, but possibly larger than the few percent level seen at lower t.

The Indiana University group, Corcoran et al.[18], used their scattering polarimeter to analyze the polarization of the recoil proton, determining the polarization parameter in pp elastic scattering over the range 20 to 200 GeV/c at Fermilab. Their results are shown in Fig. 14. Their maximum t value is $1(\text{GeV/c})^2$ but their range of energies along with other data does permit a look at the s dependence of the reaction. The asymmetry parameter appears to be down at the -15% level at the higher energies at $|t| = 1 (\text{GeV/c})^2$.

Along with the $\pi^{\pm}p$ data, Jonkheere also discussed the results for $pp_\uparrow$ scattering at 100 and 300 GeV/c, Snyder et al.[20]. Their data are shown on the top of Fig. 15. Apparently the scale on the published 300 GeV/c plot needs to be multiplied by two. Below are the results of the 150 GeV/c $pp_\uparrow$ experiment done by the CERN collaboration, presented here by M. Fidécaro.[22] Again, I have attempted to align the t axes for comparison. Both the 150 and 300 GeV/c experiments have measured values of A in the -15% range at $|t| \sim 1.2$ (GeV/c)$^2$; at larger t values the tendency of both sets of data is toward positive polarization.

## SPIN EFFECTS WITH TWO POLARIZED PARTICLES

Now I would like to discuss some experiments involving two polarized particles. In these experiments both the asymmetry parameter and the spin-spin correlation parameter are simultaneously measured. The experiments are $p_\uparrow p_\uparrow$ elastic scattering, using the ZGS polarized beam and a polarized target.

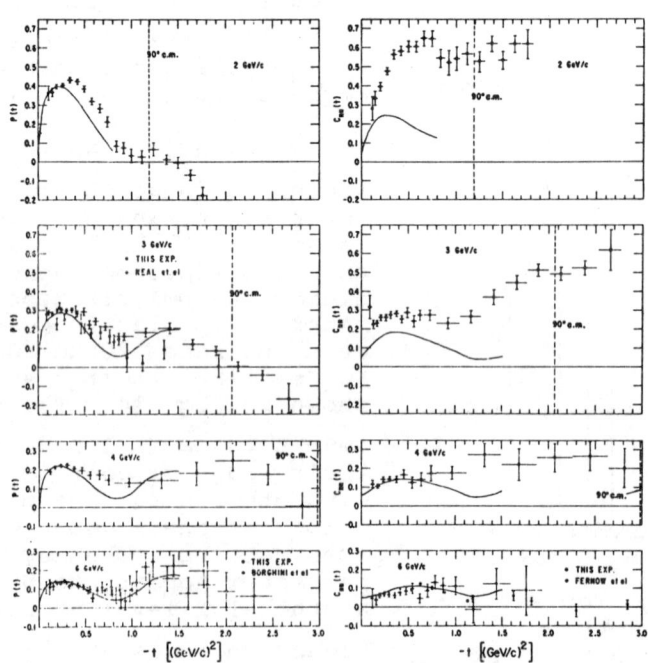

Fig. 16. A and $C_{NN}$, $p_\uparrow p_\uparrow$ elastic scattering 2 to 6 GeV/c. Ref. 23.

The first is a comprehensive experiment by the Northwestern-Argonne collaboration, Miller et al.[23], to measure A and $C_{NN}$ ($A_{NN}$ in the Ann Arbor Convention), the asymmetry and correlation parameters for beam and target polarizations normal to the scattering plane. Their measurements were at 2, 3, 4 and 6 GeV/c, with a maximum value of $|t| = 2.8(\text{GeV}/c)^2$ at nearly 90° CM at 4 GeV/c. Their results are shown in Fig. 16, along with data from other experiments. Let me concentrate on the correlation parameter, $C_{NN}$. At 2 and 3 GeV/c the values of $C_{NN}$ rise to .6 and .5, the 90° CM value near the maximum. At 4 GeV/c the maximum value of $C_{NN}$ has dropped off to ~.25 and at 6 GeV/c it is down to ~.1, quite a rapid fall off.

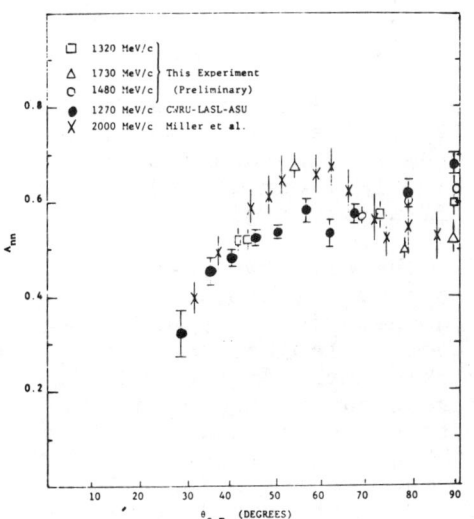

Fig. 17. $A_{nn}$ in $p_\uparrow p_\uparrow$ elastic scattering in the range 1.3-2.0 GeV/c. Preliminary data.

A Rice University group has very preliminary data on $A_{nn}$ also in the 1 to 3 GeV/c range which was presented at this conference by Mulera. A representative plot relating some of their results to other experiments is shown in Fig. 17. They are in agreement with the large value of $A_{nn} = .6$ at wide angles.

The Northwestern-Argonne group has used a solenoid and beam magnet system to rotate the beam polarization vector into the longitudinal ($\vec{L}$) and transverse ($\vec{S}$) directions in the scattering plane. The magnet for their polarized proton target is constructed to allow the target polarizations also to be longitudinal and transverse as well as normal to the scattering plane. Therefore they can measure the entire set of initial state correlation parameters. With this setup they carried out and previously reported a 6 GeV/c experiment, Auer et al.[24], the first high statistics measurement of $C_{SS}$. They found this parameter went almost linearly down from zero to -.2 at $1.2(\text{GeV}/c)^2$. In a very recent

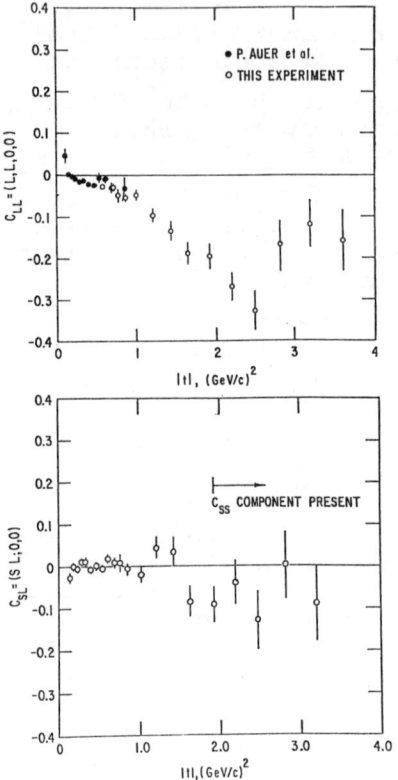

Fig. 18. $C_{LL}$ and $C_{SS}$ in 6 GeV/c pp elastic scattering. Preliminary data. Ref. 25.

experiment also at 6 GeV/c presented at Tokyo[25] they have measured $C_{LL}$ and $C_{SL}$ to $|t| = 3.5 (GeV/c)^2$. Their preliminary data are shown in Fig. 18. The $C_{LL}$ plot contains a small admixture of $C_{SL}$, seen below. The values of $C_{LL}$ are large at large $|t|$, going to $-.3$ at $|t| = 2.5 (GeV/c)^2$, larger than the 6 GeV/c values of $C_{NN}$ which were in the .1 range. Notice that the increase in $C_{LL}$ appears to begin at $|t|$ of $\sim 1$ $(GeV/c)^2$, where the elastic differential cross section curve has changed from the steep fall-off of the diffraction region.

The past two years our group, a collaboration of Argonne, Michigan, and others, also has been working here at the ZGS investigating two-spin effects in elastic p-p scattering. We have studied the momentum transfer dependence at 11.75 GeV/c and the energy dependence at 90° CM, with beam momenta from 1.5 to 11.75 GeV/c. In our experiment the beam and target spins are oriented perpendicular to the horizontal scattering plane, in the $\vec{n}$ direction, the polarization of the scattered and recoil particles are unmeasured. We determine the Wolfenstein parameters A and $A_{nn}$ and the associated parallel and antiparallel pure spin state cross sections discussed at the start of the talk.

During this two year period of running the average extracted polarized beam intensity at our target has increased from $2.5 \; 10^9$ per pulse to just over $10^{10}$ per pulse. With this increase in intensity we have been able to extend our measurements out to 90° CM at 11.75

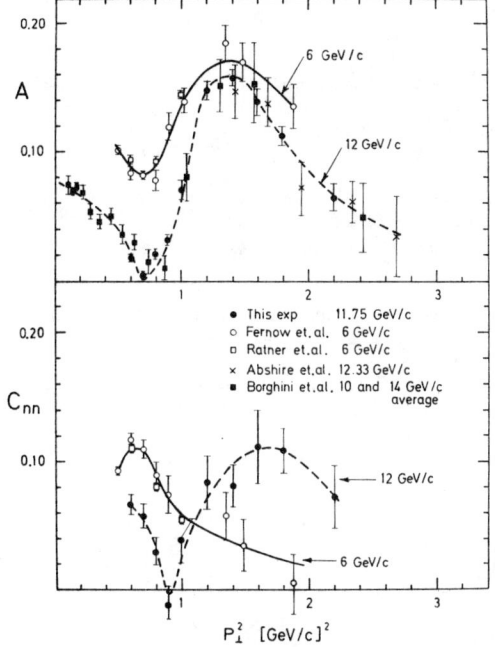

Fig. 19. A and $C_{nn}$ $p_\uparrow p_\uparrow$ elastic scattering at 6 and 12 GeV/c. Ref. 26.

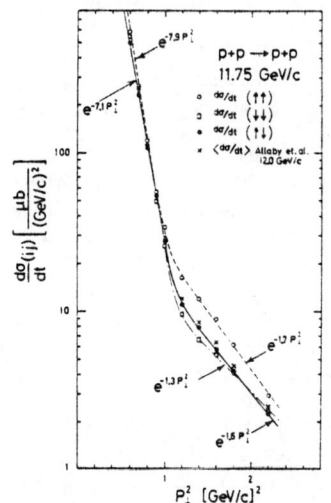

Fig. 20. Differential cross sections in pure initial spin states, $p_\uparrow p_\uparrow$ elastic scattering at 11.75 GeV/c. Ref. 26; cross section normalization, Ref. 27.

GeV/c. Our 1976 data, Abe et al.[26], is presented in the form of A and $C_{nn}$ ($A_{nn}$) in Fig. 19, along with the results of other groups and our earlier 6 GeV/c runs. Of interest here is the well known sharp dip in A at $P_\perp^2$ = .8 GeV/c$^2$ at 11.75 GeV/c and the substantial increase in the 1-2(GeV/c)$^2$ $P_\perp^2$ region. At 11.75 GeV/c $A_{nn}$ also looks remarkably similar to A. These results are reflected in the two-spin cross sections as seen in Fig. 20. The spin average cross sections are from Allaby et al.[27] As we noted in the previous Fig. 19, A and $A_{nn}$ both went to nearly zero around $P_\perp^2$ = .8 and this corresponds to the equality of the pure spin state cross sections at that $P_\perp^2$; the subsequent rise of A and $A_{nn}$ is associated with the observed substantial cross section difference. It is of obvious interest that the spin difference becomes sizeable just after the end of the diffraction peak region - where the spin averaged differential cross section slope has decreased about a factor of five, a phenomenon similar to that we noted in the $C_{LL}$ data of the Northwestern-Argonne collaboration. Our early 1977

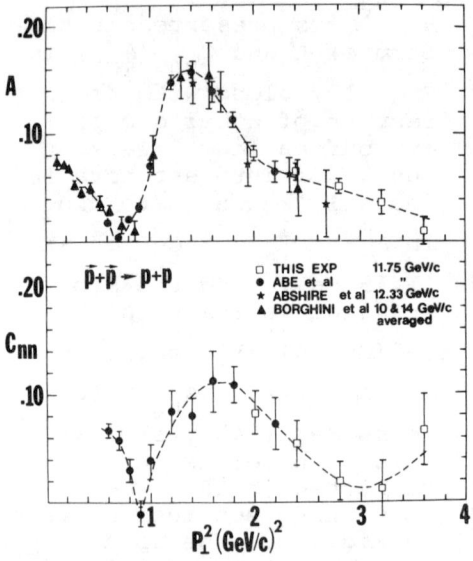

Fig. 21. A and $C_{nn}$, 11.75 GeV/c $p_\uparrow p_\uparrow$ elastic scattering to $P_\perp^2 = 3.6 (GeV/c)^2$. Ref. 28.

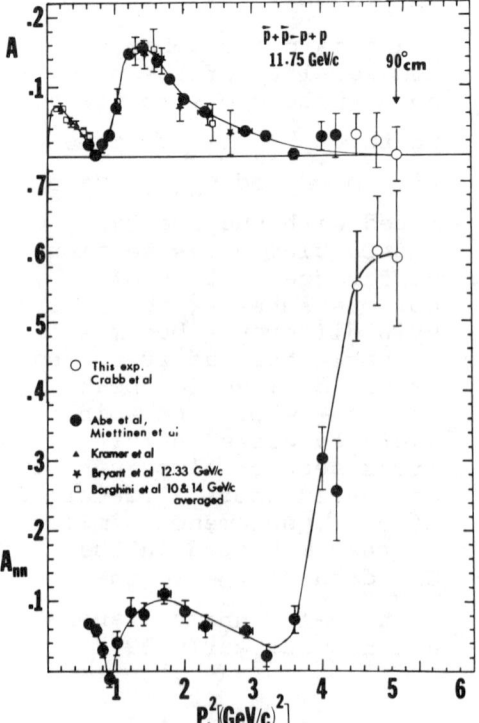

Fig. 22. Ref. 29, 30.

data increased our $P_\perp^2$ range from 2 to 3.6 $(GeV/c)^2$.[28] The A and $A_{nn}$ behavior is shown in Fig. 21. Both parameters have decreased smoothly. The dotted curve through $A_{nn}$ "to guide the eye", drawn with a dip and rise at $P_\perp^2 = 3.6(GeV/c)^2$, was obviously pure hope - a flat "guide" curve would have been equally reasonable.

Fig. 22 shows our later 1977 A and $A_{nn}$ results,[29] which went out to $P_\perp^2 = 4.2(GeV/c)^2$, and our most recent results to 90° CM, $P_\perp^2 = 5.09(GeV/c)^2$.[30] The hint of a rise of $A_{nn}$ at $P_\perp^2 = 3.6(GeV/c)^2$ in the previous run has turned into a dramatic upturn with $A_{nn}$ going to .6 - how lucky can a group be? A is consistent with zero, necessary for identical particle scattering at 90° CM. Since A is small the spin parallel up and down cross sections are essentially equal. We will call their average the spin parallel cross section. We have plotted the ratio of this average parallel cross section to the antiparallel cross section on a log scale in Fig. 23. This ratio reaches a factor of four in the 90° CM region. Clearly we will need better statistics at this

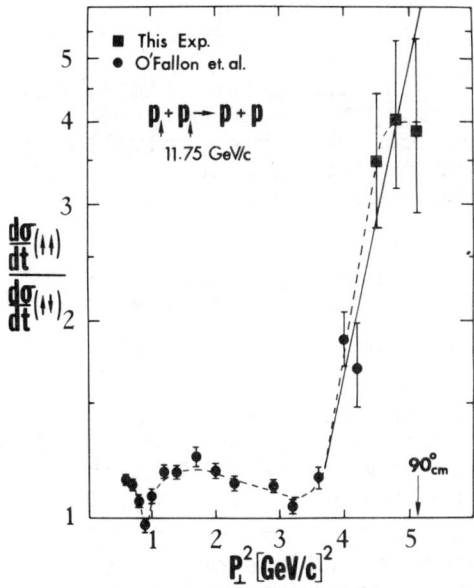

Fig. 23. The ratio of the cross sections in spin parallel to antiparallel initial spin states, $(1+A_{nn})/(1-A_{nn})$, Ref. 30.

energy or, hopefully, some running at a larger energy to test for a continuing increase or a flattening out. It is also obvious that a very exciting experiment would be the high momentum transfer measurement of the longitudinal and transverse spin parallel and antiparallel cross sections - the parameters $A_{\ell\ell}$ and $A_{ss}$; the Northwestern-Argonne group hopes to carry this out.

In Fig. 24 our spin parallel and antiparallel cross sections and the ISR data of DeKerret et al.[31] are plotted against a scaled momentum transfer scaled parameter, $\rho_\perp^2$, involving a Lorentz contraction factor and total cross section ratios, that Krisch[32] has suggested to attempt to superpose the differential cross sections over a wide range of energies. At 11.75 GeV/c it differs little from $P_\perp^2$. On this plot the antiparallel cross section appears to be falling as $e^{-2.6\rho_\perp^2}$ with no apparent change in slope from the intermediate 1 - 3 GeV/c$^2$ $\rho_\perp^2$ region. The spin parallel cross section is falling less rapidly, consistent with the same $e^{-1.6\rho_\perp^2}$ dependence of the ISR results in this range of momentum transfer. The interesting speculation is that if the antiparallel cross section continues to fade away high momentum transfer p-p scattering may end up occuring only in spin parallel states.

As mentioned earlier, we have also been studying the energy dependence of the elastic p-p spin-spin interaction at 90° CM, at the maximum $P_\perp^2$ for a given energy. Our data are from two runs, the first with beam momenta

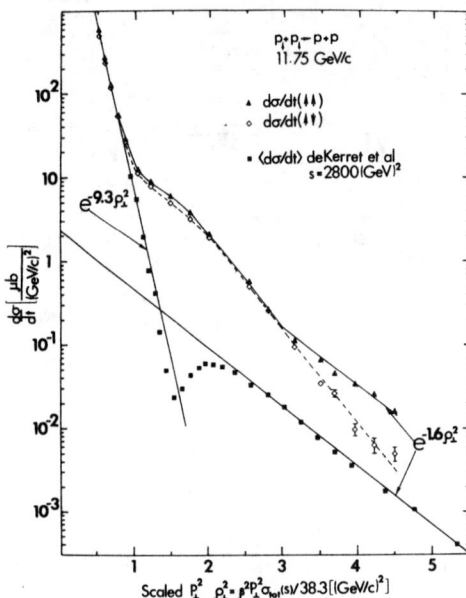

Fig. 24. Differential cross section for parallel and antiparallel initial spin states at 11.75 GeV/c. Also shown is spin-averaged ISR data, Ref. 31. The momentum transfer squared variable $\rho_\perp^2$ is discussed in Ref. 32.

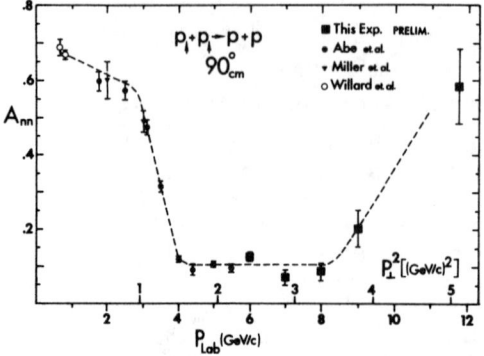

Fig. 25. $A_{nn}$ at 90° CM as a function of $P_\perp^2$ and $P_{LAB}$.

between 1.5 and 5.5 GeV/c[33] and the second up to 11.75 GeV/c.

As indicated before, A at 90° CM should be zero and our data are typically consistent with this within the errors of a percent or so.

Our 90° CM values of $A_{nn}$ are presented in Fig. 25, along with those of Miller et al.[23] discussed earlier and some new Los Alamos results at low momentum, presented at the conference by Willard. The large value of $A_{nn}$ at low momenta and the sharp drop which occurs at a laboratory momentum of ~3.5GeV/c ($P_\perp^2 \sim 1.25 (\text{GeV}/c)^2$) were clearly evident from the earlier Miller et al. data. We add an improvement in statistics and show evidence for a flattening out at the $A_{nn} \sim .1$ value above $P_{LAB} \sim 4$ GeV/c. $A_{nn}$ increases sharply at a laboratory momentum of 9 GeV/c which corresponds to $P_\perp^2$ of $3.8(\text{GeV}/C)^2$, close to the $P_\perp^2$ where we observed $A_{nn}$ increasing abruptly at 11.75 GeV/c; the first sign of a rise there was at $P_\perp^2 = 3.6(\text{GeV}/c)^2$. So at these close-by energies at least, the rapid increase of the spin parallel cross section over the antiparallel is occuring at

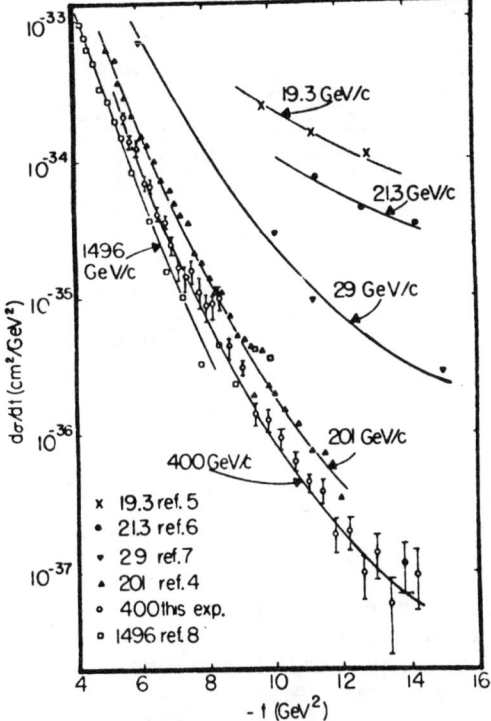

Fig. 26. Differential cross section for pp elastic scattering. Plot from Ref. 34.

similar transverse momenta. Perhaps this behavior might continue to higher energies.

To put the data in some perspective, I have to note that our momentum transfer range is small compared to the p-p spin-averaged data at higher energy. The recent results of the McGill, Northeastern, Cornell collaboration, Conetti et al.[34], at Fermilab along with the ISR and lower energy data are shown in Fig. 26. At our $P_\perp^2 \sim 5$ GeV/c$^2$ we are at the left edge of this graph. It is evident that in terms of t ($\sim P_\perp^2$ at high energies) the high momentum transfer differential cross sections have not reached an asymptotic value, and it would be a bit foolhardy to predict the spin dependences we see here will continue and that the parallel cross sections will dominate at higher energy and momentum transfers. But it is clear from the one and two spin experiments that there are dramatic spin dependences at high transverse momentum and continued investigation of these phenomena should produce further exciting results.

I would like to thank the authors who have kindly provided me with their data before publication. I also wish to acknowledge the work of the other members of our group in getting our recent data available for this conference. I want to thank Dr. Peter Hansen for his aid with some of the figures and for his helpful comments.

# REFERENCES

1. For earlier reviews and references see: H.A. Neal, Proceedings of the Conference on High Energy Physics with Polarized Beams and Targets, Argonne, 1976, M.L. Marshak, ed. (AIP, New York, 1976) p. 3; H. Spinka, Proceedings of Particles and Fields Conference, APS, DPF, Argonne, 1977, P.A. Schreiner, G.H. Thomas, A.B. Wicklund, eds. (AIP, New York, 1977) p. 167; A.D. Krisch, ibid. p. 243.
2. Convention for Spin Parameters, Proceedings of Conference on Higher Energy Polarized Proton Beams, Ann Arbor, 1977, A.D. Krisch and A.J. Salthouse, eds. (AIP, New York, 1977) p. 142.
3. L. Wolfenstein and J. Ashkin, Phys. Rev. $\underline{85}$, 947 (1952); M. Borghini et al., Phys. Rev. D17, 24 (1978).
4. K. Heller et al., Phys. Rev. Lett. $\underline{41}$, 607 (1978).
5. G. Bunce et al., Phys. Rev. Lett. $\underline{36}$, 1113 (1976).
6. A.S. Carroll et al., Phys. Rev. $\underline{D18}$, 619 (1978).
7. G. Baranko et al., XIX International Conference on High Energy Physics, 1978, Tokyo, Japan.
8. R.D. Klem et al., Phys. Rev. Lett. $\underline{36}$, 929 (1976).
9. L. Dick et al., Phys. Lett. $\underline{57B}$, 93 (1975).
10. D. Aschman et al., (to be published.)
11. J. Antille et al., XIX International Conference on High Energy Physics, 1978, Tokyo, Japan.
12. A.B. Wicklund et al., Phys. Rev. $\underline{D17}$, 1197 (1978), and private communication.
13. R. Diebold et al., Phys. Rev. Lett. $\underline{35}$, 632 (1975).
14. M.L. Marshak et al., Ref. 7 and private communication.
15. D.G. Crabb et al., Nucl. Phys. $\underline{B121}$, 231 (1977), and private communication.
16. A. Gaidot et al., Phys. Lett. $\underline{61B}$, 103 (1976).
17. I.P. Auer et al., Phys. Rev. Lett. $\underline{39}$, 313 (1977).
18. M.D. Corcoran et al., Phys. Rev. Lett. $\underline{40}$, 1113 (1978).
19. J. Pumplin and G.L. Kane, Phys. Rev. $\underline{D11}$, 1183 (1975).
20. J.H. Snyder et al., Phys. Rev. Lett. $\underline{41}$, 781 (1978).
21. G.L. Kane and A. Seidl, Rev. Mod. Phys. $\underline{48}$, 309 (1976).
22. G. Fidecaro et al., Phys. Lett. $\underline{36B}$, 369 (1978).
23. D. Miller et al., Phys. Rev. $\underline{D16}$, 2016 (1977).
24. I.P. Auer et al., Phys. Rev. Lett. $\underline{37}$, 1727 (1976).
25. A. Yokosawa et al., XIX International Conference on High Energy Physics, 1978, Tokyo, Japan, and H. Spinka, private communication.
26. K. Abe et al., Phys. Lett. $\underline{63B}$, 239 (1976).
27. J.V. Allaby et al., Nucl. Phys. $\underline{B52}$, 316 (1973).
28. H.E. Miettinen et al., Phys. Rev. $\underline{D16}$, 549 (1977).
29. J.R. O'Fallon et al., Phys. Rev. Lett. $\underline{39}$, 733 (1977).

30. D.G. Crabb et al., Phys. Rev. Lett. (to be published).
31. H. DeKerret et al., Phys. Lett. 62B, 363 (1976); 68B, 374 (1977).
32. A.D. Krisch, Phys. Lett. 44B, 71 (1973); P.H. Hansen and A.D. Krisch, Phys. Rev. D15, 3287 (1977).
33. A. Lin et al., Phys. Lett. 74B, 273 (1978).
34. S. Conetti et al., Phys. Rev. Lett. 41, 924 (1978).

## POLARIZATION IN INCLUSIVE LAMBDA PRODUCTION

K. Heller
University of Minnesota
Minneapolis, Minnesota, 55455

### ABSTRACT

At high energies the reaction $p + N \rightarrow \Lambda^0 + X$ yields $\Lambda^0$s which are strongly polarized. The behavior of this polarization should give information about quark production mechanisms. The inclusive $\Lambda^0$ polarization is also compared with $\bar{\Lambda}^0$, $\Xi^0$, and proton polarization in the fragmentation region of the incident proton at Fermilab energies.

### INTRODUCTION

The neutral hyperon beam collaboration at Fermilab is engaged in a program using polarization as a probe of the high energy production mechanisms. In this talk I will review results which have already been published and present some new data which are still preliminary. The people who have contributed to the acquisition, analysis, and understanding of this data are: R. Grobel, R. Handler, R. March, P. Martin*, L. Pondrom, M. Sheaff, C. Wilkenson: University of Wisconsin; T. Devlin, B. Edelman**, R. Edwards***, J. Norem****, L. Schachinger*****, P. Skubic+: Rutgers University; P. Cox, J. Dworkin, O. Overseth: University of Michigan; G. Bunce++, P. Yamin+++: Brookhaven National Laboratory; K. Heller+: University of Minnesota.

About three years ago this group reported that $\Lambda^0$ hyperons produced by 300 GeV protons in the interaction $p + Be \rightarrow \Lambda^0 + X$ were strongly polarized, as shown in Figure 1.[1] This was the first

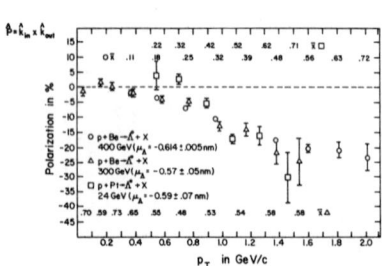

Figure 1: Polarization of $\Lambda^0$ for 3 incident proton energies plotted as a function of $p_T$. The average value of $x = P_L^*/(\sqrt{s}/2)$ is given for each point. The magnetic moment, in nuclear magnetons, $\mu_\Lambda$, for each data set is also shown.

evidence that all spin effects do not die off with energy as had been expected. Predictions available at the time were based on the triple Regge model.[2,3] These predictions called for inclusive $\Lambda^0$ polarization to be zero and the polarizations of other inclusively produced baryons to decrease with energy. In spite of its success in dealing with production cross-sections, the large $\Lambda^0$ polarization observed pointed out the failure, or at least irrelevancy, of this model.

Lambda production is an ideal tool to probe the structure of the strong interaction at high energies. Experimentally, its primary mode $\Lambda^0 \to p\pi^-$ is easily detected with a conventional charged particle spectrometer (Figure 2). Protons produce $\Lambda^0$'s

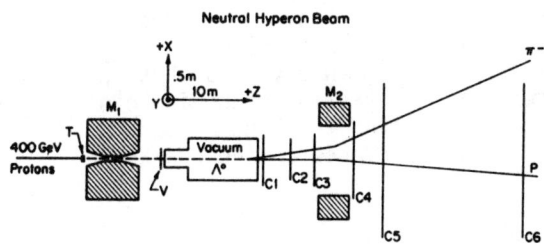

Figure 2: Fermilab neutral hyperon beam spectrometer. $M_1$ is the spin precession magnet, $M_2$ the spectrometer magnet, and $C_i$ are MWPC's. A typical $\Lambda^0$ decay is shown.

with a reasonably high probability so that the production cross-section can be measured over a large region of phase space. At the same time, the spin direction of the $\Lambda^0$ sample can be determined from the asymmetry of the daughter proton distribution in the parity violating decay. The analyzing power of the decay is 65%. There is thus no need for rescattering, which reduces statistics and has a very small analyzing power above a few GeV. The long laboratory lifetime, 7 meters at 100 GeV, means that $\Lambda^0$s can be passed through a conventional magnet to precess the spin. Because the $\Lambda^0$ is neutral, its spin can be precessed through any desired angle, up to about 150° in our magnet, without changing the direction of the particle. The ability to control the spin direction allows the elimination of most asymmetries caused by the apparatus. In addition the magnitude of the precession must give the particle's magnetic moment and is thus a powerful test of the measurement.

Theoretically, lambda production by protons can be approached from the quark viewpoint. In the interaction to be investigated one and only one valence quark has changed. An up valence quark in the proton has been exchanged for a strange quark, giving rise to a $\Lambda^0$ in the framentation region of the proton (Figure 3).

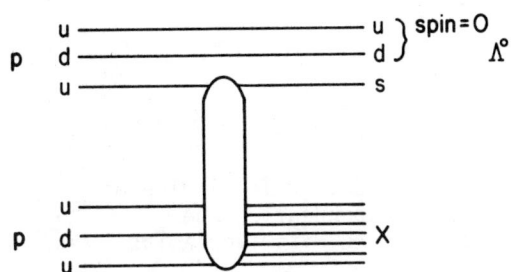

Figure 3: Quark "black box" diagram showing an incident proton (u,d,u valence quarks) becoming a lambda (u, d, s valence quarks) after an interaction with a target proton.

A strange valence quark did not exist before the interaction, so it must have been "created" by some mechanism during the interaction. The $\Lambda^0$ is not simply a rearrangement of the already existing valence quarks. Because the $\Lambda^0$ is an isosinglet, the symmetry of the standard s-wave quark wave function requires that the spin of the $\Lambda^0$ be just that of the s quark. These wave functions have been successfully used to predict baryon magnetic moments if the quarks are given masses consistent with hadron mass splittings.[4] If the $\Lambda^0$ emerges polarized, which it does, the "created" s quark must be polarized. The basic mechanism involved is thus spin dependent. The properties of the quark interaction mechanism can then be systematically investigated by finding the kinematic dependence of the polarization and its dependence on the quarks involved.

## EXPERIMENTAL RESULTS

The published polarization properties of high energy $\Lambda^0$s produced by protons[1,4,5] are:
1. The $\Lambda^0$ polarization increases monotonically with $p_t$ to over 20% at $p_t$ = 2 GeV/c. Because of the $\Sigma^0 \to \Lambda^0 \gamma$ contribution to the sample which dilutes the polarization, this is a lower limit of the $\Lambda^0$ polarization.
2. The $\Lambda^0$ polarization is perpendicular to the production plane in the direction $-(\hat{k}_{in} \times \hat{k}_{out})$.
3. The $\Lambda^0$ polarization appears to be energy independent between proton energies of 24 and 400 GeV (Figure 1).
4. The parity violating components of $\Lambda^0$ polarization are both consistant with zero to a part in $10^{-3}$.
The only other high energy inclusive polarization data published show that $\overline{\Lambda}^0$s produced by protons are unpolarized over the same kinematic region in which $\Lambda^0$s show substantial polarization (Figure 4).[6]

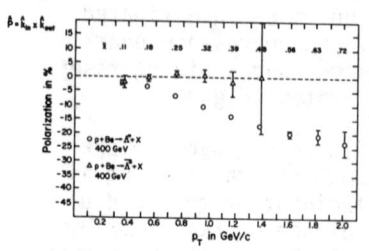

Figure 4: Polarization of $\overline{\Lambda}^0$ and $\Lambda^0$ produced by 400 GeV incident protons. The data for both sets of points were taken at the same time. At a given $p_T$, the $\overline{\Lambda}^0$ point represents the same average value of x as the corresponding $\Lambda^0$ point.

In addition our group has preliminary data which indicate:
5. The $\Lambda^0$ polarization is probably also dependent on the longitudinal momentum (Figure 5). Increasing with increasing $X = P_L^* / (\sqrt{s}/2)$.

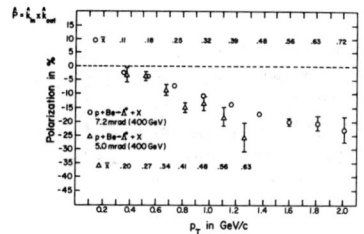

Figure 5: The magnitude of the $\Lambda^0$ polarization at 5 mrad appears systematically greater than that at 7.2 mrad. This behavior indicates that at a given $p_T$ the polarization increases with increasing x.

6. The $\Lambda^0$ polarization from $H_2$ is the same as from Be (Figure 6), indicating that the production mechanism is independent of target.

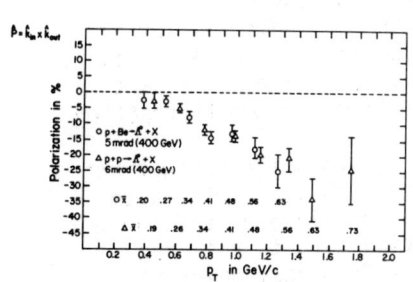

Figure 6: Polarization of $\Lambda^0$ produced by 400 GeV protons from a Be target at 5 mrad and as $H_2$ target at 6 mrad.

The $H_2$ data are still being analyzed by R. Grobel. Data exist at many incident angles so the analysis should also determine the dependence of the polarization on both transverse and longitudinal kinematic variables.

In another experiment our group examined the polarization of $\Lambda^0$s produced at $0°$ by a 320 GeV secondary proton beam. The secondary proton beam was produced at 3 mrad from a primary proton beam of 400 GeV on a Be target. The reaction was
p + Be → p + X
↳ + Be → $\Lambda^0$ + X.
An observed $\Lambda^0$ polarization would mean the produced protons were polarized. A null result on the other hand would not exclude the possibility of proton polarization. The data in Figure 7 show no $\Lambda^0$ polarization. This is inconclusive but consistent with the results of the Indiana group to be reported at this conference.[7] Their more direct measurement of inclusively produced protons at $p_t$ = 1 GeV/c shows protons are not polarized to the same extent as the $\Lambda^0$.

Our group has also taken data of $\Xi^0$ production by protons, detecting both the $\Lambda^0$ and the $\pi^0$ (via $2\gamma$ decay). These data

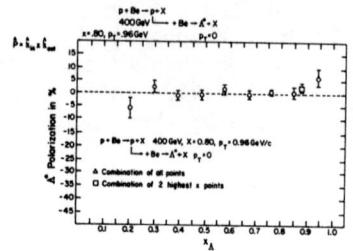

Figure 7: Polarization of $\Lambda^0$ produced by 320 GeV protons at 0 mrad as a function of x of the $\Lambda^0$. The $\Lambda^0$ polarization represents the product of the proton polarization and the spin transfer in $\Lambda^0$ production.

are undergoing preliminary analysis for polarization by P. Cox, and should yield about $10^5$ fully reconstructed $\Xi^0$s with $P_t$ above 1 GeV/c. These $\Xi^0$s will then be used to determine $P_{\Xi^0}$ from p + Be → $\Xi^0$ + X, and the magnetic moment of the $\Xi^0$.

Another approach to investigating the $\Xi^0$ polarization has been taken by our group. In our previous data on $\Lambda^0$ production, most of the $\Lambda^0$s reconstructed by our spectrometer point back precisely to the production target (Figure 8). A small fraction

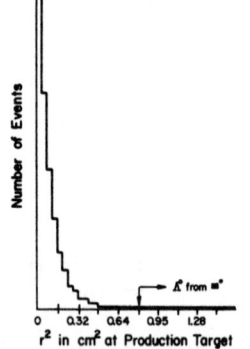

Figure 8: When the reconstructed $\Lambda^0$ trajectory is extrapolated back to the plane containing the production target, r is the distance from that point of intersection to the target center. The radius of the target is 3 mm.

do not. These large $r^2$ $\Lambda^0$s were analyzed by G. Bunce and O. Overseth to determine if they could be daughter $\Lambda^0$s from $\Xi^0 \to \Lambda^0 \pi^0$. In that decay the $\Lambda^0$ retains a substantial fraction of the $\Xi^0$ polarization in the laboratory frame. The vertex distribution of the large $r^2$ $\Lambda^0$s (Figure 9) is consistent with a parent-daughter decay such as $\Xi^0 \to \Lambda^0 \pi^0$, $\Lambda^0 \to p\pi^-$. This analysis gives the preliminary result that p + Be → $\Xi^0$ + X yields polarized $\Xi^0$s. The polarization is consistent in magnitude and direction with that of the $\Lambda^0$s, $P_{\Xi^0}$ = -0.09 ± 0.02 at $p_T$ = 0.73 GeV/c, x = .26. These data give a magnetic moment of the $\Xi^0$ of -1.20 ± 0.06 nm.

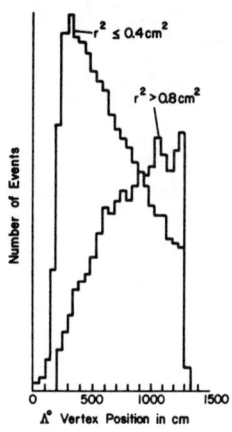

Figure 9: Vertex distribution for $\Lambda^0$s which point back to the production target (small $r^2$) and vertex distribution for $\Lambda^0$s which do not point back to the production target (large $r^2$).

## CONCLUSIONS

Lambdas and $\Xi^0$s produced by protons are polarized at $p_T > 1$ GeV/c while $\bar{\Lambda}^0$s and protons are not (Figure 10). Both

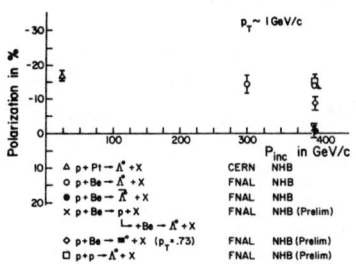

Figure 10: Summary of high energy polarization from inclusive production of baryons at $p_T \sim 1$ GeV/c by protons.

polarized particles are baryons, presumably manifestations of the incident proton with a quark change. The $\Lambda^0$ has nothing in common with the incident proton and the proton does not necessarily involve the creation of any new valence quarks.

A theory of quark interactions should explain the large polarizations observed. QCD gives promise of being such a theory. However, an analysis of Kane, Pumplin and Repko, which will be discussed at this conference, shows that a QCD perturbation expansion would not allow the large observed polarizations.[8] They conclude that if the effect does not die off at transverse momenta where the theory is expected to be valid, either QCD or their analysis method is incorrect. An experiment by our group to extend the measurement into that region will be run at Fermilab within a year.

## REFERENCES

\* Now at Lawrence Berkeley Laboratory
\*\* Now at Ford Motor Company
\*\*\*Now at Bell Laboratory
\*\*\*\*Now at Argonne National Laboratory
\*\*\*\*\*Now at University of Chicago
+ Previously at University of Michigan
++ Previously at University of Wisconsin
+++Previously at Rutgers University

1. G. Bunce et al., Phys. Rev. Lett. $\underline{36}$, 1113, 1976.
2. R. Field, Cat. Tech. report CALT-69-459.
3. F. Paige and D. Sidhu, Phys. Rev. $\underline{D14}$, 2307. 1976.
4. A. deRujula, H. Georgi, and S. Glashow, Phys. Rev. $\underline{D12}$, 147, 1975.
5. K. Heller et al., Phys. Lett. $\underline{68B}$, 480, 1977.
6. K. Heller et al., Phys. Rev. Lett., $\underline{41}$, 607, 1978.
7. G. Baranko et al., University of Indiana, preprint COO-2009-137.
8. G. Kane et al., University of Michigan, preprint UM HE 78-29.

PROTON POLARIZATION IN INCLUSIVE PROCESSES
AT 100, 200, 300, and 400 GeV

R. Polvado, G. Baranko, S. Ems, S. Gray, R. Lepore
B. Martin, H. Neal, H. Ogren, D. Rust, P. Smith, and G. Walters
Department of Physics, Indiana University, Bloomington, IN

## ABSTRACT

We have measured the polarization of the out going proton in the following high energy interactions; $p + C \rightarrow p\uparrow + x$ and $p + p \rightarrow p\uparrow + x$. Our measurements span the $p_T$ range $.5 \rightarrow 1.5$ GeV/c and the $x_f$ range $-.7 \rightarrow -.9$. Measurements were made at beam energies of 100, 200, 300, 400 GeV at the internal target area at Fermi National Accelerator Laboratory.

## DESCRIPTION OF EXPERIMENT

This series of experiments was carried out at the internal target area at Fermi National Accelerator Laboratory. Data was taken using both the warm hydrogen jet operated by the internal target group and a rotating carbon target consisting of 20μ filaments. The integrated luminosity for a single burst of the warm jet was approximately $10^{34}$ cm$^{-2}$ with $2 \times 10^{13}$ protons/pulse. Luminosity 4 to 5 times higher could be attained with the carbon target.

A complete description of the internal target spectrometer and the polarimeter have been given elsewhere[1]. The internal target spectrometer defined the incoming proton trajectory and selected the momentum and production angle. The acceptance of the spectrometer was vertically = ±2.88°, horizontally = ±.57°.

The polarimeter shown in Fig. 1 was attached to the end of the spectrometer arm. The incoming proton scattered in a two inch carbon block and was measured by a set of multiwire proportional chambers (PC1 - PC8). The entire polarimeter could be rotated about its axis allowing measurements to be taken at 0° and 180° in order to eliminate first order instrumental asymmetries. The momentum was restricted by the polarimeter chambers to $\Delta p/p = \pm 5\%$. Using a spectrometer scintillator, H1, and T2 for a time of flight measurement, a clean separation between pions and protons could be obtained.

The asymmetry ($\varepsilon$) of the projected scattering angle ($\theta x$) between 6° and 22° was then used to determine the polarization.

The polarization (P) of the incoming protons was calculated using: $-\varepsilon = PA$. A is the analyzing power of the carbon block, determined by an earlier experiment[2]. The minus sign is included to conform to standard elastic scattering sign conventions for polarization.

ISSN: 0094-243X/79/510549-05$1.50 Copyright 1979 American Institute of Physics

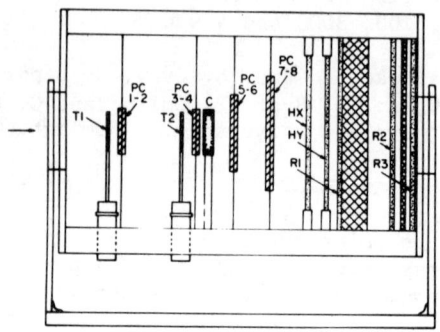

Figure 1

## DATA COLLECTION - RESULTS

The running points at which data was collected are listed in Table 1. At $p_T$ = 1.5 GeV/c all the data was taken using the rotating carbon target. Hydrogen data was taken at all other points. At $p_T$ = 1.0 GeV/c, $x_f$ = -.9 additional carbon data was collected.

The $p + p \rightarrow p\uparrow + X$ polarization data at $p_T$ = .5 GeV/c and $p_T$ = 1.0 GeV/c is shown in Figures 2 and 3. In Fig. 4 all the polarization data for $p + C \rightarrow p\uparrow + X$ is shown.

In general the polarization of the inclusive produced proton shows very little dependence on $x_f$ or $E_{beam}$. At $p_T$ = .5 GeV/c we observe a polarization of ~ +.06 from hydrogen. The corresponding hydrogen data at $p_T$ = 1.0 GeV/c exhibits no appreciable polarization. The average polarization over all $x_f$ values, and all beam energies at $p_T$ = 1.0 is $\bar{P}$ = .011 ± .005 .

The data taken with the rotating carbon target at $p_T$ = 1.0, $x_f$ = -.9 shows a negative polarization of P = -.077 ± .010 . This polarization data at $p_T$ = 1.0 GeV/c is entirely consistent with our previously reported measurements from carbon[3].

The $p_T$ = 1.5 GeV/c carbon data exhibits a much smaller polarization. However, there is some suggestion of the negative values seen at $p_T$ = 1.0 GeV/c.

TABLE 1

| $x_f$ | $P_t$ GeV/c | θ | P total GeV/c | $M_x^2$ (GeV$^2$) 400 | 300 | 200 | 100 = ELAB (GeV) |
|---|---|---|---|---|---|---|---|
| -.9 | .5  | 64.1° | .56  | 75  | 57  | 37  | 19 |
| -.9 | 1.0 | 55.8° | 1.21 | 75  | 57  | 37  | 19 |
| -.9 | 1.5 | 46.7° | 2.07 | 75  | 57  | 37  | 19 |
| -.8 | .5  | 53.3° | .63  | 149 | 112 | 75  | 37 |
| -.8 | 1.0 | 49.1° | 1.33 | 149 | 112 | 75  | 37 |
| -.8 | 1.5 | 41.7° | 2.27 | 149 | 112 | 75  | 37 |
| -.7 | .5  | 43.6° | .73  | 225 | 168 | 112 | 56 |
| -.7 | 1.0 | 42.6° | 1.43 | 225 | 168 | 112 | 56 |

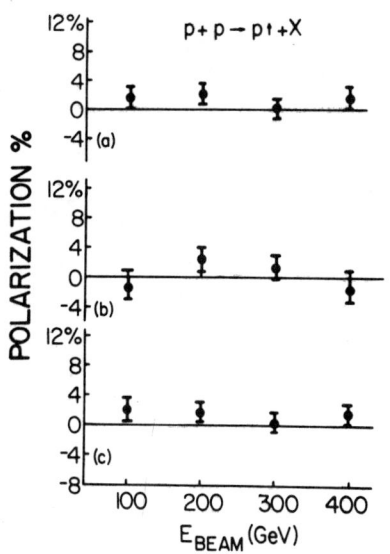
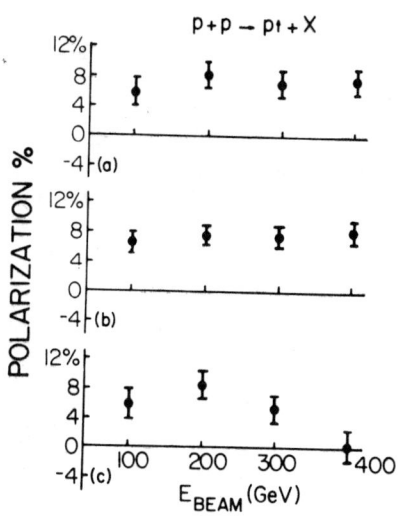

Figure 2

(a) $p_T = 1.0$ GeV/c, $x_f = -0.7$
(b) $p_T = 1.0$ GeV/c, $x_f = -0.8$
(c) $p_T = 1.0$ GeV/c, $x_f = -0.9$

Figure 3

(a) $p_T = 0.5$ GeV/c, $x_f = -0.7$
(b) $p_T = 0.5$ GeV/c, $x_f = -0.8$
(c) $p_T = 0.5$ GeV/c, $x_f = -0.9$

## DISCUSSION OF RESULTS

We can use our polarization data from hydrogen to measure the ratio of the spin dependent to spin independent amplitudes for the process $p + p \rightarrow p\uparrow + X$. A similar analysis has been done for p p elastic polarization[4] and $\Lambda$ inclusive polarization[5].

Quite generally we can express the polarization in an inclusive process as

$$P = \frac{2\,\text{Im}\, f'g^*}{|g|^2 + |f'|^2}$$

The spin flip part, $f'$, can further be written with a kinematic term $\sin\theta_{cm}$ explicitly taken out, i.e. $f' = \sin\theta_{cm}\, f$. The expression for the polarization can then be written as

$$P = \frac{2\, f/g\, \sin\theta_{cm}}{1 + \sin^2\theta_{cm}\, (f/g)^2}$$

$f_\perp$ is the spin flip amplitude component orthogonal to, $g$, the spin non - flip amplitude. Since $\sin^2\theta_{cm} \leq .03$ for all of our run points we will neglect the $\sin^2\theta_{cm}$ term and write

$$f_\perp/g \simeq \frac{P}{2\sin\theta_{cm}}$$

In Figures 5 and 6, we show these calculated ratios, at $p_T = .5$ GeV/c. The ratio $f_\perp/g$ is as large as $.8 \pm .2$, indicating the existence of an appreciable spin dependent amplitude. At $p_T = 1.0$ GeV/c however, the ratio $f_\perp/g$ is always less than $.1 \pm .1$.

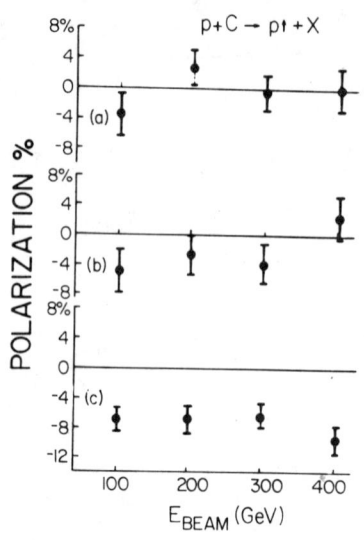
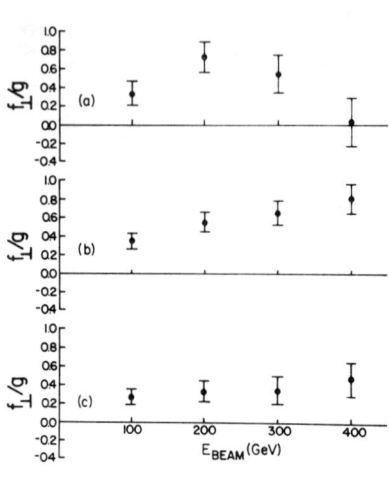

Figure 4

Figure 5

(a) $p_T = 1.5$ GeV/c, $x_f = -0.8$
(b) $p_T = 1.5$ GeV/c, $x_f = -0.9$
(c) $p_T = 1.0$ GeV/c, $x_f = -0.9$

(a) $p_T = 0.5$ GeV/c, $x_f = -0.9$
(b) $p_T = 0.5$ GeV/c, $x_f = -0.8$
(c) $p_T = 0.5$ GeV/c, $x_f = -0.7$

We can also test the predictions of the triple regge model for polarization in inclusive reactions. Using the triple regge formalism of Field and Fox[6], we consider only the contribution of the PRR amplitude to polarization in $\bar{p} + p \to p\uparrow + X$. The observed polarization will be proportional to

$$P = k \frac{d^3\sigma/dp^3 \text{ (PRR)}}{d^3\sigma/dp^3 \text{ (all terms)}}$$

at a constant $x_f$ and constant $p_T$. We then expect

$$P = \frac{k}{s^{1/2}} \quad (x_f, p_T \text{ fixed}).$$

This prediction is not consistent with the approximate s independence we observe in our polarization data at $p_T = .5$ GeV/c.

Since we measure no appreciable polarization for $p + p \to p\uparrow + X$ at $x_f = -.9$, $p_T = 1.0$ GeV/c we are forced to look for nuclear or secondary scattering effects to explain our results for $p + C \to p\uparrow + X$. We have attempted to estimate the polarization from rescattering in the nucleus using a simple model. We assume that all inclusively produced protons rescatter <u>once</u> in the carbon nucleus. We use the measured proton-carbon scattering cross-section and analyzing power[7] at $p = 1.2$ GeV/c to calculate the net polarization of the protons entering the spectrometer. This polarization is not zero since the inclusive cross-section is strongly peaked forward and cut off at the elastic limit. Using this technique we have estimated a rescattering polarization in our carbon target of $-.075$ at $p_T = 1.0$ GeV/c, consistent with our results.

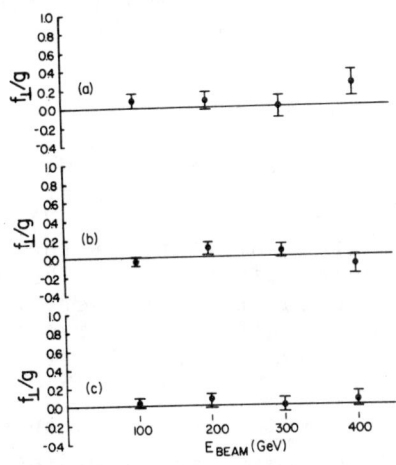

Figure 6

(a) $p_T = 1.0$ GeV/c, $x_f = -0.9$
(b) $p_T = 1.0$ GeV/c, $x_f = -0.8$
(c) $p_T = 1.0$ GeV/c, $x_f = -0.7$

## REFERENCES

1. Corcoran et al., Phys. Rev. Lett. <u>40</u>, 1113 (1978).
2. G. W. Bryant, H. A. Neal, D. R. Rust, "Proton - Carbon Analyzing Measurements for Proton Kinetic Energies Between .150 GeV and .440 GeV", Indiana University Internal Report COO-2009-102.
3. Corcoran et al., Indiana University Preprint IUHEE #10. This analysis used $+ \varepsilon =$ AP for polarization convention.
4. T. Devlin et al., To Be Published.
5. D. R. Rust et al., Phys. Lett., <u>58B</u>, 1 (1975).
6. Field and Fox Nucl. Phys., <u>B80</u>, 367 (1974).
7. Aebischer et al., Nucl. Inst. and Meth., <u>124</u>, 49 (1975).

# TRANSVERSE QUARK POLARIZATION IN LARGE $P_T$ REACTIONS AND $e^+e^-$ JETS

J. Pumplin
Michigan State University, East Lansing, MI 48824

## ABSTRACT

The polarization of quarks which are scattered at large $P_T$ in hadronic interactions or in lepto-production, or which are produced in $e^+e^-$ annihilations at high energy can be computed in quantum chromodynamics (QCD). The predicted polarizations are small, of order $\alpha_s\, m_{quark}/\sqrt{Q^2}$. This leads to a prediction that asymmetries for observable particles will be small at high $Q^2$. The prediction can be used to test QCD.

## INTRODUCTION

An extremely attractive working hypothesis is that (1) the strong interactions are manifestations of an underlying renormalizable field theory; (2) the field theory is QCD; and (3) the solutions of the field theory can be approximated reasonably well for "hard" processes by means of perturbation theory. The perturbation expansion involves a dimensionless effective coupling constant $\alpha_s$, which is analogous to the electromagnetic fine-structure constant, but which varies with $Q^2$. At large $Q^2 \approx$ large momentum transfer $\approx$ short distances, $\alpha_s$ becomes small $\approx 1/\ln Q^2$ according to "asymptotic freedom". Thus the perturbation expansion should be useful for "hard" processes.

As an example of the perturbation idea, one expects high $P_T$ inclusive reactions to be governed by underlying processes such as the one illustrated in Fig. 1.

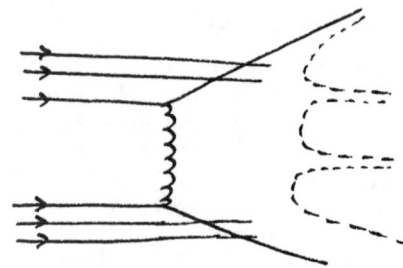

Fig. 1. One gluon exchange ($\alpha_s$) contribution to a high $P_T$ interaction.

Questions involving the initial transverse momenta of the quarks, the "dressing" of the scattered quarks into observed hadrons, and the contributions from other diagrams, which at currently available energies convert the $d\sigma/dp_T^2 \sim p_T^{-4}/\ln p_T^2$ of Fig. 1 into the observed $\approx p_T^{-8}$ are a fascinating subject. Although the dust has not yet settled on those issues, let us nevertheless proceed to ask about the next level of refinement: polarization.

## QUARK-QUARK SCATTERING

From a quantum mechanical point of view, polarization effects result from an interference between spin flip and nonflip amplitudes of the form

$$P = \frac{2 \, \mathrm{Im} \, M^*_{\mathrm{flip}} M_{\mathrm{nonflip}}}{|M_{\mathrm{flip}}|^2 + |M_{\mathrm{nonflip}}|^2} \tag{1}$$

The first order quark-quark scattering diagram (Fig. 1) is <u>real</u>, and hence by itself produces no polarization. The second order diagrams are shown in Fig. 2.

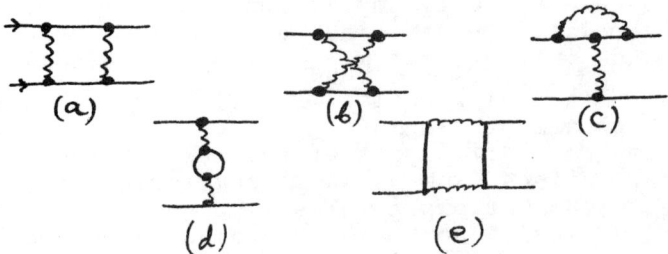

Fig. 2. Two gluon ($\alpha_s^2$) contributions to quark-quark scattering.

The revelant one is the "box" diagram 2a. It has an imaginary part due to the "physical" qq intermediate state, and can, therefore interfere with the single gluon exchange term to produce polarization. Figs. 2b-d can be disregarded because they are real. Fig. 2e can be disregarded because it is smaller than the others by a factor $m^2/s$, since it involves spin 1/2 exchanges instead of spin 1.

In order for the calculation to be meanginful, it must be "infrared-finite", i.e. well-behaved in the limit of zero gluon mass. One way to deal with that issue is as follows: Sterman and Weinberg[1] suggest that any <u>physical</u>

process will be finite in QCD. The polarization is an asymmetry in counting rates. It is free of subtleties involving the thresholds for detecting soft gluons - which are analogous to the subtleties which appear in radiative corrections to QED - and should therefore be infrared finite from the start. This is indeed the case, as can be seen directly as follows. The amplitude for the box diagram is an integral over the four-momentum q of one of the gluons. The integrand contains four propagators and a factor F(q) which results from the spins. If we write F(q) = (F(q)-F(0)) + F(0), the first term is zero at q = 0 and its integral over q is therefore finite in spite of the $1/q^2$ gluon propagator. The second term is <u>not</u> infrared finite, but it has the same spin dependence as a single gluon exchange, so it doesn't contribute to polarization.

The qualitative behavior of the predicted polarization can be determined without explicitly evaluating any integrals. The key point is the QCD assumption of a pure $\gamma_\mu$ coupling at the quark-quark-gluon vertex. This coupling makes the helicity flip vertex for single gluon exchange proportional to $m_{quark}/\sqrt{s}$. (This fact will no doubt be familiar to some readers by way of the connection between helicity amplitudes and the A and B amplitudes of $\pi N$ elastic scattering; and to other readers from $\gamma_5$ invariance in the limit $m_{quark} \rightarrow 0$.) The dominant contribution to the polarization [Eq. (1)] must come from the real nonflip part of single gluon exchange interfering with the imaginary flip part of the box diagram. This imaginary part arises from on-shell scattering of the quarks, according to the Cutkowski rules. To get an overall flip amplitude, at least one of the two gluon exchanges must involve helicity flip, and therefore fall like $m_{quark}/\sqrt{s}$. Hence

$$P \sim \alpha_s m_{quark}/\sqrt{s} \qquad (2)$$

at high energy ($\sqrt{s}$) for the quark-quark scattering, and to leading order in $\alpha_s$. My collaborators and I have calculated the box diagram explicitly, and find the coefficient of $\alpha_s m_{quark}/\sqrt{s}$ in the polarization to be of order 1 as expected. (In the $m_{quark} \rightarrow 0$ limit, the helicity flip amplitude can easily be shown to vanish to all orders in $\alpha_s$.)

The predicted quark polarization in QCD is small. The dressing ("fragmentation") of the high $p_T$ quark into physical hadrons may dilute the polarization, but is not expected to enhance it. Thus we predict <u>observable</u> polarizations to be small in the region where <u>QCD</u> perturbation theory is applicable.

To test the prediction, one can study single particle inclusive reactions. A particularly attractive candidate is the $\rho^\circ$ whose $e^+e^-$ and $\mu^+\mu^-$ decays allow a good separation from background. The density matrix for the decay contains the spin information. A second attractive candidate is the $\Lambda^\circ$. The $\Lambda^\circ$ polarization is already known to be sizable at $P_T \sim 2$ GeV/c. The prediction that this polarization should drop to at most a few percent in the region where hard scattering models are appropriate is thus a non-trivial one! Another approach would be to study asymmetries involving the transverse momenta of the jet particles[2,3].

## ELECTRON-POSITRON INTERACTIONS

The polarization of the quarks in $e^+e^- \to q\bar{q}$ is also infrared finite and calculable in QCD. The relevant diagrams are shown in Fig. 3. They imply a leading term in polarization

$$P = \left(\frac{2}{3}\alpha_s\right) \frac{p\, m\, \sin(2\theta)}{p^2(1+\cos^2\theta) + 2m^2} \qquad (3)$$

where $p = (Q^2/4 - m^2)^{1/2}$ is the quark momentum.

Fig. 3. Lowest order diagrams which produce polarization in $e^+e^- \to q\bar{q}$.

Eq. (3) predicts that the quark polarization and hence the jet polarization will be small $\sim \alpha_s m/\sqrt{Q^2}$, just like Eq. (2).

In $e^+e^-$ collisions, one could study charm production for which $m/\sqrt{Q^2}$ is not small. For example, one could measure the density matrix of the spin 1 $D^*$. However, Eq. (3) implies a polarization of the quark of only a few percent even at the most favorable energies and angles, since $\alpha_s \simeq 0.2$. So, again the QCD prediction is that the polarization is small.

In the case of light (u, d, or s) quarks, the polarization predicted by Eq. (3) can be quite large near threshold because $\alpha_s$ is presumably rather large, eg. $\simeq 1$, at such small $Q^2$. The lowest-order perturbation theory result cannot be relied upon in this threshold region, but the qualitative expectation of sizable polarization is probably correct.

In hadronic processes, $q\bar{q}$ pairs in the sea can be polarized by a "vertex correction" effect which is similar to Fig. 3, with the photon replaced by a gluon. Thus it is plausible that sea quarks will be significantly polarized.

This argument does not necessarily explain, however, the large observed $\Lambda°$ polarization at $P_T \simeq 2\text{GeV}/c$. For example, suppose the $\Lambda°$ is produced by the fusion[4] of a polarized s-quark from the sea with a spin zero di-quark from the beam proton in a relative s-wave. Even with these simple (though not implausible) assumptions, the polarization of the sea quark may be lost in forming the $\Lambda°$, because the quark polarization is perpendicular to its production plane according to parity; but that plane is in general different from the $\Lambda$ production plane because the diquark system also carries transverse momentum. If $P_T$ of the $\Lambda$ comes mainly from the sea quark, this mechanism might account for the observed $\Lambda$ polarization, because the transverse polarization of the quark is preserved by the Lorentz transformation to the $\Lambda$ rest frame. At large $P_T$, this picture is consistent with our overall conclusion that polarization must become small at large $P_T$. For in that case, either the diquark system will carry most of the transverse momentum, or the sea quark will have large transverse momentum and hence be unpolarized.

This work was done in collaboration with W. Repko (Michigan State University) and G. Kane (University of Michigan)[3]. It was supported in part by the National Science Foundation.

## REFERENCES

1. G. Sterman and S. Weinberg, Phys. Rev. Letters **39**, 1436 (1977)
2. S-Y. Pi, R. Jaffe, and F. Low, Phys. Rev. Lett. **41**, 142 (1978).
3. G. Kane, J. Pumplin and W. Repko, preprint submitted for Publication to Phys. Rev. Letters.
4. E. Lehman, M. Pratap, J. Pumplin and J. Whitmore, Phys. Rev. <u>D</u>, to be published.

PRODUCTION OF DIMUONS
FROM HIGH ENERGY POLARIZED PROTON-PROTON COLLISIONS

John Ralston and Davison E. Soper
Institute of Theoretical Science, University of Oregon,
Eugene, Oregon 97403

ABSTRACT

The high energy process $p+p \to \mu^+ + \mu^- + X$ for dimuons with large invariant mass is studied in the case that the protons are polarized. In general, the cross-section $d\sigma/dQ^2 dy d\Omega$ is determined by nine structure functions, and several interesting correlations between the polarization direction and the observed cross-section are possible. In the Drell-Yan parton model for the process, the structure of the cross-section is much simpler: there are six linear relations among the structure functions. An experimental test of these relations would provide a stringent test of the Drell-Yan model.

---

It seems appropriate at this conference to look to the future and ask what one could learn by measuring high mass muon pair production with a polarized beam and target.

We have first investigated the problem from a general point of view, assuming only that the dimuon results from the decay of a virtual photon. We have considered the cross-section $d\sigma/d^4Q d\Omega$ at $\vec{Q}_T=0$ and this cross-section integrated over $\vec{Q}_T$. In each case, the cross-section is determined by nine structure functions - functions of $P_A \cdot Q$, $P_B \cdot Q$, and S. In this talk, I will discuss the more experimentally accessible case of the cross-section integrated over $\vec{Q}_T$. (A more complete version of this work will be published elsewhere[1].)

I will then turn to the predictions of the Drell-Yan parton model for the process[2], which has enjoyed considerable experimental success in the unpolarized case. The predicted cross-section depends on certain functions that describe the number and polarization of quarks in a polarized proton. A measurement of the cross-section would then help to pin down these functions, which are only partially known at present.

Equally important, the nine structure functions just mentioned are predicted to obey several relations that are independent of the parton distribution functions. The result of these relations is that several polarization asymmetries that are allowed by the general analyses are predicted to vanish by the Drell-Yan model. If any of these asymmetries were found to be significantly non-zero at large $Q^2$ and S, the Drell-Yan model would be ruled out. Unfortunately, the converse implication is not very credible: An experimentally observed zero polarization asymmetry could indicate not that the Drell-Yan model is correct, but only that the dynamical mechanism involved is so complicated that no information about the proton polarizations reaches the dimuon. If, however, experiment were to show non-zero effects where they are expected and null effects where they are predicted, that would provide important confirmation of the Drell-Yan

ISSN: 0094-243X/79/510559-04$1.50 Copyright 1979 American Institute of Physics

model.

Let us now turn to the results. The kinematics is shown in Fig. 1. We adopt the following notation:

$P_A^\mu$ = beam momentum; along +z axis.

$P_B^\mu$ = target momentum; along -z axis.

$Q^\mu$ = dimuon momentum;
$Q^\pm = 2^{-\frac{1}{2}}(Q^0 \pm Q^3)$.

$x_A = Q^+/P_A^+$.

$x_B = Q^-/P_A^-$.

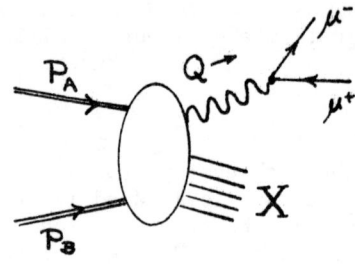

Figure 1

$\theta, \phi$ = polar angles of muon momentum in the dimuon rest frame obtained by setting $Q_z \to 0$ with a z boost, then $Q_T \to 0$ with a transverse boost.

$\lambda_A(\lambda_B)$ = average helicity of hadron A(B), normalized to $\lambda = \pm 1$ for a pure state.

$\vec{S}_A^T(\vec{S}_B^T)$ = average transverse spin of hadron A(B) (as measured in the CM frame), normalized to $|\vec{S}| = 1$ for a pure state.

$\phi_A(\phi_B)$ = azimuthal angles of $\vec{S}_A^T(\vec{S}_B^T)$.

Assuming only a one photon process as shown in Fig. 1, the cross-section integrated over $\vec{Q}_T$ can be expressed in terms of nine structure functions $\bar{W}_{0,0} \ldots, \bar{U}_{2,2}$, which are functions of S, $x_A$, and $x_B$:

$$\frac{d\sigma}{dQ^+ dQ^- d\Omega} = \frac{\alpha^2}{4Q^+Q^-S}(2\pi)^{-4}\{(2\bar{W}_{0,0} + \frac{1}{2}\lambda_A\lambda_B\bar{V}_{0,0}^L - 2\vec{S}_A^T \cdot \vec{S}_B^T \bar{V}_{0,0}^T)$$

$$+ (\frac{1}{3} - \cos^2\theta)(\bar{W}_{2,0} + \frac{1}{4}\lambda_A\lambda_B\bar{V}_{2,0}^L - \vec{S}_A^T \cdot \vec{S}_B^T \bar{V}_{2,0}^T$$

$$+ 2\cos\theta\sin\theta\cos(\phi-\phi_A) |\vec{S}_A^T|\lambda_B \bar{U}_{2,1}^A$$

$$+ 2\cos\theta\sin\theta\cos(\phi-\phi_B) |\vec{S}_B^T|\lambda_A \bar{U}_{2,1}^B$$

$$+ \sin^2\theta\cos(2\phi-\phi_A-\phi_B) |\vec{S}_A^T||\vec{S}_B^T| \bar{U}_{2,2}\}. \qquad (1)$$

In the Drell-Yan model one assumes that the incoming hadron beams can be replaced by beams of free quarks and antiquarks, two of which annihilate to form the virtual photon (see Fig. 2). We define

a = parton species label; summed over all flavors and colors of quarks and antiquarks.

$\mathcal{P}_{a/A}(x_A)dx_A$ = number of quarks of type a with momentum $k_a^+ = x_A P_A^+$ in hadron A.

$\mathcal{P}_{\bar{a}/B}(x_B)dx_B$ = number of quarks of type $\bar{a}$ with momentum $k_{\bar{a}}^+ = x_B P_B^-$ in hadron B.

$h_{a/A}^L(x)\lambda_A$ = average helicity of quarks in hadron A (similarly for B).

$h_{a/A}^T(x)\vec{S}_A^T$ = average transverse spin of quarks in hadron A (similarly for B).

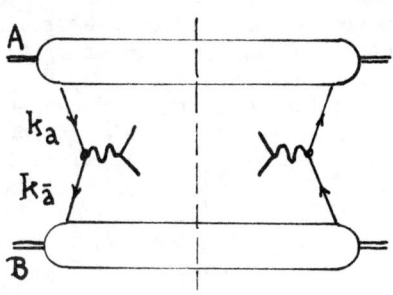

Figure 2

A straightforward calculation leads to

$$\frac{d\sigma}{dQ^+ dQ^- d\Omega} = \frac{\alpha^2}{4Q^+Q^-s} \sum_a e_a^2 \mathcal{P}_{a/A}(x_A) \mathcal{P}_{\bar{a}/B}(x_B)$$

$$\{(1+\cos^2\theta)(1-\lambda_A\lambda_B h_{a/A}^L(x_A)h_{\bar{a}/B}^L(x_B))$$

$$+ \sin^2\theta \cos(2\phi - \phi_A - \phi_B)|S_A^T||S_B^T|h_{a/A}^T(x_A)h_{\bar{a}/B}^T(x_B)\}$$

Consider first the parton model prediction for the cross-section integrated over the muon angles with longitudinally polarized protons. This case has been discussed by Close and Sivers[3]. One finds, neglecting the sum over quark types so as to simplify the discussion,

$$\frac{d\sigma}{dQ^+ dQ^-} \propto 1 - \lambda_A\lambda_B h_{a/A}^L(x_A)h_{\bar{a}/B}^L(x_B).$$

If the partons were perfectly polarized ($h_{a/A}^L = h_{\bar{a}/B}^L = 1$), the cross-section would be <u>zero</u> for protons with the same helicity. In general, one would observe an asymmetry

$$A \equiv \frac{\sigma(+,-) - \sigma(+,+)}{\sigma(+,-) + \sigma(+,+)} = h_{a/A}^L h_{\bar{a}/B}^L$$

Recent data[4] on polarized electron-proton deep inelastic scattering at SLAC suggests a quark polarization $h_{a/A}^L \sim \frac{1}{2}$ for, say, $x_A = 0.3$. Close and Sivers argue that the antiquarks in a proton are also polarized, with perhaps $h_{\bar{a}/B}^L \sim \frac{1}{5}$ at $x_A = 0.3$. Thus one might expect to observe an asymmetry $A \sim 1/10$.

Just as important as what one should see according to the parton model is what one should not see. If the target and beam are transversely polarized, the Drell-Yan model predicts a zero asymmetry in $d\sigma/dQ^+dQ^-$ as the sign of $\vec{S}_A^T \cdot \vec{S}_B^T$ is reversed, while such an asymmetry is allowed in general.

Consider next the $\theta$ dependence of the cross-section integrated over the azimuthal angle $\phi$, $d\sigma/dQ^+dQ^- d\cos\theta$. In general, the cross-section must be proportional to $1+\alpha\cos\theta$ and $\alpha$ can depend on spin: $\alpha = \alpha_0 + \alpha_L \lambda_A \lambda_B + \alpha_T \vec{S}_A^T \cdot \vec{S}_B^T$. The Drell-Yan model predicts $\alpha_0 = 1$. This has been verified to $\sim 30\%$ accuracy at Fermilab by the Chicago-Princeton-Illinois collaboration[5]. The Drell-Yan model also predicts that $\alpha$ should be independent of polarization - that is, $\alpha_L = \alpha_T = 0$.

Finally, consider the $\phi$ dependence of the cross-section. In the case of transverse polarization of beam and target, the Drell-Yan model predicts a $\sin^2\theta \cos(2\phi - \phi_A - \phi_B)$ term in the cross-section. If the two spins are parallel, for instance, the muons should be preferentially produced parallel and antiparallel to the spin direction. The effect is proportional to the product $h_{a/A}^T(x) h_{\bar{a}/B}^T(x)$ of parton transverse polarization functions, about which little information exists at present.

If one of the protons, say A, is transversely polarized and the other is longitudinally polarized, a term proportional to $\cos\theta\sin\theta\cos(\phi-\phi_A)$ is possible in general, but such a term is forbidden in the Drell-Yan model.

This work was supported in part by the Energy Research and Development Administration and in part by the Graduate School of the University of Oregon. We wish to thank J. Collins and R. Hwa for helpful discussions.

## REFERENCES

1. J. Ralston and D.E. Soper, University of Oregon preprint OITS 100.
2. S.D. Drell and T.-M. Yan, Phys. Rev. Lett. 24, 181 (1970); Ann. Phys. (N.Y.) 66, 571 (1971).
3. F.E. Close and D. Sivers, Phys. Rev. Lett. 39, 1116 (1977).
4. M.J. Alguard et al., Phys. Rev. Lett. 41, 70 (1978).
5. K.J. Anderson et al., paper submitted to the XIX International Conference on High Energy Physics, Tokyo, Japan, 1978.

# SPIN CORRELATION ASYMMETRIES OF LARGE ANGLE ELASTIC NN SCATTERING AND THE NATURE OF PARTONS

Chih Kwan Chen
Department of Physics, Purdue University
West Lafayette, Indiana 47907

## ABSTRACT

Spin correlation asymmetries of large angle elastic NN scattering are ideal for the study of the nature of the hard scattered partons. The scheme for two colored quarks to scatter by the exchange of a vector gluon is inconsistent with data, but the data rather agree with vector-type scattering of two spin-1/2 colorless fermions of small masses. The significance of extending the spin correlation measurements is also emphasized.

We will discuss the significance of spin correlation asymmetries in elastic NN scattering at 90 degrees in CM for revealing the nature of partons[1]. In the parton picture, large-angle elastic scattering takes place via a rare momentum-configuration state, in which most of the momentum of the parent hadron is carried by one energetic parton, and all the remaining partons are wee partons. The two energetic partons from the two incident nucleons undergo hard scattering, whereas the wee partons of the two incident nucleons also interact softly[2]. The intrinsic transverse motion of the partons can be ignored, owing to the presence of the soft interaction among the wee partons. In this picture, the elastic NN scattering cross section becomes

$$d\sigma/dt = \int_{1-B/P}^{1} dx Q_1(x) Q_2(x) Q_3(x) Q_4(x) W_2(d\hat{\sigma}/d\hat{t}), \quad (1)$$

where B is the upper limit of the total momentum carried by the wee partons, $W_2$ is the transition probability describing the soft interaction of the wee partons, and the Q's are the probability of observing the rare momentum-configuration state and the probability of recombination. The expression of Eq.(1) is shown (in Refs.2) to fit the existing large-angle elastic pp scattering data for the unpolarized incident protons.

For the case of polarized incident nucleons, the spin of a nucleon is assumed to be carried totally by the energetic parton if the nucleon is in the rare momentum-configuration state. This assumption is favored by recent polarized-electron-polarized proton inelastic scattering experiment[3]. Denoting the orientation of spin as $\alpha(\beta)$, and the projection of spin along $\alpha(\beta)$ direction as $\mu(\nu)$, we can write the spin dependent elastic scattering cross section

$$(d\sigma/dt)_{(\alpha,\mu)(\beta,\nu)} = \int_{1-B/P}^{1} dx Q_1(x) Q_2(x) Q_3(x) Q_4(x)$$
$$\times W_2 (d\hat{\sigma}/d\hat{t})_{(\alpha,\mu)(\beta,\nu)}. \quad (2)$$

Substituting Eq.(2) into the definition of spin correlation asymmetry parameters, we obtain

$$A_{\alpha\beta} = \frac{\int_{1-B/P}^{1} dx Q_1(x) Q_2(x) Q_3(x) Q_4(x) W_2 (d\hat{\sigma}/d\hat{t}) \hat{A}_{\alpha\beta}}{\int_{1-B/P}^{1} dx Q_1(x) Q_2(x) Q_3(x) Q_4(x) W_2 (d\hat{\sigma}/d\hat{t})}. \quad (3)$$

For the case of interest to us, we will find that $\hat{A}(\alpha,\beta)$, the asymmetry parameter of the hard scattering process is independent of the incident energy square, $\hat{s}$, if the masses of the energetic partons are small; thus Eq.(3) reduces to

$$A_{\alpha\beta}(\theta) = \hat{A}_{\alpha\beta}(\theta). \quad (4)$$

Eq.(4) implies that the measurable asymmetries are directly related to the asymmetries of the underlying hard scattering process of two partons without the complication of the unknown factors, Q's, $(d\hat{\sigma}/d\hat{t})$, B and $W_2$.

We consider the usefulness of Eq.(4) through two examples:
Example 1) Two colorless spin-1/2 fermions of small masses undergo vector-type hard scattering. At $\theta=90°$ in CM, $\hat{A}(\alpha,\beta)$ becomes explicitly computable, since two unknown structure functions in the amplitude (the second one from the identical particle effect in elastic pp scattering) become equal due to $\hat{t}=\hat{u}$, and they will cancel in the computation of $\hat{A}(\alpha,\beta)$. $\hat{A}(\alpha,\beta)$ is independent of the incident energy if the masses of the fermions are small. From Eq.(4), we then obtain

$$A_{nn}(90°) = 1/9, \qquad A_{ss}(\theta) = -A_{nn}(\theta),$$
$$A_{ll}(90°) = 7/9, \qquad A_{sl}(\theta) = 0. \quad (5)$$

For n-p elastic scattering, in which the energetic partons are not identical, we obtain

$$A_{nn}(\theta) = A_{ss}(\theta) = A_{sl}(\theta) = 0 ,$$
$$A_{ll}(\theta) = (3-2\cos\theta-\cos^2\theta)/(5+2\cos\theta+\cos^2\theta). \tag{6}$$

We note that $A_{ll}(\theta)$ used here is equal to $-C_{ll}(\theta)$, which is used in various experiments,[4] so the constraint of elastic pp scattering at 90° should read as[5]

$$A_{ll}(90°) + A_{nn}(90°) - A_{ss}(90°) = 1 \tag{7}$$

which is satisfied by the results of Eq.(5).
Example 2) We consider the extension of the previous example to the case of colored quarks. It turns out that the results will violate the exact constraint of Eq.(7), so this example will serve as a warning toward the naive combination of parton picture with colored quarks. The two energetic partons are now colored quarks, and they scatter via the exchange of a vector gluon. The presence of color quantum numbers change the results of Eq.(5) into

$$A_{nn}(90°) = -1/11, \qquad A_{ss}(\theta) = -A_{nn}(\theta),$$
$$A_{ll}(90°) = 5/11, \qquad A_{sl}(\theta) = 0. \tag{8}$$

The above results do not satisfy the constraint of Eq.(7).
The asymmetry parameter $A_{nn}(90°)$ of elastic pp scattering has been measured[6] up to $p_{lab}$ = 12 GeV/c. Below 3 GeV/c it is about +0.5, and then drops rapidly to about 1/9. Between 4 to 8 GeV.c, it is constant and is consistent with the value 1/9. Then at 12 GeV/c, it increases to about +0.6 again. In summary we conclude (a) The scheme of colored quarks, example 2, violates the exact constraint of Eq.(7). This difficulty is due to the incoherent color sum, which results in the replacement of $(d\hat{\sigma}/d\hat{t})$ in Eq.(2) by a sum over all possible color combinations. It seems that this difficulty can only be remedied by introducing some kind of coherent color sum, since the constraint of Eq.(7) is derived from a single amplitude for identical particles.
(b) In the region $p_{lab}$=4 to 8 GeV/c, the data agree with the prediction of Example 1), i.e. $A_{nn}(90°)$=1/9. Therefore in this region ($p_T$=1 to 2 GeV/c) the two energetic partons may be identified as two colorless spin-1/2 fermions of small masses and they undergo

vector-type scattering. They may be colorless quarks or just virtual nucleons.

(c) The high and positive $A_{nn}(90°)$ below $p_{lab}$=3 GeV/c seems to be caused by the presence of direct channel singularities. Therefore the rapid rise of $A_{nn}(90°)$ after 8 GeV/c may suggest the existence of another island of the direct channel singularities. It will be very significant to measure $A_{nn}(90°)$ up to higher energies, and see whether it will drop rapidly again or not.

Anyway the extension of the spin correlation measurements to other types of asymmetry parameters, to elastic np scattering, and to higher energies are significant in revealing the nature of partons further. The results discussed here also have a close connection with the spin correlations in inclusive production of two large-$p_T$ protons at high energies[5], and is an explicit example of the usefulness of spin correlation studies in high-$p_T$ processes emphasized earlier in the Symposium of High Energy Polarized Beams[6].

## REFERENCES

1. C. K. Chen, Phys. Rev. Lett. <u>41</u>, 1440 (1978); ANL report, ANL-HEP-PR-78-17 (1978).
2. D. Horn and M. Moshe, Nucl. Phys. <u>B57</u>, 139 (1973); C.K. Chen, Phys. Rev. <u>D18</u>, 3297 (1978).
3. M. J. Alguard et al., Phys. Rev. Lett. <u>41</u>, 70 (1978).
4. I. P. Auer et al., Phys. Lett. <u>70B</u>, 475 (1977).
5. Dr. J. Soffer has called the attention of the author toward this constraint.
6. A. Lin et al., Phys. Lett. <u>74B</u>, 273 (1978); D. Miller et al., Phys. Rev. <u>D16</u>, 2016 (1977); D. G. Crabb et al., Univ. of Michigan Report, UM-HE-78-36 (1978).
7. C. K. Chen, Argonne National Lab. Report, ANL-HEP-PR-78-02 (1978) (to be published on Phys. Rev. <u>D18</u>).
8. Talks by C. K. Chen and R. D. Field, Proceedings of the Symposium of Enriched Antiproton, Polarized Proton and Antiproton Beams at Fermilab Energies, edited by A. Yokosawa, Argonne National Lab. Report, ANL-HEP-CP-77-45 (1977).

Chapter 8      Symposium Summary

SUMMARY TALK

M. Borghini
CERN, Geneva, Switzerland

INTRODUCTION

This conference was presented with a wealth of interesting papers dealing with experimental results, technological advances or projects, and theoretical analyses or ideas.

Most of the experimental papers dealt with hadron beams, a few with electron scattering, but very little was said about photoproduction although quite a few groups are very active in this field.

Many experimental results presented show important spin effects, even at the highest energies used. Some of these effects were interesting because they *were* predicted -- such as parity violation in electron-deuteron and electron-proton scattering -- and thus helped clarify some theoretical issues. Other effects were interesting because they were *not* predicted -- such as the structure in the helicity-dependent proton-proton total cross-section versus energy, the sudden increase of the 90° spin-spin correlation coefficient $A_{nn}$ in proton-proton elastic scattering at 12 GeV/c, or the existence of a large polarization of the $\Lambda$'s produced inclusively, even at 400 GeV. The structure in the total cross-section may be interpreted as being due to excited and exciting dibaryon resonances, although this picture was criticized; the other effects are totally unexplained, to the best of my understanding, and are thus particularly challenging.

Most of the technological papers dealt with advances in polarized sources of protons or electrons, or in polarized beams, accelerated or not; some with progress in polarized targets; a few with electron polarimeters. Particularly noticeable were the recent developments in polarized electron beams, the improvement in polarized proton beams at Argonne, and the project for accelerating polarized electrons up to 27 GeV in the AGS at Brookhaven.

EXPERIMENTAL RESULTS

Parity is violated in electron-deuteron and electron-proton deep inelastic scattering (Prescott). This very exciting result can be used in conjunction with neutrino data and future atomic physics results to determine the weak, neutral, coupling constants of the electron and its weak isospin (Sakurai). The Weinberg-Salam-Glashow ... model of weak and electromagnetic interactions agrees with the experimental results.

An astonishingly high polarization of inclusively produced $\Lambda^0$'s and $\Sigma^0$'s was observed up to 400 GeV and transverse momenta up to 2 GeV/c (Heller), and the effect seems to be independent of energy down to 24 GeV. No polarization is observed for $\bar{\Lambda}^0$ or for protons (Heller, Polvado). Such measurements could provide a crucial test of QCD, which predicts a vanishing polarization for the $\Lambda^0$'s at momenta higher than 6 GeV/c in an apparently unambiguous way (Kane, Pumplin); the experiment will be performed in 1979 (Heller). Larger asymmetries with lesser accuracy were seen in $\pi^0$ inclusive production at 24 GeV

(Fidecaro and Terwilliger). No explanation of these effects was available.

A very strange behaviour of the spin-spin correlation coefficient $A_{nn}$ in proton-proton elastic scattering at 90° is observed as a function of incident momentum (Terwilliger): some activity is seen below 3 GeV/c followed by a plateau close to 1/9 up to 6 GeV/c, in agreement with naive quark model predictions (Chen), suddenly disturbed by an increase up to $A_{nn}$ = 0.3-0.4 at 12 GeV/c which is completely unexplained and baffling.

An important structure is seen in the total proton-proton cross-section in pure spin states (Spinka, Mulera): in particular, the difference $\Delta\sigma_L$ between the total cross-section with equal initial helicities and the cross-section with opposite initial helicities, exhibits an important structure for beam momentum values between 1 and 3 GeV/c. One phase-shift analysis claimed to require two dibaryon resonances in order to fit the data (Hoshizaki):

$^3F_2$ : m = 2.22 GeV/c$^2$ ; $\Gamma$ = 100 $\sim$ 150 MeV/c$^2$, $\Gamma_{el}/\Gamma \simeq 0.2$

$^1D_2$ : m = 1.75 GeV/c$^2$ ; $\Gamma$ = 50 $\sim$ 100 MeV/c$^2$, $\Gamma_{el}/\Gamma \simeq 0.1$ .

An opposing view (Bugg) noted that since these resonances had large branching ratios for decay into inelastic channels, the analysis should agree with available inelastic data -- which apparently it did not do -- amongst other criticisms. A lively discussion followed. It was admitted that the uniqueness of this phase-shift solution had not been demonstrated (Hoshizaki), and the situation is still undecided.

Below 500 MeV, non-surprising phase-shifts nicely describe all the proton-proton scattering data, including new accurate ones from TRIUMF (Bugg), LAMPF (Willard), and SIN (Leo). Preliminary results on $A_{nn}$ between 1 and 3 GeV/c were also presented (Mulera). Many new neutron-proton scattering data from TRIUMF were mentionned (Bugg) but no corresponding phase-shifts were available.

Polarized-electron polarized-proton deep inelastic scattering has been measured (Hughes, Schüler) and should give very important information on the spin distributions of partons within the proton (Hughes, Sivers). However, many models now fit the data, which are not yet very accurate. Better results are expected from 1979 running, and scattering from deuterium will be done with an accuracy hopefully comparable to that of the present proton data.

An impressively accurate search for parity violation in proton-proton scattering has not shown any definite effect, with standard deviations of $0.8 \times 10^{-7}$ at 15 MeV and $2.5 \times 10^{-6}$ at 6 GeV/c (Nagle). This search will continue.

The target asymmetry in elastic proton-proton scattering has been measured at 100 and 300 GeV (Jonckheere) and at 150 GeV (Fidecaro); it seems to me that the results from the two experiments disagree, with no explanation available. The polarization in elastic $\pi^{\pm}p$ scattering was also measured at 150 GeV, with no surprise (Jonckheere).

The target asymmetry in proton-neutron scattering has been measured at 24 GeV/c (Kyberd): it appears to be the mirror-symmetric of the asymmetry in proton-proton scattering, as opposed to results obtained at lower energies. These results, however, agree with predictions from a Regge pole model which fits all two-body reaction data, including the lower energy neutron-proton data (Irving).

The target asymmetry has been measured in $\pi^- p \to \pi^+ \pi^- n$ at 17 GeV/c; the results show unexpected $A_1$ exchange contribution to this reaction and interesting structures in the lower partial waves, in addition to the $\rho$, f, and g (Lutz, Rybicki).

The target asymmetry was measured in $K^+ n \to K^0 p$ at 6 and 12 GeV/c, and the results show an unexpected behaviour of the tensor-exchange amplitudes (Van Rossum, Svec).

Some examples of photoproduction experiments with polarized targets and beams and with recoil polarimeters were given (Althoff). Here also, a dibaryon resonance might be necessary in order to explain some results.

A new phase-shift analysis of $K^+ p$ elastic data at low energy showed evidence for the existence of a $Z^*$ resonance (Kelly), but did not seem to convince many people.

Small asymmetries have been seen in $\pi^-$ inclusive production at 6 and 12 GeV/c and x = 0 (Gray).

Other results on hadronic and on leptonic reactions have been obtained and collected in the very complete reviews by Terwilliger and Hughes, respectively. Schwitters reviewed experiments that can be done with polarized electron storage rings; very interesting information can be obtained.

## TECHNOLOGY

<u>Polarized proton sources</u>: These were reviewed by Haeberli: ground-state sources ($H^+$) give between 50 and 100 µA, pulsed or continuous, with about 75% polarization (Dick, Parker); Lamb-shift sources ($H^-$) give 1 µA maximum (Schmor, Suwa), but may have an efficiency comparable to that of $H^+$ sources when the possibility of multiturn injection into synchrotrons is taken into account. A few schemes for increasing the current of these sources were presented, such as the use of charge exchange reactions; a project for obtaining 10 mA of $H^-$, polarized up to 97%, was mentioned (Shatonov and Sharnovsky), which looked like a dream!

<u>Polarized proton acceleration</u>: Depolarization resonances seem to be well understood and were reviewed by Teng: they can be jumped or turned around. Ingenious schemes for avoiding depolarization by a sort of "spin echo", which involved spin rotation by magnets placed at some point of the accelerator rings ("Siberian snakes"), were presented in some details (Derbenev, Turrin). The acceleration of polarized protons at Fermilab and in the CERN SPS seems impossible; polarized deuterons might be accelerated at Fermilab. With a new $H^-$ source, polarized proton beam intensities of $10^{11}$-$10^{12}$ per pulse are expected in the ZGS at Argonne instead of a few $10^{10}$ as at present (Parker), and polarized deuterons will be accelerated in 1979. A serious proposal for injecting polarized $H^-$ into the AGS at Brookhaven has been made (Courant), with acceleration up to 27 GeV and with intensities equal to one tenth of the unpolarized beam intensity or, hopefully, more. Then injection into ISABELLE could follow as a second, and

separated, step. This exciting project would be the best way to
study proton spin effects at energies higher than that of the ZGS.
In the meantime, a polarized proton beam from $\Lambda^0$ decay has been proposed at Fermilab which would deliver a few $10^7$ protons per pulse
with 40% polarization at 300 ± 5% GeV (Underwood). A polarized beam
of $10^6$ antiprotons per pulse might be possible too. Finally, for the
first time a synchrotron, Saturne II, has been designed with the
specific purpose of avoiding beam depolarization during acceleration
(Beurtey); other future machines may have to be similarly designed.

<u>Polarized electron sources</u>: Polarized electron sources are
rapidly developing:
- At SLAC, PEGGY I, based on the photoionization of polarized lithium
atoms, has up to 85% polarization with $10^9$ electrons per pulse at
120 pps (Hughes), about 200 times less than unpolarized SLAC currents
but enough for use with polarized targets.
- At SLAC, PEGGY II, based on the photoexcitation of electrons from
gallium arsenide with circularly polarized light, has only up to 40%
polarization, but with as much as $2 \times 10^{11}$ electrons per pulse at
120 pps like the normal SLAC beam (Prescott). Acceleration in the
20 GeV linac produces no depolarization -- a good thing.
- At Bonn, a source based on the Fano effect (photoionization of unpolarized atoms with polarized light) has 65% polarization with $10^9$
electrons per pulse (von Drachenfels). Acceleration in the synchrotron does produce depolarization -- a bad thing which is being studied
for remedies.

<u>Polarized electrons in storage rings</u>: The spontaneous polarization by radiation and the depolarization of electrons in synchrotrons have been studied both theoretically and experimentally; the
simple expression which gives the rate of spontaneous polarization as
a function of the machine radii and of the energy has been very well
verified in the USSR by using the Touschek effect or the spin dependence of synchrotron radiation (Derbenev), and at SLAC by using a
polarimeter in which a laser beam is back-scattered from the electron
beam (Johnson, Schwitters). The measurement of the relative polarization as a function of the beam energy agrees with a careful theoretical calculation of the depolarization effects; absolute polarization values were not available. These measurements have been made
with one circulating beam only, and the measurements with two beams
were scarce and not reproducible. A polarimeter similar to the SLAC
one has been constructed for PETRA (Rossmanith).

Schemes for increasing the speed of spontaneous polarization
("wigglers") were discussed, as well as schemes for reducing the
amount of depolarization ("Siberian snakes") (Derbenev). However,
the latter schemes suppress the mechanism of spontaneous polarization
due to the main magnetic field of the machine, and other polarization
processes have to be used (Derbenev, Montague). "Wigglers" are being
built for PEP at SLAC (Schwitters). Electron polarization in the
proposed LEP machine in Europe (70 GeV on 70 GeV?) has special problems, as the energy spread is not negligible compared to the 440 MeV
spacing between intrinsic depolarization resonances and conclusions
cannot be reached at this time (Montague).

<u>Polarized targets</u>: Polarized targets for high-energy physics

were reviewed by Abragam. Polarized targets exist with up to some 95% proton polarization and some 40% deuteron vector polarization, using dilution refrigerators. A large dilution refrigerator is being built at CERN for muon scattering from a 1 m long, 2 l target (Niinikoski). A series of new, stable, chromium complexes (Krumpolc) are being studied; a proton polarization of 75-80% in butanol doped with "EHBA" has been obtained (Niinikoski), which makes this material very attractive; its radiation damage properties have still to be investigated. A workshop on polarizable materials, with emphasis on high radiation resistance and high hydrogen content, came up with a long list of questions but as yet few answers (Fernow). An intense polarized atomic jet could be used as a very clean and small polarized target, with reasonable luminosity if used with an internal synchrotron beam, because of the multitraversal of the jet by the beam (Dick). A target of yttrium ethylsulfate has been polarized up to 85% by rotation; it has a very poor chemical composition and the polarization has one sign only, but it can operate in rather inhomogeneous magnetic fields (Button-Shafer). An axially polarized target with a somewhat separated $^3$He liquefier and a very large solid angle has been built at RHEL (Saxon).

## THEORY

As far as I can see, the theoretical situation presented at the Conference was as follows: a Regge model can fit all the elastic scattering and charge exchange data rather well at low momentum transfer, with a limited number of parameters (Irving), but some lack of interest in such an achievement was noticeable; another interpretation was given which puts emphasis on the rapid energy variation of the real part of the non-flip amplitude (Kroll). The model of rotating hadronic matter has been mentioned again with the modification that it does not require the proton-proton total cross-section to rise with energy in order to predict spin effects (Soffer), a fact which puzzled some people; the measurements available do not provide any clue as to whether the model has any validity or not. Finally, quantum chromodynamics makes prediction for spin effects due to the strong interactions for momenta where this interaction is weak enough for perturbation theory to be applied; the corresponding experiments then suffer from low cross-sections, of course. A seemingly crucial prediction of QCD is that the large inclusive $\Lambda^0$ polarization observed below $p_T \sim 2$ GeV/c (Heller) should become zero in the momentum range where the QCD predictions agree with cross-section measurements, i.e. above 6 GeV/c (Kane, Pumplin), because of small quark masses and of quark helicity conservation; the role of gluon scattering still remains to be investigated (Sivers). Other QCD predictions (Soffer) were that the asymmetry in $p\uparrow p \to \pi^0$ ... at large $p_T$ can be related to lepton-hadron scattering, and that, even if quark-quark scattering gives 100% polarized quarks, the measured asymmetry should be smaller than 20% in absolute value. QCD cannot predict spin-spin asymmetries wtihout a knowledge of the spin distribution functions of hadrons (Sivers); these are measured by lepton-scattering but are as yet poorly known.

Spin effects in muon-pair production, $pp \to \mu^+\mu^- X$, for large muon-pair invariant masses, were investigated theoretically and could provide some stringent tests of the Drell-Yan model (Soper). Polarized proton experiments could be the only hope for extracting W bosons from hadronic background in pp collisions at very high energies (Paige). These are, however, very difficult experiments.

Asymptotic relations between spin parameters were recalled (van Hieu). The general interest of spin studies was emphasized by a theory panel (Cutkosky, Kane, Miettinen, Soffer); however, in my opinion the Conference lacked a review of current strong interaction theories and how they compare with cross-section measurements.

Unified theories of weak and electromagnetic interactions received a boost from the observation of parity-violation at SLAC, with the Weinberg-Salam-Glashow ... model being favoured. However, more has to be done, to determine the weak coupling constants and to detect intermediate bosons with or without polarization measurements (Sakurai); the interest of measuring the final polarization in $\nu p \to \nu p$, $\bar{\nu}p \to \bar{\nu}p$, $\nu n \to \mu^- p$ (Langacker) or the final density matrix in $\nu N \to \mu\Delta$ (Michel) was also mentioned.

## CONCLUSION

Much has been done, much remains to be done, and it will be interesting to see what will have been done by the time of the next Conference!

## PROGRAM

Wednesday, October 25

Registration, Building 362 Auditorium, Argonne

MORNING SESSION, 362 Auditorium

    Session Chairman: M. Goldhaber, Brookhaven National Laboratory
    Opening Remarks
      A. D. Krisch, University of Michigan
    Welcoming Address
      G. A. Smith, Argonne National Laboratory
    Polarized Targets in High Energy Physics and Elsewhere
      A. Abragam, Collège de France
    Theory Panel
      F. E. Low, MIT: Chairman
      R. E. Cutkosky, Carnegie-Mellon University
      G. L. Kane, University of Michigan
      H. I. Miettinen, Fermilab
      J. Soffer, CNRS, Marseilles

AFTERNOON SESSION, 362 Auditorium

    Session Chairman: R. L. Cool, Rockefeller University
    Parity Violations in Inelastic Electron Scattering
      C. Y. Prescott, SLAC
    Pure Spin Total Cross Sections
      H. M. Spinka, Argonne National Laboratory
    pp Phase Shifts and Dibaryon Resonances
      N. Hoshizaki, Kyoto University
    Implications of N-N Spin Measurements
      A. C. Irving, Liverpool University
    "Siberian Snake"
      Ya. Derbenev, Novosibirsk
    Polarized Target Experiments at CERN
      M. Fidecaro, CERN
    Inclusive $\Lambda$ Production
      K. Heller, University of Minnesota
    $\pi^{\pm}p$ and pp Polarization at 100 and 300 GeV/c
      A. M. Jonckheere, Fermilab

Social Evening, Oak Brook Hyatt House

Thursday, October 26

MORNING SESSION, 362 Auditorium

    Session Chairman: L. Wolfenstein, Carnegie-Mellon University
    Lepton-Hadron Scattering
      V. W. Hughes, Yale University
    Weak-EM Interference
      J. J. Sakurai, UCLA
    Spontaneous $e^-$ Polarization
      Ya. Derbenev, Novosibirsk

PARALLEL SESSION I, F108

    Session Chairman: T. H. Fields, Argonne National Laboratory

Polarized Leptons

    Polarization Physics at Bonn
      K. H. Althoff, University of Bonn
    Acceleration of Polarized $e^-$ Beams in Synchrotrons
      W. von Drachenfels, University of Bonn
    Depolarization in Storage Rings
      B. W. Montague, CERN
    Measurement of Polarization in SPEAR
      J. Johnson, University of Wisconsin
    Measurement of the Beam Polarization in the Storage Ring PETRA
      R. Rossmanith, DESY

Polarized Targets

    Recent Developments in Polairzed Targets at CERN
      T. O. Niinikoski, CERN
    Workshop Resumé
      R. C. Fernow, Brookhaven National Laboratory

PARALLEL SESSION II, 362 Auditorium

    Session Chairman: L. Rosen, LAMPF

Hadronic Interactions

    Measurements of $A_{NN}$ at LAMPF
      H. B. Willard, Case Western Reserve University
    Measurements of $A_{NN}$ at SIN
      W. Leo, University of Geneva
    Measurements of $A_{NN}$ from 1 - 3 GeV/c
      T. A. Mulera, Rice University
    Polarized Beams and Approved Experiments at SATURNE Using Polarized Probes and Targets
      R. M. Beurtey, CEN, Saclay
    Asymptotic Theorems and Polarization Phenomenon
      N. Van Hieu, Inst. of Physics, Vietnam
    KN Charge Exchange Polarization at 6 and 12 GeV/c
      L. Van Rossum, CEN, Saclay
    Tensor Exchange Amplitudes in KN Charge Exchange
      M. Svec, McGill University
    CERN-Munich Polarized Target Experiment at 17.2 GeV
      K. Rybicki, INP, Krakow

Visit to Chicago Art Institute

Symposium Banquet (Courtesy AUA)
    Sullivan Room, Chicago Art Institute

Private Viewing of Pompeii Exhibit

Friday, October 27

MORNING SESSION, 362 Auditorium

    Session Chairman: P. V. Livdahl, Fermilab

High Energy Polarized Hadron Beams
   L. C. Teng, Fermilab
Polarized Hadron Sources
   W. Haeberli, University of Wisconsin
Progress in Medium Energy N-N Scattering
   D. V. Bugg, Queen Mary College, London

PARALLEL SESSION I, 362 Auditorium

Session Chairman: B. Cork, LBL

Polarized Beams

Possibility of Polarized Beams at the AGS
   E. D. Courant, Brookhaven National Laboratory
Hyperon Beams as a Source of Polarized Protons
   D. G. Underwood, Argonne National Laboratory
Development of Polarized Gas Jets
   L. Dick, CERN
200 nA Variable Energy Polarized Proton Beam at TRIUMF
   P. W. Schmor, University of British Columbia
Studies on Accelerating Polarized Beams at KEK
   S. Suwa, KEK
Source Development at Argonne
   E. F. Parker, Argonne National Laboratory
Depolarization During Acceleration
   A. Turrin, Frascati

Contributed Papers I

Proton Polarization in Inclusive Processes at FNAL
   R. O. Polvado, Indiana University
The Polarization Parameter in Proton-Neutron Elastic Scattering at 24 GeV/c
   P. Kyberd, Oxford University
Comment on $Z_1^*(1800)$ and Spin-Rotation Parameter Measurements
   R. L. Kelly, LBL
Interpretation of the Spin Structure in Nucleon-Nucleon Scattering
   P. Kroll, University of Wuppertal
Inclusive $\pi^-$ Asymmetries Near $x = 0$
   S. W. Gray, Indiana University

PARALLEL SESSION II, F108

Session Chairman: L. Michel, Bures-Sur-Yvette

Weak Interactions and High $P_T$ Phenomena

Polarization Phenomena in $\nu$ Interactions
   P. Langacker, University of Pennsylvania
Production of Dimuons From High Energy Polarized Proton-Proton Collisions
   D. E. Soper, University of Oregon
Estimates of W Production With Polarized Protons as a Means of Detecting its Hadron Jet Decays
   F. E. Paige, Brookhaven National Laboratory
Parity Violation
   D. E. Nagle, Los Alamos Scientific Laboratory

Transverse Quark Polarization in Large $P_T$ Reactions, $e^+e^-$ Jets and Lepto-Production
  J. Pumplin, Michigan State University
Theory of Spin-Spin Asymmetries in N-N Elastic Scattering
  C. K. Chen, Purdue University

Contributed Papers II

Polarized Electron-Polarized Proton Scattering Experiments at SLAC
  K. P. Schüler, SLAC
Spin Refrigerator Target
  J. Button-Shafer, University of Massachusetts
An Axially Polarized Proton Target
  D. H. Saxon, Rutherford Laboratory

Saturday, October 28

MORNING SESSION, 362 Auditorium

Session Chairman: A. N. Skrinsky, Novosibirsk
Spin Measurements in High $P_\perp$ Scattering
  K. M. Terwilliger, University of Michigan
Polarized $e^+e^-$ Experiments
  R. F. Schwitters, SLAC
High $P_T$ Theory
  D. Sivers, Argonne National Laboratory
Symposium Summary
  M. G. Borghini, CERN

End of Symposium

## ADDITIONAL CONTRIBUTED PAPERS SUBMITTED

Electron-Positron Collisions on $1^-$ Resonances
  R. Budny, Princeton University
Proton Deuteron Polarized Cross Sections
  M. Bleszynski, LAMPF
Optimal Complete Sets of Experiments for Elastic Nucleon-Nucleon Scattering
  J. Patera, University of Montreal
Dibaryon Resonances in $\pi NN$ and $\pi\pi NN$ Dynamics
  T. Ueda, Osaka University
Inclusive Cross Sections, Structure Functions, and Scaling Variables in Nuclear and Elementary Particle Physics
  S. Frankel, University of Pennsylvania
Inclusive $\pi^-$ Asymmetries Near $x = 0$
  S. C. Ems, Indiana University
The Discrete Ambiguity Resolution and Baryon-Resonance Parameter Determination
  D. M. Chew, University of Paris
The $^6$Li Atomic Beam Polairzed Electron Source at SLAC
  J. E. Clendenin, SLAC
Highly Efficient Ionization of a Polarized Deuterium Beam in a Penning Discharge
  A. Kruger, Bonn University
A Horizontal Dilution Refrigerator with High Cooling Power
  D. A. Hill, Argonne National Laboratory

LIST OF PARTICIPANTS

| Name | Institution |
|---|---|
| A. Abragam | Collège de France |
| L. A. Ahrens | Brookhaven National Laboratory |
| K. H. Althoff | University of Bonn |
| H. H. Aly | Southern Illinois University |
| I. Ambats | Argonne National Laboratory |
| M. W. Arenton | Argonne National Laboratory |
| J. F. Arvieux | Institut Des Sciences Nucleaires |
| I. P. Auer | Argonne National Laboratory |
| D. A. Axen | Triumf, Univ. of British Columbia |
| D. S. Ayres | Argonne National Laboratory |
| S. Barkan | University of Istanbul |
| G. G. Baum | Universitat Bielefeld |
| A. F. Beretvas | Fermilab |
| R. M. Beurtey | Laboratoire National Saturne |
| A. Białas | Kracow/Fermilab |
| J. K. Bienlein | DESY |
| N. E. Booth | University of Oxford |
| G. Michel Borghini | CERN |
| R. Bornstein | Illinois Institute of Technology |
| D. V. Bugg | Queen Mary College |
| G. M. Bunce | Brookhaven National Laboratory |
| J. Button-Shafer | University of Massachusetts |
| R. D. Carlitz | University of Pittsburgh |
| A. S. Carroll | Brookhaven National Laboratory |
| E. P. Chamberlin | Los Alamos Scientific Laboratory |
| A. W. Chao | SLAC |
| G. R. Charlton | DOE/OER |
| C. K. Chen | Purdue University |
| B. T. Chertok | American University |
| J. E. Clendenin | Yale University |
| E. P. Colton | Argonne National Laboratory |
| M. Comyn | Triumf, Univ. of British Columbia |
| R. L. Cool | Rockefeller University |
| B. Cork | Lawrence Berkeley Lab |
| E. D. Courant | Brookhaven National Laboratory |
| G. R. Court | Liverpool University |
| D. G. Crabb | University of Michigan |
| R. E. Cutkosky | Carnegie-Mellon University |
| Y. S. Derbenev | Inst. of Nuclear Physics |
| M. Derrick | Argonne National Laboratory |
| L. A. Dick | CERN |
| R. E. Diebold | Argonne National Laboratory |
| Dikansky | Inst. of Nuclear Physics |
| W. R. Ditzler | Argonne National Laboratory |
| J. T. Donohue | Argonne National Laboratory |
| M. J. Eder | Zurich, Switzerland |
| A. Feltman | Brookhaven National Laboratory |
| R. C. Fernow | Brookhaven National Laboratory |

| | |
|---|---|
| W. J. Fickinger | Case Western Reserve University |
| M. Fidecaro | CERN |
| T. H. Fields | Argonne National Laboratory |
| D. H. Fitzgerald | Los Alamos Scientific Lab |
| P. Frampton | Ohio State University |
| A. Fridman | Argonne National Laboratory |
| M. Goldhaber | Brookhaven National Laboratory |
| A. A. Golestaneh | Argonne National Laboratory |
| S. A. Gottlieb | Argonne National Laboratory |
| T. D. Gottschalk | Argonne National Laboratory |
| S. W. Gray | Argonne National Laboratory |
| P. Gregory | Argonne National Laboratory |
| W. Haeberli | University of Wisconsin |
| P. H. Hansen | University of Michigan |
| J. Hauser | University of Michigan |
| K. J. Heller | University of Minnesota |
| K. Hidaka | Argonne National Laboratory |
| D. A. Hill | Argonne National Laboratory |
| E. W. Hoffman | Los Alamos Scientific Laboratory |
| N. Hoshizaki | Kyoto University |
| V. W. Hughes | Yale University |
| L. Hyman | Argonne National Laboratory |
| A. C. Irving | Liverpool University |
| K. B. Jaeger | Argonne National Laboratory |
| J. J. Jarmer | Los Alamos Scientific Laboratory |
| J. R. Johnson | University of Wisconsin |
| P. W. Johnson | Illinois Institute of Technology |
| H. Johnstad | Fermilab |
| A. M. Jonckheer | Fermilab |
| G. L. Kane | University of Michigan |
| A. Kanofsky | Lehigh University |
| M. W. Keig | Argonne National Laboratory |
| R. L. Kelly | Lawrence Berkeley Laboratory |
| R. D. Klem | Argonne National Laboratory |
| R. V. Kline | Harvard University |
| P. F. M. Koehler | Fermilab |
| V. Konig | ETH |
| A. D. Krisch | University of Michigan |
| P. Kroll | University of Wuppertal |
| B. N. Kursunoglu | University of Miami |
| P. Kyberd | Oxford University |
| P. G. Langacker | University of Pennsylvania |
| K. Lassila | Iowa State University |
| J. D. Lesikar II | Argonne National Laboratory |
| W. R. Leo | University of Geneva |
| G. M. Levman | Argonne National Laboratory |
| D. B. Lichtenberg | Indiana University |
| P. V. Livdahl | Fermilab |
| N. S. Lockyer | Argonne National Laboratory |
| F. J. Lopinto | Argonne National Laboratory |
| J. S. Loos | Argonne National Laboratory |
| F. E. Low | M.I.T. |

| | |
|---|---|
| J. L. McKibben | Los Alamos Scientific Lab |
| L. Madansky | John Hopkins University |
| M. L. Marshak | University of Minnesota |
| R. L. Martin | Argonne National Laboratory |
| E. N. May | Argonne National Laboratory |
| A. Michalowicz | LAPP - Anneiy le Vieux |
| L. Michel | Institute Hautes Etudes Scientific |
| L. P. Michelotti | Costa Mesa, California |
| H. E. Miettinen | Rice University |
| H. I. Miettinen | Fermilab |
| R. Migneron | University of Western Ontario |
| R. H. Milburn | Tufts University |
| D. R. Moffett | Argonne National Laboratory |
| D. Mohl | CERN |
| E. H. Monsay | Argonne National Laboratory |
| B. W. Montague | CERN |
| M. J. Moravcsik | University of Oregon |
| S. Mori | Fermilab |
| B. E. F. Mossberg | University of Minnesota |
| T. A. Mulera | Rice University |
| B. Musgrave | Argonne National Laboratory |
| D. E. Nagle | Los Alamos Scientific Lab |
| T. O. Niinikoski | CERN |
| P. J. O'Donnell | University of Toronto |
| J. R. O'Fallon | Argonne Universities Association |
| H. Ogren | Indiana University |
| G. G. Ohlsen | Los Alamos Scientific Lab |
| O. E. Overseth | University of Michigan |
| J. F. Owens | Florida State University |
| A. Pagnamenta | University of Illinois |
| F. E. Paige, Jr. | Brookhaven National Laboratory |
| S. Y. Park | Syracuse University |
| E. F. Parker | Argonne National Laboratory |
| S. Penselin | Institut fuer Angewandte Physik |
| A. Penzo | I.N.F.N. - Trieste |
| A. Perlmutter | University of Miami |
| E. A. Peterson | University of Minnesota |
| G. C. Phillips | Rice University |
| R. O. Polvado | Indiana University |
| C. W. Potts | Argonne National Laboratory |
| C. Y. Prescott | SLAC |
| L. E. Price | Argonne National Laboratory |
| J. Pumplin | Michigan State University |
| L. Ratner | Argonne National Laboratory |
| G. Ringland | Rutherford Laboratory |
| J. B. Roberts | Rice University |
| L. Rosen | Los Alamos Scientific Lab |
| R. Rossmanith | DESY |
| H. M. Ruck | Duke University |
| D. R. Rust | Indiana University |
| R. D. Ruth | Brookhaven National Lab/SUNY |
| K. Rybicki | Institute of Nuclear Physics |
| J. J. Sakurai | UCLA |

| | |
|---|---|
| G. L. Salmon | Oxford University |
| R. H. Sands | University of Michigan |
| J. Sauer | Argonne National Laboratory |
| D. H. Saxon | Rutherford Laboratory |
| J. L. Schlereth | Argonne National Laboratory |
| P. W. Schmor | TRIUMF, Univ. of British Columbia |
| K. P. Schuler | Yale University |
| P. F. Schultz | Argonne National Laboratory |
| R. F. Schwitters | SLAC |
| G. Shapiro | University of California, Berkeley |
| T. Shima | Argonne National Laboratory |
| J. Simmons | Los Alamos Scientific Laboratory |
| R. A. Singer | Argonne National Laboratory |
| D. W. Sivers | Argonne National Laboratory |
| A. N. Skrinsky | Institute of Nuclear Physics |
| G. A. Smith | Argonne National Laboratory |
| J. H. Snyder | Yale University |
| J. F. Soffer | Centre De Physique Theorique |
| D. E. Soper | University of Oregon |
| C. Sorensen | Argonne National Laboratory |
| H. M. Spinka | Argonne National Laboratory |
| S. Suwa | KEK |
| M. Svec | McGill University |
| E. C. Swallow | Elmhurst College |
| N. Tamura | Argonne National Laboratory |
| L. C. Teng | Fermilab |
| K. M. Terwilliger | University of Michigan |
| G. Theodosiou | Argonne National Laboratory |
| G. H. Thomas | Argonne National Laboratory |
| K. Toshioka | Argonne National Laboratory |
| T. N. Tudron | Brookhaven National Laboratory |
| A. Turrin | Istituto Nazionale di Fisica Nucl. |
| D. G. Underwood | Argonne National Laboratory |
| N. Van Hieu | Institute of Physics of Viet-Nam |
| J. Vander Velde | University of Michigan |
| L. Van Rossum | CEN-SACLAY, France |
| W. von Drachenfels | University of Bonn |
| R. G. Wagner | Argonne National Laboratory |
| P. Walden | TRIUMF, Univ. of British Columbia |
| K. C. Wali | Syracuse University |
| W. A. Wallenmeyer | U.S. DOE |
| C. E. W. Ward | Argonne National Laboratory |
| R. L. Warnock | Lawrence Berkeley Laboratory |
| H. L. Weisberg | Brookhaven National Laboratory |
| A. B. Wicklund | Argonne National Laboratory |
| H. B. Willard | CASE Western Reserve University |
| L. Wolfenstein | Carnegie-Mellon University |
| R. M. Woods, Jr. | U.S. DOE |
| A. Yokosawa | Argonne National Laboratory |

## AIP Conference Proceedings

|  |  | L.C. Number | ISBN |
|---|---|---|---|
| No. 1 | Feedback and Dynamic Control of Plasmas (Princeton) 1970 | 70-141596 | 0-88318-100-2 |
| No. 2 | Particles and Fields - 1971 (Rochester) | 71-184662 | 0-88318-101-0 |
| No. 3 | Thermal Expansion - 1971 (Corning) | 72-76970 | 0-88318-102-9 |
| No. 4 | Superconductivity in d- and f-Band Metals (Rochester, 1971) | 74-18879 | 0-88318-103-7 |
| No. 5 | Magnetism and Magnetic Materials - 1971 (2 parts) (Chicago) | 59-2468 | 0-88318-104-5 |
| No. 6 | Particle Physics (Irvine, 1971) | 72-81239 | 0-88318-105-3 |
| No. 7 | Exploring the History of Nuclear Physics (Brookline, 1967, 1969) | 72-81883 | 0-88318-106-1 |
| No. 8 | Experimental Meson Spectroscopy - 1972 (Philadelphia) | 72-88226 | 0-88318-107-X |
| No. 9 | Cyclotrons - 1972 (Vancouver) | 72-92798 | 0-88318-108-8 |
| No. 10 | Magnetism and Magnetic Materials - 1972 (2 parts) (Denver) | 72-623469 | 0-88318-109-6 |
| No. 11 | Transport Phenomena - 1973 (Brown University Conference) | 73-80682 | 0-88318-110-X |
| No. 12 | Experiments on High Energy Particle Collisions - 1973 (Vanderbilt Conference) | 73-81705 | 0-88318-111-8 |
| No. 13 | $\pi-\pi$ Scattering - 1973 (Tallahassee Conference) | 73-81704 | 0-88318-112-6 |
| No. 14 | Particles and Fields - 1973 (APS/DPF Berkeley) | 73-91923 | 0-88318-113-4 |
| No. 15 | High Energy Collisions - 1973 (Stony Brook) | 73-92324 | 0-88318-114-2 |
| No. 16 | Causality and Physical Theories (Wayne State University, 1973) | 73-93420 | 0-88318-115-0 |
| No. 17 | Thermal Expansion - 1973 (Lake of the Ozarks) | 73-94415 | 0-88318-116-9 |
| No. 18 | Magnetism and Magnetic Materials - 1973 (2 parts) (Boston) | 59-2468 | 0-88318-117-7 |
| No. 19 | Physics and the Energy Problem - 1974 (APS Chicago) | 73-94416 | 0-88318-118-5 |
| No. 20 | Tetrahedrally Bonded Amorphous Semiconductors (Yorktown Heights, 1974) | 74-80145 | 0-88318-119-3 |
| No. 21 | Experimental Meson Spectroscopy - 1974 (Boston) | 74-82628 | 0-88318-120-7 |
| No. 22 | Neutrinos - 1974 (Philadelphia) | 74-82413 | 0-88318-121-5 |
| No. 23 | Particles and Fields - 1974 (APS/DPF Williamsburg) | 74-27575 | 0-88318-122-3 |
| No. 24 | Magnetism and Magnetic Materials - 1974 (20th Annual Conference, San Francisco) | 75-2647 | 0-88318-123-1 |
| No. 25 | Efficient Use of Energy (The APS Studies on the Technical Aspects of the More Efficient Use of Energy) | 75-18227 | 0-88318-124-X |
| No. 26 | High-Energy Physics and Nuclear Structure - 1975 (Santa Fe and Los Alamos) | 75-26411 | 0-88318-125-8 |